T0398270

Time

# Presenting the Past

CENTRAL ISSUES IN MEDIEVAL AND EARLY MODERN STUDIES
ACROSS THE DISCIPLINES

*General Editor*

Nancy van Deusen

VOLUME 5

The titles published in this series are listed at *brill.com/pthp*

# Time

## Sense, Space, Structure

*Edited by*

Nancy van Deusen
Leonard Michael Koff

# BRILL

LEIDEN | BOSTON

Cover illustration: Compendium (nocturnal, vertical & horizontal sundials, compass, solar quadrant), probably Nuremberg, 1514. With kind permission of the Staatliche Kunstsammlungen Dresden Mathematisch-Physikalischer Salon
Photographer: Jürgen Karpinski, Dresden

Library of Congress Cataloging-in-Publication Data

Names: Van Deusen, Nancy (Nancy Elizabeth) | Koff, Leonard Michael.
Title: Time : sense, space, structure / edited by Nancy Van Deusen, Leonard Michael Koff.
Description: Leiden : Brill, 2016. | Series: Presenting the past : central issues in medieval and early modern studies across the disciplines, ISSN 1875-2799 ; volume 5 | Includes bibliographical references and index. | Description based on print version record and CIP data provided by publisher; resource not viewed.
Identifiers: LCCN 2015051288 (print) | LCCN 2015049884 (ebook) | ISBN 9789004312319 (E-book) | ISBN 9789004310025 (hardback : acid-free paper) | ISBN 9789004312319 (e-book)
Subjects: LCSH: Time--Philosophy. | Time--History. | Time perception. | Space and time. | Chronology. | Time--Social aspects. | Time--Political aspects. | Time in literature.
Classification: LCC BD638 (print) | LCC BD638 .T5665 2016 (ebook) | DDC 115--dc23
LC record available at http://lccn.loc.gov/2015049884

Want or need Open Access? Brill Open offers you the choice to make your research freely accessible online in exchange for a publication charge. Review your various options on brill.com/brill-open.

Typeface for the Latin, Greek, and Cyrillic scripts: "Brill". See and download: brill.com/brill-typeface.

ISSN 1875-2799
ISBN 978-90-04-31002-5 (hardback)
ISBN 978-90-04-31231-9 (e-book)

Copyright 2016 by Koninklijke Brill NV, Leiden, The Netherlands.
Koninklijke Brill NV incorporates the imprints Brill, Brill Hes & De Graaf, Brill Nijhoff, Brill Rodopi and Hotei Publishing.
All rights reserved. No part of this publication may be reproduced, translated, stored in a retrieval system, or transmitted in any form or by any means, electronic, mechanical, photocopying, recording or otherwise, without prior written permission from the publisher.
Authorization to photocopy items for internal or personal use is granted by Koninklijke Brill NV provided that the appropriate fees are paid directly to The Copyright Clearance Center, 222 Rosewood Drive, Suite 910, Danvers, MA 01923, USA.
Fees are subject to change.

This book is printed on acid-free paper and produced in a sustainable manner.

# Contents

# List of Figures

# List of Abbreviations

| | |
|---|---|
| AASS OSB | *Acta sanctorum ordinis s. Benedicti*, ed. Jean Mabillon (2nd ed. Venice, 1733–40) |
| Anal. Mon. | *Analecta monastica* (Studia Anselmiana 20, 31, 37, 41, 43, 50 [Rome, 1946–1961]) |
| Bernard, ed. Leclerc | *Sancti Bernardi opera*, Jean Leclerc, Charles H. Talbot, and Henri-Marie Rochais, eds. (Rome, 1957–1977) |
| Bernard, ed. Mabillon | *Sancti Bernardi...opera*, ed. Jean Mabillon (Paris, 1839) |
| BHL | *Bibliotheca hagiographica latina* (Subsidia hagiographica 6 [Brussels, 1898–1901]) |
| BRG | *Bibliotheca rerum germanicarum*, ed. Philipp Jaffé (Berlin, 1864–1873) |
| CCCM | *Corpus Christianorum Continuatio Mediaevalis* |
| CCSL | *Corpus christianorum. Series latina* (Turnhout, 1953ff.) |
| CHLMP | Cambridge History of Later Medieval Philosophy |
| CSEL | *Corpus scriptorum ecclesiasticorum latinorum* (Vienna, 1866ff) |
| DHGE | *Dictionnaire d'histoire et de geographie ecclesiastiques* (Paris, 1912ff) |
| Duchesne | *Histoire Francorum scriptores*, ed. Andre Duchesne (Paris, 1636–1649) |
| GC | *Gallia christiana* (Paris, 1715–1865) |
| LMA | *Lexikon des Mittelalters* |
| Mansi | *Sacrorum conciliorum nova et amplissima collectio*, ed. Giovanni Domenico Mansi (Florence/Venice, 1759–1798, repr. 1901–1927) |
| MGH | *Monumenta Germaniae historica* |
|   Briefe | *Die Briefe der deutschen Kaiserzeit* (Weimar, 1949ff) |
|   Dipl. | *Diplomata regum et imperatorum Germaniae* (Hanover/Berlin/Weimar, 1879ff) |
|   Epp. | *Epistolae* in quarto (Berlin, 1887ff) |
|   ss. | *Scriptores* in folio (Hanover, 1826ff) |
|   ss rerum Merov. | *Scriptores rerum Merovingicarum* (Hanover, 1884ff) |
|   SSRG | *Scriptores rerum Germanicarum in usum scholarum separatim editi* (Hanover/Berlin, 1871ff) |
| PG | *Patrologia graeca*, ed. Jacques Paul Migne (Paris, 1857–1876) |
| PL | *Patrologia latina*, ed. Jacques Paul Migne (Paris, 1841–1864) |

| | |
|---|---|
| PL Suppl. | *Patrologiae cursus completus: Supplementum*, ed. Adalbert G. Hamman (Paris, 1958–1974) |
| SpicFri | *Spicilegium Friburgense* |
| RS | Rolls Series (Chronicles and Memorial of Great Britain and Ireland during the Middle Ages [London, 1858–1897]) |
| VIEG | *Veröffentlichungen des Instituts für Europäische Geschichte* |
| VMO | *Vita beatae Mariae Oigniacensis* |
| VRF | *Vorreformationsgeschichtlich Forschungen* |

# List of Contributors

*Jason Aleksander*
Saint Xavier University

*Aaron Canty*
Saint Xavier University

*Peter Casarella*
University of Notre Dame

*Michael Cole*
University of California, San Diego

*Nancy van Deusen*
Claremont Graduate University

*Andrew Eschelbacher*
Portland Museum of Art

*Ken A. Grant*
University of Detroit Mercy

*Mark Howard*
Claremont Graduate University

*Gerhard Jaritz*
Central European University

*Danielle B. Joyner*
Southern Methodist University

*Ike Kamphof*
Maastricht University

*Jordan Kirk*
Pomona College

*James F. Knapp*
University of Pittsburgh

*Peggy A. Knapp*
Carnegie Mellon University

*Raymond Knapp*
University of California, Los Angeles

*Leonard Michael Koff*
University of California, Los Angeles

*Sara E. Melzer*
University of California, Los Angeles

*Vera von der Osten-Sacken*
Humboldt-Universität zu Berlin

*Hans J. Rindisbacher*
Pomona College

*Jesse W. Torgerson*
Wesleyan University

# Introduction

Time is, as Augustine implies in *De ordine*, where he is philosopher and theologian, our God-given medium of ascent, known through the ordered study of the "liberal disciplines that carry the mind to the divine (*disciplinae liberales intellectum efferunt ad divina*)": grammar and dialectic, for example, to promote thinking; geometry and astronomy to grasp the dimensions of our reality; music, an invisible substance like time itself, as an exemplary bridge to the unseen substance of thoughts, ideas, and the nature of God (theology). This ascending course of study rests on procedure, progress, and attainment—on before, following, and afterwards—whose goal is an ascending erudition that lets us finally, as Augustine says, contemplate our unseen medium—time—within time itself.

> The truly learned are those who, not allowing all the different realities to distract them, attempt their unification into a simple, true, and certain whole. Having done so, they can soar on to divine realities not rashly and by faith alone, but contemplating, understanding, and retaining them. These realities are forbidden to the slaves of pleasure, or to those hankering after perishing things. But even those who flee from such and live a clean life can be ignorant of nothingness, of formless matter, of the form of inanimate things, of body, of species, of space and time, of being in space and in time, of local motion, of change in general, of uniform motion, of eternity, of what it is to be nowhere, or to be beyond time, or to be and not to be somewhere.[1]

Augustine values disciplined study that rests on moral strength, though a "clean life" does not guarantee access to metaphysical knowledge. It is "much

---

1 *De ordine*, bk. 2, chap. 16, pp. 107, 109 (*On Order*, intro. and trans. Silvano Borruso [South Bend, IN, 2007], pp. 106, 108): "Quibus si quisque non cesserit et illa omnia quae per tot disciplinas late varieque diffusa sunt, ad unum quoddam simplex verum certumque redegerit, eruditi nomine dignissimus, non temere iam quaerit illa divina, non iam credenda solum, verum etiam contemplanda, intellegenda atque retinenda. Quisquis autem vel adhuc servus cupiditatum, et inhians rebus pereuntibus, vel iam ista fugiens casteque vivens, nesciens tamen quid sit nihil, quid informis materia, quid formatum exanime, quid corpus, quid species in corpore, quid locus, quid tempus, quid in loco, quid in tempore, quid motus secundum locum, quid motus non secundum locum, quid stabilis motus, quid sit aevum, quid sit nec in loco esse, nec nusquam, et quid sit praeter tempus et semper, quid sit et nusquam esse et nusquam non esse, et numquam esse et numquam non esse." These questions are taken up in the final essay of this volume, Leonard Michael Koff's "No-Time in Non-Places."

© KONINKLIJKE BRILL NV, LEIDEN, 2016 | DOI 10.1163/9789004312319_002

easier," Augustine says, to come to such knowledge, as well as knowledge about the purposeful use of time, by "understanding the simple and rational [*intellegibiles*] numbers," that is, by being schooled from the beginning through reason's sequenced disciplines.[2]

> Getting acquainted with the liberal arts, however, whether pursued for the sake of usefulness or for the sake of knowledge and contemplation, is extremely difficult. It is necessary to be most clever and to start from childhood with unfaltering attention and perseverance.[3]

For Augustine, goal-directed motion contrasts with vain pursuits and purposeless movement,[4] and those who are inattentive to orderly study do not, to their detriment, let reason (*ratio*) guide them step-by-step upward; for Augustine, reason is "the mental operation with the power to distinguish between, and to connect, things we learn." As Augustine says, only a "few extraordinary men have the ability to let themselves be guided by [reason] into knowing God and the soul, their own or other people's."[5] Reason gives us human purpose ("not acting without design," "not acting rashly or heedlessly" [*primum nos admonet nihil temere facere*]); it teaches us to teach the truth (*recte docere*), that is, the liberal disciplines, and to regard contemplation (*beate contemplari*) as itself a discipline in reason's ascending course of study.[6] For Augustine, reason "in

---

2  Facilius autem cognoscet ista, qui numeros simplices atque intellegibiles comprehenderit, ibid., p. 108.

3  Cum enim artes illae omnes liberales, partim ad usum vitae, partim ad cognitionem rerum contemplationemque discantur, usum earum assequi difficillimum et nisi ei qui ab ipsa pueritia ingeniosissimus instantissime atque constantissime operam dederit, ibid.

4  "I don't give two hoots about the judgment of the proud and the ignorant. They rush into reading books as into kowtowing to men. They pay no attention to human qualities, but to clothes, pomp, and the ephemeral circumstances of well being. When they read, they don't care much about what is being written about, or what the author is trying to get at, or what he explains or accomplishes" (*On Order*, bk. 1, chap. 11, p. 43).

5  Ratio est mentis motio, ea quae discuntur distinguendi et connectendi potens... (*De ordine*, bk. 2, chap. 11, pp. 90–92).

6  *On Order*, bk. 2, chap. 12, p. 99. "As reason brought grammar to perfection, it sought to take care of itself, the very power by which it produced that art. It not only analyzed and ordered itself, but further defended itself against the infiltration of falsehood. How could it then pass on to other matter, without first distinguishing, bringing out and ordering the tools of its own trade, and so creating that discipline of all disciplines known as *logic* [dialectic]? [*Quando ergo transiret ad alia fabricanda, nisi ipsa sua prius quasi quaedam machinamenta et instrumenta distingueret, notaret, digereret proderetque ipsam disciplina disciplinarum quam dialecticam vocat?*] This discipline teaches how to teach, and how to learn. In it, reason shows itself

time" arbitrates because reason is authority and the nature of time, our unseen medium, is grasped through a variety of rational projects: time is immaterial, but experienced as substantial.

The essays in this volume all explore the nature of time in light of Augustine. They look at, in this order, one early Greek project to chronicle historical time from the creation; a pope's spiritual vision of medieval Rome as present and future; a twelfth-century religious manuscript that divides time into liturgical time; the concept of *synderesis* ("simultaneous times"); a theologian's clarification of the idea of creation "in time"; the spiritual presence implicit in two modes of biblical exegesis; and the texts of no-time and not-knowing that describe an *athomus* of time. In addition, the essays here also describe time in social and political contexts: examining the boundaries separating religious life and spiritual service; charting the development of the medieval urban clock; seeing Dante's "illustrious" vernacular as the language of sustaining temporal authority; and following the eighteenth-century French debate about time and its languages—did each successive era degenerate in relation to a more perfect civilization in the past when giants and geniuses supposedly walked the earth? Finally, the essays describe time as notated in music, as embodied in memorializing stone, and as the subject and medium of consciousness.

We begin in the ninth century when George Synkellos composed a grand synthesis of the passage of time from the beginning of the world to his present day. The conception of chronological succession as described in Genesis has,

---

for what it is, what it wants, what it can do. It knows that it knows. It alone is both willing and capable of making people learned" (*On Order*, bk.2, chap. 13, pp. 101, 103, italics Borruso [*De ordine*, p. 100]). As for music, "reason wanted to take off into the heights of contemplation of divine things [*ratio ad ipsarum divinarum beatissimam contemplationem rapere voluit*]. Not to fall off those heights, though, it sought to climb in steps along a path hewn by its own devices. It sought that beauty which can exclusively be attained in simplicity without bodily eyes, but the senses stood in the way. Therefore it slowly turned its attention towards those same senses. These, staking a claim for the possession of truth, distracted it from its pursuit of higher things with their oppressive clatter. The ears were first, claiming as their possession the very words which had served as the basis of grammar, logic and rhetoric. But its immense powers of analysis, however, reason saw at once the difference between a sound and the reality it symbolized" (*On Order*, bk. 2, chap. 14, p. 103 [*De ordine*, p. 102]). From the ear, "reason stepped into the domain of the eyes. Surveying the earth and the heavens, it found that nothing but beauty pleased it: within beauty forms, within forms proportion, within proportion number. It asked itself where in the real world this or that straight, curved, or other line as conceived by the intelligence might be. It found reality far inferior. Nothing real stood comparison with what the mind could see" (*On Order*, bk. 2, chap. 15, pp. 105, 107).

of course, a Genesis-commentary tradition as early as the ninth century in Latin as well as Greek.[7] Synkellos' *Chronography*, as Jesse Torgerson describes it, sets up "a dialogue between astronomical handbooks, the canonical scriptures, logic textbooks, theological and philosophical treatises, homilies, and liturgical commentaries. Synkellos used his *Chronography* to investigate the relationship between a timeless God and a time-bound Creation" and his "investigation produced the hypothesis that humanity experienced the Trinitarian God's direct interventions into linear time as a recurring 'First-Created Day.'" Central to Synkellos' *Chronography* was, as Torgerson explains, an Aristotelian analysis of time: "Every change or movement occurs in time,"[8] and while time is not equal to motion, time is its measure. In Synkellos' *Chronography*, time was thus "not only the measure of motion, the ordering of the ages, and the progress of kingdoms, but...also bore witness to the relationship between mankind and divinity through the experience of its rupture: the past in the present and the present in the past. Synkellos took the theological principle that the Incarnation was *the* truth event and embraced its chronological paradox in a way no previous Christian chronographer had ever attempted" (Torgerson).

In *History and Memory*, Rosamond McKitterick suggested that "to examine time and its functions in the early middle ages may yield something very specific about the perception of the past, present, and even future on the part of any group."[9] In fact, Torgerson sees time as an example of "an early medieval

---

7  Among the commentators on Genesis, in both Latin and Greek, one can include Basil, *Hexaemeron*, Latin paraphrase of *Eustathius: ancienne version latine des neuf homelies sur l'Hexaemeron*, eds. Alexis Smits and Michel van Estbroeck (Paris, 1970), Sources Chrétiennes, 160; *Auctorum Incertorum, vulgo Basilii vel Gregorii Nysseni, Sermones de Creatione Hominis, Sermo de Paradis,* ed. Hadwiga Hoerner (Leiden, 1972); Ambrose, *Hexaemeron (Exameron)*, ed. C. Schenkl (Vienna, 1897) CSEL XXVII.1; Augustine, *De Genesi ad litteram*, ed. Josephus Zycha (Vienna / Prague / Leipzig, 1894) CSEL XXVIII.1; Jerome, *Hebraicae quaestiones in Genesim*, PL XXIII; Bede, *Opera de temporibus*, ed. Charles W. Jones (Cambridge, MA, 1943); *Hexaemeron*, PL XCI; John Chrysostum, *Homeliae in Genesim*, PG LIII, among others.

8  See Aristotle, *Physics*, trans. Philip H. Wicksteed and Francis M. Cornford (Cambridge, 1929, rev. and rpt. 1967), iv, chap. 14, 223$^a$.14–15 (p. 419).

9  *History and Memory* (Cambridge, UK, 2004), p. 86. On time and history, see Denis Feeney, *Caesar's Calendar: Ancient Time and the Beginnings of History* (Berkeley / Los Angeles, 2007). On perhaps the most famous attempt to re-create time and begin human history again, see Sanja Perovic, *The Calendar in Revolutionary France: Perceptions of Time in Literature, Culture, Politics* (Cambridge, UK, 2012), esp. pp. 1–2: "Sweeping away history in order to start time anew, the revolutionary calendar attempted to accomplish what had never been done before: make time express the *intentions* of history... Gone was the memory of the Sabbath, when God himself took a rest. Time rejoined the secular world. Human time and its agents became the

culture actively thinking about time in terms of the experience of worship, even as it held to a rigorous philosophical and historical time." This observation suggests that time perception is an important marker for dealing with any cultural or geographical circumstance, any group, any individual within a specific milieu. In his monumental study of what he labeled "political theology," Ernst Kantorowicz identified the imposition one on the other of "the king's two bodies," that is, the tension that arose from the fusion of corporate office with and beyond the natural lifetime, the bodily manifestation, of the one who held it.[10] Ken A. Grant extends this to place, describing Gregory VII's juxtaposition of the temporal and actual with the eternal and idealized in seeing Rome. On the one hand, Gregory viewed Rome as a city of ruins, in chaos, unable to function, a city of stinking market-places, of displaced persons, of animals roving about on muddy streets, and rampant moral and political corruption, and, on the other, as the "Eternal City," the center of Christendom, and the prefiguration of things to come, a "New Jerusalem," with the people of God's Eternal Kingdom. This was a difficult conjunction, but there is a good deal of evidence that Grant brings together supporting, in Pope Gregory's spiritual imagination,

---

material through which a break with the religious and political structures of the past was to be accomplished... The Republican calendar...aimed to create a new collective memory based on the idea of a natural equality. It was only once the collective memories of the different social classes could be conceived as belonging to one and the same time—a universal time that had now become the time of the French Republican state—that the birth of a new society could be established... If the old calendar was able to integrate celestial patterns, biological rhythms, the season and cycles of social life into *one* collective experience of time, why could a new calendar not do the same for the Revolution? In a feat of revolutionary magic, the calendar was to transform the ongoing power struggles that threatened to capsize the new republic into a *common experience of time itself*" (italics his). For an account of a "new chronotope"...where "agency, certainty, and the historical progress of mankind have faded into distant memory" so that "we are left only with unredeemed desire, uncertainty, and disorientation" and a "future that we never chose threaten[ing] us (p. 38)," see Hans Ulrich Gumbrecht, *After 1945: Latency as Origin of the Present* (Stanford, 2013).

10   Ernst H. Kantorowicz, *The King's Two Bodies: A Study in Mediaeval Political Theology* (Princeton, 1957). The king ate, slept, probably occasionally became irritated, and exercised the normal everyday routines all human beings have in common, in his own, howbeit kingly, body. The king could display the immaturity of a child, as well as its analogue at the end of his lifetime, in other words, all manner of human weaknesses. But fused with the king's humanity was the ideal, corporate, "body" of his office, his anointment within a conceptually eternal succession founded upon, and sustaining, the legal system of the land. Kantorowicz examines the fusion and simultaneity of the temporal, observable, flawed, but actual, "body" of the king, with the ideal, timeless, kingly office.

the fusion of the past, present, and ultimately the future of Rome. Gregory's idealized vision of Rome provided the impetus for Gregory's lifelong papal project toward Rome's realization as the City of God.

The narrative history of salvation, which constitutes the background for both concepts of chronology as well as simultaneity, also provides a template for a twelfth-century manuscript, the *Hortus deliciarum*, "an interpretative and interactive history of salvation," the topic of Danielle B. Joyner's essay. Hers is a multi-faceted study of the dimensions of time measurement. Time, defined and illustrated, established the order of days, weeks, months, and years within the liturgical calendar for the canonesses of Hohenbourg, presenting a program for life, as well as education, for this community.[11] Liturgical time relied upon the computation of Easter as celebrated on the first Sunday after the first full moon; this occurs on or after the vernal equinox. In spite of this, multiple systems well into the ninth century CE coexisted for advancing Lenten seasons, even over 532-year cycles. This computation of Easter and, accordingly, the Lenten seasons, a tool for discerning when Easter would occur each year, coincided with the study of the Bible, systematic theology, historical chronologies, and the four modes of interpretation—literal, allegorical, tropological, and eschatological—that constituted foundational principles of medieval monastic education. The motions of the heavens were related to daily devotions on earth within the history of salvation. The division of days into liturgical units and the division of the year into specific days, which created an ultimate and deliberate consciousness based on the passing of time, was constantly compared with the Psalms' timelessness, the Christian view of eternity, and the projection of the Christian before God, in the presence of the saints, at the end of time.[12]

---

11    Cf. Fiona J. Griffiths, *The Garden of Delights: Reform and Renaissance for Women in the Twelfth Century* (Philadelphia, 2007), a study of the *Hortus deliciarum* as "offering an education on par with that of the clerics who oversaw their pastoral care" (Joyner). The manuscript, ca. 340 folios, was compiled by Abbess Herrad (ca. 1175–1195) for the women at Hohenbourg. Although the manuscript is not extant, a facsimile edition has been published by the Warburg Institute, ed. Rosalie Green et al. (London, 1979).

12    For example, Psalm 16, 11: "Thou wilt shew me the path of life: in thy presence is fullness of joy; at thy right hand there are pleasures for evermore"; Psalm 17, 15: "As for me, I will behold thy face in righteousness; I shall be satisfied, when I awake, with thy likeness"; Psalm 23, 6: "Surely goodness and mercy shall follow me all the days of my life: and I will dwell in the house of the Lord forever"; Psalm 41, 13: "Blessed be the Lord God of Israel from everlasting, and to everlasting. Amen, and Amen"; Psalm 45, 17: "I will make thy name to be remembered in all generations: therefore shall the people praise thee forever and ever"; Psalm 48, 14: "For this God is our God forever and ever..."; Psalm 52, 9: "I will

According to Nancy van Deusen, Plato's *Phaedo*, translated from Greek into Latin at the beginning of the second half of the twelfth century, only slightly previous to the arrival, dissemination, discussion of, and commentaries on, the Latin translation of Aristotle's *Physics*, helped deepen an over-arching philosophical subject: the "harmonization within time of opposing directionalities": emotional, physical, and mental motion. In the *Phaedo*, the narrator discloses "the doctrine that the soul is a kind of harmony," and it is the concept of the composite soul, with the properties of both visible and an invisible *materia-substantia*, that established lines of reasoning, as well as disciplinary boundaries (namely the disciplines of physics as well as theology), at the nascent university at Paris. It is there that Philip, known as "The Chancellor" for his position as liaison between the newly-founded faculty of theology and the episcopal see of Notre Dame, addressed the topic of "simultaneous harmony": from a newly-constituted concept of *armonia* to the topic of *synderesis*, or the bringing together within time itself of opposite tendencies within the human spirit. *Synderesis*, a "potency" and "force," connects the "parts" of mind, spirit, and body in "a simultaneous, integrated, unity" that is exemplified by music. *Synderesis* is the "unification for which the Apostle Paul prayed, a prayer that is invoked at the onset of Philip's lengthy, complicated, discussion of what was, for him, an extremely important and useful concept" (van Deusen).

Peter Casarella's essay on the "music of creation" connects the mental milieu of the thirteenth century and the project of coordinating Augustine and Aristotle that is the agenda, for example, of Philip the Chancellor and, as we learn, of St. Bonaventure. Casarella points out that "theologians of the thirteenth century...inherited the notion that divine eternity could be equated with a form of simultaneity. They recognized a tradition of learning whereby Augustine had defined God's eternity as an eternal presence" transferred to a "scholastic tradition of the Divine 'standing now' (*nunc stans*)." But how does one think about the relation of eternity to time? How does one come to terms with duration? One response to this question is the response of Alexander of Hales; he argues that the moment of creation was not with time, or in time, or

---

praise thee for ever...I will wait on thy name; for it is good before thy saints"; Psalm 61, 8: "So will I sing praise unto thy name for ever, that I may daily perform my vows" ("King James" translation of the psalter). The incremental spiritual progress that Christians recognize in the Psalms, which culminates in the eschatological *modus* and the presence of the Christian before God—this is the end of time—is also characteristic of the liturgical sequence that occurs between the Alleluia and the Reading of the Gospel within the Mass. See Nancy van Deusen, "The Use and Significance of the Sequence," *Musica Disciplina* 40 (1986), 1–46.

without time.[13] But Bonaventure offers, to an extent, an alternative conceptualization of time itself.

As for the texts that let us speak of our presence in God's creation, Aaron Canty examines two modes of spiritual exegesis and their implicit assumptions about spiritual time and our participation in it. Hugh of St. Cher's primary interest is "in the allegorical and tropological senses [of the Psalms] and how a spiritual reading of [them] invites the reader...even propels the reader to contemplative union with God." For Hugh, the Psalms express a "mystical ascent to God, which culminates in the words of the last psalm, 'Let everything that breathes praise the LORD!'" For Thomas Aquinas, whose mode of exegesis is literal and intertextual, the Psalms are works of prayer and prophecy, and in revealing future events, "they must refer literally to some future event. This is the basis for Thomas' desire to emphasize the literal sense of the Psalms over the figurative sense, which in his view applies to David or Solomon" (Canty).

No-time and not-knowing, Walter Burley, and the "work of unknowing" (in the *Cloud of Unknowing*, of uncertain date and authorship) is the topic of Jordan Kirk's essay. The suggestion, within "a metaphysics of prayer," that possibility may be inseparable from existence is, for Kirk, a further temporal articulation in which potentiality (the possible future) can be seen to coincide with actuality (the present). Famously, Aristotle in Book 9 of the *Metaphysics* chastises the Megarians for maintaining that capability is possible only in acting,[14] or in the *Physics*, that generation is possible only in fulfillment or actualization. Within this larger discussion, there exists a concept of an *athomus* of time— an instantaneous, momentary, entirely brief, enclosed, containment of time, "exactly the duration of a syllable" which is "a minimal unit of utterance"—a discussion that leads Kirk to locate a nexus of time concepts in late medieval logic and the commentary on Aristotle's *De interpretatione* that Walter Burley produced in the first decade of the fourteenth century.

From the metaphysics of prayer, we move to an aspect of monasticism's spiritual and social boundaries. Vera von der Osten-Sacken describes a perception of eternity lived within the confines and opportunities of everyday life. The Beguines were not interested in completely withdrawing from life's realities. Rather, "according to their biographers, the Beguines were aware of the option to

---

13    Alexander of Hales, *Summa fratris Alexandri*, ed. Bernardin Klumper (Quaracchi, 1924),
      VI.1.2.4: Sic creatio—actio non est sine tempore—et comprehendo sub tempore nunc—
      quod est principium temporis—non tamen est in tempore, cum sit actio Dei prima in
      creaturam; nec cum tempore proprie dicitur, quia non ei associatur vel parificatur; sed
      non sine tempore.

14    Aristotle, *Metaphysics, Books I–IX*, trans. Hugh Tredennick (Cambridge, MA, 1933), IX.3.

live in visible and invisible spheres of society and also in the eternal world. Their intensive transchronological and antichronological mysticism went beyond the borderline between time and timelessness; biographers describe how the mystics receive access to the afterlife, where their souls could temporarily stay, while their bodies remained in the here and now" (von der Osten-Sacken).

But how was time in the "here and now" known? Gerhard Jaritz begins with Niclas Swaelbl, a clockmaker, at work in Tulln, "although there are no more surviving sources about this particular novelty of a striking clock in the small Lower Austrian town at a time when even the nearby capital of Vienna did not yet possess one." The earliest evidence of public clocks originates from northern Italy in the first half of the fourteenth century; by the mid-fourteenth century an "internationalization" of mechanical clocks had taken place in urban centers as well as residences in England, Sweden, France, and Germany—clockmakers traveled all over the continent of Europe, even to Ukraine and as far as Moscow (1404).[15] Urban clocks were marvelous, beautiful, representative of a town's status and wealth, and, of course, useful.

We should briefly mention here two other time-reckoning mechanisms: first, the astrolabe, an ancient astronomical instrument for telling ambient time, that is, time with respect to a person's position on earth in the context of the position of the sun and stars; second, the Linnaean flora clock, a natural clock that Carolus Linnaeus proposed in 1751, meant to take advantage of plants that open or close their flowers at particular times of the day, part of the Enlightenment's understanding of time as a purely natural, rather than a spiritual or historical, dimension of life—the clock was never planted in Linnaeus' day. As for the astrolabe, it made possible "vision in the round."[16] Moreover,

---

15    Gerhard Dohrn-van Rossum, *History of the Hour. Clocks and Modern Temporal Orders*, trans. Thomas Dunlap (Chicago / London, 1996), pp. 157–159 (161).

16    The most popular type of which was the planispheric astrolabe where the celestial sphere is projected onto the plane of the equator. The astrolabe "enables me to see in what psychologist James J. Gibson calls an 'ambient optic array' (see *The Ecological Approach to Visual Perception* [Hillsdale, NJ, 1986], pp. 1, 189–222). [It] is different from snapshot vision, which holds the eye still and focuses on one point. Although snapshot vision, like ambient vision, incorporates the whole cosmos, in the former I am not part of the picture; in the latter I am... Human eyes looking forward achieve a roughly hemispherical field of vision from side to side, a bit less from up to down; about 270 degrees by turning the head; full 360-degree. Ambient vision, however, represents not a snapshot series but an unfolding panorama with no vanishing point. An astrolabe allows me to see above, around, behind, and even, partially, beneath; it's not that it gives me eyes in the back of my head, it gives me omnidirectional, fish-eye vision. As observer, I am not outside the frame but inside and part of a continuous ambit of nested curves, lights, and

"what sets the astrolabe into calculative motion is an initial act of triangulation," which had profound consequences for social seeing and psychological analysis: "the medieval sense of space [which requires trigonometry to chart the heavens] puts bodies in a triangulated and hence mediated relationship with the world rather than in a subject/object dilemma."[17]

Fig. 0.1 is not, of course, a picture of an astrolabe: it is rather a picture of the Astronomical Clock now in the Sir John Soane's Museum in London, formerly the home of this neo-classical architect.[18] The 1837 Inventory of the Works of Art in the Soane Museum indicates that "the merit of the invention consists in its combining and exhibiting at one view the state of the world as acted upon in the progress of time, by the diurnal and annual revolution of the heavenly bodies immediately connected with our globe according to the Copernican system, shewing [sic] at the same time the hours and the corresponding position of those bodies in their respective orbits."[19] Unlike an astrolabe, our

---

edges—a cosmic amphitheater. There is as much happening behind me as there is in front. The nearest I can get to the spatial relation I am invoking is that of an actor acting in the round. Instead of projecting all one's presence forward toward a fourth wall of the audience you have to act with the back of your legs. In the cosmic space invoked by the astrolabe height is also depth, above is below, before behind. The circularity of time forcefully asserts itself" (Valerie Allen, "Time, Space, and the Doctrine of Triangles," paper presented at the 50th Medieval Institute Conference, Kalamazoo, Michigan, 2015, part of a larger ongoing project).

17      "Triangles are as essential to spatial figures as prime numbers are to quantity. One could even say triangles are nature's building blocks; after all, three of the five Platonic solids are formed from them (tetrahedron, octahedron, and icosahedron). Some figurate numbers are triangular. Any two legs must be longer than the remaining one... Essential to the definition of a triangle is that two collinear points, A and B, become meaningfully related once I introduce them to a third non-collinear point, C, which mediates the connection between A and B," ibid. On the implications for understanding social perception and personal awareness through visual, and figurative triangulation, see Rebecca Zorach, *The Passionate Triangle* (Chicago / London, 2013), esp. p. 23: "...perspective requires a third party—not just viewer or artwork, but a third position outside...[the] absorptive dyad." Fig. 9.8 shows a man using a sundial, not an astrolabe, but using a sundial entails establishing the same kind of triangulation that an astrolabe requires. The figure illustrates a colliding moment in the history of time-measuring mechanisms.

18      The Astronomical Clock was made by Zacharie Raingo of Paris in the early 1800s: he came to London after the Peace of Amiens in 1802 and sold six of these clocks to the Prince Regent (later George IV) for himself and his five brothers; the clock in the Soane Museum belonged to Frederick, Duke of York; Soane bought it for £75 after the Duke's death in 1827.

19      "The result is obtained as follows: To the clock is annexed a complicated, and at the same time a simply and beautifully executed kind of orrery, which is put in motion by the clock and as they perform together their several motions they shew [sic] that division of the

position with respect to the Soane Astronomical Clock does not define an angle from which we can tell both ambient and celestial time. Rather the clock has figuratively moved the entire universe inside: to a table in a nineteenth-century library, the material memory of a collector's indefatigable acquisitiveness.[20] The Soane is itself an eclectic museum that disregards the coherence of time; objects of Western historical eras are displayed one on top of another, floor to ceiling in each room, large and small, some rooms merely passageways to other rooms. Time in the Soane Museum is embodied in atemporally displayed historical objects.[21]

The Linnaean *Horologium Florae* (flower clock), where natural rhythm replaces historical time, reveals Linnaeus' understanding that flora circadian rhythms give a scientific basis to Enlightenment arguments about the orderly manifestation of Spirit in the world; a biologist's classification of flora demonstrated the presence of Spirit when it seemed purely fanciful and indeed irrational to believe that one could, even imaginatively, ascend to heaven to know earthly time from there. Andrew Marvell in "The Garden" (1678) anticipates Linnaeus' work.

How well the skilful Gardner drew
Of flow'rs and herbs this Dial new;

---

hour; the hour of the day; the day of the week; the day of the month; the month of the year; the degree and sign of the zodiac; the diurnal rotation of the earth upon its axis, producing the alternations of day and night for the different countries of the globe; the gradual progress of the earth in its annual revolution round the sun, combined with its elliptical movement which causes it to approach to and recede from the sun according to the seasons; the diurnal and annual rotation and elliptical motion of the moon round the earth as its satellite, with its phases, indicating at the same time its age and by means of a revolving dial placed above the globe, the true time and also (at will) the hour of the day or night in any given part of the world" (from the 1837 Inventory of the Works of Art in the Sir John Soane's Museum).

20    Cf. the Prague Astronomical Clock (installed in 1410), mounted on the southern wall of Old Town City Hall, whose mechanism is composed of three main components: the astronomical dial, representing the position of the Sun and Moon in the sky and displaying various astronomical details; "The Walk of the Apostles," a clockwork hourly show of figures of the Apostles and other moving sculptures, most notably Death as a skeleton striking the time; and a calendar dial with medallions representing the months. See Fig. 9.1.

21    The Museum's collections contain many important works of art and antiquities, including Hogarth's *A Rake's Progress* and *An Election*, Canaletto's *Riva degli Schiavoni looking West*, the alabaster sarcophagus of Seti I, 30,000 architectural drawings, 6,857 historical volumes, 252 historical architectural models as well as important examples of furniture and decorative arts.

Where from above the milder Sun
Does through a fragrant Zodiack run;
And, as it works, th' industrious Bee
Computes its time as well as we.
How could such sweet and wholsome Hours
Be reckon'd but with herbs and flow'rs!

As Marvell amusingly suggests, a natural clock is useful for the bee—and for us, where it tells "sweet and wholsome hours" of an imagined Edenic existence.

Dante in his *Paradiso* (17.13–27) invokes a geometric *figura* in which "earthly minds see that two obtuse angles cannot fit into a triangle" to bring to mind "contingent things before they come to be, gazing at the point to which all times are present." In the *Commedia*, "the pilgrim's interlocutors in Paradise" recognize "the temporal world's past, present, and future by gazing upon them *sub specie aeternitatis...* [By contract, the damned in the Inferno]...only know a present that is granted to them by the presence of the pilgrim." There are, in fact, two dimensions to comprehending time invoked here, as described in Jason Aleksander's essay: event and an underlying "providential" history. Thus, the greater one's intellectual capacity, the greater one's ability to see the providential structure of history, Virgil, of course, having been the most capable of all in his ability to grasp a "greater part of the truth than even that of which he himself is aware." As for the Virgil of the *Aeneid*, he sees Roman history and Roman time in the underworld and the world above, our world, on a continuum; historical time and present time are within all time itself: Roman spiritual geography is lateral (in parallel levels of being). Dantean spiritual geography is vertical. Moreover, neither Virgil nor Augustus imagines the end of Rome's presence; for just as the known world is Roman, so the liminal places that exist "outside Rome" are Roman as well: those living there exist in envious relationships with the empire. All time is Rome's: positive, eternal, where moral and political vicissitudes occur in an always Roman present; here time is the medium of hegemony.

For Dante, time, as the medium of potential salvation, enables the establishment, within providential history, of temporal authority. For such authority, an "illustrious vernacular" is a timeless first language, a "form of language created by God along with the first soul." That original and unchanging *forma locutionis* supports, within salvation history, salvific political power. Dante's understanding of the poetic and rhetorical function of the "illustrious" vernacular is tied to his political philosophy in a way that "depends upon a rich but ultimately unresolved tension" between "the demand that only an atemporal, unchanging vernacular would be suitable for the tasks of universal monarchy" and "the recognition that only a temporal, localized, and changing illustrious vernacular could possibly bring about the existence of the universal monarchy" (Aleksander).

For the eighteenth century, trying to locate itself within its classical and medieval past, history itself is an arc of change: it may have improved or it may have regressed. The issue of human time is central to the "Quarrel between the Ancients and the Moderns," a series of debates in early modern France that provides subject matter for Sara E. Melzer's essay. The topic is still with us: whether the past should be studied, emulated, and regarded as a "golden age" or whether history should be viewed optimistically as culminating with "us." Hesiod, Melzer points out, thought that the human race had deteriorated over time, degenerated into "baser" metals: from gold to silver, to bronze, to iron. As Melzer puts it: "This debate about the slope of time shaped every arena of human thought: science, philosophy, economics," with profound implications for France's intellectual life, influencing significantly, for example, the life and output of Jean-Philippe Rameau (1683–1764), his *Traité de l'harmonie* (1724), as well as *Les Cyclopes*, a violent, virtuosic piece for harpsichord about the race of one-eyed giants as portrayed in Homer's *Odyssey*, the subject of Mark Howard's essay, where time has no slope.

Howard looks at the *rondeau* in *Les Cyclopes* as the musical representation of cyclical time and space, suggesting that time as "intangible substance... expressed in the forms of measure, rhythm, duration, and repetition" and space as "unseen substance...defined through blocks or *chunks* of time with prescribed limits" are "materials that the composer may shape and manipulate in creative ways." Together they "produce *modulation* or movement over the course of a piece of music. They govern the listener's experience and cooperate with musical components such as the fundamental bass," Rameau's compositional tool "devised in order to understand harmonic particularities, relationships, and ultimately modulation or movement" (Howard).

Funerary monuments, memorials, effigies, and a "culture of memorialization through which the nation [of France] mediated its visions of the present and the future [using] objects and physical spaces that teemed with revisionist— or at least reductive—historical value" are the focus for Andrew Eschelbacher's examination of the relationship between past and present. How, for example, can an object, available and visible in itself, also contain within it unseen "substance," evoking past experience, emotional intensity, ideological passion, an entire related nexus of profound meaning only partially subject to analysis? One thinks, for example of the Holocaust memorial in Vienna as an object that draws the present momentarily into an indescribably horrifying past. Can such a past be in fact "memorialized"? Can any atonement or reconciliation be made in the present for what occurred in Vienna, and elsewhere, at the Nazi invasion and occupation? Eschelbacher points to nineteenth-century France in his useful and attentive comment: "But in funerary monuments of the nineteenth century, and the vast majority of the period's public sculpture more broadly, Dalou's *Blanqui* is remarkable in that it actively embraces the unsettled memories of the past rather than a projection of a didactic history. Rejecting the temporal fixity and boundedness of its funerary and ideological contexts, the *Blanqui* engages the ambiguities within French experience, challenging linear constructions of national histories as well as the roles of time and permanence in memorial culture."

Consciousness, both timed and untimed, is the topic of the six contributions that conclude this collection. First, Hans J. Rindisbacher's essay argues for a critical vocabulary that speaks to "temporal openness." The term *foreshadowing*, Rindisbacher says, "robs a moment of its presentness...depicting it as merely the shadow of a future event that obviously already has happened in the author's mind." The term *backshadowing* opens up the possibility that may lead the observer to "a kind of temporal egotism," which endows "our own actual present with special privilege." The backshadowing observer may even exhibit a "'tone of superiority' vis-à-vis those who failed to read literary signs in due time." Both foreshadowing and backshadowing are inimical to "eventness"; they imply that in life, mirrored in fiction, "possibilities always outnumber actualities." Rindisbacher suggests a term from Gary Morson called *sideshadowing* that "counters our tendency to view current events as the inevitable products of the past." Sideshadowing "emphasizes the here and now as the moment of decisions."

In this context, James F. Knapp and Peggy A. Knapp examine the problems of consciousness in two novels: Virginia Woolf's *Between the Acts*, which speaks, like all stream of consciousness narratives, to "the ephemerality of memory and sensation out of which the consciousness of self emerges," and Richard Powers' *The Echo Maker*, a "mystery story that explores the mental workings of its central characters, one of whom has sustained serious brain injury in a car crash and several others who try to help him."

For Raymond Knapp, "as music seems to set the terms for our experience of time, it must itself conform to the terms it sets, with every move it makes having either to fit into established temporal patterns or risk collapse into chaos. And it is, once again, in the space between the appearance and reality of music's relationship to time" that Stephen Sondheim finds "purchase for his own rich engagement" with time, "wherein music, as experienced in the 'now,' serves as a persuasive metaphor for the rhythms of life itself." Knapp follows the development of the musicals of Sondheim's middle period that, according to Knapp, provide a "particularly flexible and effective medium for the dramatic temporal confrontations that these manipulations all entail, stemming in part from the convention of the reprise, a traditional response to the felt need to replay something of the past, in affirmation, as part of an effective conclusion to a musical narrative."

In what seems counter intuitive, Michael Cole argues that the future, too, is a memory. "We are," he suggests, "all accustomed to the notion of remembering as the summoning up of past experiences in the process of dealing with the present." But as "both Augustine and Tennyson suggest, the relations between past, present, and future in human experience are a good deal more complicated than common wisdom leads us to believe." Cole moves through perhaps the most engaging description of a future remembrance that, for a reader, becomes both his present and, for Prince Myshkin in Dostoevsky's *The Idiot*, his, too. The scene Cole chooses to look at occurs when "Prince Myshkin is asked to describe a picture of an execution that illustrates in unusual detail the process by which human thought expands a present moment into an infinitely rich tapestry of [future] experience." The described scene illustrates, for Cole, the principle "that human experience of an image, such as a painting or photograph, requires the person viewing the image to place it within a temporal sequence in the act of making it interpretable" (Cole).

As for where time is as we "watch it" through a webcam is a question that falls out from the way technology changes our understanding of time's place in "double worlds." What, Ike Kamphof asks, is "real time" and the "here and now" when we "watch time"? Where does it occur? Who is in it? Is time what one "sees" in the "cast of a webcam"? Is time what one remembers seeing, what one expects, plans for and regards as "the future"? Leibniz "describes the endless stream of tiny, mostly unnoticed, perceptions that confront each individual" (Knapp and Knapp): "These...are therefore more effectual than one thinks. They make up this I-know-not-what, those flavors, those images of the sensory qualities, clear in the aggregate but confused in their parts; they make up those impressions the surrounding bodies make on us, which involve the infinite, and this connection that each being has with the rest of the universe."[22]

---

22    Leibnitz, "Preface to the *New Essays*," in *Discourse on Metaphysics and Other Essays*, trans. Daniel Garber and Roger Ariew (Indianapolis / Cambridge, UK, 1991), p. 55. See also Rocco J. Gennaro, "Leibniz on Consciousness and Self-Consciousness," in *New Essays on the Rationalists*, eds. Rocco J. Gennaro and Charles Huenemann (Oxford / New York, 1999), pp. 353–371.

We return in the final essay of this volume to the tension inherent in Augustine's famous question, "Where is law? It is in time and it is outside of time."[23] Leonard Koff suggests that it is possible to argue for "no-time," a kind of time, and he finds it in a variety of "non-places": in the real world and in the literary imagination that would describe both no-time and non-place. According to Aristotle, time past and time future depend for their reality on changes to existents perceived in time present: "So just as there would be no time if there were no distinction between this 'now' and that 'now,' but it were always the same 'now': in the same way there appears to be no time between two 'nows' when we fail to distinguish between them." Aristotle argues that time would not exist without change. But he also argues, in the form of a rhetorical question, that because "some of it [time] is past and no longer exists, and the rest is future and does not yet exist; and [because] time, whether limitless or any given length of time we take, is entirely made up of the no-longer and not-yet...how can we conceive of that which is composed of non-existents sharing in existence in any way?" Aristotle's question implies that a "now" without a discernable past or future means that time, as a measure of change, does not exist: without boundaries, there is "no time" in Aristotle's sense. His argument here, however, implied by his rhetorical question, which suggests that there is no change (hence no time), is an argument that for Aristotle isn't true. The argument is meant to catch those who, for Aristotle, reason incorrectly. A "now of indeterminate duration" has a past that, yes, "no longer exists," but that had to exist, and a future that does not "yet exist," but will exist. There always is, for Aristotle, a past and a future, hence a present, hence change, hence time, the measure of change. Nonetheless, Aristotle's argument for a "now of indeterminate duration," implied by his rhetorical question, provides the context for the argument that there can be no-time, that is, time without change, which is not the same as "no time."

For Koff, "no-time is maker-dependent that may be, or that in fact is in certain contexts—theological, for example, or utopian or dystopian—maker-independent. The content of no-time can be changeless or seen as changing (within a static indivisible instance) or changed; its content can take us out of time's flow, as Aristotle understands the 'flow of time,' and arrest that flow, conceiving it or reconceiving it in some valuable or, in some cases, necessary way. No-time can seem like—can be experienced as—an interval of time, but not in Aristotle's sense of interval as defined in the *Physics*." No-time can be

---

23    A good deal has been written on timed and untimed law, but see Nancy van Deusen, "*Ubi Lex*? Robert Grosseteste's Discussion of Law, Letter, and Time and Its Musical Exemplification," *Dayton Philosophical Review* 22 (1994), 219–232.

created or found and in a non-place one may wish to enter and from which one may not want to exit, for as Adrian Bardon puts it, *"Time is not so much a 'what' as a 'how,' and not so much a question as an answer.* Time as we know it in experience is a matter of how we adaptively organize our own experiences; in a physical and cosmological context, it is a matter of how we can most successfully model the universe of occurrences. As such, time is an answer: a solution to the problem of organizing experience and modeling events."[24]

The illustration on the cover, a Compendium, probably Nuremberg, 1514—it includes instruments for measuring time using the heavens (a solar quadrant, for example) and an instrument for guiding our movements on earth (a compass) to places where we can measure time—illustrates the essential aspect of time's nature: its dependence on our angle of vision, our angle of awareness, even when we would know time in itself.

---

24    *A Brief History of the Philosophy of Time* (Oxford, 2013), p. 175 (italics his).

FIGURE 0.1
*The Astronomical Clock by Zacharie-Nicholas-Amé-Joseph Raingo. c. 1820–1825.*
*Formerly the property of the Duke of York*
BY COURTESY OF THE TRUSTEES OF SIR JOHN SOANE'S MUSEUM

# Time and Again: Early Medieval Chronography and the Recurring Holy First-Created Day of George Synkellos

*Jesse W. Torgerson*

Ἐν ἀρχῇ ἐποίησεν ὁ θεὸς τὸν οὐρανὸν καὶ τὴν γῆν
In the beginning God created the heaven and the earth

GENESIS 1,1

Ἐγὼ τὸ Ἄλφα καὶ τὸ Ὦ, ὁ πρῶτος καὶ ὁ ἔσχατος, ἡ ἀρχὴ καὶ τὸ τέλος
I am the Alpha and the Omega, the first and the last, the beginning and the end

APOCALYPSE OF JOHN 22,13

Near the beginning of the ninth century, while residing in the environs of Constantinople, George Synkellos began composing a grand synthesis of the passage of all time—a chronography—from the Creation of the world up to his present day. Though the work, thus described, would seem to be an ideal candidate for a "Byzantine view of time," the absence of even a sketch of the author's life and career makes it particularly difficult to set George Synkellos and his text in context, let alone to posit the *Chronography* as representative. Still, we do what we can with what we have.

The little that is currently known about our author's life is extrapolated from a few fragments of data buried in the *Chronography*.[1] Many of these clues link George Synkellos to Syria-Palestine, but the only sure information is George's epithet, Synkellos.[2] Though scholars usually refer to George Synkellos as

---

1   Alexander Kazhdan, "George the Synkellos," *Oxford Dictionary of Byzantium* (Oxford, 1991) (hereafter *ODB*). The most comprehensive assessment of Synkellos' biography has just appeared in Warren Treadgold's *Middle Byzantine Historians* (New York, 2013). I am indebted to the author for advance consultation.

2   Synkellos spent significant time in Syria-Palestine; Synkellos originating from the region remains plausible. See William Adler and Paul Tuffin, *The Chronography of George Synkellos: A Byzantine Chronicle of Universal History from the Creation* (Oxford, 2002), pp. lxviii–lxix, lxxxi–lxxxiii; and most recently Robert Hoyland, *Theophilus of Edessa's Chronicle* (Liverpool, 2012), p. 11 n. 30. The key passage is Synkellos' statement: "In my journeys between Jerusalem

© KONINKLIJKE BRILL NV, LEIDEN, 2016 | DOI 10.1163/9789004312319_003

simply Synkellos, the word is not a name but an office. Between the fifth and the ninth centuries the office of *synkellos* (σύγκελλος) had developed within the Christian communities of the Eastern Mediterranean from a senior monk's attaché (the word literally means "cell mate") into an imperially appointed liaison to a patriarchal bishop. We know that the monk George Synkellos was *synkellos* to Tarasios, the patriarch of Constantinople from 784 to 806.[3] Nevertheless, attempts to further pin down Synkellos' career have resulted in little more than a series of educated guesses. In fact we only have a vague idea of the duties of late-eighth or early-ninth-century *synkelloi* in general.

To illustrate the point: on the basis of his office we might consider George Synkellos a part of the civil bureaucracy. As the *synkellos* of Tarasios, George would have attained his position through appointment by either Irene (regent 780–795; empress 797–802) or her son Constantine VI (r. 780–797). Mid-ninth-century sources on palace ceremonial rank the *synkellos* as one of the highest officials in the entire imperial hierarchy.[4] As a semi-regular at the imperial table with intimate access to his rulers, George must have been one of the most known figures in the palace. But George's office could just as well identify him as an integral part of the ecclesiastical hierarchy. According to the same ninth-century sources the *synkellos* only attended the imperial feasts for the twelve days of Christmas as a member of the *patriarch's* entourage.[5] George

---

and Bethlehem and what is known as the Old Laura [monastery] of blessed Chariton, I personally have passed by there frequently and seen [Rachel's] coffin lying there [in her tomb] on the ground." AT 153/M 122. I use this citation form throughout to refer to the *Chronography*—AT {page number}/M {page number}—with "AT" referring to Adler and Tuffin's translation of George Synkellos' *Chronography*, and "M" referring to Mosshammer's critical edition (Alden Mosshammer, *Ecloga chronographica* [Leipzig, 1984]).

3   The surest piece of information about George is found in a preface repeated in several surviving manuscripts: he was a monk and was *synkellos* under Patriarch Tarasios. Ὁ μὲν μακαριώτατος ἀββᾶς Γεώργιος, ὁ καὶ σύγκελλος γεγονὼς Ταρασίου, τοῦ ἁγιωτάτου πατριάρχου Κωνσταντινουπόλεως (Theophanes' *Preface* in Carl de Boor, *Theophanis Chronographia* (Leipzig, 1883), pp. 3.8–3.9); Ἐκλογὴ χρονογραφίας συνταγεῖσα ὑπὸ Γεωργίου μονάχου συγκέλλου γεγονότος Ταρασίου πατριάρχου Κωνσταντινουπόλεως (*Preface* to the *Chronography* at Mosshammer, *Ecloga*, pp. 1.1–1.6); Γεωργίου τοῦ εὐλαβεστάτου μονάχου καὶ συγκέλλου γεγονότος Ταρασίου τοῦ ἁγιωτάτου ἀρχιεπισκόπου Κωνσταντινουπόλεως σύνταξις ἤτοι χρονογραφία (The second *Preface* at *Ecloga*, pp. 360.1–360.4).

4   Philotheos, "Kletorologion," in N. Oikonomides, *Les Listes de préséance byzantines des IXe et Xe siècles* (Paris, 1972). See p. 163.10 and Footnote 129 for clarification.

5   Philotheos, "Kletorologion," p. 185.21. Aristeides Papadakis, "Synkellos," ODB. Constantine VII's tenth-century *Book of Ceremonies* documents the ritual appointment of these officials (II.5) and their very high rank (II.52). See J. Reiske, *De Ceremoniis Aulae Byzantinae* (Bonn,

Synkellos would have been important, but exactly when, and within which spheres of influence, we cannot say.

Even identifying Synkellos with major contemporary events is nearly impossible. If George Synkellos was in Constantinople towards the beginning of patriarch Tarasios' reign, he would have been present for an extremely significant council of the Christian churches in 787 (later canonized as the Seventh Ecumenical Council). Could George have been the *synkellos* at this time even though there is no *"synkellos* of Constantinople" mentioned in the council's acts?[6] Alternatively, was George included among the patriarchal clergy as the "deacon George" who read out an excerpt from a homily?[7] Even if one accepts that George was present for the event in one of these roles, was he returning to Constantinople for the summit or making his very first visit to the empire's capital? This, and every other proposal, remains a speculation.

We are left to elucidate the historical George Synkellos from what we can find in the *Chronography*, his one surviving work. Here, mercifully, we do find a reliable bit of biography to stand on. With a splendid piece of inductive reasoning, Richard Laqueur proposed that George Synkellos began writing his

---

1829), pp. 530.6–532.4, 713; 727. From the fifth through the ninth centuries the office's occupants were usually monks and deacons; they acquired enough influence to occasionally attain the patriarchal throne themselves. Due to our lack of specific knowledge we do not know what George's promotion signified or entailed. An audit of the Jerusalem patriarchate around 808/9 by *missi* of Charlemagne mentions a single *synkellos* of the Patriarch of Jerusalem "who manages everything under the patriarch" (*sincelo qui sub patriarcha omnia corrigit*). See Michael McCormick, *Charlemagne's Survey of the Holy Land* (Dumbarton Oaks, 2011), p. 200 ("Document 1" lnn. 7–8). According to Ignatios the Deacon's mid-840s *Life of the Patriarch Tarasios*, emperor Constantine VI appointed two *synkelloi* in approximately 796 to restrict access to the Patriarch (Chap. 47). How does the career of our George Synkellos relate to this situation? See the comments on p. 237 in Stephanos Efthymiadis, *The Life of the Patriarch Tarasios* (Birmingham, 1998). For discussion and further bibliography, see Warren Treadgold, *The Early Byzantine Histories* (New York, 2007).

6  In the 870s the Roman Anastasius Bibliothecarius also assumed George Synkellos was at the Council of Nicaea in 787, though this seems to be Anastasius' own deduction based on Synkellos' association with patriarch Tarasios. See Anastasius' letter to John the Deacon, "Epistle 7," eds. E. Perels and G. Laehr, "Anastasii Bibliothecarii Epistolae sive Praefationes," in *Monumenta Germaniae Historica Epistolae* VII, 2 (Berlin, 1928) pp. 420.5–420.11.

7  From the *Acta* of the Council of Nicaea in 787 "George, the most God-loving deacon and notary of the holy patriarchal residence" (Γεώργιος ὁ θεοφιλέστατος διάκονος καὶ νοτάριος τοῦ εὐαγὸς (sic.) πατριαρχείου) read from a sermon by Bishop Antipater of Syrian Bostra: J.D. Mansi, *Sacrorum Conciliorum Nova et Amplissima Collectio* v. 13 (Florence, 1767), col. 13D–E.

chronography in AD 808 and stopped merely two years later in AD 810.[8] In this short time Synkellos managed to accomplish a great deal: he covered nearly six thousand years of the past: from the Creation of the world on Day 1 in the first "Year of the World" (conventionally written "AM 1" for the Latin *anno mundi*),[9] up to the reign of the Roman Emperor Diocletian in AD 284 (by Synkellos' reckoning AM 5777, or 5,777 years from the Creation). There is good evidence that Synkellos bequeathed the completed portion of his proposed master work, along with drafts, notations, or excerpts of what still remained to be written, to the abbot Theophanes the Confessor.[10] Theophanes stated that Synkellos did so because he was physically incapable of fulfilling the original plan; he was dying.

Theophanes disseminated a continuation of the *Chronography* under his own name as the *Chronicle* (believed to have been completed by AD 814).[11] The *Chronicle* of Theophanes completed Synkellos' project, recording the period from the reigns of Diocletian and Constantine the Great (AD 284 or AM 5777) up to his present day (AD 814 or AM 6305). It is worth comparing, for a moment, the reception of these two halves of what was intended as a single work. Theophanes' *Chronicle* has received a great deal of scholarly attention as arguably the single most important surviving source for the early medieval past; without it we would not possess any continuous contemporary account of East Mediterranean events from the mid-seventh century to the late eighth.[12] The *Chronicle*'s perceived importance is also due to the fact that it was composed of excerpts from an array of unattributed sources, many of which would

---

8    Richard Laqueur, "Synkellos," *Paulys Real-Encyclopädie* (Stuttgart, 1932), col. 1398. Synkellos quite possibly retained his position through the death of Tarasios until he began writing his *Chronography* in AD 808. In 810 Synkellos appears to have updated some but not all of all the references to "the current year" in his work. See AT 3/M 2 (dating the present as both AD 808 *and* AD 810), AT 8/M 6 (dating the present as AD 808), and AT 301/M 244 (dating the present as AD 810).

9    Though scholarly convention demands the Latin "AM," Synkellos calculated by Κόσμου ἔτη ("in the year of the universe").

10   Theophanes was abbot of Megas Agros near Constantinople in Bithynia. Alexander Kazhdan, "Theophanes," *ODB*. Theophanes stated that Synkellos "both bequeathed to me (who was his close friend) the book he had written, and provided materials with a view to complete what was missing" (Mango and Scott, *Chronicle*, p. 1).

11   The argument dating the work's completion is tidy: the chronicle ends with March 813 and seems to give a positive view of the Emperor Leo V (r. 813–820), who in 815 would re-impose iconoclasm. Theophanes, a supporter of the use of icons, was unlikely to portray the emperor positively unless the ban was not yet official.

12   See the representative comments in George Ostrogorsky's textbook, *History of the Byzantine State*, trans. Joan Hussey (New Brunswick, 1969), pp. 87–89.

otherwise be unknown. Scholars have spent decades on retracing the author's steps in compiling the work. The attention currently being devoted to whether one of these sources is the lost *Chronicle* of Theophilos of Edessa gives an idea of the importance of this task to the common historical record.[13] The *Chronicle* of Theophanes remains a bottomless well for scholarly curiosity.[14]

The importance of Theophanes' *Chronicle* for the early medieval history of the Eastern Mediterranean stands in stark contrast to the relative neglect of Synkellos' *Chronography* among historians. While Synkellos' *Chronography* also contains excerpts from many texts that would otherwise be lost, these texts are primarily of interest to classicists and students of chronography.[15] With the exception of an article by Ihor Ševčenko, the *Chronography* has never been studied for its relevance to the ninth century milieu in which it was composed.[16] There are now two compelling reasons to rectify this imbalance.

---

13   If the "Eastern Source" behind Theophanes' narration of events in the Near East is Theophilos' account, produced by a Christian community under the Umayyad rulers of Damascus, we can reconstruct much more of the text than otherwise. For an early discussion see Andrew Palmer, *The Seventh Century in the West-Syrian Chronicles* (Liverpool, 1993). See now Robert Hoyland, *Theophilus of Edessa's Chronicle* (Liverpool, 2011) and an opposing point of view in Maria Conterno *"Palestina, Siria, Costantinopoli: la «Cronografia» di Teofane Confessore e la mezzaluna fertile della storiografia nei «secoli bui» di Bisanzio"* (Ph.D. dissertation, Università di Firenze, 2011).

14   Jakov Ljubarskij warned that critical analysis of the text, especially under the gaze of P. Speck, would destroy any possibility of reading the text as a composite whole. See his *"Quellenforschung* and/or Literary Criticism: Narrative Structure in Byzantine Historical Writings," *Symbolae Osloenses* 73/1 (1998), 10–11. Nonetheless, work on Theophanes as an author continues: Panayotis Yannopoulos, *Théophané de Sigraine* (Bruxelles, 2013).

15   For a recent example, see Luca Arcari, "Are Women the *aition* for the Evil in the World? George Syncellus' Version of 1 Enoch 8:1 in Light of Hesiod's Theogony and Works and Days," *Henoch* 34 (2012), 5–20. Joseph Scaliger set the modern precedent of using Synkellos' *Chronography* to get at his sources: Anthony Grafton, *Joseph Scaliger: A Study in the History of Classical Scholarship* II (Oxford, 1993), pp. 580–591. The ongoing work of Alden Mosshammer and William Adler is indispensible for assessing Synkellos' accuracy as a compiler and adjudicator between ancient sources. Besides Mosshammer's edition and Adler's translation (with Paul G. Tuffin), see also Adler's *Time Immemorial* (1989) and Mosshammer's recent "The Christian Era of Julius Africanus with an Excursus on Olympiad Chronology," in *Julius Africanus und die Christliche Weltchronistik*, ed. M. Walraff (Berlin, 2006), pp. 83–112; and Ibid., *The Easter Computus and the Origins of the Christian Era* (Oxford, 2008).

16   Ihor Ševčenko, "The Search for the Past in Byzantium around the Year 800" *DOP* 46 (1992), 279–293; see also George L. Huxley, "On the Erudition of George the Synkellos," *Proceedings of the Royal Irish Academy* 81c (1981), 207–217. Adler and Tuffin discuss the issue on pp. lxxxi–lxxxiii.

First, according to recent reassessments of the manuscript evidence, the *Chronography* and its continuation, the *Chronicle*, were originally placed back-to-back in the same codices: they circulated together and so would have been read together, apparently just as Synkellos had intended.[17] If medievals did not read Theophanes' *Chronicle* apart from Synkellos' *Chronography*, neither should medievalists.

Second, despite how little we know about his life and career, there is enough circumstantial evidence to insist that Synkellos' *Chronography* cannot be set aside as the faint ivory-tower whispers of an obscure antiquarian. According to the account of AD 808 in the *Chronicle* of Theophanes—the same year that Laqueur deduced Synkellos had begun the *Chronography*—the *synkellos* of the Patriarch of Constantinople was accused of conspiracy against the emperor and was punished with "lashes, banishment, and confiscation."[18] It remains most likely that this was a subtle reference to none other than Theophanes' "close friend," our own George Synkellos.[19] This supposition, combined with what we do know of the office of *synkellos* in the ninth century, strongly suggests Synkellos was very active politically and that his work was composed in the aftermath of a high-stakes political gamble. If, upon his exile, Synkellos immediately turned to writing the *Chronography*, it would appear that he believed there was present meaning to the archaic past, that the study of time mattered a great deal.

My analysis of Synkellos' *Chronography* takes this premise—and this premise only—from our scanty knowledge of Synkellos' biography: Synkellos wrote because he sought to communicate something of import to his contemporaries. What was his message? With biography failing to illuminate the issue, we must turn to the text itself. In the following argument, I focus in particular on the conceptual clues and generic cues in Synkellos' programmatic statements,

---

17     Filippo Ronconi, *"La première circulation de la « Chronique de Théophane » : Notes paléographiques et codicologiques,"* and Jesse W. Torgerson, "From the Many, One? The Shared Manuscripts of the Chronicle of Theophanes and the Chronography of Synkellos," in *Colloque Théophane: Travaux et mémoires 19* (Paris, forthcoming).

18     "In the month of February (808) many officials planned a revolt... [Nikephoros] punished [them] with lashes, banishment, and confiscation, not only secular dignitaries, but also holy bishops, and monks, and the clergy of the Great Church, including the *synkellos*, the *sakelarios*, and the *chartophylax*, men of high repute and worthy of respect." Trans. Mango and Scott, *Chronicle*, p. 664 from de Boor *Chronographia*, pp. 483–484.

19     The other possible known historical figure is John, who would become metropolitan bishop of Sardis. See Stephanos Efthymiades, "John of Sardis and the *Metaphrasis* of the *Passio* of St. Nikephoros the Martyr," *Rivista di Studi Bizantini e Neoellenici* 28 (1991), 25–26.

especially in the first pages of the *Chronography*. In doing so I suggest how Synkellos might have communicated the significance of his ideas about time to his ninth-century audience. I ground my argument in a brief survey of two late antique texts to which Synkellos explicitly referred: Eusebius' *Chronological Canons* and Ptolemy's *Handy Tables*. Synkellos' *Chronography* held a number of premises in common with these texts, such as the centrality of Aristotelian logic to the analysis of time and the close relationship between the present political order and the organization of past time.

I argue that the uniqueness of Synkellos' *Chronography* derives from its transgression of strict generic boundaries, setting up a dialogue between astronomical handbooks, the canonical scriptures, logic textbooks, theological and philosophical treatises, homilies, and liturgical commentaries. Synkellos used his *Chronography* to investigate the relationship between a timeless God and a time-bound Creation. His investigation produced the hypothesis that humanity experienced the Trinitarian God's direct interventions into linear time as a recurring "First-Created Day." It is not immediately apparent what Synkellos meant by the neologistic term he used for the idea, but it amounts to his most original contribution. I hope that what follows will not only serve as an argument for the intellectual milieu of George Synkellos, but will also contribute to cross-disciplinary interest in what early medieval elites throughout the Mediterranean world might have expected their chronographies to do.

### Synkellos' Chronography as an Early Medieval Universal Chronicle

Since the popularity of chronicles in general is unique to the middle ages, it is important to define the genre. Modern scholars use the designation "chronicle" to describe texts ranging from multivolume masterworks of all past time (chronographies) to a couple of pages devoted to brief historical notices in order of occurrence (annals).[20] Thanks to forward-thinking studies by a number of scholars, it has become increasingly accepted that chronicles had direct political implications in the social and cultural contexts in which they were written,

---

20   There are of course many exceptions. For instance, see Rosamond McKitterick, *History and Memory* (Cambridge, 2004), esp. pp. 97–99 on distinguishing generic differences in the Carolingian context; for what a chronicle is not, see Karl F. Werner's contextualized description of medieval *historiae*: "L'Historia et les rois," *Religion et culture autour de l'an Mil*, eds. D. Ionga-Prat and J.-Ch. Picard (1990), pp. 135–143.

copied, and disseminated.[21] Nevertheless, function alone cannot define form; if we are more willing to accept that chronicles mattered, we do not completely agree on a definition of the subject. Issues of genre, terminology, and audience are (rightly) topics of strong disagreement among those trying to decipher a form of literature we no longer read or write.[22]

Textbooks and dictionaries tell us that an inquiry into causation is considered the primary goal of history, and so we routinely distinguish "proper histories" from chronicle-type texts.[23] Having been denigrated for lacking narrative history's critical inquiry into causation, chronicles are (perversely) defined not by the nature of their own inquiry, but by the nature of their structure.[24] The structure of a chronicle, annal, or chronography is characterized by short narratives covering relatively brief time periods, most often one year, which give the appearance of independence from one another. This structure, *prima facie*, directly inhibits the pursuit of causal connections between past events. We tend to read chronicles disingenuously, as though they are trying to be histories but happen to have this annalistic structure in the way. We could excuse ourselves—in a field plagued by a dearth of source material, the

---

21      Sarah Foot surveyed the foundational impact of Hayden White in "Finding the Meaning of the Form: Narrative in Annals and Chronicles," in *Writing Medieval History*, ed. N. Partner (London, 2005), pp. 88–108. For ground-breaking monographs on chronicles' socio-political contexts see Gabrielle Spiegel, *The Past as Text* (Baltimore, 1999) and Brian Croke, *Count Marcellinus and His Chronicle* (Oxford, 2001). Recent work has trended towards incorporating the manuscript tradition, such as Simon MacLean's study of the twelfth-century English afterlife of a tenth-century Frankish chronicle, "Recycling the Franks in Twelfth-Century England," *Speculum* 87 (2012), 649–681.

22      See R.W. Burgess and M. Kulikowski, "Medieval Historiographical Terminology: the meaning of the word *annales*," *The Medieval Chronicle* VIII (2013), pp. 165–192. One contemporary equivalent could be Wilson Alvarez's Chrono-Zoom project (http://www.chronozoom.com/), created by researchers at U.C. Berkeley, and Moscow State University as an interactive visualization of time's order from the formation of the universe.

23      See Neville Morley's perceptive discussion in *Writing Ancient History* (1999), pp. 50–52, distinguishing "history" as the past in general from the professional historian's particular "way of talking about the past."

24      The Oxford English Dictionary reflects the entrenchment of chronicles' subjugation: "*Chronicles*, or *annals*, are simpler or more rudimentary forms of history in which the events of each year, or other limited period, are recorded before passing on to those of the next year or period, the year or period being the primary division; whereas in a *history*, strictly so called, each movement, action, or chain of events is dealt with as a whole, and pursued to its natural termination, or to a convenient halting-point, without regard to these divisions of time."

temptation to extract "facts" for a basic historical narrative is beyond endur-
ance—but is it possible to read chronicles more responsibly? Could medieval
authors yet suggest to us how they supposed their texts would be read and interpreted? There is not space to be comprehensive, but I will attempt to sketch
a paradigm, and contextualize my own approach to Synkellos' ninth-century
*Chronography*.

The chronicle that exerted the most influence in the early middle ages was
the fourth-century two-volume *Chronological Canons* of Eusebius of Caesarea,
originally written in Greek and completed ca. 325.[25] By the end of the century
the work's first volume, a discussion of "pre-history" before the birth of
Abraham, was declining in popularity, while the second volume was widely
read across the Mediterranean world. The second volume of the *Canons*
presented, in a single codex, all known history from Abraham to the First
Ecumenical Council under Constantine I. These post-Abrahamic *Canons* were
translated into Syriac and Armenian anonymously and into Latin by St. Jerome
(ca. 382).[26] For the early medieval world, both East and West, Eusebius' *Canons*
were both paradigmatic and definitive. Medieval readers would have
approached Synkellos conditioned by reading Eusebius. Synkellos anticipated
this situation, carrying on an explicit methodological debate with Eusebius
throughout the *Chronography*.[27]

Though it is clearly pertinent to determine how Eusebius' project was
understood, we have few sources that can establish this context. Rather than

---

25    On the history of chronicle writing to Synkellos see Adler *Time Immemorial*, and Alden
      Mosshammer, *The Chronicle of Eusebius and the Greek Chronographic Tradition*
      (Lewisburg, 1979). See the survey of Western early medieval universal chronicles in
      Michael I. Allen "Universal History 300–1000: Origins and Western Developments," in
      *Historiography in the Middle Ages*, ed. D. Deliyannis (Leiden, 2003), pp. 17–42; and for
      specific studies Erik Kooper's *The Medieval Chronicle* (currently to Volume VIII) is now
      essential reading.

26    Eusebius' *Canons* consisted of two books, both making use of the graphic potential of a
      codex, as opposed to papyrus. In the first book, not discussed here, Eusebius laid out a
      large horizontal table of the various accounts of this period of the past. Besides the frag-
      ments preserved by Synkellos, this work survives in an Armenian recension. See
      Mosshammer, *Greek Chronographic*, pp. 65–66. For the layout, see Anthony Grafton and
      Megan Williams, *Christianity and the Transformation of the Book* (Cambridge MA, 2006),
      pp. 136–146; and on the reconstruction of Eusebius' *Canons*, see Brian Croke, "The
      Originality of Eusebius' Chronicle," *American Journal of Philology* 103/2 (1982),
      195–200. St. Jerome copied the second book and seems not to have substantially altered
      Eusebius' text, though he added a preface and a continuation of the chronicle up to AD
      378 (likely for the Roman Synod of 382: Mosshammer, *Greek Chrongraphic*, pp. 67–68).

27    There are many examples but see esp. AT 222/M 180; AT 244/M 197–198; and AT 333/M 271.

using Synkellos' reading of Eusebius to interpret Synkellos, I will draw upon the seventh-century Latinate polymath Isidore of Seville to function as our Virgilian guide to the expectations early medieval *litterati* might have brought to chronicles. In Book I of his encyclopedic work the *Etymologies*, Isidore clearly distinguished *chronica* from *historia*; he did not discuss the more common later medieval term—the singular form *chronicon*—at all. Book I of the *Etymologies* described *historia* as a broad category for *all* narrative accounts of the past: "*historia* is a *narration* of deeds accomplished; through it what occurred in the past is sorted out."[28] Isidore did not only conceptually distinguish *chronica* from *historia*, but he also physically separated them, placing his description of *chronica* in Book V, "On the Laws and Times." Isidore placed the chronicle-like annals (*annales*) in Book I along with *historia*. Though a perplexing decision from our point of view, Isidore associated annals and histories based on their similar scope: annals organized an account of a particular time and place by temporal units, as opposed to narrative coherence; neither annals nor histories attempted to account for the *entire* past.[29]

Isidore's *chronica* must be separated from the mass of annalistic texts we loosely call "chronicles." Isidore's *chronica* was a very specific group of texts we call "universal chronicles," or less often, "chronographies." In other words, it is necessary to translate Isidore's *chronica* as "*universal* chronicles," since Isidore's *annales* are what modern scholars usually mean by "chronicles." For Isidore, only *chronica* discussed the entire past from the creation of the world, and the most widely circulating example of this sort of text would have been Eusebius' *Chronological Canons*. In terms of scope, Isidore's definition also fits Synkellos' *Chronography*. If Eusebius' *Canons*, Isidore's *Chronica*, and Synkellos' *Chronography* are identical in terms of scope: does this similarity extend to purpose and method?

To answer this question we might note that Isidore's discussion of *chronica* proceeded didactically. He first explained how to order gradually increasing amounts of time: from moments and hours, days and nights, to weeks, months, solstices, equinoxes, seasons, years, Olympiads, Jubilees, and finally to eras and

---

28    Emphasis mine. Historia est narratio rei gestae, per quam ea, quae in praeterito facta sunt, dinoscuntur (I.xli.1) in Wallace M. Lindsay, *Isidori Hispalensis episcopi Etymologiarum sive originum* (Oxford, 1911); trans. Barney, Beach, Berghoff, and Lewis, *The Etymologies of Isidore of Seville* (Cambridge, 2006).

29    While "*historia* is of those times that we have seen,"..."*annales* are of those years that our age has not known " (Ibid., I.xliv.4). Isidore places *historia* among rhetorical works within his explication of an education in the *trivium* of Grammar, Rhetoric, and Dialectic, after laying out types of literature in a series of contrasting pairs (that is, prose vs. verse at I.38–I.39), Isidore has *historia* (I.41–I.44) oppose fable (I.40).

ages (*saecula et aetates*).[30] Isidore then stated that a *chronica* was the organization of "the succession of times" (*successio temporum*), where "times" denoted all measured lengths, from passing instants to ages of the world. He concluded with an example: his own epitome of time's six ages, from the Creation to his present day.[31] Isidore's chronographer could not assume, as did the diarist, calendrist, historian, or annalist, that the correct reckoning of time was a given. In Isidore's potentially representative view, the composition of a *chronica* was the conclusion of a scholar's categorization of time itself, a philosophical and scientific investigation into the nature and division of time.[32] Even if the genres of *historia* and *chronica* shared the same basic material—past events—a *historia*'s logic was plausible narrative, while a *chronica*'s logic was the order of events in time. Causation was not relevant to the organization, reckoning, and periodization of events in *chronica*; the chronographer investigated *when* an event occurred, not *why*.[33]

If this is a valid reading then, at least for Isidore, the more natural sister science to early medieval chronography was not history but astronomy: the measuring of time's passage by the motions of the heavens.[34] In fact, this supposition can be supported with additional evidence.[35] The parallels between

---

30    Ibid., V.xxviii–xxxviii.

31    Sam Koon and Jamie Wood. "The *Chronica Maiora* of Isidore of Seville: An Introduction and Translation," *e-Spania: Revue interdisciplinaire d'études hispaniques médiévales et modernes* 6 (2008), 3–5 (accessed 3-6-12).

32    Correct chronology was also the focus of Bede the Venerable's (d. AD 735) *De Temporum Ratione* though he was more motivated by didactic concerns: his "World Chronicle" is buried as Chapter 66 within his instruction on the calculation of Easter. See Faith Wallis, *Bede: The Reckoning of Time* (Liverpool, 1999).

33    Sarah Foot, "Finding the Meaning," p. 90, blames narrative demands: histories describe events as "one thing *because of* another"; a chronicle places "one thing *after* another...a conjunction of non-causal singular statements."

34    There is convincing evidence of continued interest in astronomy between the seventh and the ninth centuries. See Anne Tihon, "L'Astronomie à Byzance à l'Epoque Iconoclaste (VIII–IXᵉ Siècles)," in *Science in Western and Eastern Civilization in Carolingian Times*, ed. P. Butzer (Basel, 1993), esp. on the Solar Diagram in *Vat. gr. 1291*, pp. 193–201. See also the survey on astronomy in this period in Paul Magdalino, *L'Orthodoxie des astrologues: La science entre le dogme et la divination à Byzance (VIIᵉ–XIVᵉ siècle)* (Paris, 2006), pp. 17–32.

35    Scholars continue to uncover just how central the *Handy Tables* and other astronomical texts were to the enterprise of early Byzantine chronography. Besides *Vat. gr. 1291*, the *Chronographeion Suntomon* or Abridged Chronography (*Vat. gr. 2210*, AD 854/5) is another manuscript that contained texts from both genres. See Mossman Rouéché, "Stephanus the Alexandrian Philosopher, the *Kanon* and a Seventh Century Millenium," *Journal of the Warburg Institute*, 74 (2011), pp. 11–17. Paul Magdalino discussed two surviving

chronography and astronomical handbooks are both conceptual and codico-logical.[36] As we shall see, Eusebius alluded to astronomical works via his arrangement of the text on the page; Synkellos directed his readers to the period's most important astronomical handbook on multiple occasions.

We can elucidate both the explicit and the allusive by examining the astronomical text to which Synkellos referred. The polymath known as Ptolemy (Claudius Ptolemaeus, fl. AD 146–170) had composed his masterwork in second-century Alexandria: the Μαθηματικὴ Σύνταξις (most commonly known by its Latinized Arabic title, *Almagest*). Ptolemy excerpted and reas-sembled tables charting astronomical data and calendrical cycles from the *Almagest* into a collection called the *Handy Tables* (Πρόχειροι Κανόνες).[37] The work was considered essential for centuries: One surviving eighth-century manuscript—*Vaticanus graecus 1291*—was used through Synkellos' lifetime.[38]

Ptolemy's tables did not only convey information, they depicted a method. In the excerpted image from *Vat. gr. 1291* (Fig. 1.1), Ptolemy's synchronization of daily calendars exemplifies both his system of organization, and its implicit hierarchization of the content. Here Ptolemy coordinated the Nones of the Roman month of May with (left to right) the calendar of the Hellenes, of the cities of Alexandria and Tyre, then the calendar of the Arabs, and those of Sidon, Gaza, Askalon, Heliopolis, Lycia, Cappadocia, Bythinia, Seleucia, Asian Pamphylia, Cyprus, and Crete. The Roman calendar (far left) was the constant to which the others were normalized. The order communicated the message: Rome had long since conquered the Hellenistic kingdoms and so Ptolemy's table subjugated Hellenistic calendars to a universal Roman time as kingdoms

---

anonymous astronomical manuscripts, and whether Synkellos could be the author (*L'Orthodoxie des astrologues*, p. 55); Magdalino thinks not, suggesting that Synkellos was not, so to speak, a *hapax legomenon* (p. 56).

36    See Synkellos' direct citation at AT 381/M 314–315 (in discussing Alexander the Great).

37    G.J. Turner. "Ptolemy: Claudius Ptolemaeus," in *The Oxford Classical Dictionary* 3rd rev. ed., eds. S. Hornblower and A. Spawforth (Oxford, 2003). The *Handy Tables* are preserved today as in Byzantium: with the commentary of Theon of Alexandria (fl. AD 364). See Adler and Tuffin on Synkellos' use of, but relative independence from, these sources (pp. lxiv–lxix), and Synkellos' reference to Ptolemy's *Almagest* and *Handy Tables* in his discussion of Nabonasar (AT 73–75, 299–300).

38    Based on its list of Roman-Byzantine Emperors, David H. Wright proved that this manu-script was compiled under Constantine V (r. 741–775), "The Date of the Vatican Illuminated Handy Tables of Ptolemy and of its Early Additions," *Byzantinische Zeitschrift* 78 (1985), 355–362. Based on the variation of hands in the list of Roman Emperors it was likely in active use for well over a century through the reign of Basil I (d. 886).

under the empire.[39] If Ptolemy's work is representative, then an astronomer's synchronization of calendars conveyed political content via its system of organization. What would it tell us if this conceptual paradigm was shared between chronography and astronomy? If an astronomer's synchronization of local calendars with a universal calendar is similar to a chronographer's synchronization of local historical events with universal time, did chronographies also convey overt political content in their system of organization? Synkellos' own direct references to the *Handy Tables* would seem to confirm this hypothesis.

Though Synkellos' method relied on following the Septuagint's record, as the *Chronography* approached the Incarnation, Synkellos increasingly turned to Ptolemy's *Handy Tables* and specifically to the *Handy Tables*' lists of kings.[40] These king lists were occasionally transmitted independently as a "Royal Canon" (Κανὼν Βασιλέων).[41] The lists were relevant to Ptolemy's astronomy because astronomical cycles and specific events, such as the appearance of a comet, were always dated by the reign of a local king who, when synchronized with his contemporary kings, served to cross-reference and confirm the observations. These harmonized observations in turn rendered the synchronized king lists, as a collective whole, the authority of a universal political-historical time. The "Royal Canon" is the textual go-between that instantiates the conceptual link between the sciences of astronomy and chronography: the political narrative embedded in both means of ordering time.[42]

Eusebios' *Canons* presented synchronized regnal successions with a political message even more explicit than that identified in Ptolemy's table (Fig. 1.1). The *Canons*' organization of time on the manuscript page also visualized

---

39  See A.E. Samuel, *Greek and Roman Chronography* (Munich, 1972), pp. 186–188 on an "Eastern" Julian calendar.

40  Synkellos needed to establish the concurrent rulers of the Babylonians, Persians, Greeks and Latins both at the conquest of Troy, and at the death of Alexander the Great, but would not accept the histories of the Maccabees as having the authority of older scriptures: "Written by Josephus (sic), this book does not belong to divinely inspired scriptures; it is, however, extremely useful" (AT 398/M 329). A. Mosshammer, *Easter Comptus*, pp. 17–18; 25–26.

41  Mossman Roueché explains: "the *Royal Canon* is a table whereby historical data (the commencement and length of a ruler's reign) are correlated with the time series underlying the mathematical tables" of the *Handy Tables*. "Stephanus," 11. I am extremely grateful to the author for sharing this research prior to publication.

42  Ptolemy's "Royal Canon" was authoritative. "Just as the mathematical *Handy Tables* could be used to calculate the positions of heavenly bodies for a given date, they could also be used to check the accuracy of chronological information in the Royal Canon, by reference to the same celestial phenomena" (Roueché, "Stephanus," 14).

political hierarchy. Eusebius synchronized the regnal chronologies of multiple kingdoms by aligning them with Olympiads and a tally of years from the life of Abraham. The appearance of Eusebius' original Greek manuscripts must be imagined on the basis of its anonymous Armenian translation and of St. Jerome's Latin translation. While specific details in the layout may be only tenuously original, surviving copies all testify to the same overall concept. The ninth-century Latin manuscript *Merton College 315* provides an example synchronous with Synkellos' composition of his *Chronography*, giving us an idea of the text's visual appearance during his milieu. For the pre-Incarnation period, Eusebius stretched multiple columns across both pages, each filled with the history of a different kingdom or people group. The scribe of the tri-colored ninth-century copy replicated this intent by stretching the text across both pages of the codex, with each horizontal row indicating a passing year, and each vertical column a different kingdom (Fig. 1.2). In Eusebius' *Canons*, plotting time meant synchronizing the rulers of particular kingdoms with each other.

In Fig. 1.2, featuring the life of Moses, a reader would have reckoned down the far left column, where Eusebius used decades from Abraham as a baseline for his universal chronology. Column 1 identifies decades 460 (CCCCLX) and 470 (CCCCLXX) with green ink. Column 2 notes the Assyrians (red ink), Column 3 the Jews (black ink), and Column 4 Sikyonian Greeks (red ink). On this folio Column 3 also initiates Eusebius' second objective chronological tally in green ink: the number of years since God's covenant with the seventy-three year-old Abraham. The right hand page in this opening continues with the Argives (black ink) in Column 5, introduces the Athenians (red ink) half way down Column 6, and ends on the far right with the Egyptians (black ink) in Column 7. Like Ptolemy, Eusebius created a universal time by organizing diversity. Unlike the astronomer, however, the chronographer's visual multiplicity followed a plot: Eusebius made seven columns dwindle down to one. To organize time was to structure political power, but to calculate time's passing was to give political power a plot.

Just as Isidore had described, Eusebius organized the succession of kingdoms into eras and epochs. As time advanced line-by-line down each page of the *Canons*, the format gradually simplified: the multi-column two-page spread of separate kingdoms (Fig. 1.2) ultimately devolved into a single column of prose under a single empire, the Roman *imperium* (Fig. 1.3). Imperial Roman time was not only political but religious. The unity of reckoning achieved through Roman political universality was enumerated with a Christian formula: years from the Incarnation. The *Canons* made the triumphs of Roman hegemony and of the Incarnation essentially co-terminus. The scribe of our ninth-century copy of the *Canons'* Latin translation presented this unification symbolically: a bold Christogram stretching from margin to margin (Fig. 1.3). Eusebius' system of reckoning made time's progress teleological. The succession of years led inevitably to a universal temporal order under Christ and Rome; the Incarnation was the goal of the ancient past and the *raison d'etre* of the Roman Empire.[43]

## The Structure of the Chronography of George Synkellos

The manner in which astronomers and chronographers arranged and organized the progression of time stemmed from their view of the relationship between past time and the present moment. The tabular grid created by Eusebius for his *Canons* illustrated a progression from diversity to universality that reflected the contemporary political ideology under his lord, the Emperor Constantine I (r. 306–337). Synkellos did not present his readers with a systematic tabulation of time in an immediately comprehensible visual format. In fact just the opposite: medieval readers of the *Chronography* looking for ordered synchronicity between past figures and events, as achieved so elegantly by Eusebius, may well have been overwhelmed by the great mass of chronological and historical information compiled by Synkellos, much of it in lists.[44] That is, while Eusebius integrated various kingdoms' systems of reckoning, Synkellos separated each kingdom's records into distinct subsections whose dates often did not even match.

Consider the layout of the *Chronography* in our oldest nearly-complete manuscript, *Paris BnF Grec 1764* of the tenth century. The pages in the reproduced image contain Synkellos' discussion of the nineteenth and twentieth Generations of the Patriarchs: those of Abraham's grandfather, Nachor, and his

---

43  History's division into periods progressing towards the apocalyptic has a long history and a vast literature. Common eschatological frameworks include the "Four Kingdoms" of the prophet Daniel and the "Six Days of Creation." The most popular early medieval apocalyptic, the apocalypse of Pseudo-Methodius (extant in Syriac, Latin, and Greek), divided history into seven epochs. See the still-essential work by Paul Alexander, *The Byzantine Apocalyptic Tradition* (Berkeley, 1985), translation and commentary by Benjamin Garstad, *Apocalypse of Pseudo Methodius and an Alexandrian World Chronicle* (Washington, 2012). On the relation between eschatology and history see Paul Magdalino, "The history of the future and its uses: prophecy, policy, and propaganda," *The Making of Byzantine History*, eds. R. Beaton and C. Roueché (Aldershot, 1993), esp. pp. 3–5.

44  So, in solidarity, Daniel J. Thornton: "...endless (to the casual reader at least) tables of monarchs, emperors, bishops, and the years of their reign." BMCR 2004.10.27.

father, Terah (Fig. 1.4). According to Synkellos, the lives of these two patriarchs covered 150 years—from AM 3163 to AM 3313—from the birth of Nachor to the birth of Abraham.[45] The entry reproduced here grouped Generation Nineteen and Twenty together and brought Synkellos' chronology up to the time of the birth of Abraham's Generation Twenty-One (AM 3313).[46] After stating the dates he assigned to Nachor and Terah—AM 3163 to AM 3313—Synkellos provided the lists of rulers for four different kingdoms.

One might expect that each of these groups would contain the kings who were exactly contemporaneous with Nachor and Terah. On the contrary, the chronological information provided by Synkellos is eclectic at best: despite the neat hierarchical appearance of the manuscript (Fig. 1.4), the actual years covered in each kingdom's short list of rulers do not align with each other. Under "The Egyptians" Synkellos noted rulers for the years AM 3117–3315; under "The Assyrians" Synkellos listed rulers for the years AM 3216–3403; under "The Sikyonians" Synkellos recorded the reigns of the first Sikyonian-Greek rulers as AM 3239–3290; and, finally under "The Thebans" Synkellos ennumerated kings for the years AM 3053–3231.

These four lists of rulers do not coordinate for even a single year, and in fact the Thebans and Sikyonians do not even overlap at all: there is an eight-year gap between the last Theban ruler (who died in AM 3231) and the first Sikyonian (who ruled from AM 3239). It must have been truly cumbersome for the early medieval, no less twenty-first century, reader to make cross-references between lists. Anyone interested in identifying rulers synchronous with Nachor's birth, for instance, would have had to flip back and forth through other entries in the *Chronography* in order to hunt down all relevant figures. Presuming that the surviving arrangement of the text was both original and intentional, Synkellos' ambivalence suggests that his central argument did not depend on making connections between historical figures.

Let's zoom out from the manuscript page and consider the work as a whole. Since Synkellos never completed his *Chronography* we cannot be entirely sure of his overall vision for its organization. We are fairly certain that Synkellos originally divided his *Chronography* into two portions: one leading up to, and

---

45   At the mention of tangible "dates" by Synkellos, it is tempting to check how "accurate" our author was. Synkellos' dates largely followed what might be called the "Alexandrian school" of chronography, synthesized by the fifth-century chronographer Annianos (see below, pp. 39–40). Comparing these and other efforts would miss the point. In chronography all dates are basic additions and subtractions from "fixed points" (such as the Incarnation): the dates are necessitated by the pre-determined hierarchy of sources.

46   AT 129/M 104.

the other following from, AM 5434 (63 BC), Pompey's conquest of Jerusalem.[47] The portion before AM 5434 was revised and is relatively polished and coherent. If we focus on this portion of the text, it appears to have been planned out as four distinct periods, or epochs.[48] My reader should be aware that these divisions are *not* explicitly indicated in the text as such. Nevertheless, I believe there are enough clues to indicate Synkellos intended his own attentive medieval readers to deduce them.[49] Each of Synkellos' epochs synchronized the royal successions of a distinct set of neighboring kingdoms with the records of the leaders of the Hebrew people (whether patriarchs, prophets, priests, judges, or kings); each epoch displayed its chronological data distinctively, forcing a reader to approach the text in a different way; each epoch was both initiated and completed by comprehensive summaries and editorial discussions; and, each epoch was unified by one or two key synchronizations between

---

47    J.W. Torgerson, "From the Many, One?" Synkellos divided all time at AM 5434 to emphasize a prophecy concerning the end of the rule of Jewish priests and the rule of a non-Jew over Judea: "At that time also, the 'anointed ones who rule' prophesied by Daniel came to an end" (M 373,24-25/AT 446); and "Herod, being an Idumaean Arab, was the first Jewish ruler of foreign stock" (M 383,16/AT 457). Nevertheless, Roman triumphalism is not absent from the *Chronography*: "The first to be monarch, [Julius Caesar] proved by far the most humane of all the kings who have ever ruled" (M 365,8-9/AT 436).

48    I use the term "epoch" here in its general English sense, "a period of time." The Greek word ἐποχή has the more technical meaning of "a fixed point," which in chronology, astronomy, or *computus* is the past point from which one calculates years and dates. I use "epoch" to refer to these periods of time because today Byzantine historians conventionally use the term "era" in the technical sense just described (ἐποχή). Thus an "era" refers to a particular dating system that calculated the "Year 1" from a distinct point (*ODB* s.v. "Antiochene Era," "Byzantine Era," or "Alexandrian Era"). Synkellos, for instance, largely adopted the "Minor Alexandrian Era," basing his calculations upon the "fixed point" (ἐποχή) set by the fifth-century Alexandrian Annianos who dated the Incarnation to March 25th, 5500 (on whom see below, pp. 39–40).

49    To prevent misunderstanding: though Synkellos must have at very least *planned* his chronicle with something resembling these epochs, he never explicitly referred to them as distinct "sections" of the work. Their existence is contingent upon the reader's acceptance of my analysis of the text and manuscripts. Synkellos provided summaries between the first and second epochs (AT 111–125/M 87–101), and between the second and third (AT 160–177/M 128–142). Between the third and fourth epochs there is a treatise but no chronological summary (AT 318–339/M 259–278). An early marginal note (σχόλιον) did correct this "omission," and by the tenth-century this σχόλιον had been incorporated into the text proper (*Paris BN Grec 1764*; see AT 318/M 259). The dramatic parallel between Nebuchadnezzar leading Jechonias to Babylon (AT 319–320/M 260) and Pompey leading the Jewish prisoners to Rome (AT 429–431/M 357–359) at least suggests a planned division.

particular figures or events for which it was essential to establish chronological congruence.[50]

Synkellos' division of the past into four evenly spaced epochs presumed that the past was equally knowable. Synkellos used his first epoch to demonstrate this point explicitly, arguing that time was quantifiable from the first moment of Creation. The years of the first epoch could be tallied just as well as those of the fourth, and so all of time could be analytically divided into overarching historical periods. This agenda perhaps explains Synkellos' ambivalence towards detailed organization, as noted above (Fig. 1.4).

The division of *all* time into epochs distinguishes the *Chronography* from the *Canons*. Eusebius began the *Canons* proper at the point when he could compare multiple historical records from Abraham on. Eusebius' temporal system presumed that, for the chronographer, time in the early history of the world was dissolute and amorphous. As it came to be calibrated by fewer and fewer kingdoms, time truly "came together" at the Incarnation; universal time finally emerged with the conjunction of Augustus and Christ.

The first portion of Synkellos' *Chronography*, his first epoch, covered the ancient period which Eusebius had left unorganized. This portion introduced Synkellos' method and clarified his thesis. I will focus on the method first, and then return to the philosophical basis for his reckoning below. A key premise of Synkellos' approach was the compatibility between traditional chronological methodology and a literal reading of the Septuagint translation of the Hebrew Scriptures. Synkellos linked his use of the Septuagint for chronology to the text's authority as scripture in the tradition of the Greek-speaking Church.[51]

Synkellos pointed out that the most respected chronographers of Late Antiquity—Julius Africanus (d. ca. 240) and Eusebius of Caesarea (d. 339/40)— agreed with him: for the world's earliest period, the Septuagint's record of 2,242 years from Adam to the Flood was more accurate than either the Hebrew scriptures themselves or their Samaritan translation.[52] Even though the Septuagint

---

50 Unfortunately, Adler and Tuffin treat the organization of the *Chronographia* as a constant (p. lvii). Against Scaliger's assertion that our manuscripts of Synkellos represent the work of dishonest scribes who "piled the historical entries indiscriminately in a random heap" (*librarii vero eas historias* σποράδην *in tumultuarium congeriem concesserunt*, as quoted in Grafton, *Scaliger*, II, pp. 540–542), I affirm the premise, as I understand Mosshammer, too—that Synkellos' alternating and evolving format was intentional.

51 Sacred tradition: ἱερὰ (or ἱερατικὴ) παράδοσις. See Adler and Tuffin, *Chronography*, pp. liii–lv.

52 AT 27/M 20. Adler and Tuffin note Synkellos' preference for the Septuagint, but do not explain it as a key methodological principle (*Chronology*, pp. xxxvi and xlix). Africanus actually tallied 2,262 but, as Synkellos explains, this is only because he was working

was a late Greek translation from the Hellenistic period, "the LXX [Septuagint] translation was translated from, so it would seem, an ancient and a strictly accurate Hebrew text of scripture."[53] Synkellos further exalted the Septuagint by refuting the non-Scriptural records of Egyptian and Chaldaean kings claiming to predate Noah's Flood.[54] No other text was a viable alternative.

Synkellos then turned and sailed into the wind, demonizing the authorities he had just cited: Africanus and Eusebius. The venerable late antique chronographers had attempted to harmonize these same non-Scriptural sources with the Septuagint. At moments of apparent discrepancy they had even abandoned the Septuagint in favor of the other records: they had reneged on their own principles. Africanus and Eusebius were unreliable dissemblers.[55] The conclusion is clear: only Synkellos could be consistently relied upon to defend canonical chronology from the definitive text of the Septuagint. The argument had the rhetorical impact of all but equating Synkellos' chronological schema with the canonicity of Holy Scriptures.

This discourse sets Synkellos' *Chronography* apart: it was not primarily a reference, but an apology for a particular method of chronography. Synkellos was far more bent on proving his authority to order time than on synchronizing the lives of historical figures for his readers to consult, as in the *Canons*. The *Chronography*'s original titles seem to support this conclusion. "Selection of Chronography" (Ἐκλογὴ χρονογραφίας) and "Abridged Chronography" (Σύντομος Χρονογραφία),[56] evoke Synkellos' mode of argumentation via quotation and then refutation.

---

with a faulty manuscript that stated Methuselah was 187 years old when his son Lamech was born; the more accurate reading is that he was 167 years of age. The Hebrew scriptures and their translations in Greek and Samaritan were wildly divergent, with the Hebrew falling 586 and the Samaritan 935 years short of the Septuagint reckoning. See AT 118/M 94.

53   Πανταχόθεν τοιγαροῦν τῆς τῶν ὁ ἑρμηνείας ἐκ παλαιᾶς, ὡς ἔοικε, καὶ ἀδιαστρόφου Ἑβραίων γραφῆς μεταβεβλῆσθαι συνισταμένης (AT 125/M 100).

54   Esp. Manetho's second or third-century B.C. Ptolemaic *Aegyptica* and Berossus' *Babyloniaca* (from the same period, though likely written under the patronage of the Syria-based successors to Alexander, the Seleucids). For a discussion of Synkellos' access to these, see Adler, *Time Immemorial*, pp. 148–157. If Manetho is taken literally, he recorded 11,985 years before the Flood; Berossus proposed 432,000 years. See Gerald Verbrugghe and John Wickersham, *Berossos and Manetho, Introduced and Translated* (Ann Arbor, 1996).

55   AT 95/M 74.

56   Adler and Tuffin, *Chronography*, p. xxix. In the chronicle's first half Synkellos describes his work as a χρονογραφία, the classical Greek term for a chronological record. Not incidentally this is also the term for a method of reckoning in an astronomical treatise (See AT 125/M 100). In a presumably intentional linguistic switch Synkellos used the variants

Why was Synkellos so eager to make an authoritative claim about time's order? What was the ideological payoff if a reader granted Synkellos' assertions about chronological method? Eusebius had shown a progression in time—from the dissolution of the ancient past, to the clarity of chronology at the Incarnation of Christ under the Romans—to make a connection between Empire and Church. By contrast, Synkellos began his reckoning of time from the very first moment of Creation, which he insisted was a chronological point fully accessible to a chronographer's investigations. Synkellos argued from the same premise as Eusebius—the events of Christ's life were the primary fixed points of chronography—but built up his own unique hypothesis of the Creation as a chronological event.

### Synkellos' First-Created Day as a Date

Synkellos argued that since the entire temporal order of the universe stood on the chronological point (ἡ ἐποχή) of Christ's Resurrection, the temporal event of the Creation was knowable. Synkellos' argument is paradoxical, but not illogical. Synkellos held that the Resurrection occurred on the day that in contemporary solar calendars was the twenty-fifth of the Roman month March and the twenty-ninth of the Egyptian month Phamenoth. Synkellos also claimed that in the year the Resurrection occurred (AM 5534), this day was marked by the date 1 Nisan, an apparent reference to the Jewish lunar calendar. Near the end of the *Chronography*, Synkellos succinctly stated the implications of precisely dating the day God rose from the dead:

> And after his burial, [Christ] arose on the third day, on 29 Phamenoth, that is 25 March, when the Lord's day, the first day of the week, was dawning, on the eighth day before the Kalends of April, the first day of the first-created Hebrew and Christian month of Nisan, concerning which it was said: 'In the beginning, God created the heaven and the earth' (Gen. 1.1.), and again, 'This is the book of the creation of the heaven and the earth, on which day God created.' (Gen. 2.4)[57]

---

χρονογραφεῖον and χρονογράφιον, the *koine* and Byzantine Greek terms for a chronicle in charts and in the second portion of the text (AT 115/M 91; AT 118/M 94; AT 121/M 96 and AT 473/M 396).

57   καὶ ταφεὶς ἀνίσταται τῇ γ′ ἡμέρᾳ, Φαμενὼθ κθ′ ἤτοι Μαρτίου κε′, ἐπιφωσκούσης κυριακῆς μιᾶς σαββάτων, πρωὶ καλανδῶν Ἀπριλλίων, α′ τοῦ πρωτοκτίστου μηνὸς Νισὰν παρ᾽ Ἑβραίοις καὶ Χριστιανοῖς, περὶ ἧς εἴρηται· "ἐν ἀρχῇ ἐποίησεν ὁ θεὸς τὸν οὐρανὸν καὶ τὴν γῆν," καὶ πάλιν· "αὕτη ἡ βίβλος γενέσεως οὐρανοῦ καὶ γῆς, ᾗ ἡμέρᾳ ἐποίησεν ὁ θεός" (AT 463/M 389).

Note that Synkellos did not restart his sentence when the subject transitioned from the day of the Resurrection to the day of the Creation. He did not even restate the subject noun "day" (ἡ ἡμέρα), but he conjoined a statement about the First-Created Day in AM 5534 and the First-Created Day in AM 1 with the relative clause "concerning which" (περὶ ἧς); the day was single. If the days were identical, then, according to classical logic, statements about the day of Creation were statements about the day of the Resurrection. If the day of the Creation and the day of the Resurrection was the same day, then March 25 in AM 5534 was both the date of the Resurrection, and the date from which the Resurrection was tallied. For Synkellos' *Chronography*, the Resurrection would be both the premise and the conclusion.

Synkellos then went on to argue that Christ's Incarnation—the archangel Gabriel's announcement of the descent of the Holy Spirit into the womb of the Mother of God—had also taken place on the exact same calendrical alignment, the exact same day, in the year AM 5500:

> We have committed all our labour on this work to demonstrate the premise that this First-Created Day corresponds with the day of the divine proclamation and the miraculous conception of the only-begotten son of God from the holy Virgin; and with the day of the life-bringing Resurrection from the dead, a day which for those made worthy to celebrate it in spirit and truth is both more divine than the other days and the source of all light.[58]

Synkellos linked three events—the Creation, the Incarnation, and the Resurrection of Christ—with a formulation that is, so far as I can ascertain, unattested before the *Chronography*: the First-Created Day (ἡ ἁγία πρωτόκτιστος ἡμέρα).[59] Synkellos later included a fourth day of divine intervention, bringing the instances of 1 Nisan, 29 Phamenoth, and 25 March to a total of four: (1) the Creation in AM 1; (2) the post-Flood drying of the earth in AM 2243; (3) the Incarnation of Christ in AM 5500; and, (4) the Resurrection of Christ in AM

---

58    Περὶ ταύτης καὶ ἡμῖν ὁ πᾶς τοῦδε τοῦ γράμματος πόνος καταβέβληται, δεῖξαι τὴν αὐτὴν καὶ μίαν πρωτόκτιστον ἡμέραν σύστοιχον τῇ τοῦ θείου εὐαγγελισμοῦ καὶ τῆς ὑπερφυοῦς ἐξ ἁγίας παρθένου συλλήψεως τοῦ μονογενοῦς υἱοῦ τοῦ θεοῦ ἡμέρᾳ καὶ τῇ τῆς ζωοποιοῦ ἐκ νεκρῶν ἀναστάσεως θεοειδεστέρᾳ καὶ ὁλοφώτῳ τοῖς ἀξίοις ἑορτάζειν αὐτὴν ἐν πνεύματι καὶ ἀληθείᾳ (AT 463–464/M 389.20-25).

59    Supported by searches in the Thesaurus Linguae Graecae (www.tlg.uci.edu). George Monachos, Synkellos' successor in universal chronicling who wrote around the 840s, used the phrase, though it did not play a central role in his conception of time. C. de Boor, *Georgii monachi chronicon* (Leipzig, 1904), pp. 129.3, 177.27.

5534.[60] Synkellos was explicit: the alignment of the *dates* indicated recurrences of the *same day*: the First-Created Day.

What does it mean to date the "First-Created Day"? Presumably Synkellos was drawing on the chronographic tradition, and so we dutifully turn to earlier works in the genre. Synkellos presented the *Chronography* in the tradition of the fifth-century Alexandrian Annianos who had, in turn, positioned himself as heir to Julius Africanus, the third-century godfather of Christian chronography.[61] Annianos' calculation of years is nearly the same that we find reproduced in Synkellos, for Annianos reckoned that both the Incarnation and the Creation had occurred on a 25 March, the latter exactly 5500 years after the former.[62] Synkellos expressed complete agreement with the calculations of Annianos, reproving Annianos only for his attempt to synchronize the records of Babylonian and Egyptian kings with pre-Flood chronology.[63]

Synkellos did, however, add to Annianos. His fourth First-Created Day—the earth's drying while Moses was in the Ark in AM 2243—was his own invention. In his discussion of this fourth day, Synkellos distanced himself from Annianos' chronological claims in subtle but profound ways. Synkellos specified that all four recurrences of the First-Created Day were not only 25 March and 29 Phamenoth, but also the first day of the week (Sunday) and 1 Nisan.[64] In describing this day, Synkellos went beyond the already specific Genesis narrative by adding "lunar days" (κατὰ σελήνην), stating, in particular, that the water subsided on "Luna 12."[65] It is not possible, however, for a day to be the moon's

---

60    Creation: AT 4/M 3; Ark on dry earth: AT 32/M 23–24; Incarnation: AT 449–450, 454–455/M 376–377, 380–382; Resurrection: AT 462–463, 465/M 388–389, 390 and AT 472–473/M 394–395.

61    On Annianos and his relationship to his scholarly predecessors, see Victor Grumel, *Traité d'études byzantines I: La chronologie* (Paris, 1958), pp. 92–94.

62    On the development of the importance of 25 March, see Grumel, *Chronologie*, pp. 27–30. Synkellos closely adheres to Annianos' calculations, and likely simply copied statements from his works and the now-lost works of Panodoros (AT 46–48 & 474/M 35–36 & 396). These chronicles were, unlike the *Chronography*, interspersed with calculations and Paschal Tables. Synkellos also consulted Maximus the Confessor on the computation of Easter (AT 455/M 382), and an "Ecclesiastical Computation" related to the "Astronomical Tables" (AT 301–304/M 245–247). This maybe have been "Annianos' attempt to bring Panodoros' Astronomical Canon into conformity with biblical chronology and the traditional dating of Christ's Incarnation" (Adler and Tuffin, pp. lxiv–lxix; see AT 455/M 381–382 and AT 46–47/M 35–36).

63    Grumel, *Chronologie*, p. 95. AT 46–47/M 35.

64    See Grumel, *Chronologie*, p. 93 n. 1.

65    The Septuagint version of Genesis relates that Noah entered the Ark on the twenty-seventh day of the second month, Iyar (the Hebrew text states the seventeenth). Nearly a

twelfth (Luna 12) and also the first day of a particular lunar month, 1 Nisan. 1 Nisan is, by definition, Luna 1.[66] Synkellos then generalized this new and problematic idea: these additional dating specifications were also true of the First-Created Day of the Resurrection, which he now dated as 25 March, 29 Phamenoth, *and* 1 Nisan.

The application of the idea to the Resurrection created yet another chronological impossibility. Though the Jewish lunar calendar at the time of Christ was not intercalated to ensure that Passover (14 Nisan) fell after the vernal equinox (21 March), it usually did so; theoretically 1 Nisan of AM 5534 could have also been 25 March.[67] However, the *historic* Resurrection could not possibly have occurred on 1 Nisan. As recounted by all four canonical gospels, Christ's historical *passio* occurred during the celebration of Passover, his Resurrection just after. Passover might fall on a range of solar calendar dates, but in the Jewish calendar Passover was always 14 Nisan.[68]

Before we attribute these apparent problems to Synkellos' ignorance, note that elsewhere he correctly defined the date of Passover as "the fourteenth of the first month at evening."[69] Thus, the contradiction just elucidated does not

---

year later on 1 Nisan the flood waters finally dried up, and exactly one year after embarkation, the Ark was emptied on 27 Iyar (Genesis 8, 13–19).

66    As the Venerable Bede succinctly explained to his students: "Whenever Holy Scripture… indicates a day of the month on which something was said or done, it signifies nothing other than the age of the Moon." *De Temporum Ratione* 11.313, trans. Faith Wallis, *Bede*, p. 42.

67    Sacha Stern, *Calendar and Community* (Oxford, 2001), pp. 34–46.

68    The key passage is Exodus 12, 18: "In the first month, on the fourteenth day of the month at evening, you shall eat unleavened bread, until the twenty-first day of the month at evening." Nisan was kept generally in the springtime by the occasional addition of a thirteenth lunar month (Adar II) but ranged across the Roman solar months March, April, and May. It was not until the tenth century that Jewish lunar reckoning was universally "fixed" to the solar calendar so that the first full moon after the vernal equinox always belonged to Nisan. On the repeating 19-year cycle of the lunar calendar, see Grumel, *Chronologie*, 31–56, in particular pp. 41–48. Sacha Stern has shown convincingly that the "fixed" Jewish lunisolar calendar became accepted only very gradually over the course of the fourth to ninth centuries, and not universally until the tenth (*Calendar and Community*, pp. 155–181, 197–200).

69    AT 207/M 168. As this passage continues, Synkellos gives even more specific information, stating that based on a tradition dating back to the year of the Resurrection: "even to this day one can see in Jericho at the vernal equinox new grain being harvested early in the warmer locations. From this grain, the most holy church in Jerusalem customarily offers the bloodless offering [the Eucharist] during the anniversary of the life-bringing Resurrection of Christ our God."

alert us to Synkellos' incompetence, but to the fact that by "Nisan" he must have meant something other than what we assumed. Indeed, when Synkellos referred to dates using months of the Jewish calendar, he was in fact referring to a calendar entirely distinct from the rabbinic lunisolar calendar in all respects save for the month names: "Let anyone who reads this [chronography] reckon the first of the first Hebrew month of Nisan as the beginning of every year in this chronicle, and not the first of the Egyptian month Thoth, or the first of the Roman month of January, or some other beginning-point used by some other nation."[70] Synkellos went on to define a 365-day solar calendar that matched Hebrew month names with the Roman and Egyptian solar calendars: "Nisan" was a 30-day month, from "25 March up to 23 April, and from 29 Phamenoth up to 28 Pharmouthi," and so on.[71]

In other words, Synkellos was using a Hebrew *solar* calendar. Synkellos' calendar has been recognized as a distinct system of reckoning,[72] and merits further study in the context of "Romanizing" calendars created and used by various local cultures in Mesopotamia, Syria, and Palestine.[73] Synkellos' apparent coordination of multiple calendars to the First-Created Day was, rather, a translation of the date March 25 into three different (but parallel) solar calendars.[74]

Synkellos' innovative tripartite dating of the First-Created Day was not a claim to chronological synchronization so much as it was a statement of cultural universality. Synkellos' statement of the dates for the First-Created Day played the same unifying role in his conceptualization of time as the visual presentation of the Incarnation did in Eusebius' *Canons*. If Eusebius used a graphic depiction of the dawn of universal time under Roman rule to *show* Christian providence coordinated a universal chronology, Synkellos used chronological terminology to *state* the same idea in regards to his First-Created Day.

The chronographer Annianos does not seem to have employed anything like this concept, and it remains unclear what Synkellos meant by calling this universally dated day "First-Created." We could compare the *Chronography* with the Byzantine universal chronicle closest in scope and date of composition, the

---

70    AT 8/M 6.

71    AT 9–10/M 6–7.

72    Jürgen Tubach, "Synkellos' Kalendar der Habräer," *Vigilae Christianae* 47 (1993), 379–389.

73    Discussed as the *hemerologia* by Samuel, *Chronology*, pp. 172–178, 186–188. See also Stern, *Calendar and Community*, pp. 211–275; and the recent work of Jonathan Ben-Dov who suggests discussing these surviving texts in the context of a regional culture of exchange and influence, *The Head of All Years* (Leiden, 2008), pp. 266–270.

74    See Tubach's table of the three in "Synkellos' Kalendar," p. 381.

seventh-century *Chronicon Paschale*.[75] The *Chronicon Paschale*'s influence is questionable since it only survives in one tenth-century manuscript (*Vat Gr. 1941*).[76] Even if we grant that Synkellos had read this text, the anonymous chronicler's "paschal" focus is very different from his own. The paschal chronicler's achievement was to comprehensively bring together a chronographer's reckoning by annual increments, with a computist's reckoning by solar and lunar cycles. The *Chronicon Paschale* unlocked the chronological potential of the 19-year lunar cycle, the 28-year solar cycle, and their product the 532-year paschal table, to project days and dates into the past when the sources had not recorded such specificity.[77] To this end, the paschal chronicler made precise calculations that relied on a strictly linear conception of time, never stating that days with the same date were in any way the same day.[78]

## Synkellos' First-Created Day as a Concept

Synkellos' idea was unprecedented in chronography: a day that cinched up the linear thread of time like a drawstring, gathering together temporally disparate historical events as though through a loophole in the fabric of time itself. Synkellos had no actual chronological need for his assertion: the thesis that God first set matter in motion on March 25 was not *chronologically* significant for any of the calculations or synchronizations in the *Chronography*. Synkellos' use of terminology from multiple calendars does not indicate an interest in

---

75    Analysis and recent bibliography in Mary Whitby, "The Biblical Past in John Malalas and the *Paschal Chronicle*," *From Rome to Constantinople*, eds. H. Amirav and H. Romeny (Leuven / Paris, 2007), pp. 279–302. *Chronicon paschale ad exemplar vaticanum* 11, ed. Ludwig Dindorf (Bonn, 1832), trans. Mark Whitby and Mary Whitby, *Chronicon Paschale: 284–628 AD* (Liverpool, 1989).

76    Passages similar to those in Synkellos or Theophanes seem to have come from a common source, rather than from Synkellos' reading the *Chronicon Paschale* directly (see Whitby and Whitby, *Chronicon*, p. xiv).

77    For instance, under AD 609: "And so from the death of Constantine until now there are 272 years, while from his twentieth anniversary, 284 complete years. Easter indeed fell on the third of April 272 years ago in year 13 of the moon's cycle, in the second year of Olympiad 279." Whitby and Whitby, *Chronicon*, pp. 147–148; Dindorf, *Chronicon*, p. 698. The anonymous paschal chronicler seems to have suggested subtle typologies in correspondence between days of the week, such as Christ's baptism occurring on a Wednesday, the same day God created the waters. See Treadgold, *The Early Byzantine Historians*, p. 343, for other examples.

78    The paschal chronicler adhered to a strictly historical and linear time even in the entry for AD 562, at the completion of the first 532-year cycle on a date that was demonstrably the same astronomical day as Christ's resurrection (see Dindorff, *Chronicon*, p. 684; Whitby and Whitby, *Chronicon*, pp. 134–135.)

cross-cultural chronology: Synkellos' equation of 25 March, 29 Phamenoth, and 1 Nisan was in fact simply an equivalence of three different, but compatible, solar calendars. As such one would have to presume that those dates would align *every* year.

Synkellos' chronological arguments, such as his dates for Abraham and Moses, were limited to harmonizing the *years* of various rulers' reigns. Synkellos did not attempt to prove that his *dates* aligned with celestial events, such as the appearances of comets, and he hardly mentioned days of the week. The idea of the First-Created Day was clearly central to Synkellos' ambitions. The meaning of this phrase, however, remains far from apparent. How might Synkellos have expected his readers to understand his novel formulation? What did it mean?

Work in critical theory has pointed out that generic expectations are communicated from author to reader through a series of cues or references which "make present...the text's presence in the world," a presence shared within the community of author and audience.[79] That is, textual cues do not simply alert a reader to a single genre and then step aside: they continue to negotiate with the reader's expectations and so situate the text in relationship to multiple genres.[80] Synkellos' project is not entirely comprehensible as pure chronography. If we seek to take Synkellos on his own terms and to trace the experience of his medieval readers, we must follow his generic cues, line by line.

Let us return to the beginning. Synkellos began the *Chronography* by quoting the Septuagint's first sentence, the instantiating moment of Creation ἐν ἀρχῇ:

> In the beginning (ἐν ἀρχῇ) God created the heaven and the earth.[81]

With his next words, Synkellos provided commentary on his first noun:

> The beginning (ἡ ἀρχὴ) of all chronological movement of the visible creation subject to time...[82]

One reason to begin with the same words as the Septuagint might have been to set up the argument that absolute adherence to the canonical translation set

---

79    John. Frow, *Genre* (London / New York, 2006), p. 109, citing Gérard Gennette's *Seuils* (Paris, 1987).

80    Ibid., pp. 114–123.

81    AT 1/M 1. "Εν ἀρχῇ ἐποίησεν ὁ θεὸς τὸν οὐρανὸν καὶ τὴν γῆν."

82    ἀρχὴ πάσης χρονικῆς κινήσεως τῆς ὑπὸ χρόνον ὁρατῆς κτίσεώς ἐστιν. Ibid., though substituting the more literal "all chronological movement" for Adler and Tuffin's "whole chronological process."

the *Chronography* on a pedestal of canonicity. In the immediate context, however, Synkellos used the first line from the Book of Genesis to introduce a philosophical discussion and a distinction. Synkellos' prologue argued that, properly speaking, the beginning (ἡ ἀρχὴ) must be temporal, the beginning of matter, of motion and, therefore, by definition, of time. "It is abundantly clear" that "heaven and the earth, the light and the darkness, the spirit and the abyss"—all created matter—came into existence with "the first-created 24-hour day itself...this no one of sound mind will oppose."[83] Then:

> Moses, the beholder of God, learnt naturally and through divine instruction that it was also the first day of the first month of Nisan and commenced his narrative from it, saying 'In the beginning God created the heaven and the earth.' For it is abundantly clear that a day is at the head of every monthly and yearly chronological cycle.[84]

According to Synkellos, Moses, as the author of Genesis, must have meant the "beginning" of Creation as the beginning of a (solar) calendrical cycle.

In his gloss Synkellos avoided discussion of both the equinox and the plenitude of the moon by asserting that, inductively, Moses must have meant time to be reckoned from (solar) 1 Nisan. Synkellos was not concerned with the astronomy or historicity of the matter, but the principle.[85] Synkellos added a philosophical proof. If, as had all other chronographers, he were to reckon the beginning of time from the Creation of the moon on the fourth day there would be two beginnings: one "of the heaven and earth earlier in time" and, then, a second, "later, during which the First-Created Day began its existence." This is "opposed to divinely inspired-utterances and to the natural order of things." Creation and time must be co-terminus: "This Holy First-Created Day is incontrovertibly proved to be a chronological beginning."[86] What were the stakes in making this claim?

---

83   πρόδηλον γὰρ ὅτι...ὁ οὐρανὸς καὶ ἡ γῆ καὶ τὸ φῶς καὶ τὸ σκότος τὸ πνεῦμά τε καὶ ἡ ἄβυσσος καὶ αὐτὸ τὸ πρωτόκτιστον νυχθήμερον ὅπερ ἀρχὴ τῆς χρονικῆς κινήσεως πέφυκεν...οὐδεὶς ἀντιφράσοι τῶν εὖ φρονούντων (AT 2/M 2).

84   οὗ χάριν καὶ πρώτην τοῦ πρώτου μηνὸς Νισὰν φυσικῶς αὐτὴν καὶ θεοδιδάκτως ὁ θεόπτης Μωϋσῆς παραλαβὼν ἐξ αὐτῆς ἤρξατο τῆς συγγραφῆς λέγων "Ἐν ἀρχῇ ἐποίησεν ὁ θεὸς τὸν οὐρανὸν καὶ τὴν γῆν." πρόδηλον γὰρ ὅτι παντὸς μηνιαίου καὶ ἐνιαυσιαίου χρόνου (AT 2/M 2).

85   Grumel, *Chronologie*, pp. 87–88, 95.

86   εἰ γὰρ μὴ τοῦτο δῶμεν, ἔσται μὲν ἄλλη τις ἀρχὴ οὐρανοῦ καὶ γῆς κατὰ τὸν χρόνον πρεσβυτέρα καὶ ἄλλη νεωτέρα, καθ' ἣν ἡ πρωτόκτιστος ἡμέρα τοῦ εἶναι ἤρξατο, ὅπερ ἐναντιοῦται ταῖς θεοπνεύστοις φωναῖς καὶ τῇ φυσικῇ τῶν πραγμάτων ἀκολουθίᾳ. ...ἀναγκαίως οὖν ἐκ πάντων δείκνυται χρονικὴ ἀρχή. AT 3/M 2.

To my knowledge no previous chronographer had attempted to defend the assertion that the creation of matter on the very first day meant the beginning of time. Christian chronographers and computists ubiquitously began their calculations from the "fourth day," the day on which the Book of Genesis had said God created the sun and the moon.[87] Synkellos confidently asserted that no one of sound mind could continue to propose this premise without offending basic logic. This was a cue to Synkellos' readership that his reasoning was based on Aristotle's standard definition of time: while time is not equal to motion, time is the measure of motion.[88] By referring to Aristotle in the context of a discussion of the Creation, Synkellos not only grounded his argument in textbook logic, but also placed himself in line with widely accepted philosophical and theological treatises on the world's origins.[89]

Using Aristotle's logic to ground an exegesis of Genesis 1 resonates with the philosophical work of Synkellos' near-contemporary, John of Damascus (d. 749–754).[90] The Damascene was a theologian-philosopher who, like Synkellos, wrote in Greek and had ties to Umayyad Syria.[91] John of Damascus had begun his magnum opus, the *Fount of Knowledge* (Πηγὴ Γνώσεως), with an excursus— the *Dialectica*, or "Philosophical Chapters" (Κεφάλαια Φιλοσοφικά)—on the Aristotelian terminology he would apply to his theology of the Trinitarian God.[92] Synkellos could have read John's text while in Palestine or Constantinople; the

---

87    See Grumel, *Chronologie*, p. 88, on Panodoros' argument for this position (presumably repeated in Annianos' lost works). For most chronographers, the creation of matter fell on the (theoretical) 19th of March, the 21st being then the "fourth day," the beginning of astronomical time, the vernal equinox, and the eventual date of the Resurrection.

88    It is this definition that lends time its universality. As Aristotle put it: "Every change and every motion is in time" (πᾶσα μεταβολὴ καὶ πᾶσα κίνησις ἐν χρόνῳ ἐστίν: *Physics* 4.14: 223a.14–15). Translation from Glen Coughlin, *Aristotle: Physics or Natural Hearing* (South Bend, 2005), p. 92.

89    Through the early middle ages Aristotelian logic never lost its position in Greek pedagogy. See Klaus Oehler, "Aristotle in Byzantium," *GRBS* V (1964), 133–146; and Mossman Roueché, "A Middle Byzantine Handbook on Logical Terminology," *JÖB* 29 (1980), 71–98.

90    Alexander Kazhdan, "John of Damascus," *ODB*.

91    John of Damascus belonged to a family, the Manṣūr, who were native to Syria and had likely headed the Umayyad financial administration into the eighth century. See Andrew Louth, *St. John Damascene: Tradition and Originality in Byzantine Theology* (Oxford, 2002), pp. 3–7. Louth argued that the monastery John of Damascus retired to was Mar Chariton, which Synkellos visited many times, as above p. 18, n. 2, "St. John Damascene: Preacher and Poet," *Preacher and Audience: Studies in Early Christian and Byzantine Homiletics*, eds. P. Allen and M. Cunningham (Leiden, 1998), pp. 248–249.

92    Louth, *Damascene*, pp. 38–46. Louth notes the works' lasting import as "[scholastics'] principle resource for the Trinitarian and Christological doctrines defined by the

chronographer's use of Aristotle may well have been in imitation of the *Fount of Knowledge*.[93] John of Damascus' work explicitly relied upon the same standard Aristotelian definition of time—"time is the measure of motion"—so key to Synkellos' reasoning.[94]

Thus, Synkellos' chronological assertion was in part the harmonization of an accepted philosophical commonplace with the practice of chronography. This conceptual cross-pollination supported the controversial assertion that the beginning of time was coterminous with the creation of matter. In working out his harmonization, Synkellos' chronological rendering of time's beginning went where no philosopher had. As we have seen, Synkellos not only asserted this basic relationship between matter and time, but his First-Created Day was a claim that dates thousands of years apart were a single day. Where did he get this idea, and how did he expect his readers to understand it?

### Synkellos' First-Created Day as a Revelation of Grace

Another philosopher-theologian, the fourth-century bishop Basil of Caesarea, known in patristics as one of the three great fourth-century "Cappadocian Fathers," also wrote a work on the Creation, but framed his account, the *Hexaemeron*, as a series of homilies.[95] In Homily 2 on the phrase "the earth was

---

Oecumenical Synods of the early Church, and continuing up through the Reformation era and the period of Protestant scholasticism, ...[into] systematic theology" (p. 3).

93   Scholars have found evidence of familiarity with the Damascene's writings in Constantinople in the early decades of the ninth century during debates over the religious use of icons at very the time Synkellos was working on his *Chronography* (808–810). See Mansi, *Sacrorum Conciliorum Nova et Amplissima Collectio*, XIII, cols. 356C–364E, trans. in Daniel Sahas, *Icon and Logos* (Toronto, 1986), pp. 168–175; and Louth, *Damascene*, pp. 13; 197–198. It was becoming customary to cite Aristotle in ninth-century debates over the religious use of icons. See Kenneth Parry, *Depicting the Word: Byzantine iconophile thought of the eighth and ninth centuries* (Leiden, 1996) esp. pp. 19–20; 56–63; and the careful correctives in Thalia Anagnostopoulos "Object and Symbol: Greek Learning and the Aesthetics of Identity in Byzantine Iconoclasm" Ph.D. dissertation (University of California, Berkeley, 2008), esp. pp. 77–81.

94   Χρόνος ἐστὶ μέτρον κινήσεως. P. Bonifatius Kotter, *Die Schriften des Johannes von Damaskos* I: Dialectica (Berlin, 1969), Chap. 68, ln. 18 (p. 141). For further discussion, see also Ibid., Chap. 62, lnn 45–51 (p. 131), and Kotter, *Schriften* II: Expositio Fidei (Berlin, 1973), Chap. 15, lnn 9–13 (p. 43). On the importance of Aristotle to John of Damascus, see Klaus Oehler, "Aristotle in Byzantium," *GRBS* V (1964), 143–144.

95   Synkellos mentions that a manuscript attributed to Basil of Caesarea (d. 379) solved "the question of chronological agreement between the two kingdoms of the Hebrews (Israel and Judah)" (AT 295/M 240).

invisible and unfinished,"[96] Basil demonstrated that God was not merely a craftsman who arranged pre-existing matter, but that He created all matter from this first moment, which included the beginning of time.[97] Basil located this issue in a discussion of scripture's use of "one day" (ἡμέρα μία) as opposed to the "first day" (πρώτη ἡμέρα).[98] Though there was no sun, the point of specifying a twenty-four hour day-and-night period was "in order that through the term it might be related (τὸ συγγενὲς) to eternity."[99] Basil turned to an idea strikingly similar to the First-Created Day to explain that this meant the day was *both* eternal and temporal: "In order that you might carry the idea on to the future life, [Scripture] specifies [this] icon of eternity as "one," the first-fruit (ἀπαρχὴ) of days, equal-in-age to light, the Holy Lord's Day, which has been honored by the resurrection of the Lord."[100] Basil's concept is similar to Synkellos', but still maintained the line between theological typology and historical chronology.[101]

Basil's choice to communicate these ideas through sermons suggests that Synkellos could also have intended that his First-Created Day invoke the context of liturgical worship. In fact, we have already seen Synkellos make this same generic reference himself. Synkellos did not defend his idea of a First-Created Day mathematically, by providing, for instance, extensive tables charting five-and-a-half millennia of calendrical cycles. Rather, Synkellos claimed that his knowledge of universal time was a prerogative shared by those who were within the fold of Christian orthodoxy, who were granted access to divine grace. This claim was initially made in the conclusion of his first statement of the thesis:

> It is abundantly clear *for those deemed worthy of divine grace* that the first Pascha of the Lord also began on this holy first-created day.[102]

Only one "worthy of divine grace" could know or perceive that this alignment of dates occurred on the Holy First-Created Day. Divine grace provided

---

96    Homily 2: "Περὶ τοῦ ἀόρατος ἦν ἡ γῆ καὶ ἀκατασκεύαστος," in Stanislas Giet, *Basile de Césarée. Homélies sur l'hexaéméron*, 2nd ed. (Paris, 1968), pp. 138–187.

97    Ibid., sec 2, 33A, lnn 2–6 (p. 148).

98    Ibid., sec 8, 49A, lnn 9–10 (p. 178).

99    Ibid., sec 8, 49C, lnn 10–11 (p. 182).

100   Ἵνα οὖν πρὸς τὴν μέλλουσαν ζωὴν τὴν ἔννοιαν ἀπαγάγῃ, μίαν ὠνόμασε τοῦ αἰῶνος τὴν εἰκόνα, τὴν ἀπαρχὴν τῶν ἡμερῶν, τὴν ὁμήλικα τοῦ φωτός, τὴν ἁγίαν κυριακήν, τὴν τῇ ἀναστάσει τοῦ Κυρίου τετιμημένην. Homily 2, 8.74–8.77.

101   See the treatment of Byzantine chronicles in a cultural and literary context, by Alexander Kazhdan, *Byzantine Literature* I–II (Athens, 1999, 2006).

102   Πρόδηλον δὲ ὅτι καὶ πρῶτον κυριακὸν πάσχα τοῖς καταξιωθεῖσι τῆς θείας χάριτος κατὰ ταύτην ἤρξατο τὴν ἁγίαν πρωτόκτιστον ἡμέραν. AT 2/M 2 (emphasis mine).

Synkellos with the date of the Resurrection; the same grace gave him the date of the Creation, of the opening of the Ark, and of the Incarnation.

Following his prefatory discourse on creation and time Synkellos emphasized the relevance of his First-Created Day to Christ's life:

> On this day also (ἐν ταύτῃ [ἡμέρᾳ] καὶ)
>> Gabriel foretold the divine conception...
> on this day also (ἐν ταύτῃ [ἡμέρᾳ] καὶ)
>> the only begotten Son arose from the dead...
> on this same holy day (κατὰ τὴν αὐτὴν ἁγίαν...ἡμέραν)
>> of the life-bringing Resurrection,
>> the 5534th year from the creation of the universe commenced.[103]

Synkellos capitalized on the ambiguity of the word day (ἡ ἡμέρα) as both "date" and "present day" in order to make the assertion that when these dates align this is, somehow, a recurring now, "this same holy day." The poetic syntax smooths the conceptual paradox.

The concept of a recurring "same holy day" appears in the homilies of John Chrysostom, Patriarch of Constantinople at the turn of the fifth century. Chrysostom's recurring holy day would have been familiar to Constantinopolitan churchmen of the ninth century from copies of his homilies on the feast of the Resurrection.[104] In Chrysostom's paschal homilies Christ's Resurrection and the yearly feast celebrating that event partook of the same present moment:

> This is the very day (Αὕτη ἡ ἡμέρα)
> on which Adam was freed,
> on which Eve was released from grief,
> on which brutal death shuddered,
> on which the power that burst from the mighty stones was let loose,
>> and the barriers of the tombs which were torn asunder were undone, ...

---

103   AT 1/M 1. I have arranged the text to highlight the repeated phrases.

104   Chrysostom was one of the most emulated homilists throughout the Byzantine period. Synkellos cites his commentary on Matthew in relation to his discussion of the Creation (AT 5/M 4). The surviving manuscripts indicate interest in his paschal homilies in the ninth and tenth centuries in particular: ninth century: Moscow, Gosudarstvennyj Istoričeskij Musej (GIM) Cod. Sinod. Gr. 284 (Vlad. 215); Escorial, Real Biblioteca, Cod. Chi IV.6 (Andrés 401); tenth century: Athens, Mouseio Benaki, Cod. T A 319, (110); Jerusalem, Patriarchikē bibliothēkē, Cod. Panagiou Taphou 6; Oxford, Bodleian Library, Cod. Barocci 174 and Cod. Barocci 199.

on which grew the abundance and fruitfulness of the resurrection,
    as in the garden inhabited by the race of men,
on which the lilies of the newly-illumined were made to spring up...
on which the multitude of the Jews was put to shame,
on which the ranks of the faithful are made glad,
on which the wreaths of the martyrs are made afresh.
"This, then, is the day that the Lord has made, let us rejoice and be glad in it."[105]

In a strictly chronological sense these events did not all occur on the actual date of the Resurrection, for there were as yet no newly-baptized neophytes when Christ exited the tomb, nor were Jews yet feeling any shame. Synkellos, too, put grammar at the service of theology.

At the mention of "the faithful" in the above quotation, a grammatical shift from the past into the present tense occurs without a break in the syntactical cadence (*are* made glad; *are* made afresh).[106] A historical treatment of Christ's resurrection would render these phrases in a past tense, denoting those faithful to Christ at that time, perhaps the faithful group of disciples huddled in the Upper Room. Chrysostom was well aware that at the Resurrection the martyrs could not yet have testified to their faith. In these temporal contradictions, Chrysostom seems to have sought to enjoin the "ranks of the faithful" gathered with him at the close of the fourth century to consider these acts in the past as part of the present reality. Chrysostom's point was that all of the events he described were called into being by the act of Resurrection. Embedded in the grammar of the rhetorical flourish was the assertion that in the subsequent liturgical life of the church, specifically at the yearly celebration of the Resurrection, these past events existed in a unified present, "this very day."

---

105    One example among many. "In resurrectionem domini" in M. Aubineau, *Homélies pascales* (Paris, 1972), p. 324. Translation mine. Αὕτη ἡ ἡμέρα ἐν ᾗ ὁ Ἀδὰμ ἠλευθερώθη, ἐν ᾗ ἡ Εὖα ἀπηλλάγη τῆς λύπης, ἐν ᾗ ὁ ἀνήμερος θάνατος ἔφριξεν, ἐν ᾗ τῶν κραταιῶν λίθων ἡ δύναμις παρελύθη ῥαγεῖσα καὶ τὰ τῶν μνημείων κλεῖθρα διασπασθέντα ἀνέθη...ἐν ᾗ τὸ τῆς ἀναστάσεως εὐθαλὲς καὶ εὔκαρπον ὡς ἐν κήπῳ τῇ οἰκουμένῃ τῷ γένει τῶν ἀνθρώπων ἐβλάστησεν, ἐν ᾗ τὰ τῶν νεοφωτίστων ἀνεφύησαν κρίνα...ἐν ᾗ τὰ τῶν Ἰουδαίων κατῃσχύνθησαν πλήθη, ἐν ᾗ τὰ τῶν πιστῶν εὐφραίνονται τάγματα, ἐν ᾗ τὰ τῶν μαρτύρων ἀναθάλλουσι διαδήματα. "Ταύτην τοίνυν τὴν ἡμέραν ἐποίησεν ὁ κύριος, ἀγαλλιασώμεθα καὶ εὐφρανθῶμεν ἐν αὐτῇ." Interestingly, Synkellos concluded a key passage with the same citation: "Concerning which [day] it was said: 'In the beginning God created the heaven and the earth, on which day God created.' Concerning this, David the ancestor of God, as a prelude to universal salvation, has sung: 'This is the day that the Lord created; let us rejoice and be glad in it.'" (AT 463/M 389).

106    I am grateful to Alexandre M. Roberts for first bringing this shift to my attention.

A supra-chronological salvific time had been described as an experiential aspect of not only yearly but daily worship in a surviving text written much closer to Synkellos' own milieu: the *Ecclesiastical History and Mystical Contemplation* (Ἱστορία Ἐκκλησιαστική καὶ Μυστικὴ Θεωρία) attributed to Patriarch Germanos (r. 715–730).[107] Like John of Damascus, Germanos was an iconophile condemned by the iconoclast Council of 754. He had then been post-humously exonerated by the iconophile Council of 787; this council had been led by patriarch Tarasios of Constantinople, the very patriarch under whom Synkellos himself eventually served. Though this connection is intriguing, it cannot be assumed that Synkellos would be familiar with writings attributed to a patriarch from an earlier era simply because of their doctrinal agreements.

Fortunately there is a direct textual connection between Synkellos' *Chronography* and the *Ecclesiastical History*: the texts share a ninth-century translator.[108] Anastasius Bibliothecarius, an emissary for the Carolingian Louis II, visited Constantinople in 870 and there selected a number of works for translation into Latin. Besides translating excerpts from Synkellos' *Chronography* and Theophanes' *Chronicle*, Anastasius also made a translation of Germanos' *Ecclesiastical History* for the Carolingian Charles the Bald.[109] It is entirely plausible to suppose that Anastasius found these two texts in close physical proximity.

Of all the texts we have surveyed, the liturgical commentary of the *Historia Ecclesiastica* offers the closest conceptual parallels to Synkellos' claim that a cosmos bound by linear temporality experienced the action of the timeless eternal God as a recurring First-Created Day. The author of the *Historia Ecclesiastica* also dissolved the line between human temporality and divine eternality in his description of the liturgical experience of the Church.[110]

---

107   Mango and Scott, *Theophanes*, pp. 563–565; de Boor, *Theophanis*, pp. 407–409. Germanos (Patriarch of Constantinople from 715–730) fought against imperial religious policies: he opposed Philippikos' revival of monotheletism in 712, and then Leo III's ostensible icono-clastic policies in the 720s. Germanos was deposed by the emperor in 730.

108   Liturgical variants place the commentary no earlier than the eighth century, and not much later than the early ninth (thus, inclusive of Synkellos' time as the *synkellos*). Germanos is only the most likely candidate for authorship of the *Historia Ecclesiastica*. See René Bornert, *Les Commentaires Byzantins de la Divine Liturgie* (1966), pp. 132–160; and Robert Taft, "The Liturgy of the Great Church" *DOP* 34/5 (1980/1), 47–58.

109   See Anastasius' dedicatory letter, *Monumenta Germaniae Historica: Epistolae VII* (Munich, 1978), pp. 434–435. A ninth-century manuscript of the translation survives as Codex 711 at the Bibliothèque Municipale de Cambrai.

110   See Paul Magdalino's discussion of the text as part of a dialogue that intertwined icono-clasm, eschatology, liturgy, and politics in "The History of the Future and Its Uses:

The church is earthly heaven (ἐπίγειος οὐρανός), in which the heavenly (ἐπουράνιος) God dwells and walks about, typifying (ἀντιτυποῦσα) the crucifixion, burial and resurrection of Christ.[111]

In the act of performing the liturgy, the celebrants and the people became a part of the whole of salvation history, spanning the Old and New Testaments, as they assembled with the saints in the "kingdom of Christ."[112] The priest did not merely contemplate figures and symbols of Christ, but actually entered the heavenly kingdom and divine splendor:

> Then the priest, leading everyone into the heavenly Jerusalem, to His holy mountain exclaims: Behold, let us lift up our hearts! ...Then the priest goes with confidence to the throne of the grace of God and...speaks to God. He converses...with uncovered face seeing the glory of the Lord... 'one-to-one' he addresses God...contemplating the heavenly liturgy, [he] is initiated even into the splendor of the life-giving Trinity.[113]

Finally, the congregation was invited to partake of the Eucharist, "so that it might be fulfilled that 'Today I have begotten you.'"[114] They join fully in this experience and become "eye-witnesses of the mysteries of God, partakers of eternal life, and sharers in divine nature."[115] By this participation in the divine life and reality,

---

prophecy, policy and propaganda," in *The Making of Byzantine History: Studies Dedicated to Donald M. Nicol on his Seventieth Birthday*, eds. R. Beaton and C. Roueché (Aldershot, 1993), pp. 22–23.

111   Ed. and trans by Paul Meyendorff, *St. Germanus of Constantinople On the Divine Liturgy* (Crestwood, NY, 1984). Sec. 1, p. 56. Slightly altering Meyendorff's translation of: Ἐκκλησία ἐστὶν ἐπίγειος οὐρανός, ἐν ᾧ ὁ ἐπουράνιος Θεὸς ἐνοικεῖ καὶ ἐμπεριπατεῖ, ἀντιτυποῦσα τὴν σταύρωσιν καὶ τὴν ταφὴν καὶ τὴν ἀνάστασιν Χριστοῦ.

112   Ibid., Sec. 41, pp. 100–101.

113   Ibid., Sec. 41, pp. 90–91. Εἶτα πάντας ἀναβιβάζων ὁ ἱερεὺς εἰς τὴν ἄνω Ἰερουσαλὴμ εἰς τὸ ὄρος τὸ ἅγιον αὐτοῦ καὶ βοᾷ· Βλέπετε ἄνω σχῶμεν τὰς καρδίας·...Εἶτα πρόσεισιν ὁ ἱερεὺς μετὰ παρρησίας τῷ θρόνῳ τῆς χάριτος τοῦ Θεοῦ...ἀπαγγέλλων τῷ Θεῷ...ἀλλὰ ἀνακεκαλυμμένῳ προσώπῳ τὴν δόξαν Κυρίου κατοπτεύων·...καὶ μόνος μόνῳ προσλαλεῖ Θεοῦ...τε καὶ λαμπρότητα τὴν ἐπουράνιον λατρείαν νοερῶς ὁρῶν καὶ μυεῖται καὶ τῆς ζωαρχικῆς Τριάδος τὴν ἔλλαμψιν τοῦ μὲν Θεοῦ καὶ Πατρὸς τὸ ἄναρχον καὶ αγέννητον.

114   Ibid., Sec. 41, p. 96. καὶ πληρωθήσεται τό· "Ἐγὼ σήμερον γεγέννηκά σε."

115   Ibid., Sec. 41, pp. 98–99. Ὅθεν γενόμενοι τῶν θείων μυστηρίων αὐτόπται καὶ μέτοχοι ζωῆς ἀθανάτου καὶ κοινωνοὶ θείας φύσεως.

the souls of Christians are called together to assemble with the prophets, apostles, and hierarchs in order to recline with Abraham, Isaac, and Jacob at the mystical banquet of the Kingdom of Christ. ...We are no longer on earth but standing by the royal throne of God in heaven, where Christ is.[116]

In the *Ecclesiastical History* the "very day" of the Resurrection captured the experience of the eternal present moment of divine life bestowed through grace upon the gathered faithful. The idea that a reality joining earth and heaven was revealed on the basis of faith resonates with Synkellos' claim that on the First-Created Day "the new creation begun in Christ ushered from death to life all those with a correct belief in Him."[117] The paradigm in which Synkellos constructed his *Chronography* is incomprehensible apart from the *Ecclesiastical History*'s ecclesiology.

Synkellos claimed that the orthodox believer knew the eternal God through his experience of divine grace in faith. In the same way Moses, not present at the moment of the Creation, could know the date of the creation of the world because of his experience of God's grace. Furthermore, Synkellos too, though not present at the Creation or during Moses' vision, had been "deemed worthy of divine grace" through his correct belief. Synkellos could use authoritative tradition concerning the date of the Incarnation and the Resurrection to interpret Moses' vague statements with chronological exactitude. Synkellos' philosophically and theologically astute vision of time, encapsulated in his First-Created Day, was a claim to objective knowledge of a universal chronology through subjective experience of divine truth.

### Conclusion

A chronographer's conception of time was the same as a philosopher's: the measure of motion. Nevertheless, chronographers did not pursue an "objective" or an apolitical tally of time. How could they when the established chronological method was to reckon past time by the successions of kingdoms? So long as

---

116   Ibid., Sec. 41, pp. 100–101. καὶ συγκαλοῦνται μετὰ προφητῶν καὶ ἀποστόλων καὶ ἱεραρχῶν τῶν χριστιανῶν αἱ ψυχαὶ συνελθεῖν καὶ ἀνακλιθῆναι μετὰ Ἀβραὰμ καὶ Ἰσαὰκ καὶ Ἰακὼβ ἐν τῇ μυστικῇ τραπέζῃ τῆς βασιλείας Χριστοῦ. ...οὐκ ἔτι ἐπὶ γῆς ἐσμεν ἀλλ᾿ ἐν τῷ θρόνῳ τοῦ Θεοῦ τῷ βασιλικῷ παρεστηκότες· ἐν οὐρανῷ ὅπου ὁ Χριστός ἐστι.

117   πρωτόκτιστος ἡμέρα τοῦ πρωτοκτίστου μηνὸς ὑπάρχουσα, καθ᾿ ἣν ἡ ἐν Χριστῷ καινὴ κτίσις ἀρξαμένη πάντας εἰς ζωὴν ἐκ θανάτου μετήγαγε τοὺς ὀρθῶς εἰς αὐτὸν πιστεύοντας. AT 465/M 390.

chronology meant ordering time according to the rise and fall of kingdoms, every chronographer would find a plot embedded in his methodology. In the post-Constantinian Empire, that historical plot was perforce providential, an imperial providence dominated by the simultaneity of imperial Rome's rise and Christ's birth.

Synkellos created an innovative vision of the arc of providence by applying the epistemological implications of the Incarnation to the earliest periods of human history. This essay has argued that his project resulted in a unique conception of time itself. In Synkellos' *Chronography* time was not only the measure of motion, the ordering of the ages, and the progress of kingdoms, but time also bore witness to the relationship between mankind and Divinity through the experience of its rupture: the past in the present and the present in the past. Synkellos took the theological principle that the Incarnation was *the* truth event and embraced its chronological paradox in a way no previous Christian chronographer had ever attempted.

How significant was this achievement? Is Synkellos' struggle with the idea that the key to linear time was the intervention of a timeless Divinity also a revelation of "a Byzantine" contemplating man's experience of time in general? Caveats and cautions are easy to muster. Synkellos' system of reckoning never gained widespread currency, whether because of, or in spite of, its sophistication. We have pointed out at length that Synkellos' obscure biography makes it very difficult to understand him as a historical figure. Furthermore, it is unclear whether we should associate Synkellos, his work, and his ideas with Constantinople or with the intellectual milieu of Greek learning outside the Roman Empire. If Synkellos is "byzantine," how are we defining Byzantium? Does Synkellos, if he permits us any generalities, in fact tell us something about Syria-Palestine whence he gathered much of his material and perhaps received his intellectual formation?[118]

Even if we are left with only these ambiguities and the marvel of the surviving text, it is impossible to ignore Synkellos' authorial voice. Synkellos' all-encompassing goal was to defend and promote his chronological thesis: the date of the Resurrection in AM 5534 and the Incarnation in AM 5500 was the same First-Created Day as the Creation in AM 1. To this end Synkellos argued for two central ideas concerning time's order. First, Synkellos believed that a truly universal calendar should reflect the philosophical beginning of universal

---

118    On the Syrian milieu in relationship to Constantinople, see Marie-France Auzepy, "De la Palestine à Constantinople (VIIIe–IXe siècles): Étienne le Sabaïte et Jean Damascène," *Travaux et Mémoires* 12 (1994), 183–218. On the Syrian milieu in relationship to Baghdad, see Sidney Griffith, *Christianity under Islam* (Princeton, 2008), pp. 40–48; and Dimitri Gutas, *Greek thought, Arabic Culture* (New York / London, 1998) esp. pp. 22 and 155.

time: the movement of matter from the moment of creation ἐν ἀρχῇ.[119] Second, knowledge of the earliest period of time was possible, but only via the inspired scriptures as interpreted by those with access to a supra-temporal divine grace, itself accessible through the liturgical worship of the Church. There "those deemed worthy" experienced the Creation, the Incarnation, and the Resurrection of Christ as the ever-present "life-giving Trinity." The *Chronography* of George Synkellos aimed to prove that a true reckoning of all time, the entire past and present, unfolded from a Holy First-Created Day of and for the People of God, the Church of Christ.

Several years ago the medievalist Rosamond McKitterick suggested that to "examine time and its functions in the early middle ages may yield something very specific about the perception of the past, present, and even future on the part of any group."[120] Thus, it may not be too grandiose to claim that in the foregoing discussion we have glimpsed an early medieval culture actively thinking about time in terms of the experience of worship, even as it held to a rigorous philosophical and historical time. In this way can George Synkellos serve as a *homo byzantinus* set to thinking about the nature and meaning of time?[121]

Having made an effort to underscore Synkellos' creativity, it seems disingenuous to argue that we should make a generalization out of him. Synkellos, however, is not the only character in this story. George Synkellos created an ecumenical and therefore canonical measure of time that gave present meaning to the past, to knowledge of the stars, planets, the successions of kings, and the very temporal progression of the universe from the celestial to the quotidian. He claimed time universal and eternal for the Church of Constantinople where all was in the present as it was in the beginning and ever would be. For Synkellos, to experience God was to know time unto the ages of ages. If Synkellos intended his readers to follow his hypothesis of the First-Created Day, then it is in the cultural logic attributed to these imagined Byzantines that we can posit our larger cultural group. Synkellos may not have *been* a *homo byzantinus*, but he was writing *for* one.

---

119   Similarly, since the invention of the atomic clock (TAI) we have discussed whether to disassociate the reckoning of time from the movement of the earth in space. See D. Feeney, *Caesar's Calendar* (Berkeley, 2007), p. 294 n. 120.

120   McKitterick, *History and Memory*, p. 86.

121   On Alexander Kazhdan's *Homo Byzantinus*, see *People and Power* (1991) and *Homo Byzantinus: Papers in Honor of Alexander Kazhdan: DOP* 46 (1992), especially J. Ljubarskij, "Man in Byzantine Historiography," pp. 177–186, and Ihor Ševčenko, "The Search for the Past in Byzantium around the Year 800," pp. 279–293.

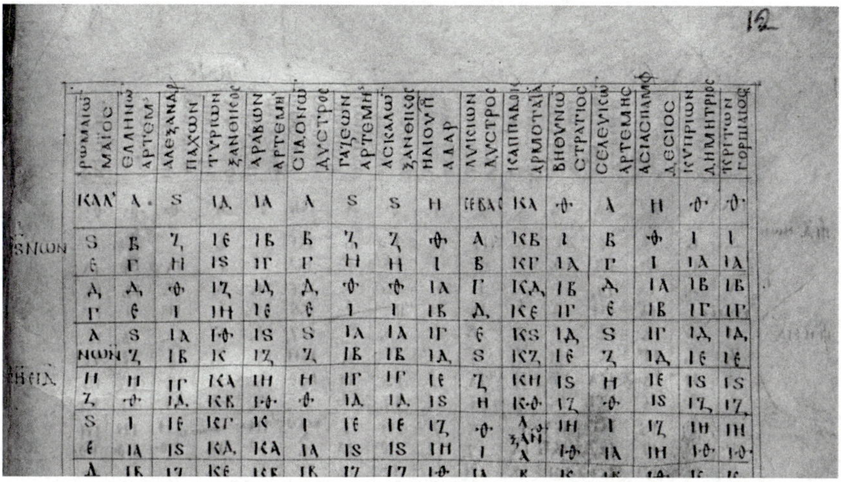

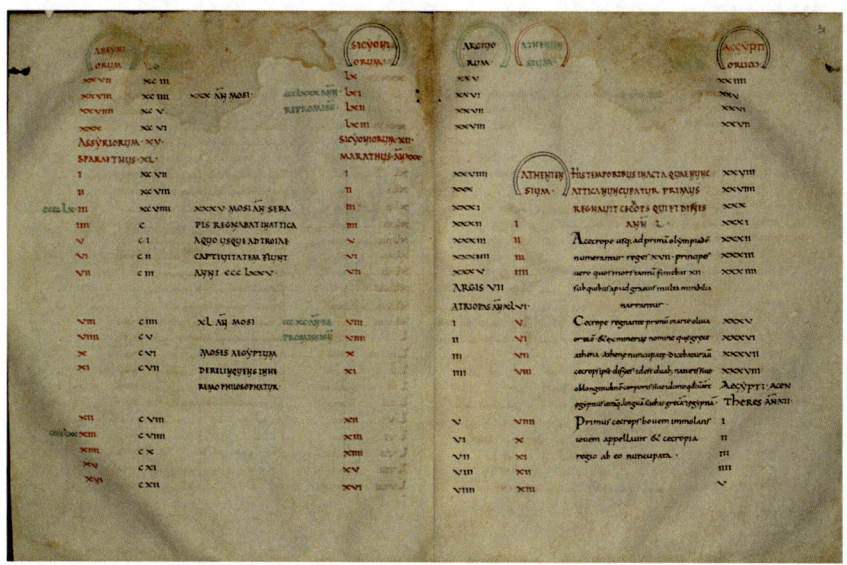

FIGURE 1.1 *Synchronization of various calendars from around the Hellenistic world to the Roman month of May (left-hand side) as in the copy of Ptolemy's* Handy Tables *in Vaticanus graecus 1291 (s. viii), f. 12r*
COURTESY OF BIBLIOTHECA APOSTOLICA VATICANA

| Column 1 | Column 2 | Column 3 | Column 4 | Column 5 | Column 6 | Column 7 |
|---|---|---|---|---|---|---|
| Decade of Abraham | – Assyrians – | Hebrews – | Sicyonian Greeks | – Argives – | Athenians – | Egyptians |

FIGURE 1.2 *Layout of the Chronicle of Eusebios-Jerome synchronizing Moses and Cecrops the Athenian* Merton College MS 315 ff. 30v–31r
COURTESY OF THE WARDEN AND FELLOWS OF MERTON COLLEGE, OXFORD

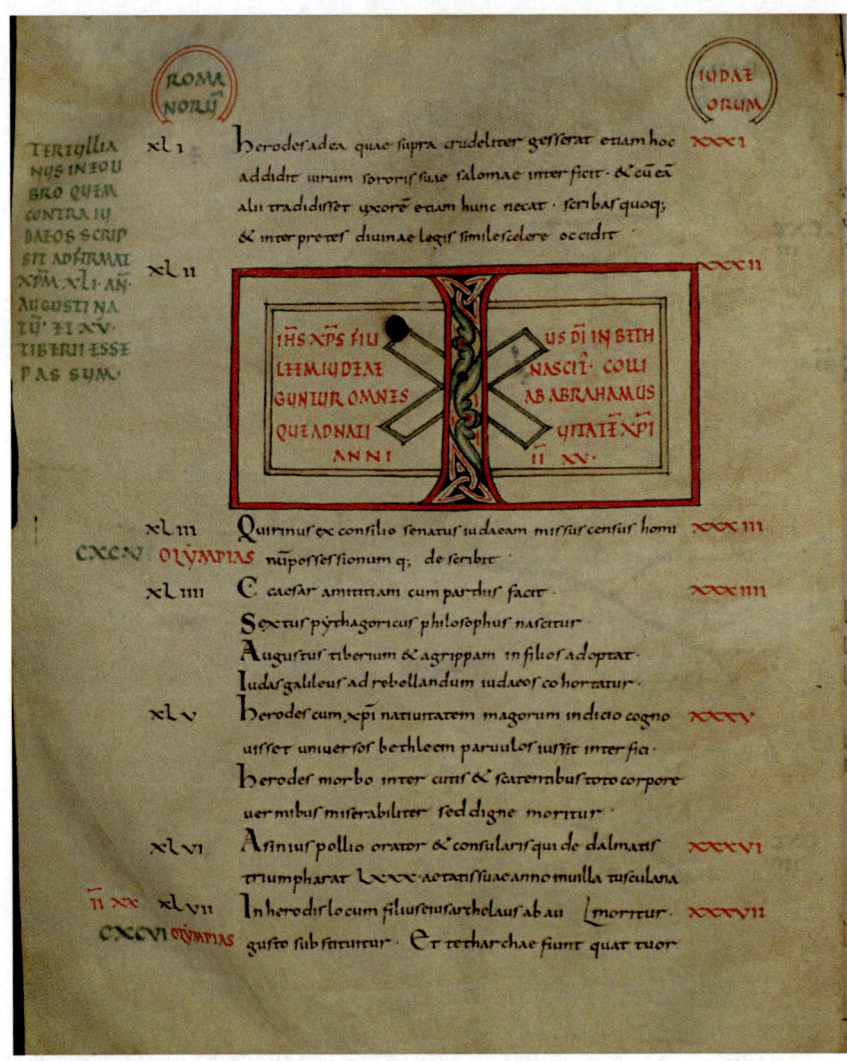

FIGURE 1.3   *The Incarnation of Christ as the kingdom of the Jews comes under the Roman Empire* as in the copy of Eusebius-Jerome's *Chronicle* in Merton College 315, *f 125v*
COURTESY OF THE WARDEN AND FELLOWS OF MERTON COLLEGE, OXFORD

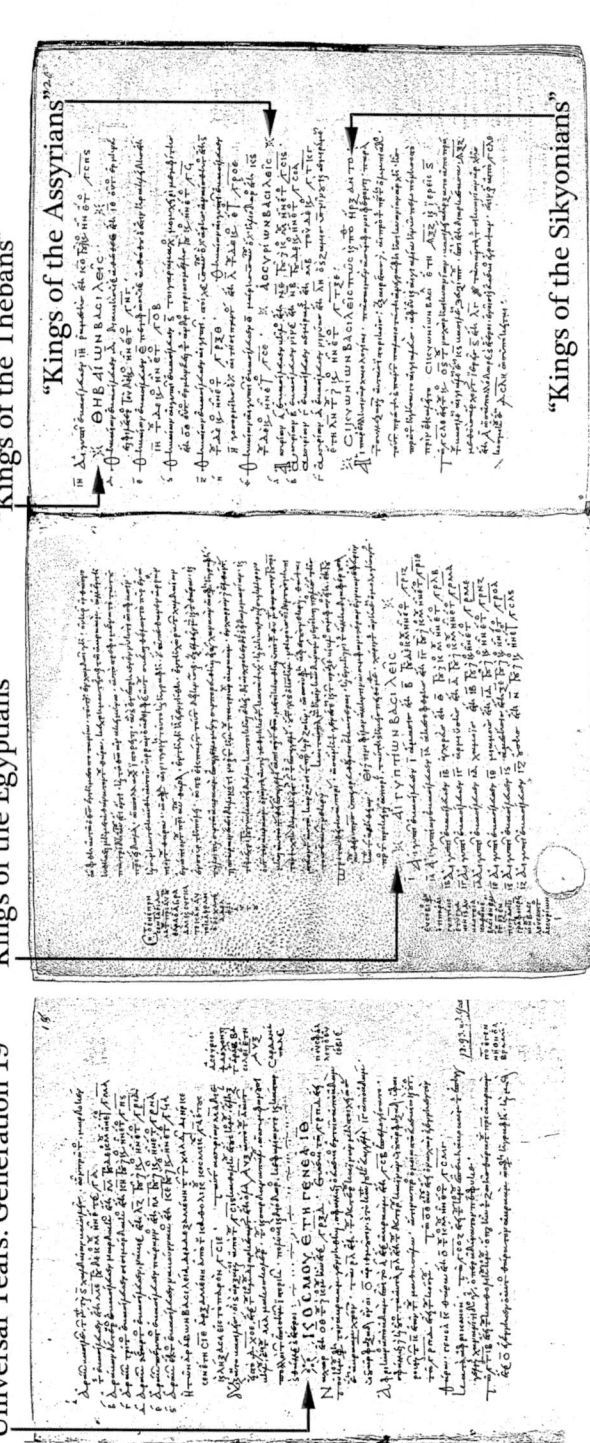

FIGURE 1.4 BnF Parisinus Graecus 1764 (*left: f 18r; center: f 19v; right: f 20r*)
COURTESY OF BIBLIOTHÈQUE NATIONALE DE FRANCE

Organizational scheme for an entry in the "second epoch" of George Synkellos' *Chronographia* as in *BnF Parisinus Graecus 1764*. Here the entry is initiated by "Universal Years: Generation 19 (and 20)" (left: f 18r) which is palaeographically emphasized by block capitals. This heading is followed by several pages of prose narrative and discussion of the nineteenth and twentieth Biblical generations (not shown: ff 18v and 19r). The entry concludes with headings (in regular capital script) of lists of the kings of the Mestraia Egyptians (center: f 19v), Thebans, Assyrians, and Sikyonian Greeks (right: f 20r). Each organizational division is also marked by a pair of line-and-dot asterisks.

# CHAPTER 2

# Registering Rome: The Eternal City through the Eyes of Pope Gregory VII

*Ken A. Grant*

When Pope Gregory VII (1073–1085) saw Rome, the physical city, a city of monuments and ruin, market and field, tower and river, he saw chaos, a city mirroring the world around it, a world he believed to be teetering on the brink of catastrophe. What he wanted to see and labored to bring about so that all could recognize Rome's splendor and its centrality, was a vision of Rome as a city of concord and order, the heart of Christendom, structuring and ordering the lives of the people so that all might gain admittance to the heavenly city. Rome was accordingly two places for Pope Gregory VII: the tumultuous, chaotic temporal city, a place of strife, intrigue, vocal support, and loud condemnation; and a bastion of order, an idealized reality created as the heart of the Church, the stable center of a crumbling world in need.

Each of these Romes is revealed to us through Gregory's own words in his *Register* and through the words of his supporters, Paul of Bernried and Bishop Bonizo of Sutri. Gregory's dual vision is important in understanding the man who shaped and attempted to guide the late eleventh-century reform movement that often bears his name, serving as a lens by which to examine why Gregory acted with such passion and determination.[1] His dual vision is also an important glimpse into what the city of Rome meant in the late eleventh century. As Richard Krautheimer notes, Rome was in the process of recreating itself, if not in stone and steeple, which came later, then at the very least in concept and ideal.[2] Pope Gregory VII's vision was to bring to fruition that process of recreation, shaping an intention to build the orderly, eternal Rome of his hopes out of the Rome of his experience. As we shall see, however, Gregory was not wholly

---

1   Whether Gregory's passionate pursuit of reform during his pontificate was constructive or destructive (or, most likely, somewhere in between) is not central to the argument here; my purpose is to continue to explore motivations for Gregory's actions. Cf. Maureen Miller, "The Crisis in the Investiture Crisis Narrative," *History Compass* 7/6 (2009), 1570, 1577: [It is time to] "undertake a radical reconsideration of the papacy from a truly post-confessional perspective" and to "set aside questions of whether a belief or development was right or wrong, good or bad, and concentrate on how and why ideas and practices developed."

2   Richard Krautheimer, *Rome—Profile of a City, 312–1308* (Princeton, 1980, rpt. 2000), p. 141ff.

© KONINKLIJKE BRILL NV, LEIDEN, 2016 | DOI 10.1163/9789004312319_004

successful; the physical Rome he left behind was as chaotic as ever, yet the idealized Rome, the Rome Gregory believed ought to be, was an idea that started to
take root in the thoughts and actions of those who followed him.

•••

Chaos was a constant in the life of Rome, at least in the two centuries proceeding the Gregorian era. Thus, it would not have been difficult for Gregory to see
Rome only as a place of crumbling ruins, not only in her physical space, but
also in her institutional structure. Rome was a city reduced in population and
embroiled in local political infighting, a pale and hazy reflection of its former
glory and influence.[3] All that remained of Rome's former station in the thoughts
of the people in the Mediterranean and Western European worlds was its continued destination as an important pilgrimage site, as H.E.J. Cowdrey notes,
"Such wealth as it had came largely from the pilgrims, penitents, and suppli-
ants who came from all over Latin Christendom and beyond."[4]

During the tenth and first half of the eleventh century, what might be called
the Papacy's own *saeculum obscuram*, or dark age, the papacy reached its
nadir: the Formosan/Anti-Formosan battles in the late ninth and early tenth
centuries, the rise and fall of the corrupt, but efficient, Theophylact family in
the early and mid-tenth century, the rule of Alberic II, the self-titled patrician
and senator of all Romans who directly influenced papal elections in the mid-
tenth century, the control of the papacy by the Crescentii family in the late
tenth and early eleventh centuries, and the control of the papacy by the
Tusculan family in the period from 1024 to 1046 (with a very brief Crescentii
family papacy in early 1045).

Responding to this long period of degradation and parochialism, the Salian
king Henry III attempted to restore the papacy by removing the warring
Roman families from the papal election process. Henry III intervened in the
papal election process for reasons that focused as much on his own future
needs as those of the church, for he needed a pope that had broad support and
legitimacy so that his crowning as Emperor would be viewed as above reproach.
This necessitated not only the ending of the feuding between the Crescentii
and Tusculan families, but also the removal of Pope Gregory VI (1045–1046)
from office, since he was tainted with charges of simony.[5] Hildebrand, the

---

3   H.E.J. Cowdrey, *Pope Gregory VII 1073–1085* (Oxford, 1998), p. 1.

4   Cowdrey, *Pope Gregory VII*, p. 2.

5   *The Papal Reform of the Eleventh Century—Lives of Pope Leo IX and Pope Gregory VII*, ed. and
    trans. I.S. Robinson (Manchester, 2004), p. 188.

future Pope Gregory VII, traveled with Gregory VI as his chaplain, thus situating Hildebrand as an eye-witness to the problems and potential reforms of both city and institution.

Four popes (Clement II [1046–1047], Damasus II [1048], Leo IX [1049–1054], and Victor II [1058–1061]) began the process of reforming the church and the papacy. Pope Leo IX made the greatest strides towards this reform as he embraced a Cluniac reform of the clergy, with its attendant calls for the extirpation of simony and clerical marriage and a renewal of "certain canonical conditions; that is, that his appointment should be ratified by his election by the clergy and people of Rome."[6] With the help of Humbert of Moyenmoutier, Peter Damian, Anselm of Lucca, and Hildebrand, Pope Leo IX centralized the organizational structure of the church, starting a process that Gregory VII would later advocate as a way of attempting to realize his vision of Rome as center of the church. Leo IX expanded the demands for loyalty and obedience to the pope throughout the church, even including a demand for the patriarch of Constantinople to accept the primacy and authority of Rome and bishop of Rome, which led, in part, to the finalization of the split between East and West.

The process toward creating a more orderly center of the church accelerated under Leo IX, as two important decisions deeply affected the papacy and the wider church: one, the promulgation at the Easter Synod of the Papal Election Decree, which stipulated that the cardinal clergy would take a direct hand in electing the Pope;[7] two, the Treaty of Melfi that allied the papacy with Robert Guiscard and the Normans who controlled much of Southern Italy and Sicily. The first provided the church with the ability to shape and control the papal elections with minimal secular influence; the second provided the Pope with a protector that was not beholden to the Emperor. In both cases, this allowed the church and the pope to act with greater freedom in continuing the reforms initiated by Henry III and the German popes and continued to consolidate control in the hands of the pope in Rome. Rome started to begin to shape itself as the heart, the center of all spiritual life, not only of the Church, but also of Christendom. Hildebrand, and the other reformers, started to see Rome as

---

6   Walter Ullmann, *A Short History of the Papacy in the Middle Ages*, (London, 1972; rpt. New York, 2003), p. 129.

7   *Select Historical Documents of the Middle Ages*, trans. Ernest F. Henderson (London, 1910), pp. 361–364. "1. That, when the pontiff of this Roman universal church dies, the cardinal bishops, after first conferring together with most diligent consideration, shall afterwards call in to themselves the cardinal clergy; and then the remaining clergy and the people shall approach and consent to the new election. 2. That—lest the disease of venality creep in through any excuse whatever—the men of the church shall be the leaders in carrying on the election of a pope, the others merely followers."

something other than the place of intrigue and plots and more as a place of centralized structure and order.

The election of Pope Alexander II in 1061 revealed, however, the lingering resentments of the formerly powerful Roman families and the incomplete nature of the reform process, as the old, parochial Rome fought against the reformers. After Alexander's election, Cadalus, linked either to the Crescentii or Tusculan families, was put forth as a rival claimant to the papacy. The resulting Cadalan Schism devastated Rome as fighting raged on and off over the next three years. Rome's allies, the Normans, were only of superficial help. The city received greater assistance from an alliance that Hildebrand developed for the church with Duke Godfrey and his wife, Beatrice, from Tuscany. The newfound Tuscan allies tipped the balance of power in favor of Alexander II, and he successfully defended his right to the papacy. Equally important, the people of Rome remembered Hildebrand's role in the resolution of the conflict, and his continuing efforts on their behalf.

On 22 April 1073, after Pope Alexander II was buried at the Lateran Basilica, the scene that unfolded at St. Peter *in vincoli*, with its assemblage of "cardinal-clerks of the holy Roman catholic and apostolic church—acolytes, subdeacons, deacons, priests, in the presence of the venerable bishops and abbots, with the clerks and monks consenting, and with the acclamation of many crowds of both sexes and different orders,"[8] challenged the letter if not the spirit of the Election Decree of 1059. This crowd took it into its own hands to render a decision regarding who the next pope should be and elected Hildebrand by acclamation. Even in the midst of the order that the reformers were trying to create, chaos still erupted.

After decades of papal elections dictated by the Emperor and the first faltering steps of the papal elections after the Election Decree of 1059, the people of Rome reclaimed their role, and installed the person who they believed to be the one who could solve the various crises of outside interference and need for moral reform. While Hildebrand may have reasoned that he was chosen so that the canonical abuses of clerical marriage, simony, and lay investiture could be extirpated for the benefit and spiritual autonomy of the church, it may have been that the Romans simply wanted their version of "order," namely that

---

8 H.E.J. Cowdrey, *The Register of Pope Gregory VII 1073–1085 An English Translation*, (Oxford, 2002), p. 1. Erich Caspar, *Das Register Gregors VII*, Monumenta Germaniae Historica [MGH] (Berlin, 1920), p. 2: ...nos sanctae Romanae, catholicae et apostolicae ecclesiae cardinales clericis, acoliti subdiaconi diaconi presbyteri, presentibus venerabilibus episcopis et abbatibus, clericis et monachis consentientibus, plurimis turbis utriusque sexus diversique ordinis acclamantibus...

Romans would decide who would sit in St. Peter's chair. Hildebrand, a well-known and respected son of Rome, was their choice.[9]

• • •

The chaos of the temporal Rome, and the potential for the eternal, orderly Rome, Gregory VII's dual vision of the city, is seen in the way Gregory described his ascension to the Apostolic See. His description of the chaotic scene that unfolded in Rome in the days immediately after Pope Alexander II's death in 1073 is noted with great emotional weight in Gregory's letter to Abbot Desiderius of Montecassino.[10]

> Our lord Pope Alexander is dead. His death has fallen upon me, and has stricken and utterly dismayed my whole heart. For upon his death, the Roman people was so unwontedly quiet and passed the reins of counsel into our hand, that this seemed clearly to have happened by the mercy of God. Therefore, after taking advice, we decided that after a three-day fast and after litanies and the prayer of many supported by alms, upheld by God's aid we would determine what seemed best regarding the election of the Roman pontiff. But suddenly, when our lord the pope whom we have mentioned was being given burial in the church of the Saviour, there arose a great tumult and uproar of the people; they surged up before me as in a frenzy, so that I could say with the Prophet, "I came into the depth of the sea and the tempest overwhelmed me; I was in distress saying, 'My throat is dry,'" and "Fear and trembling have come upon me and darkness has covered me."[11]

---

9    Hildebrand served as economus for the monastery at St. Paul Outside the Walls, as legate and chaplain for Pope Gregory VI, as archdeacon, and as a part of the inner-circle of advisors to popes such as Leo IX and Alexander II. While Hildebrand's influence on the reformers is debatable, his continual presence in the circle of reformers is assured.

10   Copies of the letter, or similar variants, were sent to Prince Gisulf of Salerno, Archibishop Guibert of Ravenna, Countess Beatrice, Abbot Hugh of Cluny, Archbishop Manasses of Rheims, King Sweyn Erthrithson of Denmark, and the Abbot of Marseilles.

11   Cowdrey, *Register*, p. 2. Erich Caspar, *Das Register Gregors VII*, p. 3: Dominus noster papa Alexander mortuus est. Cuius mors super me cecidit et omnia viscera mea concutiens penitus conturbavit. Nam in morte quidem eius Romanus populus contra morem ita quievit et in manu nostra consilii frena dimisit, ut evidenter appareret ex Dei misericordia hoc provenisse. Unde accepto consilio hoc sta tuimus, ut post triduanum ieiunium, post letanias et multorum orationem elemosinis conditam divino fulti auxilio statueremus, quod melius de electione Romani pontificis videretur. Sed subito, cum predictus

Gregory VII's description of the scene that unfolded outside of *San Pietro in vincoli* is filled with angst, dread, and uncertainty, a version of events quite different in tone than the far more tempered narrative offered in the official record of the election event.[12] Gregory's description of the crowd erupting "in a frenzy"[13] is decidedly different from the official depiction of the events, one of a very small number of inclusions in the *Register* not directly dictated by Gregory, which offers the gentile call, "Does it please you?" with the equally sedate response, "It pleases."[14] Indeed, Gregory is seeing a moment when order dissolves; he recognizes a moment of terror when he is thrust into a situation that seems beyond his capabilities. In Gregory's letter to Abbot Desiderius, we see his torment and self-doubt, particularly through his frequent use of Scripture to describe inner feelings and to rationalize behavior. Not only does Gregory see the physical, chaotic moment when the crowd surged up to him, but also came to understand it through his own lens of struggle and ideology. He likewise recognized that the events were also a moment pregnant with

---

dominus noster papa in ecclesia Salvatoris sepulturae traderetur, ortus est magnus tumultus populi et fremitus et in me quasi vesani insurrexerunt, ita ut cum propheta possim dicere: ' Veni in altitudine maris et tempestas demersit me, laboravi clamans, raucae factae sunt fauces meae'; et: 'Timor et tremor venerunt super me et contexerunt me tenebrae'). I draw on Cowdrey's translation of the *Register* into English for quoted passages.

12    "We have elected for ourselves as shepherd and supreme pontiff a religious man, well versed in both branches of knowledge, a most pre-eminent lover of equity and righteousness, strong in adversity, moderate in prosperity, and (according to the Apostle's words) endowed with good character, honourable, modest, sober, chaste, hospitable, managing well his own household, who was from boyhood most admiringly brought up and through the bosom of this mother church, and who for the merits of his life has to this day been exalted in the dignity of the archiaconate—namely Archdeacon Hildebrand. From now and for all time, we wish and approve that he should both be called Gregory, pope and successor of the Apostle. 'Does it please you?' 'It pleases.' 'Do you desire him?' 'We desire him.' 'Do you applaud him?' 'We applaud him.'" (Cowdrey, *Register*, p. 1). Caspar, p. 2: Eligimus nobis in pastorem et summum pontificem virum religiosum, gemine, scientiae prudentia pollentem, aequitatis et iustitiae, prestantissimum amatorem, in adversis fortem, in prosperis temperatum et iuxta apostoli dictum bonis moribus ornatum, pudicum, modestum, sobrium, castum, hospitalem, domum suam bene rezentem, in gremio huius matris ecclesiae a pueritia satis nobiliter educatum et doctum at que pro vitae merito in archidiaconatus honorem usque hodie sublimatum, Heldibraudum videlicet archidiaconum, quem ammodo et usque in sempiternum et esse et dici Gregorium papam et apostolicum volumus et approbamus. ' Placet vobis?' 'Placet.' 'Vultis eum?' 'Volumus' 'Laudatis eum?' 'Laudamus.'

13    Cowdrey, *Register*, p. 2 (Caspar, p. 3).

14    Cowdrey, *Register*, p. 1 (Caspar, p. 1).

potential in that he saw the possibility of order, and thus asked for the prayers of those with whom he corresponded.

> I accordingly ask you through the Almighty Lord that you urge, and from true charity, invite the brothers and sons whom you are nurturing in Christ to pray to God for me, so that the prayer which should have freed me from falling into danger may at least safeguard me now that I am in danger.[15]

This sense of chaos in the temporal world also surfaced in other letters that are part of the *Register*. After offering reassurance to King William I of England (4 April 1074) regarding the death of Pope Alexander II and the transition to his pontificate, Gregory related to William the grim state of the world and the dangers that he and church face:

> We have ascended unwillingly a ship which through a billowing sea, is cast into uncertain courses by the battering of the winds and by the force of tornados and by waves rising up to the very skies; it is avoiding, although with peril, rocks that are hidden and other things that appear in the deep from afar—yet also near exhaustion. For the holy Roman church, over which we though unworthy and unwilling preside, is continually and daily shaken by diverse trials, by very many harassments of hypocrites, and by the snares and guileful reproaches of heretics; she is, in truth, for different causes rent asunder secretly and openly by this world's powers.[16]

Quite interestingly, prior to Gregory's dire description, he adds a word of hope: William "stands alone among kings as the one whom...we believe to cherish the things that are written above,"[17] that is, eternal things. Even as he commented

---

15   Cowdrey, *Register*, p. 2 (Caspar, p. 4: Te itaque per omnipotentem Dominum rogo, ut fratres et filios, quos in Christo nutris, ad exorandum Deum pro me provoces et ex vera caritate invites, quatenus).

16   Cowdrey, *Register*, p. 74 (Caspar, pp. 101–102: Navem inviti ascendimus, quae per undosum pelagus violentia ventorum et impetu turbinum et fluctibus ad aera usque insurgentibus in incerta deicitur; saxis occultatis et aliis a longe in altum apparentibus, licet cum periculo, obviat tamen et ex animo. Sancta quippe Romana ecclesia, cui licet indigni et nolentes presidemus, diversis temptationibus quam plurimis persecutionibus ypocritarum et hereticorum insidiis et dolosis obiectionibus continue et cotidie quatitur, mundanis vero potestatibus occulte et evidenter per diversa distrahitur).

17   Cowdrey, *Register*, p. 74 (Caspar, p. 102).

on the destructiveness of the temporal realm, Gregory pointed to the eternal as a sign of hope, holding the two visions, what is and what should be, simultaneously.

At this point, though, it is the chaotic version of Rome that Gregory saw most clearly. Gregory's description of the world is even more explicitly stated in his letter addressed to the faithful of St. Peter calling on them to support his decision to excommunicate and depose Henry IV at the Lenten Synod of 1076.

> You have heard, brothers, the novel and unheard-of presumption, you have heard the atrocious loquacity and boldness of schismatics and blasphemers of the name of the Lord in blessed Peter, you have heard the pride which has reared up to the insult and shame of the holy and apostolic see, *such as your fathers have not at any time seen or heard nor does the sequence of things written teach that the like has ever come forth* from pagans or heretics.[18]

Gregory's commentary on the unprecedented nature of the attacks on the church is an indictment of the deteriorating condition of the world. He sought to restore concord and order, but those who worked against him, those who "have hated, and do hate, me for my fidelity to you (Peter),"[19] seemed to have the upper hand. Gregory clearly recognizes the grave severity of the condition of the world, yet was continually working toward attempting to realize his vision of what should be.

• • •

Two of Gregory VII's ardent supporters, Bonizo of Sutri, a contemporary of Gregory, in his *Letter to a Friend*, and Paul of Bernried, an early twelfth century proponent of the Gregorian reforms, in his *Life of Gregory*, describe the problems of temporal Rome, in particular the chaos that attended the Christmas Eve mass Gregory VII celebrated at Santa Maria Maggiore in Rome in 1075.[20]

---

18    Cowdrey, *Register*, p. 182, emphasis mine (Caspar, pp. 254–255: Audistis, fratres, novam et inauditam presumptionem, audistis sceleratam scismaticorum et nomen Domini in beato Petro blasphemantium garrulitatem et audaciam, audistis superbiam ad iniuriam et contumeliam sanctae et apostolicae sedis elatam qualem vestri patres nec viderunt nec audierunt unquam nec scripturarum series aliquando a paganis vel haereticis docet emersam).

19    Cowdrey, *Register*, p. 181 (Caspar, p. 253: ...qui me pro tua fidelitate oderunt et odiunt...).

20    Paul of Bernried, a member of a "south German friendship circle devoted to reform," likely completed his "Life of Gregory" in 1128. Bishop Bonizo of Sutri's "Letter to a Friend," likely

The events, of course, had their origin in the years prior to the events of December 1075. The old Roman partisan fights, mentioned earlier, resurfaced from time to time, and Gregory's election clearly raised the ire of particular rival factions. Cencius Stephani, whom Paul Bernreid describes as, "the son of perdition, the most wicked and most dangerous of all men,"[21] conspired to take revenge for the sentence laid against him by Pope Alexander II for the murder of his godfather and supporting the group who resisted the reformers by playing a part in the Cadalan schism during the 1060's.[22] Bonizo of Sutri, in his *Letter to a Friend*, notes that Cencius was arrested in early 1075 and imprisoned for a capital offense, but gained his freedom soon thereafter.

> He was undeservedly allowed to live and was set free, however, at the behest of the glorious Matilda, who was present at that time, and of many Roman citizens, after he had surrendered hostages into the hands of the pope, together with the tower by means of which he strove to ascend to heaven. The tower was utterly destroyed.[23]

Paul of Bernreid, who also noted this turn of events, stated that Cencius' freedom quieted the city and the "faction of the wicked was troubled with great fear and fell silent."[24] But this did not last long, as Cencius soon began to conspire to "capture him and betray him [Gregory] to wicked people who would kill him."[25] Wibert, Archbishop of Ravenna (the future anti-pope Clement III (1080–1100)) was included in this plot, and while Paul states that Henry IV was likewise involved there is no clear evidence to support this claim.[26] On Christmas Eve, 1075, Cencius and his co-conspirators carried out

---

written in late 1085 to early 1086, "is perhaps the best known of the polemics of the Investiture Contest" (I.S. Robinson, *The Papal Reform*, pp. 36–44, 64–72).

21    Ibid., p. 292. PL 148, p. 52: ...perditionis filius, omnium hominum sceleratissimus et inquissimus...

22    I.S. Robinson, *Papal Reform*, pp. 292–293.

23    Robinson, *Papal Reform*, p. 231. *Liber ad Amicum*, MGH, ed. Ernestus Dummler (Berlin, 1891), p. 605: Sed precibus gloriosae Matildis, que ibi aderat illis diebus, et multorum Romanorum civium vix emeruit, ut vivus dimitteretur, datis obsidibus in manu papae et turri, per quam ad celum ascendere nitebatur; que funditus destructa est.

24    Robinson, *Papal Reform*, p. 294. PL 148, p. 58: ...et iniquorum factio, timore magno concussa, conticuit.

25    Robinson, p. 295. PL 148, p. 58: ...exspectans nimirum opportunitatem ut cum caperet, et iniquorum populo ad necandum traderet.

26    Cowdrey, *Pope Gregory VII*, p. 327: "Cencius's outrage attracted widespread and enduring attention in friend and foe alike in both Italy and Germany. This attention is largely to be

their plan to abduct Gregory. He was celebrating the midnight mass at the manger altar of Santa Maria Maggiore on this most solemn of occasions; this fact may have prompted the attack on that particular night because of the proximity of Cencius' holdings to Santa Maria Maggiore.[27] During the distribution, loud shouts erupted from the people as armed troops "ranged through the church with drawn swords, striking at those whom they could reach,"[28] and, "then they laid hands on him and seized him."[29] One knight raised a sword to decapitate him, but as Berthold recounts in the *Chronicon*, he was struck by fear and collapsed.[30] Gregory was wounded in the melee and dragged from the church, yet "he, like a gentle lamb, lifted up his eyes to heaven and made no reply to them; he did not cry out nor did he resist nor pray them to show some mercy."[31] He was stripped of the "pallium and chasuble, of his dalmatic and tunicle, they dragged him like a thief, clad only in his alb and his stole, and mounted him behind a certain sacrilegious man."[32] The knight, who attempted to kill him, was "seized by a demon before the entrance hall of that church. For a long time he rolled about, foaming at the mouth and, his horse fleeing with him, he was never found again."[33]

Gregory was taken to Cencius' compound in Parione, and placed in the tower at the compound. Over the course of the night, a great many Romans descended on Cencius' establishment with fire and siege weapons; in the morning

---

explained by the synchronism between Censius's outrage and the rapid deterioration during the winter of 1075–6 of relations between Gregory and Henry IV of Germany which led to their sentences upon each other. The suddenness of the breach between Gregory and Henry renders more than improbable the allegations of later writers that Cencius in collusion with persons outside of Rome, whether Henry himself, Duke Robert Guiscard, or Archbishop Guibert of Ravenna. His attempted coup was a local Roman event."

27  Robinson, *Papal Reform*, p. 232.
28  Ibid., pp. 296–297. PL 148, p. 60: Tunc undique lustrantes ecclesiam, evaginatis mucronibus percutientes quos poterant...
29  Robinson, *Papal Reform*, p. 297. PL 148, p. 60: Tunc injecerunt in eum manus, et tenuerunt.
30  Robinson, *Papal Reform*, p. 297.
31  Ibid., p. 297. PL 148, p. 60: Ille vero, ut agnus innocens et mansuetus, ad coelom oculos erigens, nullum eis dedit responsum, non reclamavit, non reluctatus est, neaque ut sibi in aliquot miserentur rogavit.
32  Robinson, *Papal Reform*, p. 297. PL 148, p. 60: ...exutum Pallio et Casula seu Dalmotica et Tunica cum Camisia, relictis ei tantum amictu et stola, ut furem tractum, post dorsum cujusdam sacrilegi posuerunt.
33  PL 148, p. 60: ...arreptus a daemonio, ante ejusdem ecclesiae atrium equusque ejus fugiens nequaquam ulterius inventus est.

they surrounded the tower and began systematically to tear it apart.[34] A Roman noble couple "rushed"[35] in to aid Gregory, tending his wounds and treating his exposure (due to spending the night in the cold tower). The noblewoman's actions were recognized as mirroring those of Mary Magdalene. She praised Gregory's heart and affections, cleaned his wounds, and excoriated those that attacked him.[36]

After Gregory was brought out of the tower, Cencius' sister and another of the knights who participated in the conspiracy are also mentioned roundly abusing Gregory. The knight, during his tirade in which he proclaimed that he would cut off Gregory's head, was killed when a spear was hurled out of a window of the tower and "pierced his throat."[37] Cencius, realizing that he was on the losing side of this battle, begged Gregory for clemency. Gregory promised restoration if Cencius made a pilgrimage to Jerusalem. When, or rather if, he returned he would need to submit himself to Gregory's advice in order to make certain of his renewal.[38]

This extraordinary event received no direct attention in the *Register*. The lack of commentary on such an event may be understandable in light of the events of the Lenten Synod of 1076, resulting in the excommunication of the bishops of Lombardy, the excommunication of the bishops in a variety of locales across the Alps, and, most importantly, the excommunication and deposition of Henry IV. The Christmas Eve incident, as evidenced by the Bonizo of Sutri and Paul of Bernried accounts, was of great interest to the Romans, and to those who sought to use the story to strengthen sympathy and support for Gregory VII. Yet Gregory did not recall the story to any of his regular correspondents, as the tide of events likely swept it away following the decisions of the Lenten Synod. The reaction of the Romans within the narrative highlights the complex relationship between Gregory and the inhabitants of the city, denoting a strong bond between them, regardless of their loyalties. This bond, though, does not translate into additional references to the people of Rome. Whatever the reason may have been for the episode in the *Register*, it represents yet another moment depicting Rome as a chaotic and disorderly.

The only possible reference to the event in the *Register* is found in Gregory VII's April 1076 letter to Wifred, a Milanese knight, and supporter of Gregory,

---

34    Robinson, *Papal Reform*, pp. 297–298.
35    Ibid., p. 298.
36    Ibid., pp. 298–299.
37    Ibid., p. 299. PL 148, p. 61: ...gutter ejus...obtruncans...
38    Robinson, *Papal Reform*, p. 300.

a vague allusion to the Christmas Eve abduction in the context of Gregory's encouragement of the knight.

> But he who said, 'Have confidence, for I have come to overcome the world', gives us the most assured faith that he will speedily bring aid to his church and utterly confound the devil and his members.[39]

Gregory adds:

> Do you, therefore, dearest son, be strong in the Lord and in the power of his might, and strengthen those whom you know to be standing fast in the Christian faith; but as for those who by their works have denied the Christian faith, bid them they should repent, and that they should be covered in shame to live in the devil's slavery.[40]

Gregory's clemency towards Cencius may be reflected in the final call to bring those who have denied the faith to repentance. The Church forgives those who legitimately and authentically seek forgiveness through repentance. "Have confidence, I have come to overcome the world,"[41] Gregory quoted directly from John 16, 33, noting that Jesus will restore order to the chaotic world, that he will overcome that which harms the faithful. It is a call for the faithful to see the ordered reality that could be, as opposed to chaotic reality.

• • •

When Gregory was not referencing the chaos of the temporal city, his statements about Rome or Romans reveal either an understanding of the pragmatic needs of the city or a sense of muted wistfulness for past kindnesses. A reference to the people of Rome in 1076 is found in a very brief note sent to the emir an-Nasir in which he mentioned that the emir's good treatment of the Christians in Mauretania had duly impressed the nobility of Rome, who of

---

39  Cowdrey, *Register*, p.197 (Caspar, p. 277: ...sed qui dixit: 'Confidite, quia ego vici mundum', dat nobis certissimam fidem festinanter se ecclesiae suae succurrere et diabolum et membra eius omnino confundere).

40  Ibid., p. 277: Tu itaque, karissime fili, confortare in Domino et in potentia virtutis eius et eos conforta, quos in christiana fide cognoveris permanere, eos autem, qui fidem christianam operibus negaverunt, ut resipiscant ammone, ut erubescant in servitute diaboli vivere.

41  Ibid., p. 277.

whom, Alberic and Cencius, wished to travel to the emir's land to act possibly in the commercial and financial interests of Rome.[42] The reference is decidedly mundane, which reinforces the reality that while the city of Rome as a city was not unimportant, it was certainly not the central focus for Gregory's actions and ideals.

The Rome he wanted to see was always superimposed on what actually was before him; the churning chaos may have been his present reality, but what Rome could be was what drove him in his reforming actions. Interestingly enough, the seeds of the eternal Rome could be seen, periodically and momentarily, in the temporal Rome. Even the disastrous Christmas Eve assault of 1075, as clear a sign of the temporal, chaotic Rome as possible, offered a vision of what might be—a Rome, whose inhabitants rushed to Gregory's defense in loving obedience and support for the head of the church. Even in the midst of the chaos, Gregory's last mention in the *Register* of the city and her people struck a note of affection. The May 1081 letter to Abbot Desiderius of Montecassino detailed the current political state of affairs, which had been deteriorating due to the continued fight against Henry IV. Here Gregory spoke positively regarding the people of Rome and their support:

> You should know, moreover, that the Romans and those who are near to us are in all respects prepared with faithful and ready heart for the service of God and of ourself.[43]

This may be the last moment of mutual admiration and support, as the relationship began to fail as Henry's incursions into the city intensified, and the chaos of the temporal city once again overwhelmed any other vision of the physical city. The support collapsed almost completely in 1085 when the confrontation flared into an intense street battle between Henry's troops and Gregory's rescuers, namely the Norman troops of Robert Guiscard. However, we do not read about this collapse in the *Register*. The final impression of Rome and her people is that of the weary Pope expressing gratitude regarding the support for what was to come. An impression taken from the *Register* shows Rome to be a place of support for Gregory. Gregory's gratitude was muted, though, as he continued to seek a better future, focusing his sight on what was yet to be, or what should be. The physical Rome receded as the idealized, spiritual Rome became Gregory's main reference. The people of Rome, no

---

42    Cowdrey, *Register*, pp. 204–205 (Caspar, pp. 287–288).

43    Cowdrey, *Register*, p. 412 (Caspar, p. 589: Scias preterea, quod Romani et qui circa nos sunt fido et prompto animo Dei et nostro servitio parati per omnia existent).

matter how supportive, were too much the creatures of the temporal Rome, their support finally wavering and collapsing when Gregory needed it most.

When he does reference Rome as an urban, temporal center, he called it simply, the City.[44] When he referred to the people of Rome, they were simply the Romans or the Roman people. When referencing individuals passing through the lands directly held by the papacy, he offered the generic term, "territory of our land." The tone of these various usages is matter of fact, and carries little of the gravitas that weights his references when he mentions the need to pledge obedience to Rome, the center of the church and foundation of all order. The references are functional, used infrequently in the first years of Gregory's papacy, and rarely in the later years.

In light of the chaos of the diminished physical and temporal Rome, how could one begin to restore order? For Gregory, obedience to Rome was the key, not obedience to the temporal Rome of the turbulent election or the Christmas Eve debacle, but the Rome of Gregory's idealism, the Rome of Peter and Paul, the idealized Rome of Gregory's mental construction. Gregory articulated this Rome by describing it as the center of Christendom, the see of St. Peter, the place from which flows the balm for the world. The references to the idealized center of the church are clearly seen in the use of four particular stock phrases that Gregory employed with great regularity throughout the *Register*. Gregory is consistent in utilizing Rome as the symbol of something much greater than the physical city of Rome, an ideal deserving of obedience and devotion, not an eleventh-century central Italian town.[45]

While many components of Gregory's reform activities and his own perspectives on the reforms, the church—and the world—are altered by the rupturing events of the Lenten Synod of 1076 where Gregory excommunicates and deposes Henry IV, and the Drama at Canossa (January 1077), when Gregory forgives and restores Henry to the community of the church (but only conditionally lifts the deposition), his references to Rome remain constant. Gregory consistently defines Rome as the idealized center of the church and Christendom. Only one particular reference receives additional use by Gregory after the Lenten Synod of 1076, variations on the depiction of the Roman Church as 'Mother'.

---

44    Cowdrey, *Register*, p. 183: "When I received your majesty's letter, those persons were far distant from the City, mainly on account of the bad air...." (Caspar, pp. 256–257: *Quando a litteras tuae magnitudinis accepi, longe ab Urbe maxime causa infirmitatis aberamus....*).

45    Gerd Tellenbach, *Church, State and Christian Society at the Time of the Investiture Contest*, (Oxford, 1959), p. 1, argues that Gregory was attempting to create the "right order of the world."

When Gregory referenced Rome as symbol, center, and ideal, he used a collection of phrases that speak to this much broader, symbolic, and abstract reality. Gregory employed various iterations of Holy Church, Roman Church, Holy
Roman Church, and Mother of the Holy Roman Church. Additionally, Gregory
referenced *ad limina* visits directly on five occasions, all prior to the Lenten
Synod of 1076. These demands for an appearance at the 'threshold of the apostles' add to the understanding of Rome as the center of the church, a view
supported in the apostolic age.

The first reference to one of the variations on the title of the church, in the
23 April 1073 letter to Abbot Desiderius of Montecassino, subtly connects the
greater church to Gregory.

> As for yourself, do not fail to come to us as quickly as may be, for you are
> not unaware of how greatly the Roman church needs you and has confi-
> dence in your sagacity.[46]

Gregory developed multiple layers to his request, noting a desire for Desiderius
to come to Rome to support the church, which to him represented both the
church at large as well as its personification, Gregory. While Gregory and
Desiderius had a contentious relationship, mostly due to their difference of
opinion on the matter of Robert Guiscard and the Normans,[47] the implication
of Gregory's plea is clear; he was asking all to support the church, regardless of
what the various recipients of his letters might think of the individual who
held the office. Gregory elaborated on this point in the 26 April 1073 letter to
the future anti-pope Clement III, then Archbishop Guibert of Ravenna:

> I therefore ask you through Almighty God that, even if not for my own
> merits then at least from love for the apostles, you will now be at pains to
> show towards me the charity that you promised to bear toward the
> Roman church particularly at this time, and as you are bound to remem-
> ber, especially towards myself, seeing that the time and the character its
> affairs call for it to be proven.[48]

---

46  Cowdrey, *Register*, p. 2. (Caspar, p. 4: Tu autem ipse quantotius ad nos venire non preter-
    mittas, qui, quantum Romana ecclesia te indigeat et in prudentia tua fiduciam habeat,
    non ignores).

47  H.E.J. Cowdrey, *The Age of Abbot Desiderius—Montecassino, the Papacy, and the Normans
    in the Eleventh and Early Twelfth Centuries*, (Oxford, 1983), pp. 122–123.

48  Cowdrey, *Register*, p. 4 (Caspar, p. 6: Rogo itaque vos per omnipotentem Deum, ut caritatem,
    quam erga Romanam ecclesiam maxime hoc tempore et, ut meminisse debetis, erga me

Gregory situated the seat of the authority of the church not in himself, but in Christ and the apostles, thereby circumventing any personal animosity from others by noting that he was simply carrying out those the call of God, the prophets, and the apostles, especially St. Peter.

In a letter dated 14 October 1073 to four prefects of Sardinia, Gregory continued to explicate the role of the Roman church for people throughout Europe, reinforcing the idea that Rome represents the wider church and that, therefore, Sardinia needed to offer the obedience due to the Roman church.

> It is well known to you and all who venerate Christ that the Roman church is the universal mother of all Christians. Although in consideration of her duty she should be vigilant for the welfare of all peoples, for you it is, however, right for her to exercise a special and so to speak private care.[49]

The language of church as mother is evocative, indicating a profound and universal concern for its constituents. Gregory continued by referencing the history of the relationship between the Roman church and Sardinia, especially noting Sardinian failings:

> In truth, because by the negligence of our own predecessors the charity has grown cold that existed in ancient times between the Roman church and your people, you have so greatly estranged yourselves from us even more than the peoples who are at the ends of the earth that amongst you the Christian religion has come to the utmost harm.[50]

Gregory then offered a remedy to this dire situation and in so doing reinforced the primacy and authority of the Roman church.

> Wherefore it is highly necessary for you that from now on you should take more earnest thought for the salvation of your souls, recognizing like

---

specialiter vos gerere promisistis, quoniam quidem tempus et rerum qualitas eius probari postulat, nunc quidem in me, etsi non meis meriti, saltim amore apostolorum...).

49  Cowdrey, *Register*, p. 33 (Caspar, p. 46: Vobis et omnibus, qui Christum venerantur, cognitum est, quod Romana ecclesia universalis mater sit omnium christianorum. Quae licet ex consideratione officii sui omnium gentium saluti debeat invigilare, specialem tamen et quodammodo privatam vobis sollicitudinem oportet eam impendere).

50  Cowdrey, *Register*, p. 33 (Caspar, pp. 46–47: Verum quia neglegentia antecessorum nostrorum caritas illa friguit, quae antiquis temporibus Inter Romanam ecclesiam et gentem vestram fuit in tantum a nobis plus quam gentes, quae sunt in fine mundi, vos extraneos fecistis, ut Christiana religio inter vos ad maximum detrimentum devenit).

legitimate sons the Roman church as your mother; the same devotion
that your forebears of old paid here, you too should pay.[51]

Expanding on the developing theme of the primacy of the Roman church in
the *Register*, Gregory, in a 20 March 1074 letter to King Sancho I of Aragon, and
a 19 March 1074 letter to King Alphonso VI of Leon-Castile, provided concrete
steps that the two kings must take in order to reclaim and reinforce their connection to the church. In each case the renewed use of the Roman liturgical
rite is the first step. To King Sancho I of Aragon, Gregory writes:

> Of a certainty, in so far as you assert that in your domain by your zeal and
> order the office of the Roman order is in force, you may recognize that
> you are a son of the Roman church and that you have the same concord
> and the same friendship with us that the kings of Spain used of old to
> have with the Roman pontiffs.[52]

Gregory, in the letter to King Alphonso VI of Leon-Castile, repeated this
demand for liturgical unity, and tied to it an explicit assertion of universality,
providing a direct and indirect use of Matthew 16:16–19 to support and justify
these demands and claims.

> I accordingly exhort and warn you as most dear sons that, like a good
> posterity, even after long-standing divergences you recognize the Roman
> church as truly your mother, in whom you may also find ourself to be
> your brother; and that you receive the order and office of the Roman
> church, not of the Toledan or any other, but that, like the other kingdoms
> of the west and north, you hold to that which has been founded through
> Christ by Peter and Paul upon the firm rock and consecrated by their
> blood, against which the gates of hell, that is, the tongues of heretics,
> have never been able to prevail.[53]

---

51  Cowdrey, *Register*, pp. 33–34 (Caspar, p. 47: Unde multum vobis necessarium est, ut de
    salute animarum vestrarum studiosius amodo cogitetis et matrem vestram Romanam
    ecclesiam sicut legitimi filii recognoscatis et eam devotionem, quam antiqui parentes
    vestri sibi impenderunt, vos quoque impendatis).
52  Cowdrey, *Register*, p. 67 (Caspar, p. 92: In hoc autem, quod sub ditione tua Romani ordinis
    officium fieri studio et iussionibus tuis asseris Romanae ecclesiae, te filium ac eam con-
    cordiam et eandem amicitiam te nobiscum habere, quam olim reges Hyspanie, cum
    Romanis pontificibus habebant, cognosceris).
53  Cowdrey, *Register*, p. 68 (Caspar, p. 93: Quapropter ut filios karissimos vos adhortor et
    moneo, ut vos sicut bone soboles etsi post diuturnas scissuras, demum tamen ut matrem

In the 15 October 1079 letter to King Harold Hein of Denmark, Gregory again reinforced the universal primacy of the church and Rome as the symbolic center of that church.

> With a sincere disposition of charity, beloved, we congratulate you that, although placed at the furthest end of the earth, you nevertheless vigilantly seek to find out the things that are known to belong to the observance of the Christian religion, and because, recognizing the holy Roman church to be your mother and that of all men, you desire for yourself and ask for her instructions.[54]

To strengthen the connection between these far flung kingdoms and the heart of the church, Gregory encouraged kings across Europe to send representatives to Rome, so that they, "might be in a position fully to inform us about the customs and bearing of your people and would be able when more fully instructed to bring back to you the instruction or prescriptions of the apostolic see."[55] Combining the papal declarations of Roman primacy and authority with the reinforcements representatives would receive when they journeyed to Rome, Gregory developed the structure for an understanding of "Rome" as the center of the church, from which everything flowed and to which everything returned.

The issue of centralized control is apparent even in the language allowed in the liturgy, as Gregory warned Duke Wratislav of Bohemia that the use of vernacular languages could be dangerous. Gregory conflated the use of the vernacular in the liturgy with reading the scriptures in a vernacular language, noting that, "if it (scripture) were transparent to all, it might perhaps be cheapened and subject to contempt, or being wrongly understood by incompetent

---

re vera vestram Romanam ecclesiam recognoscatis, in quo et nos fratres reperiatis; Romanae ecclesiae ordinem et officium recipiatis, non Toletanae vel cuiuslibet alliae, sed istius, quae a Petro et Paulo supra firmam petram per Christum fundata est et sanguine consecrata, cui porte inferni, id est linguae hereticorum, nunquam prevalere potuerunt, sicut caetera regna occidentis et septemtrionis teneatis).

54  Cowdrey, *Register*, p. 328 (Caspar, pp. 464–465: Sincero caritatis affectu dilectioni tuae congratulamur, quia, licet in ultimis terrarum finibus positus, ea tamen, quae ad christianae a religionis cultum pertinere noscuntur, vigilanter studes inquirere et quod sanctam Romanam aecclesiam matrem tuam et universorum recognoscens ipsius documenta tibi exoptas et exposcis).

55  Cowdrey, *Register*, p. 329 (Caspar, p. 465: ...qui et vestrae gentis mores seu continentias sciret nobis pleniter intimare et apostolicae sedis documenta sive mandata plenius eruditus ad vos posset perferre).

person might lead into error."[56] In a 15 October 1080 letter to the people of Tuscany and Fermo, and the exarchate of Ravenna, Gregory reminded the people that Roman control over all aspects of the Christian life was not innovative or novel, even for those areas and churches who wished to display some form of independence.

> We do not doubt that it is known to your understanding with how great an observance of religion the holy church of Ravenna has been accustomed to flourish and with how great stores of things needful it has been accustomed in years past to abound, and also—what is of more concern for her—with what special love she has also cleaved to her mother, the holy Roman church, from the very beginning of the Christian faith.[57]

Gregory reiterated and justified this conceptualization of Rome as center and foundation of the church in the famed 15 March 1081 letter to Bishop Hermann of Metz. This letter was Gregory's grand apologia for the second excommunication and deposition of Henry IV and contains his most organized and thorough defense and justification for his motives and actions. In the process of mounting this defense, Gregory elaborated on the rationale for his authority and the primacy of Rome, and offered that what truly elevates the Roman church to its exalted status is the divine will, the writings of the holy fathers, and the actions of the councils.[58] This Rome is the idealized and symbolic center of the faith and the church, and bears little resemblance to the city at the bend in the Tiber River, as it is rooted in a far more structured and orderly reality and sense of purpose.

In each case, Gregory attempted to create a sense of structure and order, developing a city that more closely resembled the heavenly city of the New Jerusalem, a place of concord and order, a place of God's presence, a place where all lived in full obedience to the will of God. Gregory's letters continually reinforce this image. For Gregory, the grand construct of the ordered and idealized Rome overshadowed and overwhelmed the chaotic, domestic reality. In

---

56   Cowdrey, *Register*, p. 336 (Caspar, p. 474: ...ne, si ad liquidum cunctis pateret, forte vilesceret et subiaceret despectui aut prave intellecta a mediocribus in errorem induceret).

57   Cowdrey, *Register*, p. 377 (Caspar, p. 531: Prudentiae vestrae notum esse non dubitamus, quanto religionis cultu sancta Ravennas aecclesia pollere quantisque rerum necessariarum copiis solita sit preteritis a annis affluere, quodque magis illius interest, quam speciali dilectione matri suae sanctae Romane aecclesiae ab ipso fidei christianae principio semper adheserit).

58   Cowdrey, *Register*, p. 388 (Caspar, pp. 544–562).

the Summer 1080 letter to the "faithful of St. Peter," Gregory highlighted this stark differentiation between the concrete and conceptual realities in a statement regarding an upcoming military assault on the city of Ravenna.

> They [the Normans], indeed, unanimously promise that, as they have sworn, they will render aid for the defence of the holy Roman church and of our honour against all mortal men. Also the princes who are both far and wide in the environs of the City and also in Tuscany and other regions firmly promise the same to us.[59]

Physical space could not contain all that Gregory intended when he spoke of the Roman church, for the physical Rome, the Rome of the Romans and their often decidedly parochial or mundane issues could not provide the necessary authority or structure that the world so desperately needed. An idealized Rome needed to be constructed, one that guided and directed the people and brought them to a place of salvation. The means by which Gregory developed his conceptualization of Rome as center and heart of the Church was a strict demand for obedience to the papal see. The physical reality of the city faded, as it served only as a fading backdrop to his sense of a greater reality of the church. Gregory focused on the idealized and symbolic Rome as center of the church to reinforce his exploration of the power and authority to act as he thought necessary for the protection of the church and society.

Did Pope Gregory VII succeed in creating the Rome that he wanted to see, the idealized, eternal Rome, center of Christendom? No, for the physical Rome that Gregory left behind was as chaotic and disorderly as when he ascended the Apostolic See, with sections of the city suffering horribly from Guiscard's "rescue" of Gregory in 1084. Likewise, the Rome of Gregory's vision, a Rome as idealized spiritual and authoritative center of Christendom, was not yet realized, as many bishoprics and cities around Europe clearly did not fully embrace complete obedience to that Rome. Yet while the city was not what Gregory wanted it to become in his lifetime, the vision of the idealized, eternal Rome had certainly been planted, a vision embraced, encouraged, and nurtured by other advocates of reform, as well as Gregory's successors.

---

59   Cowdrey, *Register*, p. 372 (Caspar, p. 525: Qui profecto unanimiter promittunt se, sicut iurati sunt, ad defensionem sanctae Romanae ecclesiae nostrique honoris contra omnes mortales auxilium impensuros. Id ipsum quoque nobis et qui circa Urbem longe lateque sunt et in Tuscia caeteris que regionibus principes firmiter pollicentur).

# CHAPTER 3

# Building Block of Times, Knowledge and Wisdom in the *Hortus deliciarum*

*Danielle B. Joyner*

The *Hortus deliciarum* was a twelfth-century manuscript that combined images and texts to narrate an interpretive and interactive history of salvation.[1] Produced under the aegis of Abbess Herrad for the canonesses in her care at the Augustinian foundation of Hohenbourg, this manuscript merged monastic traditions, such as biblical and liturgical exegesis and the workings of the computus, with the complex narrative and didactic imagery becoming more frequently employed in the twelfth century.[2] This essay examines ideas about time, specifically how time was defined and pictured as an essential component of the salvation narrative recounted in this deeply pedagogical and carefully constructed manuscript. Although the topics of time and the computus establish the liturgical calendar and naturally complement one another, neither topic is frequently discussed in relationship with women.[3] In her 2007 study, Fiona Griffiths persuasively argued that the *Hortus deliciarum* was created to offer the Hohenbourg canonesses an education on par with that of the clerics who oversaw their pastoral care.[4] I agree with her conclusions and suggest that a thorough understanding of time and its related concepts and processes was an important component of that education.

---

1  Herrad of Hohenbourg, *Hortus deliciarum*, eds. Rosalie Green et al., 2 vols (London, 1979). For a more complete discussion of this argument, see Danielle B. Joyner, *Painting the Hortus deliciarum: Medieval Women, Wisdom, and Time* (University Park, 2016).

2  Examples of this include the Floreffe Bible (London, British Liteary, add. Ms. 17738), St. Albans Psalter (Hildesheim, Dom-Museum Ms. DS 37), and Speculum Virginum (London, British Library, Ms. Arundel 44). See Anne-Marie Bouché, "Vox Imaginis: Anomaly and Enigma in Romanesque Art," in *The Mind's Eye: Art and Theological Argument in the Middle Ages*, eds. Jeffrey F. Hamburger and Anne-Marie Bouché (Princeton, 2006), pp. 306–335; Jochen Bepler, *Der Albani-Psalter: Kommentarband* (Simbach, 2008); and *Listen, Daughter: The Speculum Virginum and the Formation of Religious Women in the Middle Ages*, ed. Constant J. Mews (New York, 2001).

3  The topic of time intersects most frequently with women in studies of Books of Hours, as in Kathryn A. Smith's study, *Art, Identity, and Devotion in Fourteenth-Century England: Three Women and their Books of Hours* (London, 2003).

4  Fiona J. Griffiths, *The Garden of Delights: Reform and Renaissance for Women in the Twelfth Century* (Philadelphia, 2007).

Studies focusing on ideas about time in the Middle Ages often turn to philosophical texts for their definitions.[5] Following Herrad's cue, this study begins instead with the computus. In the preface of his second book on "the nature, course and end of times,"[6] written ca. 725 in Northumbria, Bede credited the impetus of its composition to the pleas of his students, who desired additional help with the abstruse complexities of computus reckonings. The venerable teacher cited patristic authors as authorities for the process of determining the annually shifting Lenten seasons and Easter dates, but he acknowledged as his greatest source of benevolent support the Creator of the seasons, ages, and unstable cycles of times.[7] In her valuable edition of this influential text, Faith Wallis translated Bede's prefatory passages with an ear attuned more to a smooth English rendering than to a strictly literal translation of times in the plural. This plurality, though awkward in English, directly corresponds with Bede's etymologically based definition of times in the opening of his second chapter.

> Times take their name from "measure," either because every unit of time is separately measured, or because all the courses of mortal life are measured in moments, hours, days, months, years, ages and epochs.[8]

In other words, time exists in the plural because it is composed of individual units that, like a set of building blocks, embody discrete quantities. No less an

---

5  For example, see Charlotte Gross, "Augustine's Ambivalence about Temporality: His Two Accounts of Time," *Medieval Philosophy and Theology* 8 (1999), 129–148; Richard Cross, "Absolute Time: Peter John Olivi and the Bonaventurean Tradition," *Medioevo, Rivista di Storia della Filosofia Medievale* 27 (2002), 261–300; and Rory Fox, *Time and Eternity in Mid-Thirteenth-Century Thought* (Oxford, 2006). More diversity exists in *Time and Eternity: The Medieval Discourse*, eds. Gerhard Jaritz and Gerson Moreno-Riaño (Turnhout, 2003).

6  All Latin quotations of this work are cited from PL 90. See PL 90:294A, *De temporum statu, cursu, ac fine*.

7  PL 90: 295A, *Quibus concitus parens, perspectis venerabilium Patrum scriptis, prolixiorem de temporibus librum edidi, prout ipso largiente potui, qui aeternus permanens, tempora quando voluit constituit, et qui novit temporum fines: imo ipse labentibus temporum curriculis finem cum voluerit imponet.*

8  *Bede: The Reckoning of Time*, trans. Faith Wallis (Liverpool, 1999), p. 13. I would like to thank Faith Wallis for her candid response to my questions about this, as well as for her ongoing help. PL 90:298D, *Tempora igitur a temperando nomen accipiunt, sive quod unumquodque illorum spatium separatim temperatum sit: seu quod momentis, horis, diebus, mensibus, annis, saeculisque et aetatibus omnia mortalis vitae curricula temperentur.*

authority than Genesis 1, 14 proffers a similar plural sense of times, *Dixit autem Deus fiant luminaria in firmamento caeli ut dividant diem ac noctem et sint in signa et tempora et dies et annos.*[9] The Douay Rheims translation takes *tempora* as "seasons": "And God said: Let there be lights made in the firmament of heaven, to divide the day and the night, and let them be for signs, and for seasons, and for days and years."[10] Some medieval authors made other choices, as seen in Book 2, Dist. XIV, Ch. 11 of the *Sententiae*, where Peter Lombard noted that Augustine promoted an interpretation of *tempora* as "seasons,"[11] whereas Bede opted for "times" in his commentary on Genesis, which interprets *tempora* as "daily times."[12]

Bede's students were not the only voices lamenting the computus; indeed the reconciliation of solar and lunar cycles baffled and fascinated countless authors over the centuries.[13] The sacramental nature of Easter required that it be celebrated on the same day that Christ was resurrected, but the determination of that day was problematic in light of the two calendar systems mentioned in the Gospels' accounts of Christ's Passion. In the days leading to his death, Christ and the Apostles celebrated Passover, a Jewish feast regulated by lunar cycles. Christ's resurrection, however, occurred on a Sunday, a weekday designated by the Roman solar calendar. With both lunar and solar calendars

---

9     Latin biblical quotations are cited from *Biblia sacra iuxta vulgatam versionem*, eds. B. Fischer et al., 4th ed. (Stuttgart, 1994).

10    English translations of the Vulgate are taken from *The Holy Bible, Douay Rheims Version*, Bishop Richard Challoner, rev. 2nd ed. (Rockford, IL, 2000).

11    Peter Lombard, *Peter Lombard: The Sentences, Book 2, On Creation*, trans. Giulio Silano (Toronto, 2008), p. 63: "And so we must take the times marked by the stars not as units of time marked by duration, but as climatic changes, because these occur by the movement of the heavenly bodies, as do the days and the years to which we are accustomed. For they are *as signs* of calm weather and of storms; and *for the fixing of times* because through them we distinguish the four seasons of the year, namely spring, summer, fall, and winter."

12    Silano, *The Sentences*, 63: "Or they are *as signs and for the fixing of the times*, that is, to keep distinct the hours of time, 'because, before they were made, the order of time was not marked by any indicators, whether of the noon hour or of any other.'"

13    No single study examines the computus tradition over the entire Middle Ages, but for more focused studies see especially Arno Borst, *Schriften zur Komputistik im Frankreich von 721 bis 818* (Hannover, 2006); Immo Warntjes, *The Munich Computus: Text and Translation: Irish Computistics between Isidore of Seville and the Venerable Bede and its Reception in Carolingian Times* (Stuttgart, 2010); and *Computus and its Cultural Context in the Latin West, AD 300–1200: Proceedings of the 1st International Conference on the Science of Computus in Ireland and Europe, Galway, 14–16 July, 2006*, eds. Immo Warntjes and Dáibhí Ó Cróinín (Turnhout, 2010). Faith Wallis, "What a Diagram Shows: A Case Study of Computus," *Studies in Iconography* 36 (2015), 1–40.

in play, a fixed annual date would not always correspond with Passover, and a celebration aligned with Passover would not always fall on a Sunday. In 312, the Council of Nicaea likely mandated that Easter should be celebrated on the first Sunday after the first full moon that occurs on or after the vernal equinox. Nevertheless, multiple systems plotting Lenten seasons across nineteen, eighty-five, and eventually 532-year cycles coexisted during the first eight centuries of Christianity.[14] Aspects of these formulations were still debated during Bede's lifetime, and he did not resolve all of the problems.[15] What he crafted in his handbook was a clear and helpful explanation of the 532-year Great Paschal cycle that also greatly elaborated on its theological, historical, and salvific implications. This clarity, along with Bede's obvious regard for his students, insured the ongoing usefulness of his handbook for Carolingian reformers, tenth-century schoolmasters, and twelfth-century monastic foundations.[16]

The liturgical calendar relies on formulations established by the computus tradition, but by no means was the computus relegated to being a mere technical tool. Rather, computus formulations converged with biblical exegesis, theological speculation, historical records, and artistic creation to become an integral component of monastic education. This vast and totalizing system aligned the motions of the heavens with events on the earth; it prescribed daily devotions and connected annual celebrations with greater spans of history. Although Bede largely relied on verbal explication in his handbook, this material prompted a stunning array of textual, numerical, and visual mnemonic devices, aids, and shortcuts for its multi-step processes.[17] In the centuries after Bede's life, authors such as Helpericus of Auxerre[18] and Hermanus Contractus

---

14      A clear and succinct history of the early computus appears in Alden Mosshammer, *The Easter Computus and the Origins of the Christian Era* (Oxford, 2008).

15      Jennifer Moreton, "Doubts about the Calendar: Bede and the Eclipse of 664," *Isis* 89/1 (1998), 50–65, and Faith Wallis, "Bede and Science," in *The Cambridge Companion to Bede*, ed. Scott DeGregorio (Cambridge, 2010), pp. 113–126.

16      Charles W. Jones, *Bede, the Schools, and the Computus*, ed. Wesley M. Stevens (Aldershot, 1994), and Wesley M. Stevens, *Cycles of Time and Scientific Learning in Medieval Europe*, (Aldershot, 1995).

17      See for example, John Hennig, "Versus de Mensibus," *Traditio* 11 (1955), 65–90; Rolf Max Kully, "Cisiojanus," *Schweizerisches Archiv für Volkskunde* 70 (1974), 93–123; and Faith Wallis, "Images of Order in the Medieval 'Computus,'" in *Ideas of Order in the Middle Ages*, *Acta* XV, ed. Warren Ginsberg (Binghampton, 1988), pp. 45–68.

18      There is limited bibliography on Helpericus. See A. Cordoliani, "Les traités de comput du haut Moyen Âge (526–1003)," *Bulletin Du Cange*, 17 (1943), 62–63; Patrick McGurk, "Computus Helperici: Its Transmission in England in the Eleventh and Twelfth Centuries," *Medium Aevum*, 43 (1974), 1–5; and Helperic, *De computo*, PL 137, 15–48.

at Reichenau,[19] followed his example to compose handbooks that simplified the process, whereas Abbo of Fleury embraced the complexities in his composition of an inventive poem that was both a short-cut for certain reckonings and a virtuoso display of verbal and numerical ingenuity.[20] That the computus tradition absorbed exegetical, cosmological, theological, historical, and artistic activities conveys its foundational relevance. That computus material survives in approximately 9,000 manuscripts reveals its flexible applicability to multiple projects, even those created for women, such as the *Hortus deliciarum*.[21]

Bede's text and definition of times in the plural occupy a central place in the computus tradition, as later readers copied, refined, and paraphrased his works, but this should not suggest that Bede's definition was ubiquitous. Richard Sorabji's study of philosophical questions about time in the late antique and early Christian centuries reveals the range of descriptions given to this elusive phenomenon.[22] More recently, Margot Fassler has remarked on the variety of ways that time was characterized in the Middle Ages by Augustine and liturgical practices. She writes, "...time is not either linear or cyclical, but both, and various units move in different ways—forward, backward, some simultaneously, some synchronized, some neither. There are expectations for lines and cycles and for closure, there are places where time will be unraveled and where structures will be open-ended."[23] From the thirteenth-century on, many philosophers had adopted Aristotelian definitions of time as an accident of quantity or as "the number of motion in respect of before and after."[24] Although no single definition of time was ubiquitous, Bede's description of the

---

19    Arno Borst, "Ein Forschungsbericht Hermanns des Lahmen," *Deutsches Archiv für Erforschung des Mittelalters* 40 (1984), 379–477; Werner Bergmann, "Chronographie und Komputistik bei Hermann von Reichenau," in *Historiographia Mediaevalis: Studien zur Geschichtsschreibung und Quellenkunde des Mittelalters, Festschrift für Franz-Josef Schmale zum 65. Geburtstag* (Darmstadt, 1988), pp. 103–117; and Nadja Germann, *De temporum ratione: Quadrivium und Gotteserkenntnis am Beispiel Abbos von Fleury und Hermanns von Reichenau* (Leiden, 2006).

20    Michael Lapidge, and Peter S. Baker, "More Acrostic Verse by Abbo of Fleury," *The Journal of Medieval Latin* 7 (1997), 1–27.

21    Stevens, *Cycles of Time and Scientific Learning*, p. 46.

22    Richard Sorabji, *Time, Creation and the Continuum: Theories in Antiquity and the Early Middle Ages* (Chicago, 1983).

23    Margot Fassler, "The Liturgical Framework of Time and the Representation of History," in *Representing History, 900–1300: Art, Music, History*, ed. Robert A. Maxwell (University Park, 2010), p. 151.

24    *Hoc enim est tempus: numerus motus secundum prius et posterius* (from Aristotle's *Physics* cited, translated, and discussed in "The Language of Time," in Fox, *Time and Eternity*, chap. 1, p. 11).

plurality of times remained a central component of computus and liturgical traditions. His plural definition was also an important concept for the salvation narrative portrayed in the *Hortus deliciarum*.

•••

This magisterial manuscript, compiled by Abbess Herrad between ca. 1175–1195 for the women at Hohenbourg, recounted in 340+ folios an expansive history of salvation. Unfortunately, the original manuscript no longer survives, but a partial facsimile edition published by the Warburg Institute in 1979 provides a standard edition of the lost masterpiece.[25] The manuscript begins with the days of Creation, then Old Testament stories progress chronologically to fol. 66v, where transitional texts and images precede the New Testament portion of the manuscript. An image of the Annunciation to the Zacharias on fol. 84v commences the lengthy section devoted to Christ's life, which concludes on fol. 167r with a depiction of the Ascension into Heaven. The next thirty folios chronicle various evangelizing acts of the Apostles, then history shifts to allegory on fol. 199v, where a visual rendering of the Virtues and Vices battling with each other introduces an allegorical section that loosely alternates between themes of Ecclesia and the soul. Following this, a visual account of the Antichrist story signals the prophetic end times, and then the Last Judgement according to Matthew 25 unfolds, supplemented by several images from the Book of Revelation. On fol. 263v, a full-page image of the Bosom of Abraham signals both the end of the historical/visual narrative and the desired eternal abode. The manuscript, however, continues with a purely textual recapitulation that also touches on church synods and liturgical practices, and this leads to the computus material on fol. 316r. An unusual and provocative portrait of the foundation's past and present coupled with several moralizing poems concludes the manuscript.

---

25    In addition to the commentary volume that accompanies the facsimile, studies focusing on individual images include Gérard Cames, *Allegories et symboles dans l'Hortus deliciarum* (Leiden, 1971); Michael Curschmann, "Texte-Bilder-Strukturen: Der Hortus deliciarum und die frühmittelhochdeutsche Geistlichendichtung," *Deutsche Vierteljahresschrift für Literaturwissenschaft und Geistesgeschichte* 55 (1981), 379–418; Karl Morrison, *History as a Visual Art in the Twelfth-Century Renaissance* (Princeton, 1990); Caroline Walker Bynum, *The Resurrection of the Body in Western Christianity, 200–1336* (New York, 1995); Katrin Graf, *Bildnisse schreibender Frauen im Mittelalter, 9. bis Anfang 13. Jahrhundert* (Basel, 2002); and Otto Gerhald Oexle, "Relind und Herrad von Hohenburg und die Entstehung des 'Hortus deliciarum,'" *Retour aux sources: Textes, études et documents d'histoire médiévale offerts à Michel Parisse* (Paris, 2004), pp. 551–563.

The *Hortus deliciarum* was composed of quotations from over fifty identifiable and anonymous sources combined with remarkably diverse images that include biblical and allegorical narrative sequences, figural diagrams, linear cosmological diagrams, and computus tables. As a salvation history—an account of Ecclesia in the world from Creation to beyond the Last Judgement—the project is steeped in issues relating to temporal and eternal domains, to history and prophecy, and to the calendar and devotional practices. As a compilation, many of the texts and images contained therein not only touch on aspects of times, but deriving from different traditions, they represent varied points of view on the topic. Finally, although women are not often associated with the computus, the integral place in this manuscript of times and associated concepts demonstrates that the canonesses were expected to master this important material.

Overt references to times and temporality crystallize at five specific points in the *Hortus deliciarum* as clustered images and texts. Four of the five examples designate points in the narrative when temporal and eternal realms intersect: in the Creation and Last Judgement sequences, in the transition between the Old and New Testament sections, and at Christ's Crucifixion. The fifth appears at the end of the manuscript, where Herrad set five folios of computus material. Although studies of the *Hortus deliciarum* have tapped the computus folios in attempts to date the manuscript with greater precision, they seldom integrate this material into interpretations of the salvation narrative.[26] Despite its placement at the end of the manuscript, the computus material provides a point of departure for the following analysis.[27]

A papal list spanning from Christ and Peter to Gregory VIII (1187) and Clement III (1187–1191) opens this section with an unbroken line of religious authority extending across nearly 1200 years, or, as the canonesses could attest, just over two consecutive 532-year cycles. The Great Paschal cycle of 532 years represents the amount of time that must pass before the dates and celestial configurations associated with Easter start their cycle anew. The number derives from multiplying the twenty-eight-year solar cycle by the nineteen-year lunar cycle. In computus handbooks, as on pages 3–20 in the ninth-century Cod. Sang. 250 made at the monastery of St. Gall, the information associated

---

26    Ferdinand Piper, *Die Kalendarien und Martyrologien der Angelsachsen: so wie das Martyrologium und der Computus der Herrad von Landsperg* (Berlin, 1862).

27    I have discussed this material in a different context in "Counting Time and Comprehending History in the *Hortus deliciarum*," *Was zählt: Ordnungsangebote, Gebrauchsformen, und Erfahrungsmodalitäten des 'numerus' im Mittelalter*, ed. Moritz Wedell (Cologne, 2012), pp. 105–118.

with this cycle is often written across columns that fill multiple pages.[28] In the *Hortus deliciarum*, however, two small tables on fol. 319r condense much of this information into more manageable figures (Fig. 3.1). The thirty-five possible dates for Easter range from March 22 (XI Kalends April) to April 25 (VII Kalends May), and the table on the right assigns to these dates Paschal Letters, which span from B. (B punctus) through V. (V punctus) and .A (punctus A) through .Q (punctus Q). Arranged in the center column from the top down, the progressive Easter dates are framed between two columns orienting them against the fixed feasts of Christmas and the first Sunday of Advent. The left column lists the weeks and days between Christmas and Quadragesima Sunday, the first Sunday of Lent. The right column similarly posts the weeks separating the octave of Pentecost, the last of the shifting feasts, from the first Sunday of Advent. Thus, on the latest possible date for Easter, April 25 (VII Kalends May), the Paschal Letter is .Q (punctus Q), there are eleven weeks and two days between Christmas and the first Sunday of Lent, and twenty-three weeks separate the octave of Pentecost from the first Sunday of Advent. By organizing the dates of Easter in this fashion, not only does the table portray the swing of the temporale cycle against the stable backdrop of the sanctorale cycle, but it defines their relationship in terms of countable blocks of times.

The thirty-five Paschal letters reappear in the accompanying table in a 133-box grid composed of seven columns divided into nineteen rows (Fig. 3.1). On the right, Roman numerals from I–XVIIII label the rows according to the nineteen-year lunar cycle. Written on the left is the corresponding sequence of lunar epacts, the age of the moon on March 22 for each year. Roman numerals I–VII at the top assign to each column the concurrent, that is, the weekday of March 24. Set between these corresponding axes, the box in the upper left corner represents the first year of a nineteen-year cycle, when the lunar epact is zero, March 24 is a Sunday (I) and Easter falls on April 7, as the S. (S punctus) denotes. This pattern continues for all 133 boxes. An accompanying text explains how to expand this table fourfold to include leap years and to establish the full 532-year Paschal cycle. Presumably additional information, now lost, correlated this table with specific chronological years.

Folio 319v contains a more unusual grid that reconfigures these schematized sequences back into annular order (Fig. 3.2). Each box in the grid represents a

---

28   St. Gall Stiftsbibliothek, Cod. Sang. 250. A digital version of this manuscript can be accessed at the website Codices Electronici Sangallenses, http://www.cesg.unifr.ch/en/. These tables are similar to Appendix 2 in Wallis, *Bede: The Reckoning of Time*, pp. 392–404. See also Anton von Euw, *Die St. Galler Buchkunst vom 8. bis zum Ende des 11. Jahrhunderts* 2 (St. Gall, 2008).

single year and frames Paschal letters, small dots and dashes indicating the
times separating Christmas from Quadragesima Sunday, and larger dashes signifying the weekdays of Christmas. Additionally, crosses designate each leap
year, rubricated capitals mark the first year of the nineteen-year lunar cycle,
and non-capitalized red initials refer to the first year of the fifteen-year Roman
Indictions. An accompanying paragraph dates the first box in the upper left
corner as the year 1175, when Easter fell on April 13 (.D), the preceding Christmas
was celebrated on a Wednesday (I), and Quadragesima Sunday followed
Christmas by nine weeks and four days.[29] Each successive box continues in
this manner, and the grid extended to the bottom of fol. 320v to include all 532
years of the Paschal cycle.

This information is standard for Paschal tables, but an unusual poem on fol.
321 offers an intriguing reconfiguration.[30] Rather than presenting numbers or
tabular information, this 532-word hexameter poem praising God and virtues
is annotated with a series of small dots that appear over most of the words.
Additionally, the letter "B" marks every fourth word. A short paragraph on the
folio explains that each word of the poem denotes a single year, and the "B"
designates leap years. The number of letters in that word correspond with the
weeks between Christmas and Quadragesima Sunday, and the dots above each
word are the days added to the weeks. Simply put, the 532 words of this poem
encode the Great Paschal cycle into a rhythmic, rhyming puzzle that prompts
another counting of times.[31] Although the *Hortus deliciarum* is the earliest
appearance of the poem, as well as the only complete surviving copy, Abbess
Herrad is not believed to have composed it, likely because of the reference to
"brothers" (*fratres*) in line eight.[32]

What Herrad included in these five folios represents only a fraction of the
computus material circulating in the late twelfth century, and the list is long for
information she did not include, such as lunar patterns, the zodiac sequence,

29   *Hortus deliciarum* vol. 2 #1160. This refers to the explanatory paragraph that accompanies
     the table in the Warburg facsimile.
30   *Hortus deliciarum* vol. 2 #1161. See also, Bernard Bischoff, "Ostertagtexte und Intervalltafeln,"
     *Mittelalterliche Studien, Ausgewählte Aufsätze zur Schriftkunde und Literaturgeschichte
     Band II* (Stuttgart, 1967), pp. 192–227.
31   It is interesting that the date included in the explanatory paragraph is 1159, but the dated
     words begin with 1171 at the second line.
32   Johannes Autenrieth, "Einige Bemerkungen zu den Gedichten im Hortus deliciarum
     Herrads von Landesburg," in *Festschrift Bernhard Bischoff zu seinem 65. Geburtstag*, eds.
     Johannes Autenrieth and Franz Brunhoezl (Stuttgart, 1971), pp. 307–321, and Fiona
     Griffiths, "Herrad of Hohenbourg and the Poetry of the Hortus deliciarum: Cantat tibi
     cantica," in *Women Writing in Latin*, ed. Laurie Churchill (New York, 2002), pp. 231–263.

solstices and equinoxes, and embolismic years. Whether or not Hohenbourg had additional manuscripts with more expansive computus contents for now remains unknown.[33] None of this material directly quotes Bede's handbook, yet it clearly relies on a sense of the plurality of times—individual units that are measured and counted. Individual days and weeks could be plotted against a simple calendar template drawn on fol. 318v, with twelve stacked horizontal lines representing the months and vertical dashes marking the days.[34] Different combinations of years—four-year leap years, fifteen-year Roman Indictions, nineteen-year lunar cycles and twenty-eight-year solar cycles—could be counted across the grid and the poem. All of these formulations abstract and systematize the motions of the sun, moon and stars in the heavens, and it is with the creation of the heavens that Herrad first defined the plurality of times.

In the Creation sequence at the beginning of the manuscript, Herrad portrays the connections between the heavens and times with a combination of biblical and cosmological traditions.[35] Folios 2–8 blend figural narrative sequences with interpretive texts to recount the first five days of Creation, and fols. 9–16 supplement the biblical sequence with cosmological texts, diagrams and classicizing figures. The story of Adam and Eve recommences on fol. 17r after this cosmological interlude. Departing from the more literal sequences commonly found in twelfth-century bibles,[36] Herrad opted to portray the First Day as a provocatively moralizing story about Lucifer's damning choices and expulsion from heaven.[37] She maintains this

---

33    No other codices are known to survive from Hohenbourg's library or scriptorium. Speculation about this issue appears in Griffiths, *The Garden of Delights*, pp. 64–81, and Susann El Kholi, *Lektüre in Frauenkonventen des ostfränkisch-deutschen Reiches vom 8. Jahrhundert bis zur Mitte des 13. Jahrhunderts*, in *Würzburger Wissenschaftliche Schriften* (Würzburg, 1997).

34    *Hortus deliciarum* vol. 2 #1157, *Martyrologium in sequenti pagina per circulum notatum idem ipsum in hac pagina ut lucidius perspectum intelligatur per lineas notatur* (The martyrology in the following page is denoted as a circle, but on this page it is denoted as lines so that the view can be understood more clearly).

35    Studies on Herrad's Creation imagery include Rosalie Green, "The Adam and Eve Cycle in the Hortus deliciarum," in *Late Classical and Medieval Studies in Honor of Albert Mathias Friend, Jr.*, ed., Rosalie Green (Princeton, 1955), pp. 340–347, and Gérard Cames, "La Création des animaux dans l'Hortus deliciarum," *Cahiers Archéologiques fin de l'Antiquité et Moyen Âge* 25 (1976), 131–142.

36    See Walter Cahn, *Romanesque Bible Illumination* (Ithaca, 1982), and Conrad Rudolph, "In the Beginning: Theories and Images of Creation in Northern Europe in the Twelfth Century," *Art History* 22 (1999), 3–55.

37    I have discussed this sequence in greater detail in "All that is Evil: Images of Reality and Figments of the Imagination in the *Hortus deliciarum*," in *Imagination und Deixis,*

interpretive approach by pairing a representation of the Trinity with aspects of the Second and Third Days, likely to suggest the active presence of the Trinity in the acts of Creation as described by Augustine and others.[38] Compared with these thoughtful sequences, the events of the Fourth Day in the upper register of fol. 8v (Fig. 3.3) are remarkably literal illustrations of Genesis 1, 14, "And God said, 'Let there be lights made in the firmament of heaven, to divide the day and the night.'" On the left, God represented as Christ stands on a simple groundline and gestures toward the right where an anthropomorphized sun and moon pair shines down from the semi-circular heavens onto personifications of Day holding aloft torches and Night clasping a billowing veil. Genesis continues, "And let them [the lights] be for signs and times, and for days and years."[39]

In the *Hortus deliciarum*, Day and Night stand as embodied times in the plural, though admittedly as rather ambiguous quantities, since their respective lengths expand and contract through the year. The ensuing cosmological folios include diagrams detailing how and why these changes occur. Two of the four rota diagrams contained in fols. 9–15 are variations of common images deriving from late antique texts such as Macrobius's *Commentary on the Dream of Scipio*.[40] On fol. 10r, a rota representing the basic structure of the cosmos depicts the earth in the center of seven concentric rings signifying the orbiting planets (Fig. 3.4). The outermost ring is divided into twelve segments and labeled with the names of the twelve zodiac signs. In Cicero's allegorical *Dream of Scipio*, a conversation between two men occurs amidst the stars, where they enjoy an extraterrestrial view of the cosmos. In his commentary, Macrobius takes advantage of their heavenly position to draw this diagram and clarify a point of confusion about the cosmos.[41] It should be noted that the cosmos is described as spherical and the two-dimensional diagram depicts a cross-section of the sphere. Macrobius explains that a person standing on the ground and looking up into the sky sees a foreshortened view that collapses the distances among the planets and stars. This creates the optical illusion, reiterated by different literary works, that the planets of the lower spheres

---

  *Imagination und Deixis: Studien zur Wahrnehmung im Mittelalter,* eds., Kathryn Starkey and Horst Wenzel (Stuttgart, 2007), pp. 105–125.

38  See for example, *Trinity and Creation: A Selection of Works of Hugh, Richard and Adam of St. Victor,* trans. Boyd Taylor Coolman, and Dale M. Coulter (Turnhout, 2010).

39  This is the literal translation rather than the Douay-Rheims translation.

40  *Commentary on the Dream of Scipio by Macrobius,* trans. William Harris Stahl (New York, 1990).

41  Ibid., pp. 174–175.

"pass through" the signs of the zodiac.[42] If the vantage point is shifted from a terrestrial to a celestial position, the vertical distance between the planets and stars reappears. This diagram became a well-known cosmological image during the following centuries, as Bruce Eastwood has demonstrated, and it reappeared in numerous contexts.[43] In the *Hortus deliciarum*, Herrad paired her image not with a quotation from Macrobius, but instead with an anonymous allegorizing text.[44]

The rota on fol. 11v, also a variation of a Macrobian diagram, expands on the first cosmological image by shifting the perspective and introducing new elements (Fig. 3.5).[45] Rather than representing a cross section of the cosmos with the zodiac ring as the border, this diagram rotates the image 90 degrees forward then tilts the right side up so the zodiac band becomes a diagonal belt across the circle. Five horizontal lines distinguish the five climate zones of the earth and cosmos, with the frigid zones at the top and bottom, and the temperate zones on either side of the hot equator.[46] As the sun orbits the earth in circles parallel to the climate zones, it follows an annual course up and down the diagonal zodiac path. The lowest and highest points on the zodiac belt mark the winter and summer solstices respectively, and the sun crossing the equator corresponds with the vernal and autumnal equinoxes. As a pair, these two diagrams demonstrate the basic cosmological mechanics that generate the different blocks of times associated with a year: days and nights, months, and the four seasons. Finally, a titulus that survives without its text on fol. 13v articulates the plurality of times, *De divisione temporum*, that is, "On the division of times."[47] Perhaps the now missing paragraph quoted or paraphrased Bede's definition of times in the plural.

Abbess Herrad adhered to a literal biblical interpretation that times were created on the Fourth Day and she portrayed this with figural illustration, technical diagrams, and textual excerpts. She maintained this close connection between times and the heavens in the Last Judgement sequence at the end of the narrative, though in this prophetic account she emphasized the rupture

---

42  Ibid.

43  Bruce S. Eastwood, "Planetary Diagrams for Roman Astronomy in Medieval Europe, ca. 800–1500," *Transactions of the American Philosophical Society* 94 (2004), 1–157, and Bruce Eastwood, *Ordering the Heavens: Roman Astronomy and Cosmology in the Carolingian Renaissance* (Leiden, 2007).

44  *Hortus deliciarum* vol. 2 #39.

45  *Commentary on the Dream of Scipio*, trans. Stahl, pp. 150–151.

46  There is a slight discrepancy in the *Hortus deliciarum* image which might indicate some confusion on the part of the original artist, or the nineteenth-century copiest.

47  *Hortus deliciarum* vol. 2 #47.

that will occur between them. Beginning on fol. 247v, a series of narrative registers remarkably similar to the mosaic at Sta Maria Assumpta in Torcello depict the Judgement according to Matthew 25.[48] A single angel in both the mosaics and on fol. 251r of Herrad's Judgement imagery (Fig. 3.6) stands and rolls up the scroll of the heavens to illustrate the prophecy of Isaiah 34, 4, that the heavens will be tucked away as time and history reach their conclusion.[49] As if this motif were too easily overlooked, Herrad reiterates the idea with the inclusion of two registers of imagery, slightly out of order, on fol. 247v (Fig. 3.7).[50] In the central register, beneath the line of penitent awaiting Judgement, a massive conflagration burning across lands and sky is labeled with a quotation of Luke 21, 33, "Heaven and earth pass away, but my words shall not pass away." Contrasting with this violent immolation is the calm stillness of the lower register where two medallions representing New Heaven and New Earth suggest a post-apocalyptic rupture of the spatial-temporal unity characterizing the created world. Paradisiacal flowers bloom on the medallion of New Earth, and shining resplendently from the center of New Heaven, like a prefiguration of the soon-to-be popular Veronica visage, is Christ's face as the true and unconquered sun.[51] The sun and moon pair make their final appearance amidst the stars above Christ's face, though an inscription explains that, now unmoving, the sun and moon pale in comparison with Christ's brilliant clarity.[52] With its use of medallions, this register is similar to the Last Judgement image in Vision 12, Book III of Hildegard's *Scivias*.[53] More importantly for this study, by emphasizing the disjuncture between the mutable temporal heavens and immutable eternal heavens, the image signals a definitive end of times in the salvation narrative.

---

48    Among various studies see Otto Demus, "Studies Among the Torcello Mosaics—III," *Burlington Magazine* 85/497 (1944), 195–200, and Irina Andreescu, "Torcello. III. La chronologie relative des mosaïques pariétales," *Dumbarton Oaks Papers* 30 (1976), 245–341.

49    Inscriptions around Herrad's angel read, *Angelus involvit celum quasi rodale*, and *Ysayas. Celi ut liber complicabuntur*, which paraphrase Isaiah 34, 4, *et conplicabuntur sicut liber caeli* and Revelation 6, 14, *et caelum recessit sicut liber involutus*.

50    An inscription on the image notes that these registers appear out of sequence, *Ista conflagratio erit post judicium*.

51    Jeffrey F. Hamburger, "*Frequentant memoriam visionis faciei meae*: Vision and the Veronica in a Devotion Attributed to Gertrude of Helfta," in *"The Holy Face": Proceedings of the International Colloquium, Bibliotheca Hertziana Rome and Villa Spelman, Florence*, eds. H.L. Kessler, G. Wolf, and E. Cropper, (Baltimore / Bologna, 1998), pp. 229–246.

52    *Hortus deliciarum* vol. 2 #840, *Christi claritas prefulgida precellit omnia novi celi luminaria.*

53    Lieselotte E. Saurma-Jeltsch, *Die Miniaturen im 'Liber Scivias' der Hildegard von Bingen: die Wucht der Vision und die Ordnung der Bilder* (Wiesbaden, 1998). In this image, different components of the eternal world are also framed in separate medallions.

Creation and the Last Judgement demarcate the extreme end-points of times and history, when temporal and eternal domains abut against one another. Herrad's salvation narrative reaches beyond temporal boundaries to include eternity, which is figured in these folios, among other ways, as a full-page image of Hell on fol. 255r and the Bosom of Abraham as heaven on fol. 263v. Between Creation and the Last Judgement, most of the historical, allegorical and prophetic events comprising this history are portrayed as figural sequences framed in horizontal registers stacked vertically on the page.[54] It is tempting to interpret these registers as a visual equivalent of the nature of time—a time that is cohesive, singular, and linear.[55] This interpretation aligns neatly with assumptions scholars often make about time. Take, for example, Eric Auerbach's influential chapter *"Figura,"* which traces the use of the term *figura* through literary and theological texts to plot a developing Christian hermeneutic. In describing the Christian exegetical practice of combining figure with fulfillment, Auerbach wrote,

> In the modern view [of historical development], the provisional event is treated as a step in an unbroken horizontal process; in the figural system the interpretation is always sought from above; events are considered not in their unbroken relation to one another, but torn apart, individually, each in relation to something other that is promised and not yet present.[56]

Auerbach speaks of history as if it were an unbroken swathe of fabric, like the Bayeux Tapestry, or the horizontal registers in the *Hortus deliciarum* linked end-to-end, with typological exegesis renting the fabric to introduce a vertical sensibility as individual people or events are extracted and paired together despite their historical i.e. linear distance. More recently, Alexander Nagel and Christopher S. Wood offer a variation on this description in their *Anachronic Renaissance* when they note,

> With its power to compel but not explain a folding of time over onto itself, the work of art in the fifteenth and sixteenth centuries was able to

---

54 This narrative structure is frequently referred to as "strip narrative," see Götz Pochat, *Bild—Zeit, Zeitgestalt und Erzählstruktur in der bildenden Kunst von den Anfängen bis zur frühen Neuzeit* (Vienna, 1996), and Moshe Barasch, *The Language of Art* (New York, 1997).

55 Stephen G. Nichols, *Romanesque Signs: Early Medieval Narrative and Iconography* (New Haven, 1983).

56 Eric Auerbach, "Figura," *Scenes from the Drama of European Literature* (Minneapolis, 1984), p. 59.

lay a trail back to Europe's multiple pasts, to the Holy Land, to Rome—
monarchial, Republican, Imperial, or Christian—and sometimes to
Rome's Byzantine legacy.

and then continue,

Artifacts played an indispensable role in the overall cultural project of
time management, not simply as beneficiaries or participants, but as the
very models of the time-bending operation.[57]

Neither study is unusual in assuming a stable fabric of time that is singular and
cohesive. That assumed definition of time, however, corresponds neither with
the definition circulating with so much computus material nor with the definition crafted by Herrad in the *Hortus deliciarum*.

With the plurality of times accepted as the base definition, the horizontal
registers in the *Hortus deliciarum* project a narratival rather than a temporal
function. Admittedly, aspects of time and narrative often parallel and overlap
one another, but ultimately they remain distinct and separate concepts.[58] As
seen in the Creation sequence, times were created on the Fourth Day as an
event in this biblical narrative; in the Last Judgement sequence, the heavens
were stilled and eternity reigned as part of the narrative. This should not suggest, however, that this visual framework solely defines the salvation narrative
in the manuscript. Many other images and texts exist outside of its regular
borders. These images, such as the cosmological diagrams, the computus
tables, and images separating the Old and New Testament sections of the
manuscript, serve a different purpose than the figural narrative sequences. As
seen in three of the transitional images, these compositions construct over-
views of the narrative by prompting an interpretive analysis of, among other
issues, a passage of times.

On fol. 67r/v, a pair of figural rota diagrams contrast ritual Temple sacrifice
with the Christian Eucharist commemorating Christ's Crucifixion.[59] The first
image depicts a puzzling two-headed figure seated on a seven-armed candela-
bra and surrounded by bust-length figures proffering animals for sacrifice

---

57   Alexander Nagel and Christopher S. Wood, *Anachronic Renaissance* (New York, 2010),
     pp. 10–11.

58   For a fascinating discussion of this, see Paul Ricoeur, *Time and Narrative* 3 vols, trans.
     Kathleen McLaughlin (Chicago, 1984).

59   Annette Krüger, and Gabriel Runge, "Lifting the Veil: Two Typological Diagrams in the
     Hortus deliciarum," *Journal of the Warburg and Courtauld Institute* 60 (1997), 1–22.

(Fig. 3.8). An altar with a goat immolating and a veil drawn suggestively aside establishes the location as inside the Jewish Temple but in front of the veil partitioning off the Holy of Holies. A length of cloth wraps around the heads and torso of the two-headed figure, and an inscription explains that this is the Old and New Testaments joined together.[60] A turn of the page corresponds with a progression past the Temple veil and into the Holy of Holies, where a cloth-draped altar, a cross, and the outlines of a church building rest upon the Ark of the Covenant (Fig. 3.9). Christ as Priest and King stands above the cross and he is surrounded by bust-length portraits of the Virtues, who have replaced animal sacrifices. Drawn at a conclusion of the Old Testament histories, this pair of diagrams signals a shift in the narrative of the manuscript. Drawn on opposite sides of the folio, these two images can not be viewed simultaneously. Just as both rotae portray a Christian meaning allegorically hidden beneath the literal Old Testament text, the visual separation of the two images reinforces the necessary process of seeking hidden truths in one text in order to truly understand the significance of the other. A turning of this page also recreates, in a much smaller block of time, the temporal progression that occurred between the eras of the old and new covenants, thus aligning the pages of the manuscript itself with a sense of times passing.[61]

A very different strategy for visualizing times appears on fol. 80v in the Tree of Abraham, a variation on the well-known Tree of Jesse iconography (Fig. 3.10).[62] In this full-page image, Abraham stands between the boughs of a tree that is planted into the ground by God. As described in Genesis 15:5, Abraham was rewarded for his obedience with a blessed and abundant lineage.[63] These many generations appear as the rows of heads neatly arranged between the trunks of the tree. Branches coiling off either side frame groups of faithful Jews from the Old Testament, who complement Abraham's blood lineage with his spiritual family.[64] As the generations pass, the tree grows steadily

60    *Vetus et novum testamentum in simul junctum.*

61    The division of world history into six or seven ages, or three eras, is discussed in Elizabeth Sears, *The Ages of Man, Medieval Interpretations of the Life Cycle* (Princeton, 1986), and J.A. Burrow, *The Ages of Man, A Study in Medieval Writing and Thought* (Oxford, 1988).

62    Among many studies of the Tree of Jesse iconography, see esp. Arthur Watson, *Early Iconography of the Tree of Jesse* (London, 1934); Margot Fassler, "Mary's Nativity, Fulbert of Chartres, and the *Stirps Jesse*: Liturgical Innovation circa 1000 and its Afterlife," *Speculum* 75 (2000), 389–434.

63    Genesis 15, 5, "And he brought him forth abroad, and said to him: Look up to heaven and number the stars, if thou canst. And he said to him: So shall thy seed be."

64    Jérôme Baschet, "Medieval Abraham: Between Fleshly Patriarch and Divine Father," *Modern Language Notes* 108 (1993), 738–758.

upward and outward until Christ appears at its pinnacle. The flower-strewn green length of new ground stretching on either side of Christ strongly contrasts with the barren dirt mound at the bottom of the page. Occupying this verdant field are the ranks of Christian faithful who succeed the Jewish faithful below, from popes and hermits to bishops and virgins. Reconfigured into a compelling organic metaphor, the passage of times is implied by the succession of generations mapped onto the ramified growth of this most blessed tree.[65] Its placement in the transitional section preceding Christ's life parallels the common use of the Tree of Jesse as an introduction to the Gospel of Matthew and its opening account of the earthly lineage of Christ.[66]

This arboreal outgrowth of times is distilled into a slender, delicate strand in the fascinating image of God Fishing for the Leviathan (Fig. 3.11).[67] In the upper left corner of fol. 84v, God stands and holds out a fishing pole, the line of which drops down the entire page. Medallions framing the heads of prophets and patriarchs lead down the line like a series of sinkers to the hook at the bottom. A cross is drawn onto the hook, where Christ's fully-clothed body is suspended against its wooden beams. The written source for this image is likely a passage on the Annunciation to the Blessed Virgin in Honorius's *Speculum ecclesiae*.[68] Although excerpts from this section appear in the *Hortus deliciarum*, the exact descriptive passage does not. Honorius describes Matthew's genealogy of the Virgin as a line ending with a hook that is the wood of the cross and her son Christ. Perhaps the Divine Angler is privy to knowledge beyond the ken of ordinary fishermen, for when he dropped Christ into the world like a worm-bearing hook into the sea, the Leviathan was lured from his depths and swallowed the bait.[69] In this image, the fishing line equates with a passage of times as it drops from God into the abyss below, and that passage acquires a weighted urgency

65    See also Gabrielle Spiegel, "Genealogy: Form and Function in Medieval Historiography," in *The Past as Text: The Theory and Practice of Medieval Historiography* (Baltimore, 1997), pp. 99–110, and Christiane Klapisch-Zuber, *L'ombre des ancêtres, Essai sur l'imaginaire médiéval de la parenté* (Fayard, 2000).

66    Musée Condée-Chantilly, MS 9 olim 1695. The Ingeborg Psalter, made ca. 1195–1200 for the Danish Queen of French King Philip II, provides a stunning example. Florens Deuchler, *Der Ingeborgpsalter* (Berlin, 1967).

67    For interesting related motifs, see Lois Drewer, "Leviathan, Behemoth and Ziz: A Christian Adaptation," *Journal of the Warburg and Courtauld Institutes* 44 (1981), 148–156; Jessie Poesch, "The Beasts from Job in the Liber Floridus Manuscripts," *Journal of the Warburg and Courtauld Institutes* 33 (1970), 41–51.

68    PL 172:901D.

69    This elaborates on Job 40, 20–21, "Canst thou draw out the leviathan with a hook, or canst thou tie his tongue with a cord? Canst thou put a ring in his nose, or bore through his jaw with a buckle?"

in the struggle between good and evil. The pair of typological diagrams, the Tree of Abraham, and God Fishing for the Leviathan all give different interpretive shapes to times, and they all portray Christ's life and death as the turning point of history.

Befitting its significance, the depiction of Christ's Crucifixion combines two narrative registers into a single vertical composition on fol. 150r (Fig. 3.12).[70] Historical events recorded in the Gospels are drawn with allegorical and interpretive figures and numerous inscriptions to designate this moment not only as the fulfillment of prophecy, but also as latent with prophecy. At the base of the cross, Adam's desiccated bones recall Genesis and the connection between Adam the first man and Christ the new man. Juxtaposed on either side of the cross, personifications of Ecclesia and Synagoga recall the pair of Typology Rotae on fol. 67r/v and remind the canonesses of the progression from the era under the law to the era of grace. Ecclesia's four-headed mount echoes words that extend backward and forward in times, from Ezekiel 1:5–10 to the Book of Revelation 4:6–9. Inscriptions on the image quoting the Song of Songs reiterate this dual turn, foretelling a future time when the Sulamite, that is Synagoga, will reunite with Ecclesia and Christ.[71] Above the cross, the inconsolable sun and moon are standard features of Crucifixion imagery, but here they also echo the celestial bodies that served as signs for the beginning and end of times in the Creation and Last Judgement sequences.

Mounted on the tetramorph, Ecclesia lifts a chalice toward Christ's side wound to capture the flowing blood, with a gesture that repeats Christ's movement in the second Typology Rota, to announce the birth (or baptism) of Ecclesia.[72] Furthermore, this gesture implicates times contemporary with the

---

70　Studies on Crucifixion iconography include Stanley Ferber, "Crucifixion Iconography in a Group of Carolingian Ivory Plaques," *Art Bulletin* 48 (1966), 323–334; C.W. Marx, "Aspects of the Iconography of the Devil at the Crucifixion," *Journal of the Warburg and Courtauld Institutes* 42 (1979), 233–235; Celia Chazelle, "An 'Exemplum' of Humility: The Crucifixion Image in the Drogo Sacramentary," in *Reading Medieval Images, The Art Historian and the Object*, ed. Elizabeth Sears (Michigan, 2002), pp. 27–35.

71　Wolfgang Seiferth, *Synagogue and Church in the Middle Ages: Two Symbols in Art and Literature*, trans. Lee Chadeayne and Paul Gottwald (New York, 1970); Jeremy Cohen, "'Synagoga conversa': Honorius Augustodunensis, the Song of Songs, and Christianity's 'Eschatological Jew,'" *Speculum* 79 (2004), 309–340; Elizabeth Monroe, "'Fair and Friendly, Sweet and Beautiful': Hopes for Jewish Conversion in Synagoga's Song of Songs Imagery," in *Beyond the Yellow Badge: Anti-Judaism and Antisemitism in Medieval and Early Modern Visual Culture*, ed. Mitchell B. Merback, (Leiden, 2008), pp. 33–62.

72　Rudolf Suntrup, "Te Igitur-Initialen und Kanonbilder in mittelalterlichen Sakramentarhandschriften," in *Text und Bild, Aspekte des Zusammenwirkens zweier Künste in Mittelalter und früher Neuzeit*, eds. Christel Meier and Uwe Ruberg (Wiesbaden, 1980),

Hohenbourg canonesses, who commemorated this moment in their prayers and liturgy. Herrad strengthens this connection between Christ's Crucifixion and the Hohenbourg canonesses by including in the following folios several paragraphs referring to the calendar and computus. A paragraph on fol. 151r raises the difficulty of pinpointing with certainty the actual dates of Christ's death and resurrection, noting that one tradition names 8 Kalends April (March 25), whereas others list either 6 or 5 Kalends April (March 27–28).[73] The oldest and most authoritative martyrology, it notes, uses overlapping dates of the Annunciation and Crucifixion as evidence to designate 8 Kalends of April, the sixth day of the week and the 14th lunar day, as the correct date. Unfortunately, only the tituli of the computus paragraphs survive, but they refer to a tantalizing array of technical topics.[74]

Herrad's decision to include paragraphs describing computus elements after the Crucifixion supports the idea that the computus material at the end of the manuscript was an important component of her salvation narrative. The *Hortus deliciarum* was not created for passive reading; rather it was designed to challenge, compel, and guide the canonesses through multiple processes imparting knowledge about the Church in the world. Ultimately, it was then up to each individual to transform that knowledge into a deeper wisdom. To teach those lessons and prompt those inner journeys, Herrad carefully blended visual, textual, and numerical elements in her salvation narrative. Thus, figural and diagrammatic images of Creation correspond with the Crucifixion scene just as much as the transitional images preceding the New Testament section complement the computus tables. As Herrad portrays, times are generated by the motions of the heavens, they are recorded in biblical histories, schematized in computus formulations, and defied by the miraculous events associated with Christ's life, death and resurrection. A full understanding of the nature of times allowed the canonesses to discover how and why their daily devotions corresponded with the greater history of salvation. With this knowledge at hand, the canonesses could then acquire a deeper wisdom regarding the relationship between their temporal lives and eternal souls.

• • •

pp. 278–331. According to Rupert of Deutz, whom Herrad quoted on fol. 153r, this was the baptism and not the birth of the church. *Hortus deliciarum* vol. 2, #543, quotes from Rupert's *De divinis officiis*, bk. 6, chap. 35.

73    *Hortus deliciarum* vol. 2, #528.

74    *Hortus deliciarum* vol. 2 #530, *De inventione paschalis temporis*; #531, *Quod typicum sit pascha*; #532, *Item de pascha, in Gemma anime*; #533, *De embolismo*.

Although this essay has focused on visual aspects of these processes, a number of texts demonstrate the close correlation of textual and visual strategies in the manuscript. The confluence of times portrayed in the image of Christ's Crucifixion is clearly articulated in the liturgical commentaries of Rupert of Deutz and Honorius of Augustodunensis that Abbess Herrad favored. Rupert of Deutz, a Benedictine closely associated with reform efforts in the early twelfth century, explained in the introduction of his *Liber de divinis officiis* that since all lands are colored purple with Christ's blood, the heavens flower with his victories, the times are stamped with his mysteries, and the days and hours are ornamented with memories of him, that a study of liturgical devotions leads one towards divine mysteries.[75] Rupert's clear and systematic commentary on the hours, mass, and liturgical year was a text that generally circulated without imagery in the twelfth century.[76] Herrad drew most of her quotations from Book III on the liturgical year, and she focused especially on the Lenten season, Passion week, and Easter. Honorius lived and worked roughly a decade later than Rupert, and though his *Speculum Ecclesiae* and *Gemma Animae*, along with other works, became widely admired in the twelfth century, much of his life remains a mystery.[77] All three of these works similarly apply strategies of biblical exegesis to the Divine Office and liturgy.

Both men structure their commentaries according to the organization of different liturgical cycles, such as the hours of the Divine Office or the progression of feast days in the liturgical year. Both men also employ analytical strategies that correspond with computus formulations and a plural sense of times. For example, Septuagesima Sunday, the third Sunday before Ash Wednesday, once marked the beginning of the pre-Lenten season.[78]

---

75    Latin quotations from Rupert's text are taken from PL 170, *Libro de divinis officiis*. PL 170:13. *Visum est autem hoc opusculum sic ordinari ut primum ea quae pene quotidiana sunt, deinde quae certis temporibus et causis variantur, digeram, ipso adjuvante Christo, cujus pio sanguine universa terra purpurata est, cujus victoriis caelum floret, cujus mysteriis tempora insignita sunt, cujus memoria dies et horae ornatae sunt.*

76    Rhabanus M. Haacke, *Programme zur bildenden Kunst in den Schriften Ruperts von Deutz*, *Siegburger Studien* 9 (Siegburg, 1974); John van Engen, *Rupert of Deutz* (Los Angeles, 1983); Maria Lodovica Arduini, *Rupert von Deutz (1076–1129) und der "Status Christianitatis" seiner Zeit: Symbolisch-prophetische Deutung der Geschichte* (Cologne, 1987).

77    Among many, see esp. Valerie I.J. Flint, such as "The Place and Purpose of the Works of Honorius of Autun," *Revue Benedictine* 87 (1977), 97–127.

78    An interesting consideration of the important days of the Lenten season can be found in Essay III, "The Lenten Agon: From Septuagesima to Good Friday," in O.B. Hardison, Jr., *Christian Rite and Christian Drama in the Middle Ages* (Baltimore / London, 1967), pp. 80–138.

In anticipation of Lenten fasting, the white cloth was removed from the altar, the alleluia was excluded from the Mass, and the Gospel passage was Matthew 20, 1 – 8, the parable of the Laborers in the Vineyard. The readings assigned to this Sunday return to Genesis, to the beginning of history, and during the weeks leading to Easter, the readings progress forward chronologically through the Old Testament. Both authors note that Septuagesima Sunday is about 70 days before Easter, and it was certainly no coincidence that the Babylonian captivity lasted 70 years. Furthermore, the Israelites' historical exile is analogous with an individual's exile while captive in the secular world and wandering from God. Rupert and Honorius both remark upon parallels between the progression of Lenten weeks toward Easter and the successive Ages of the World, but Rupert maintains a more methodical pairing of weeks and ages until the seventh Sunday, the fourth Sunday of Lent and the week prior to Palm Sunday, which he describes as the period of rest that follows Judgement.[79]

Honorius seems to delight in the interpretive possibilities associated with these times. In the *Speculum ecclesiae*, he elaborates by first connecting these 70 days with the 7,000 years comprising the entire history of humanity, then he aligns the seven millennia with the seven days of Creation.[80] In a chapter on the Lenten fast in his *Gemma animae*, Honorius constructs a different configuration by aligning the first age of the world not with Septuagesima Sunday, but with Quadragesima Sunday, the first Sunday of Lent. With this shift in place, he briefly notes figures from the six ages of the world that correspond to the progressive Lenten Sundays, such as the fourth Sunday and the fourth age when great peace flourished during Solomon's reign.[81] Honorius takes a different

---

79  PL 170:101B. In bk. 4, chap. 13, Rupert writes, *Dominica haec ab ea, quae septuagesima dicitur, septima est, Sabbatum mundi, id est requiem significans, in qua sanctorum et electorum animae, deposito carnis onere, et velut post bella victores, depositis armis feriatae laetantur, et nunc iterum iidem electi singulis stolis, id est animae beatitudine munerati resurrectionem, in qua binas stolas, scilicet tam animarum quam corporum immortalem gloriam recipient, praestolantur.*

80  Latin quotations from the *Speculum Ecclesiae* are taken from PL 172. Here, PL 172: 855B, Honorius writes, *Per septuaginta annos quibus populus in Babylone affligebatur septem milia annorum intelliguntur, quibus genus humanum in hac vita peregrinatur. Sicut enim septem primis diebus omnis creatura disponitur, ita per septem milia annorum hic mundus extendi creditur.*

81  Latin quotations from the *Gemma Animae* are also taken from PL 172. Here, PL 172:658B, *Totum tempus hujus vitae ab initio usque in finem per Quadragesimam intelligitur, quia est tempus afflictionis, sicut Paschae tempus significat futurum tempus gratulationis. Sed hebdomadae sunt sex aetates mundi.*

approach in the *Speculum ecclesiae* discussion of the Lenten fast. Here, he aligns the week of Ash Wednesday with the fourth age of world when Solomon built the Temple, and he reconfigures different blocks of time to seek new connections. For example, Solomon built his temple in seven years and, following its destruction by the Babylonians, it was rebuilt by the priest Jesus in 46 years. Solomon can be understood as Christ, and the temple as Ecclesia, thus the temple originally took seven years to build because the Holy Spirit blessed Christ with seven gifts, and the 46 years taken for its reconstruction are the 46 days of Lent that lead to Easter.[82]

These examples are fairly typical not just for Rupert and Honorius, but for any number of biblical exegetes from Origen and Augustine on. Scholarship has labeled their methods of analysis as typological exegesis and also as numerology. Oftentimes, however, scholarship has conceived of time as a singular, unified entity. When times are conceived in the plural, this analytical process acquires a different sensibility. Auerbach, Wood and Nagel described the process as extracting events from a unified swathe of history, or folding history over to make connections. If time is understood as a plural phenomenon, then there is neither a unified swathe nor a linear history. There are instead endless blocks of time. Rather than hunting and picking, the exegete actively assembles and examines different configurations of times in order to uncover the mysteries hidden beneath literal and temporal strictures. When narratival and chronological rules do not apply, the exegete who desires access to secret knowledge is free to wander through history, stacking, shuffling, rearranging, and comparing blocks of varying sizes in the pursuit of divine mysteries.[83]

Sometimes vastly different blocks are compared, like the alignment of millennia with individual days; other times the blocks are similar, as when Rupert

---

82   PL 172:878A, *Quarta aetate mundi aedificavit Salomon Domino templum septem annis, quod destructum a Babyloniis, reaedificatum est a Jesu sacerdote XL et VI annis. Salomon est Christus, templum Ecclesia. Hoc templum VII annis a Salomone aedificatur, quia Ecclesia VII donis Spiritus sancti a Christo informatur.*

83   A visual equivalent of this exegetical strategy appears in the pair of rota diagrams on fol. 32r in Lambert of St. Omer's *Liber Floridus*, an early twelfth-century compilation (Wolfenbüttel, Herzog-August Bibliothek, Cod. Guelf. 1 Gud. lat., fol. 32r). Six petal-like lobes that surround central medallions framing personifications of the *Mundus maior* and *Mundus minor*, contain inscriptions that describe a layering of multiple time-spans. See Penelope C. Mayo, "Concordia Discordantium: A Twelfth-Century Illustration of Time and Eternity," in *Album Amicorum, Kenneth C. Lindsay, Essays on Art and Literature* (Binghampton, 1990), pp. 29–57, and Veronika Pirker-Aurenhammer, "Modelle der Zeit in symbolischen Darstellungen des Mittelalters," *Das Münster* 53/2 (2000), 98–119.

aligns the Passion and Creation weeks. In this comparison, Rupert notes that God began to create the world on a Sunday and the Savior enters into the labor of his passion on a Sunday.[84] Forging additional connections, Rupert points out that on the fifth day of the week, Thursday, Christ and his Apostles had their Last Supper together and Christ washed their feet. On the Fifth Day of Creation, God made some creatures that would be submerged beneath waves and others that would rise into the air. Tellingly, some liturgical practices on Holy Thursday are weighed down with sadness, such as the silence commemorating Christ's betrayal, whereas others prompt elation and joy, such as the consecration of oil, which is a sign of the Holy Spirit.[85] Working toward ever smaller blocks, Rupert remarks on the individual hours of Friday; man was formed on the third hour of the Sixth Day, and the third hour on Good Friday sees Christ sentenced to death. At the sixth hour man disobeyed God's command and ate the forbidden fruit, and Christ was crucified. By the ninth hour, man was expelled from Paradise and Christ died.[86]

All three commentaries are organized according to various liturgical cycles, but in the *Hortus deliciarum*, Herrad inserts quotations of the commentaries into a historical or allegorical context. For example, Honorius' discussion of Septuagesima Sunday appears at three different places in the *Hortus deliciarum*. A short paragraph in the Old Testament recapitulation on f. 69v describing Moses leading the Israelites through the Red Sea and out of Egypt compares this event with baptism and the hopeful return to the *patriam paradisi*.[87] A second

---

84    Herrad included this quotation in the manuscript. *Hortus deliciarum* vol. 2 #448, bk. 5, chap. 9, *Dominica die que prima est, in qua creationem mundi cepit Deus operari, ingreditur salvator ad laborem passionis et per totam illam ebdomadam que major ebdomada dicitur salutem nostram operatus, die septima cessavit et in sepulchro requievit.*

85    *Hortus deliciarum* vol. 2, #456, bk. 5, chap. 14, *Feria quinta mundane creationis genus ex aquis ortum partim gurgiti remissum, partim levatum in aera, ut stirpe una prodita diversa raperent loca, sic feria quinta, quam cenam nominamus, sacramenta unius ejusdem salutis partim nos deprimunt in tristitiam, sicut in regularibus horis in quibus neque festivum neque more usitato integrum Deo debitum reddimus servicium, quia capita, id est inicia precidimus, fines quoque earum sub silentio abscondimus....*

86    *Hortus deliciarum* vol. 2, #500. Only the title survives of a passage from Rupert that comments on the liturgical acts performed at the ninth hour, bk. 6, chap. 4, *De eo quod hora nona convenimus adorare.* Honorius incorporates a similar framework into his discussion of the Divine Hours in the *Gemma Animae*, as in bk. 1, chap. 113 where he aligns the daytime hours of Terce, Sexts, and Nones with events leading to Christ's death.

87    *Hortus deliciarum* vol. 2, #245, *Per Egyptum hic mundus significatur, in quo genus humanum a paradisi patria peregrinatur, et diabolo gravi oppressione subjugatur. Cui ad Deum per justos vociferanti Christus ad liberandum mittitur, per quem mundus multis miraculis*

excerpt on f. 222r seems to have inspired the three scenes of Odysseus and the Sirens, which Honorius retells with a Christian twist of the obedient men slaying the evil Sirens.[88] The third passage works smoothly into Christ's life as a lengthy inscription accompanying the representation on fol. 108r of the Parable of the Laborers in the Vineyard, the reading for Septuagesima Sunday.[89]

By setting these interpretive passages back into a historical context, Herrad unravels the layered times, that is to say disassembles the configured times, that comprise the Divine Office and Liturgy. The Victorines, an Augustinian foundation in Paris, advocated learning the literal history of the Bible before advancing to more complex interpretations.[90] In the *Hortus deliciarum*, Abbess Herrad recounted literal histories and supplemented their basic narrative with interpretive tools and strategies to reveal how history becomes the basis of daily prayers, weekly devotions, and annual liturgical cycles. In the folios of this fascinating manuscript, the canonesses could see those histories and, more importantly, figure out what to do with them. They could learn that times were created on the Fourth Day in conjunction with the creation of the sun, moon and stars in the heavens. They could trace in the cosmological diagrams how the basic mechanics of the cosmos generate days and nights, seasons and years. They could work through metaphors and abstractions of greater spans of time to ponder broader historical trajectories. They could count through complex heavenly cycles of nineteen, twenty-eight, and 532 years to marvel at the work of the divine hand. They could discover connections between celestial patterns and historical events, and then discern their own place in those patterns. They could apply a deeper analysis to break through the linearity of historical narrative as they assembled, reassembled, and analyzed blocks of times. Finally, they could embrace the belief that the Incarnation of Christ and grace of God would one day prompt a complete dissolution and cessation of temporal confines, at which point they could all enjoy eternal wisdom in the Garden of Delights.

---

*concutitur. Populus per mare Rubrum, id est per baptismum, ad patriam paradisi reducitur, ubi mel et lac exundat, id est affluentia omnium bonorum exuberat.*

88   *Hortus deliciarum* vol. 2, #756.

89   This tradition extends back to Origen, and was popularized by both Augustine and Gregory, a tradition that is briefly discussed in Burrow, *The Ages of Man*, pp. 59–68, and Sears, *The Ages of Man*, pp. 81–90.

90   Beryl Smalley, *The Study of the Bible in the Middle Ages* (Oxford, 1941, rev. ed, 1952, rpt. 1983) and Boyd Taylor Coolman, *The Theology of Hugh of St. Victor: An Interpretation* (Cambridge, 2010).

**Decennovenalis ciclus.**

| Epactae | I | II | III | IV | V | VI | VII | Anni |
|---|---|---|---|---|---|---|---|---|
| Nullae | S. | R. | .C | .B | .A | V. | T. | I |
| XI | L. | K. | I. | H. | G. | F. | M. | II |
| XXII | .E | .L | .K | .I | .H | .G | .F | III |
| III | S. | R. | Q. | P. | O. | V. | T. | IIII |
| XIIII | P. | C. | I. | H. | G. | F. | E. | V |
| XXV | .E | .D | .C | .B | .H | .G | .F | VI |
| VI | L. | R. | Q. | P. | O. | N. | M. | VII |
| XVII | .M | .L | .K | .Q | .P | .O | .N | VIII |
| XXVIII | .E | .P | .C | .B | .A | V. | T. | VIIII |
| VIIII | L. | K. | I. | H. | O. | N. | M. | X |
| XX | .M | .L | .K | .I | .H | .G | .N | XI |
| I | S. | R. | Q. | .B | .A | V. | T. | XII |
| XII | L. | K. | I. | H. | G. | F. | E. | XIII |
| XXIII | .E | .P | .K | .I | .H | .G | .F | XIIII |
| IIII | S. | R. | Q. | P. | O. | A. | T. | XV |
| XV | P. | C. | B. | H. | G. | F. | E. | XVI |
| XXVI | .E | .P | .C | .B | .A | .G | .F | XVII |
| VII | L. | K. | Q. | P. | O. | N. | M. | XVIII |
| XVIII | .M | .L | .K | .I | .P | .O | .N | XVIIII |

**Paschales litterae**

| | Ebdae cum diebus a natali domini usq. in XL mam. | Dies paschae. | Ebdae ab oct. pent. usq. ad advent. dni. |
|---|---|---|---|
| B. | Ebd. VI. Dies III. | XI K. April. | XXVIII |
| C. | — VI — IIII | X — | XXVIII |
| D. | — VI — V | VIIII — | XXVIII |
| E. | — VI — VI | VIII — | XXVIII |
| F. | — VII | VII — | XXVIII |
| G. | — VII — I | VI — | XXVII |
| H. | — VII — II | V — | XXVII |
| I. | — VII — III | IIII — | XXVII |
| K. | .— VII — IIII | III — | XXVII |
| L. | .— VII — V | II — | XXVII |
| M. | — VII — VI | Kal. April. | XXVII |
| N. | — VIII | IIII N. Ap. | XXVII |
| O. | — VIII — I | III — | XXVI |
| P. | — VIII — II | II — | XXVI |
| Q. | — VIII — III | Non. Ap. | XXVI |
| R. | — VIII — IIII | VIII Id. Ap. | XXVI |
| S. | — VIII — V | VII — | XXVI |
| T. | — VIII — VI | VI — | XXVI |
| V. | — VIIII | V — | XXVI |
| .A | — VIIII — I | IIII | XXV |
| .B | — VIIII — II | III — | XXV |
| .C | — VIIII — III | II — | XXV |
| .D | — VIIII — IIII | Idus Ap. | XXV |
| .E | — VIIII — V | XVIII K. Mai. | XXV |
| .F | — VIIII — VI | XVII — | XXV |
| .G | — X | XVI — | XXV |
| .H | — X — I | XV — | XXIIII |
| .I | — X — II | XIIII — | XXIIII |
| .K | — X — III | XIII — | XXIIII |
| .L | — X — IIII | XII — | XXIIII |
| .M | — X — V | XI — | XXIIII |
| .N | — X — VI | X — | XXIIII |
| .O | — XI | VIIII — | XXIIII |
| .P | — XI — I | VIII — | XXIII |
| .Q | Ebd. XI. Dies II. | VII K. Mai. | XXIII |

FIGURE 3.1  Hortus deliciarum, *fol. 319r, Computus tables*
PHOTO: WARBURG INSTITUTE

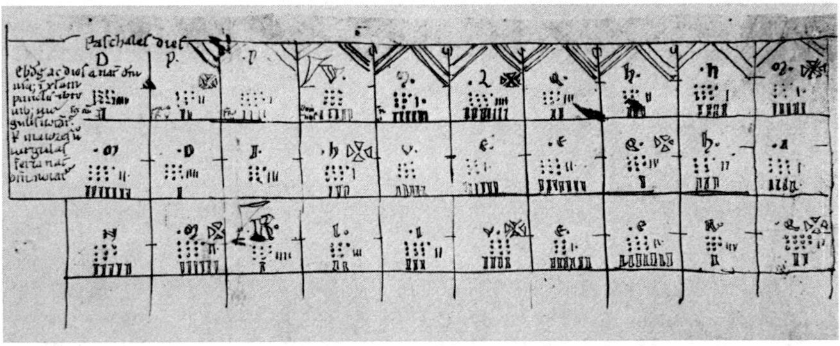

FIGURE 3.2  Hortus deliciarum, *fol. 319v, Computus grid*
IMAGE COURTESY OF THE BIBLIOTHÈQUE NATIONALE DE FRANCE

FIGURE 3.3   Hortus deliciarum, *fol. 8v, Fourth Day of Creation*
IMAGE COURTESY OF THE BIBLIOTHÈQUE NATIONALE DE FRANCE

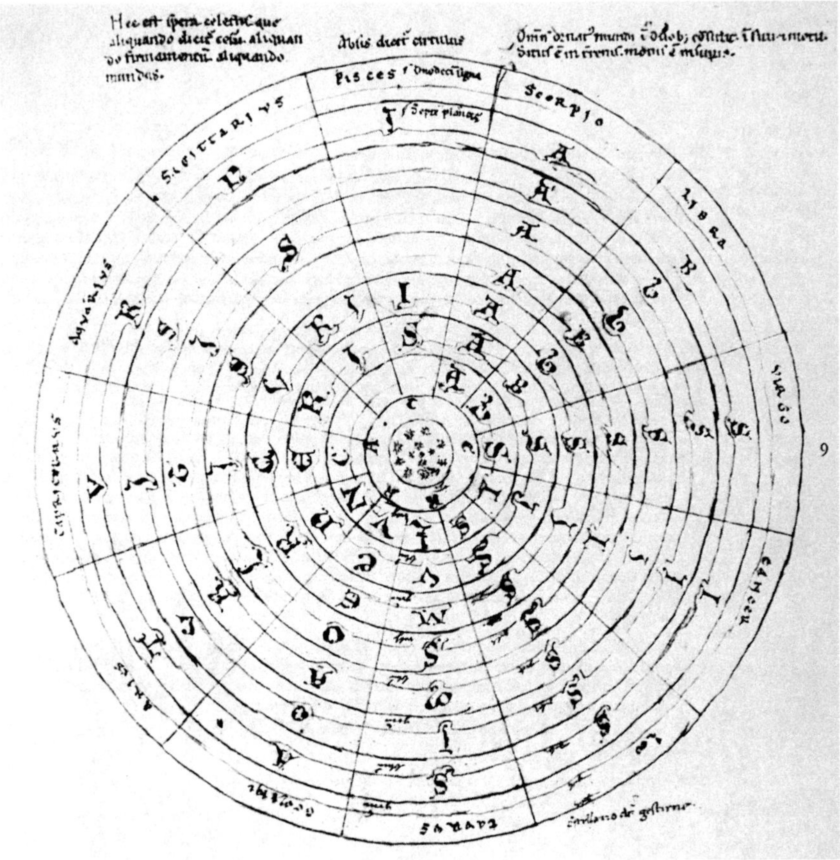

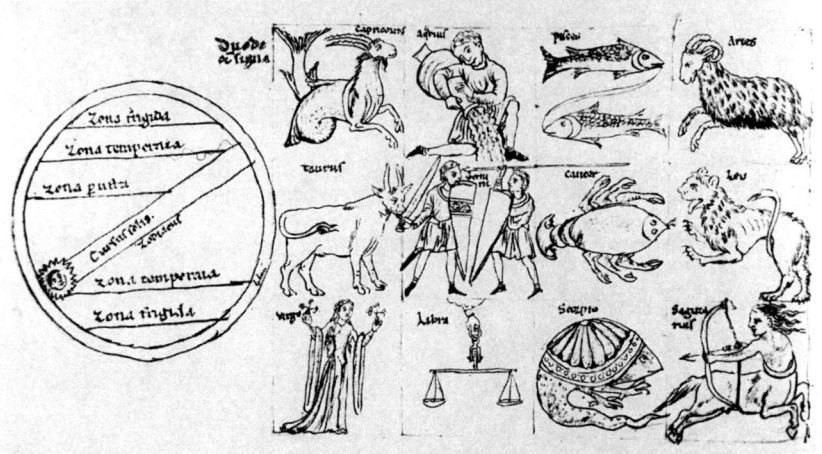

FIGURE 3.4　Hortus deliciarum, *fol. 10r, Diagram of the planets and zodiac*
　　　　　　PHOTO: WARBURG INSTITUTE

FIGURE 3.5　Hortus deliciarum, *fol. 11v, Diagram of the climate zones*
　　　　　　PHOTO: WARBURG INSTITUTE

FIGURE 3.6   Hortus deliciarum, *fol. 251r, Angel rolling up the scroll of the heavens*
IMAGE COURTESY OF THE BIBLIOTHÈQUE NATIONALE DE FRANCE

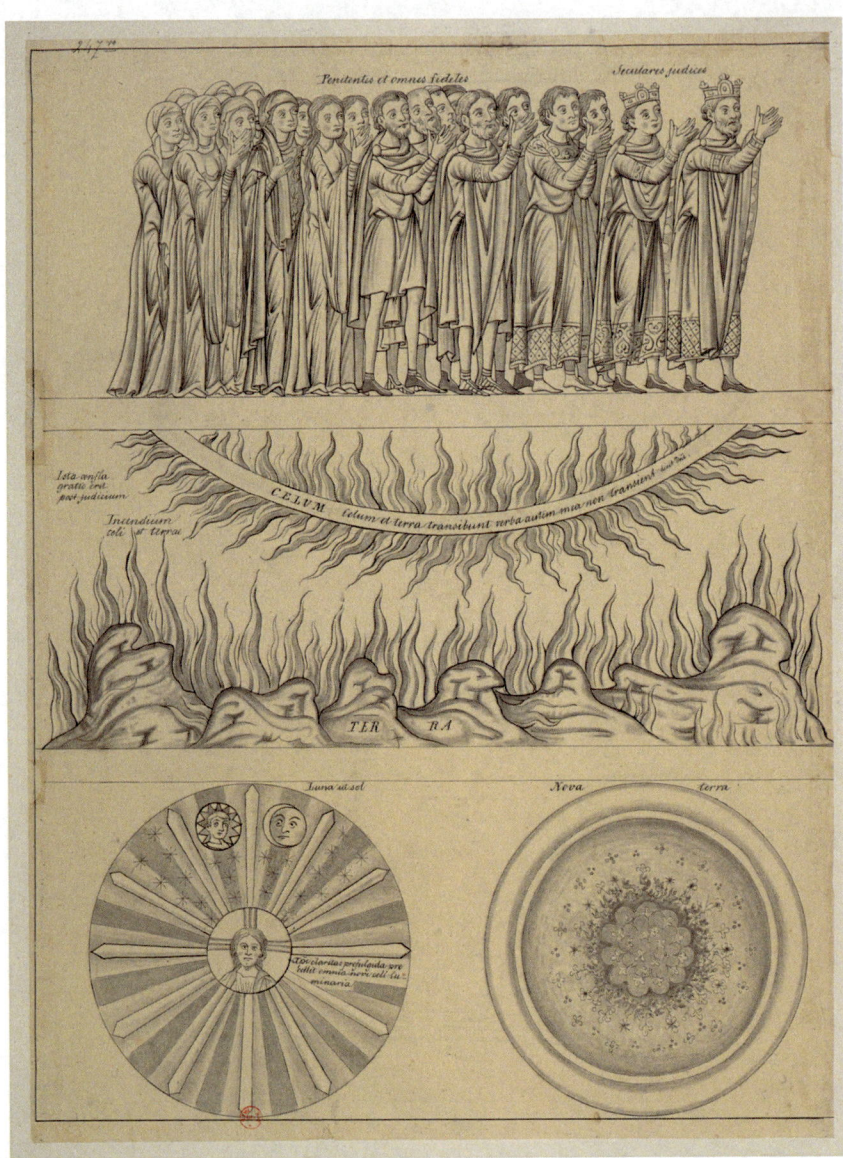

FIGURE 3.7   Hortus deliciarum, *fol. 247v, Conflagration; new heaven and new earth*
IMAGE COURTESY OF THE BIBLIOTHÈQUE NATIONALE DE FRANCE

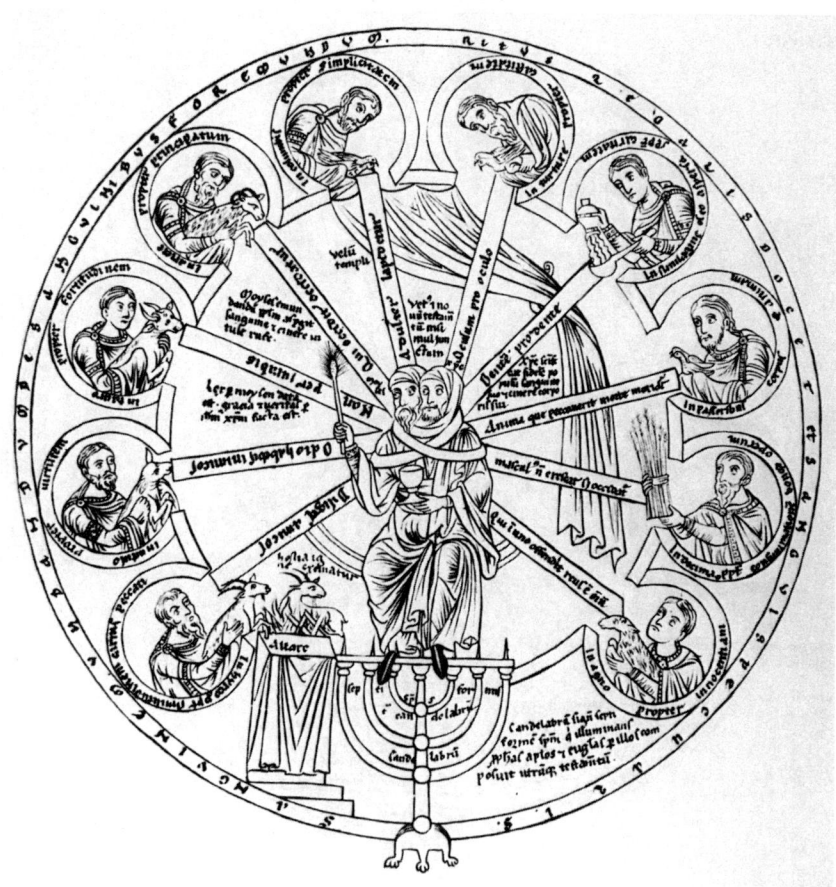

FIGURE 3.8   Hortus deliciarum, *fol. 67r, Old Testament typological rota*
IMAGE COURTESY OF THE BIBLIOTHÈQUE NATIONALE DE FRANCE

FIGURE 3.9   Hortus deliciarum, *fol. 67v, New Testament typological rota*
             IMAGE COURTESY OF THE BIBLIOTHÈQUE NATIONALE DE FRANCE

FIGURE 3.10    Hortus deliciarum, *fol. 8ov, Tree of Abraham*
IMAGE COURTESY OF THE BIBLIOTHÈQUE NATIONALE DE FRANCE

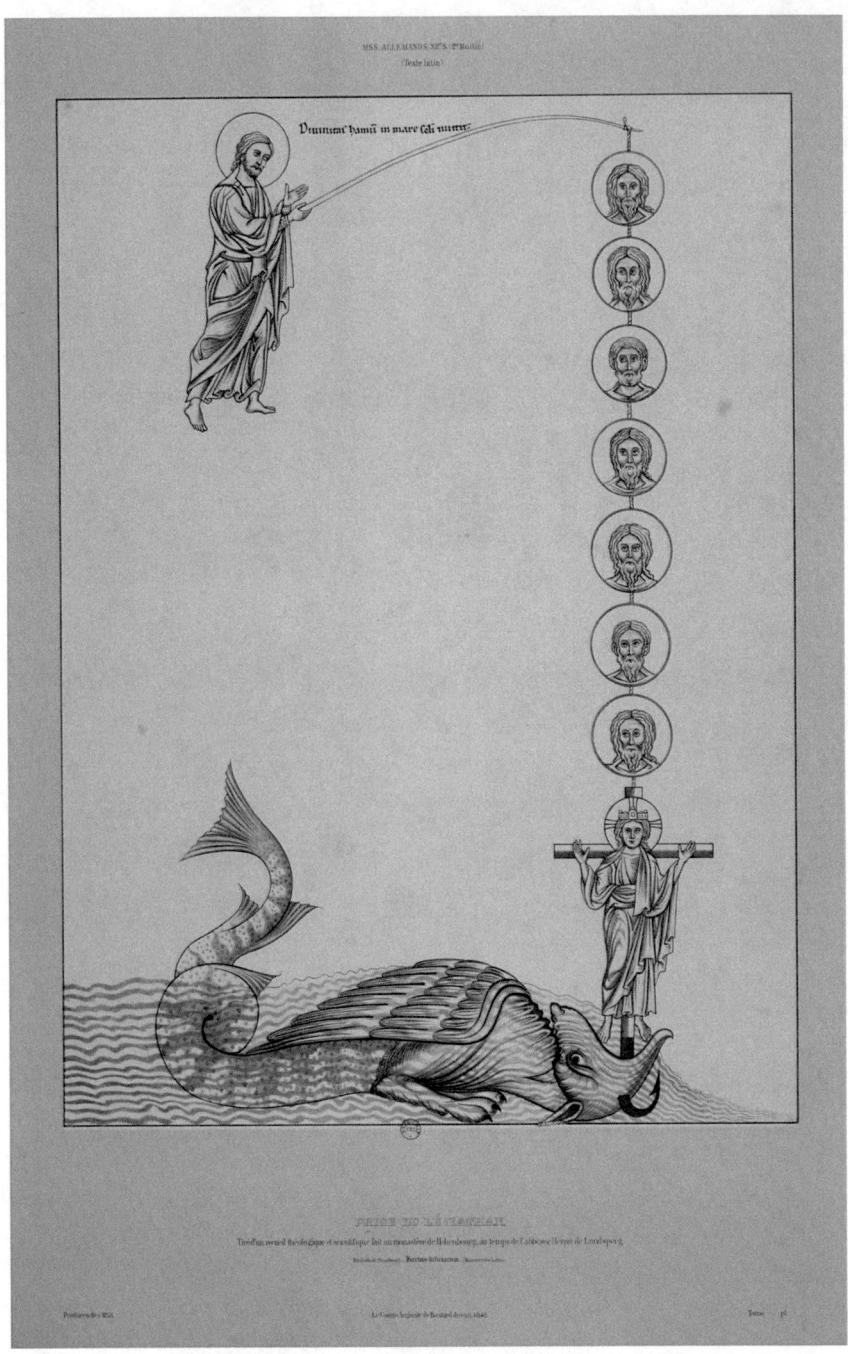

FIGURE 3.11    Hortus deliciarum, *fol. 84v, God Fishing for the Leviathan*
IMAGE COURTESY OF THE BIBLIOTHÈQUE NATIONALE DE FRANCE

FIGURE 3.12    Hortus deliciarum, *fol. 150r, Crucifixion*
IMAGE COURTESY OF THE BIBLIOTHÈQUE NATIONALE DE FRANCE

# CHAPTER 4

# Simultaneous Times: *Synderesis* and Its Musical Exemplification

*Nancy van Deusen*

Plato, in the *Phaedo*, translated from Greek into Latin ca. 1157, causes the narrator, Phaedo himself, to state, during the final discussion of Socrates' disciples just before their teacher's impending death: "In the first place, my own feelings at the time were quite extraordinary.... I felt no sorrow at all, and at the same time, I felt no pleasure at being occupied in our usual philosophical discussions. I felt an absolutely incomprehensible emotion, *a sort of curious blend of pleasure and pain combined*, as my mind took it in, in that in a while my friend was going to die." The subject matter is powerful, the situation intense, and both the priorities presented as well as the final outcome influential upon the Latin-reading public of the thirteenth century, as the translation of the *Phaedo* became available and eventually copied, and incorporated into principal libraries of Europe, the Sorbonne, Paris, and what is now the Vatican Library, Rome.

From the above passage, Plato continues, "For the doctrine that the soul is a kind of harmony has always had, and has now a wonderful hold upon me, and your mention of it reminded me that I myself believed in it before." To reinforce the point made at its onset, this "curious blend" of contraries, the "harmony" as "the last of all to be composed and the first to perish," the *Phaedo* contained an image, that of the harp/*lyra*, which incorporated, in its resolution, both *visible* material (the instrument itself made presumably of wood), as well as an *invisible* substance, that is, its "soulish" *harmonia*, which, as the *Phaedo* stated, was the last to be incorporated into the physical, visible, instrument, and the first to perish.[1] It is this concept of composite, including the

---

1  See Raymond Klibansky, *The Continuity of the Platonic Tradition during the Middle Ages. Outlines of a Corpus platonicum medii aevi* (London, 1939, with new prefaces and supplement, rpt. Munich, 1981), and Nancy van Deusen, "The Harp and the Soul: The Image of the Harp and Trecento Reception of Plato's *Phaedo*," in *The Harp and the Soul. Essays in Medieval Music* (New York, 1989), pp. 384–418; on the twelfth-century translation of the *Phaedo* and its influence, see. van Deusen, "Composite Harmony: An Aspect of the Conceptual Background to the Problem," in *Theology and Music at the Early University* (Leiden, 1995), pp. 113–126, esp. pp. 113–114. Cf. Plato's *Phaedo*, English translation, Hugh Tredennick (*The Collected Dialogues of Plato, Including the Letters*, eds. Edith Hamilton and Huntington Cairns [Bollingen Series

properties of both visible as well as invisible *materia-substantia* that is also the emphasis of Aristotle's *Physica*, newly available in multiple translations into Latin ca. the end of the twelfth beginning of the thirteenth centuries.[2] The two works, in many respects, reinforce one another, perhaps one of several reasons why they were both of such importance during the first two centuries of what is commonly viewed as the early reception of the Latin Aristotle.

Philip, known as "The Chancellor," for his position as liaison between the newly-founded faculty of theology at the University of Paris, and the episcopal see of Notre Dame, in his *Summa de Bono*, during the first decades of the thirteenth century, takes on this topic *of harmonia*, newly forged by conceptual impetus from the *Phaedo* and the *Physics*, eventually channeling this newly-constituted concept of *armonia* into a long-standing tradition of writing concerning *synderesis*.[3] He has much to say on this subject. Distinct from a

---

LXXI, Princeton, 1989]. Klibansky stated, in his 1939 publication, that the influence of Plato in the Middle Ages had yet to be assessed. This influence is difficult to ascertain, since Plato is very awkward to quote, due to the dialogue format. Hence, the reception of the *Phaedo* is more subtle than, for example, Aristotle's *Metaphysics*, beginning with its clarion-call, "All men by nature desire to know." The reception of Plato's *Phaedo* is discussed in both publications mentioned above.

2   See introduction to *Roberti Grosseteste Episcopi Lincolniensis Commentarius in* VIII *Libros Physicorum Aristotelis*, ed. Richard C. Dales (Boulder, CO, 1963), esp. "The Reception of the *Physics* by the Latins," pp. vi–ix, at p. vii: "The earliest citations of the *Physics*, though very slight and apparently not from first-hand knowledge, seem to have been made by Adelard of Bath and Thierry of Chartres before the middle of the twelfth century. By the end of that century it had been translated into Latin at least four times, three times from the Greek and once from the Arabic, and references to it became increasingly frequent," as well as Bernard G. Dod, "Aristoteles latinus," in *The Cambridge History of Later Medieval Philosophy* (CHLMP), eds. Norman Kretzmann, Anthony Kenny, Jan Pinborg with Eleonore Stump (Cambridge, UK, 1982), pp. 45–79, esp. "Table of Medieval Latin Translations of Aristotle's Works and of Greek and Arabic Commentaries," pp. 74–79. John E. Murdoch, in his contribution to CHLMP, "Infinity and Continuity," pp. 564–591, states that the *Physics* was the most commented upon of Aristotle's natural philosophical works through the first half of the fourteenth century (p. 565). It is important to observe that the *Physics*, especially in the early decades of the thirteenth century, makes its appearance in other ways than within the commentary format.

3   Cf. *Philippi cancellarii parisiensis Summa de bono*, edita studio et cura Nicolai Wicki, 2 vols (Corpus philosophorum medii aevi, Bern, 1985). Prior to the publication of this edition, also of much value because of its identification of Philip's quotations, and extensive bibliography, one finds very little pertaining to this seminal figure of the early thirteenth century (d. 1236), as the sparse bibliography, CHLMP, p. 878, indicates. Although Philip's *Summa* is not a commentary on the *Physics*, it is clear from the copious use he makes, particularly of the *Physics* and *De anima*, that one of his main intentions is to collate and reconcile Aristotle with Augustine, often also accessing Peter Lombard. The office of "chancellor" of the episcopal see

view of *synderesis* as "conscience," "*conscientia*," Philip describes *synderesis* as the "inner eye of the eagle," resolving contrary tendencies and directionalities into a convincing whole, composed of parts, which, in application, "with good effect" could also constitute a "work," "opus," or "*operatio*." This contribution to a volume on the topic of time and its structures will examine Philip's view of *synderesis*, which can only be described as a break-through at the time of his writing; but which also brings together, in Philip's extensive, skillful, and carefully-considered, treatment, conflicting opinions and tendencies covering many centuries, into *concordantia*. His concept of *synderesis* could also be best exemplified within music, the material and measurement discipline with the ministry of "making difficult things plain." And this was the case, in what was at that time radical new music exemplification of the first half of the thirteenth century.

The passages referred to above, taken from the *Phaedo*, bring up the subjects of the substance of the soul, the issue of contrary motions of impulse and intention, as well as the analogy of conflicting ideologies as expressed by the various and diverse characters who deliver their points of view throughout the dialogue—all of which finding resolution or harmonization within motion and time. Philip's attention to, and input into, the subject of harmonization within motion and time did not come from nowhere. It is not a chance coincidence that the Latin *Phaedo* became available to the reading public at approximately the same time that translations from Greek into Latin of the *Physics* also entered the main stream of discussion. It is true that Plato's pupil, Aristotle, is easier to quote than Plato himself; reference to Aristotle's works in many cases can be spotted almost immediately, and the continuation of discussion of the *Physics* concerning Plato's priorities elicited more commentaries during what is known as the high middle ages than any other of Aristotle's natural philosophical works. Aristotle's *Physics* takes up where the *Phaedo* left off, in an entirely different manner of presentation that made it amenable to lectures and learning, both used as a textbook, and with its text interspersed with commentary.

What insights can be gained from the *Physics*? How did these particular insights provide new dimensions to an older, or perhaps less nuanced, topic, that of *conscientia*? Why was the term, *conscientia*, not sufficient for Philip's

---

of Notre Dame, together with the office of chancellor of the emerging university of Paris (although a good deal of the bibliography given, and comments made, deal with this office during the later half of the thirteenth century), can be found in J.M.M.H. Thijssen, *Censure and Heresy at the University of Paris 1200–1400* (Philadelphia, 1998), especially pp. 8–11, p. 112, para. 2; pp. 125–126, n. 36.

project of bringing together Augustine, Chalcidius' translation of, and commentary on, Plato's *Timaeus*, as well as the coalescence of Plato with the newly-available Latin translations of Aristotle? How did Philip the Chancellor filter a past use of the term through what would have been an exciting new lens of Aristotle's treatises? Aristotle, in the *Physics* is concerned primarily with *materia/ substancia* and had this to say about what, by his own admission, was most difficult to understand.[4] *Materia* that has not been delimited, in other words, inchoate, contains motion, has similar propensities and potentiality, whether unseen or seen, and contains within itself diverse motions that move in contrary directionalities to, and against, one another. These concepts are not at all easy to understand, at least not on the surface of things, particularly taken in the abstract without exemplification—even when mulled over by generations of writers, as was, in fact, the case. The Aristotelian commentary tradition appears to extend nearly into infinitude, as evidence for the enormous interest in, and influence of, this particular writer and the importance of the topics he raises. This is true throughout the centuries, but, especially, from the early thirteenth century, when the *Physics* in its Latin translations began to receive increasing attention that continued to the printed books of the late sixteenth century.

From a discussion of *materia*, both seen and unseen, its innate *motus*, its *proprietates* and *potentia* that can be actualized and then brought to conclusion as *perfectio*, one moves on, in the *Physics*, to the next point of departure, namely from contrary motions, the harmonization of conflicting, otherwise mutually exclusive, intentions to *harmonia*, or simultaneous coalescence achieved by dint of will and work through, and within, time-lapse.[5] *Harmonia*,

---

4  Since Latin terms that were chosen to express the Greek are a focus here, the Latin terms are included. Another reason for doing this, rather than including the common English cognates, is that in many cases English cognates have lost their expressive power, are no longer as pungent, nor as comprehensive, as the Latin terms. Often the English term contains very little interest for a reading audience today. By contrast, an English term may be loaded with ideological accretion, such as material/matter/materialism. Harmony/*armonia* is an example as well of an English cognate that is now mostly devoid of its specific significance, which is not the case for the Latin term as it is introduced with particular focus in the thirteenth century. It would seem that thirteenth-century writers from Philip the Chancellor to Robert Grosseteste, Thomas Aquinas, and Roger Bacon—as well as others—were all, one way or another, concerned with understanding, translating, and adequately explaining this term. Cf. also van Deusen, "Roger Bacon on Music," in *Roger Bacon and the Sciences*, ed. Jeremiah Hackett (Leiden, 1997), pp. 223–241.

5  Aristotle discusses substance in Book I, proceeding to the notion of contraries (I.4.187[a] 20–189.[a]1) in which musical examples are crucial, as I.188[a]32–I.188[b]8: "But we must see how this can be arrived at as a reasoned result. Our first presupposition must be that in nature

emphasized in the *Phaedo*, is a preoccupation that is emphasized emphatically, not only by Aristotle himself, but by commentators on the *Physics* who sought to understand what The Philosopher meant by this term, and what conclusions could be drawn from it. It is probably no exaggeration to state that writers of the entire thirteenth century were set upon dealing with this concept.

Philip is no exception. Writing what appears to be his life's work on the topic of "goods," a *Summa de bono*, engaging himself in an early response to the availability of the Latin *Physics*, he begins his work with a short narrative, from the Old Testament book of Ruth: Vadam in agrum et colligam spicas que fugerunt manus metentium, ubicumque clementis in me patris familias reperero gratiam,[6] to which Philip responds that all of this is significant. As his commentary then unfolds, we are repeatedly introduced to terms and concepts that will signal the collection process the commentator has undertaken in bringing together a comprehensive Latin heritage with early patristic authors such as Jerome, and, especially, Augustine, in order to provide a background for understanding new tools for thought and articulation presented by Aristotle. Writing as Philip did, around the beginning of the second decade of the thirteenth century, the *Physics* would have just begun to cause the sensation it continued for centuries to generate.[7]

---

nothing acts on, or is acted on by any other thing at random, nor may anything come from anything else, unless we mean that it does so accidentally. For how could white come from musical, unless musical happened to be an attribute of the not-white or of the black? No, white comes from not-white—and not from any not-white, but from black or some intermediate. Similarly, musical comes to be from non-musical, but not from any thing other than musical, but from unmusical or any intermediate state there may be. Nor again do things pass away into the first chance thing; white does not pass into musical...but into not-white—and not into any chance thing which is not white, but into black or an intermediate, musical passes into not-musical—and not into any chance thing other than musical, but into unmusical or any intermediate state there may be." I.9.192.ª21: "For contraries are mutually destructive" to motion within substance discussed in Book II, especially 192ᵇ13–14, and change, VI.4, in which commensurability is discussed: "A difficulty may be raised as to whether every motion is commensurable with every other or not. Now if they are all commensurable and if things that move an equal distance in an equal time have an equal speed, then we may have a circumference equal to a straight line, or, of course, the one may be greater or less than the other."

There are three crucial points here: first, *substantia/materia* contains motion, second, motions may be, and often are, contrary to one another, and third, *natura/substantia/materia* are simply present, at the disposal of the one who possesses the capacity for, and will to, work with what is there.

6   Ruth 2, 2: "Let me now go to the field, and glean ears of corn after him in whose sight I shall find grace." "King James" English translation used, edition I, p. 3.

7   Cf. van Deusen, "On the Usefulness of Music: Motion, Music, and the Thirteenth-Century Reception of Aristotle's *Physics*," *Viator* 29 (1998), 167–187.

Philip is collecting or gathering together "grains of corn" from the Latin classical heritage as well as those from the Old and New Testament scriptures, and thus brings immediately into his discussion what will become a priority for him, as it was for Aristotle, namely, the nature and properties of *materia*, as well as how its properties could be worked with and brought to appropriate perfection. In effect, Philip's large-scale *Summa de bono* constitutes an excellent example of this very concept of collection, or gathering, of material as Philip brings together a large spectrum of sources. His own input into, and work with, this collection is in bringing together a convincing *concordancia* of allusions, passages, quotations, all expertly joined together by Philip who is obviously an expert at his craft. Philip was a relationship-builder, a joiner. His life's work was that of serving as liaison between the episcopal see of Notre Dame cathedral, still in the beginning stages of physical and structural construction, and the theological faculty of the newly-founded University of Paris. A good deal here was new: the cathedral building, the theology faculty, and the university, Philip's position, as well as the Latin translations and their reception of Aristotle. There is a sense of excitement, as well as a good deal of tension, that one can perceive in reading Philip's work; a spirit of inquiry and intellectual foment that underlie his writing. His was an extremely important task, specifically, of harmonization during some of the most crucial early years of that university's existence.

*Materia* becomes the underlying structure of Philip's panoramic work. Along the way, *in viam*, as he writes, Philip presents a tool-chest of terms as his priorities, instruments or *organa* for dealing with this topic of seen, unseen *materia/substancia*. The reason why this topic of *materia/substancia* is brought up time and time again is two-fold, one, as he writes, the Old and New Testament scriptures, to which Philip makes repeated reference, constitute a *colligata*, a stockpile of *substancia*. Secondly, *materia/substancia* is crucial to all of the major Aristotelian texts under consideration, that is, especially, the *Physics*, unseen substance and instruments for dealing with it a focus of the *Metaphysics*, and unseen substance in terms of "soulish substance" in Aristotle's treatise on the soul, *De anima*. As a background to, and reinforcing these treatises, Augustine's two treatises on the soul, concerning the qualities and quantities of the soul, his treatise, *De anima et eius origine*, as well as a treatise, *De spiritu et anima*, no longer thought to have been authored by Augustine, are also brought repeatedly into the discussion. (Philip's knowledge of nearly all of Augustine's writings was phenomenal, as the abundance of quotations evidence.) In view of its importance, also as a platform for further exploration, we meet the topic of *materia/substancia* repeatedly with a variety of treatments, orientations, directionalities, and goals, all brought together by Philip's immense reading—a project, that completed, gives the reader the impression, centuries after the fact, that it undoubtedly represented his life's work; but that it also brought together the two missions intrinsic to his career, namely, pastoral responsibility, undoubtedly also preaching activity, and his connection to the students in the newly-reconstituted and systematized discipline of theology at the Parisian university.

Next, Philip goes on to the topic of "goods," both silver and gold with their visible as well as unseen values. The invisible attracts Philip as he places it in a comparison. Silver pertains to intelligence, gold to wisdom; silver to faith, gold to loving kindness; silver to the perfection of the speculative intellect, gold to the perfection of what he names practical intellect, that is, the affections.[8] "Gold" is the place where these qualities come together. This gold has its source in principles, not in conventions or habits. One is able to acquire the desired unseen substance, given by analogy, of gold, only by conscious, knowledgeable, patient, perseverant, disciplined, work—attention to detail and perseverance that also overcomes the contrary impulses of laziness, arrogance, complacence, it would seem, and which also overcomes the heavy weight and the onerous task of slogging through "*materia*."

Now, Philip himself at this precise spot in the introductory section of his treatise does not in fact furnish the explicit connection and write, "As we all know, music as a material and measurement discipline that uses as its unseen substance motion, time, and sound, can therefore be understood as an example, standing in a proportional relationship to 'gold' since it is knowledge according to underlying principles, put into, and refined by, experience and assiduous practice." This was not necessary, and, I believe that he would have unpardonably insulted, or, at the very least been inexcusably banal, with respect to his reading audience with this reminder of what they all very well knew, since music had had its place as the illustrative, exemplary, discipline between the material and measurement disciplines and unseen substance since Roman education, as elucidated by Augustine in his *De ordine*, as well as

---

8   Edition I, p. 4; Argentum pertinet ad intelligentiam, aurum ad sapientiam, argentum ad fidem, aurum ad caritatem, argentum ad perfectionem speculativi intellectus, aurum ad perfectionem practici intellectus, id est affectus. De sapientia ergo que pertinet ad mores dicitur: "Auro locus est in quo conflatur," de intelligentia que pertinet ad questiones dicitur: "Argentum venarum suarum habet principia" (quoting from Job 28). "Silver" refers to the capacity of the mind to inquire, consider, remember, and articulate what has been considered (*anima*); "gold" refers to the unseen, often even unarticulated, "substance" of the soul, what exists on the deepest level of the personality (*spiritus*). This becomes clear in the biblical passages he chooses to make his points; but one of the problems in dealing with this entire nexus of considerations is the lack of consistency with which modern translations treat of what would have been rendered in Latin, *anima/spiritus*—a crucial distinction that Philip strove to maintain.

*De musica.* But, furthermore, most importantly, he would have violated expected and justified disciplinary methodology that placed basic principles first, fulfilled and informed by the material and measurement disciplines, culminating in the applications he was then making as he applied them all to scriptural interpretation and the commentaries of so many authors, read and absorbed by himself. As he states, "The good that we intend principally to treat pertains to theology."[9]

From the observation that gold is good, Philip goes on to what all "goods" have in common. Human beings like goods. They have an appetite for them. In fact, one can observe what is almost an insatiable appetite for *materia/substancia* in nearly every circumstance in which it presents itself. What then, is to be desired above all? He asks a timely and important question, and from this he proceeds to the ultimate interrogation of goods compared to, and coherent with, truth. Some apparent goods are truly good, others are not, and that is a major difference between them, it would seem. Goaded on by his reading of the Latin *Metaphysics*, Philip then delves into a comparison of "good" to "truth," where these two come together, as well as going into the distinctions between them. Here, as in many cases, Philip begins with Augustine, in this case, the *Soliloquies*, and Augustine's refreshingly succinct statement, "Truth is that which is."[10] Since such a statement is not entirely unequivocal, Philip has, by

---

9    Amos Funkenstein in his *Theology and the Scientific Imagination from the Middle Ages to the Seventeenth Century* (Princeton, 1986), pp. 3, 6, calls attention to the fact that deep-rooted differences in education and intellectual culture can be attributed to what is connected and what is separated within disciplines. Cf. also van Deusen, *Theology and Music,* pp. xii–xvi. In Philip's case here, he had gone beyond the input of the "material and measurement disciplines," that is, those that explain particularity (arithmetic), relationship (geometry), and the nature of motion (physics), as well as the exemplification of particularity, relationship, and the nature of motion specifically in music. His lengthy treatise concerns itself then with the next progression, the application of particularity, relationship, and motion to the unseen substance of the nature of God (theology). Funkenstein rightly points out categorical distinction within, and between, Aristotle's treatises, each one of which treating of a separate discipline, but has failed to see that each one of the disciplines exemplified basic principles, such as Aristotle's *theoria* model presented in the *Metaphysics.* Ultimately a unified system of learning emerged. A foundational principle articulated in the *Metaphysics* was accordingly viewed through, and understood by, a distinctive disciplinary window, such as through material and motion in the *Physics,* within the structure of a dramatic time-lapse or theater piece in the *Poetics,* and through the course of human action in the *Nichomachean Ethics.* This structure is, I believe, central to the understanding of Aristotle's influence, also in determining the organization of the early university, points made more extensively in the introduction to *Theology and Music,* pp. ix–xvi.

10   Edition, I, p. 9: Sequitur de comparatione boni ad verum, in quo conveniant et in quo differant. Augustinus enim de hoc multum agit in libro De vero et bono. Videtur quod non

taking on this particular tack, thrust his *Summa* into the mainstream historically of the most basic considerations of the human spirit.

This was obvious, as well, to Philip himself. He follows Augustine's statement with several other points of view, those of Hilary, Anselm of Canterbury, then again, what he thought Aristotle had included in the *Metaphysics*: "Truth is indivisible, and is what it is."[11] And so, writes Philip, we will see how all of these definitions convene, and what manner of movement (*modus*) can be assigned to them. This is precisely what Philip then proceeds to do within several hundred pages of his *Summa de bono*. He begins with the concept of the indivisibility of "truth," quoting Boethius that truth is simple, but also a composite, in which truth is also inseparable from potentiality, with possible contingency then to its goal or final end. In this, he brings up the question of movement (*passio/modus*) within physics, as either activity or of being acted upon, concluding with the statement that that which exists in God as maximally indivisible, as well as maximally truthful, therefore maximally good, also as indivisible from what is potential, is indivisible from its end.[12] Philip

---

sit differentia, quia verum et bonum convertuntur. Omne enim quod est, in eo quod est verum est, et in eo quod est bonum est. Ergo si est verum, est bonum et converso.

Item, Augustinus in libro Soliloquiorum: "Verum est, inquit, id quod est." Cf. this entire section, pp. 3–9, as well as the succession of quotations carefully documented, not only by Philip himself within his text, but by the editor: the Old Testament Book of Ruth, Jerome, Seneca's letters, Rhabanus Maurus, Paul's second epistle to the Corinthians, the Psalms, Isaiah, Job, Ecclesiastes, the New Testament books of 1 Timothy, Romans, John's Gospel, Genesis, Exodus, Luke's Gospel, the Old Testament prophet, Amos, Aristotle, *Nichomachean Ethics*, the *Ethica nova*, Pseudo-Dionysius, *De divinis nominibus* in John the Scot's translation, Aristotle, *Metaphysics*, and Augustine's *Psalm commentaries, Soliloquies, Enchiridion*—all, several times more than once, within the first introductory passages of his treatise, which, nevertheless, in spite, or because, of their panoramic expanse, provide the foundation for his subsequent argumentation.

11   This section closes with a discussion of appetite versus fear concluding with the above quotation (p. 10): "Verum est id quod est" (Augustine), as well as "Verum est declarativum aut manifestativum esse" (Hilarius), "Veritas est rectitudo mentis sola mente perceptibilis" (Anselm of Canterbury), concluding with the *Metaphysics*: "Veritas est indivisio esse et eius quod est," which, the editor notes, is actually not to be found in the *Metaphysics*.

12   Edition, I, p. 11: Illa autem diffinitio: veritas est indivisio esse et eius quod est explicatur per hoc quod dicitur a Boethio in libro De hebdomadibus: "Omni simplici idem est esse et quod est, omni vero composito aliud est esse et quod est." ...Sic ergo summa veritas est indivisio actu et potentia, re et ratione. Unde dicitur in ea esse est id quod est. Unde Hilarius: "Esse non est accidens in Deo, sed subsistens veritas." In aliis vero in quibus esse et quod est differunt, est veritas circa compositionem. Sed in quibusdam est re separare, in quibusdam ratione, in quibusdam potentia. In quibusdam est re separare, ut ubi

here is referring to Aristotle's *theoria* model, found within the *Metaphysics*, exemplified by motion within material in the *Physics*, namely, that what is constitutes a continuity with what it is becoming, eventually arriving, step by step, increment by increment at a *finalis*, an end-goal, prepared one step at a time by what had happened previously. Philip has much more to say about both motion and the *theoria* model, but what is interesting here is that an exemplification within the study of physics (*quod passio in phisico motu*) leads to comprehension of a godly principle, God's definition of truth and good. In this opening discussion Philip brings up Latin terms/concepts such as *convenientior* (I,10), *consonat* (I,11), *simplici* (I,11), *coniunctio, finalis, mensuratur, per participationem, medium, sicut unum est tempus quo omnia temporalia mensurantur, proportionaliter* (I,12); *consonare, planum, secundum proprietatem, secundum proportionem, secundum proportionem diversis, convenientius in simplicibus, simplex, affectus* (I, 13), *partes* (I,14). These are concepts that had become important in reading and attempting to understand newly-available translations especially of Aristotle's *Metaphysics*, and *Physics*, that would also receive exemplification within writing concerning music. One can observe, furthermore, that convening and measuring specifically, of *times*—measurable, simultaneous, temporality—has followed directly upon his presentation of "truth" and "good," arguably during the most important opening foundational phase of Philip's work.

Time, motion, and concurrence are Philip's priorities, having begun his *Summa* with the concept of movement *in via*, motion within *materia*, these considerations apparently occupied his mind.[13] In the first volume of his

---

utrumque per se subsistit vel alterum sine altero, ut lux est aer vel actio est id quod passio in phisico motu, agendo enim patitur; in quibusdam ratione, ut in diffinitione et diffinito; in quibusdam potentia, ut in omnibus contingentibus que sunt possibilia ad finem. Propter hoc cum idem sit quod est et esse in Deo, maxime ibi est indivisio et maxime veritas, sicut maxima bonitas quia indivisio potentie et finis.

13    The number of times this topic is raised is too vast to include all cases. He writes of: *modus* divisible according to *modi* (I, 68), triplex modus cognitionis (I, 81), motus sed mensuratur dilectio secundum (I, 91), diverse motions (I, 77, 79), mobilia diversa (I, 108), multis intentionibus (as motions, I, 172, 233–235) motus respondens (I, 92), Queritur de eo quod dicitur substantia mobilis...omnis enim motus fit propter carentiam...mobiles sint natura...Sicut motus octava (I, 106), circulariter moventur circa thronum maiestatis (I, 107), movetur vel contraria in eodem; quies (I, 107), movetur...secundum partem, in loco (I, 108), in loco, sed in loco non est proprie nisi corpus, mobile in tempore (I, 108), secundum accessum et recessum, in motu locali (I, 109), moventur composita ex partibus quantitativus (I, 109), hoc modo moventur (I, 109, 110), per accessum et recessum ad contrarium (I, 110), de divisibilitate secundum motum localem (I, 110), moto subiecto (I, 110), mobilia

*Summa* alone, there is divisible, yet measurable motion, *motus contrarietis*, *motus respondens*, *substantia mobilis*, contrary motion, contrary motion according to part(s), in place, motions of time as a contained body within place between boundaries (*mobile in tempore*), composite motion according to quantified parts, rising and falling contrary motion, divisible motion, proportional motion according to what is first and what follows (principles of prior/posterior), simultaneous times, motion/time relationship, measurable motion, motion within time, composition/*motus*—and many other cases of the bringing together of motion within, and coincident with, time throughout both volumes of Philip's lengthy *Summa de bono*. Although the *Summa* is not a commentary on the *Physics* of Aristotle, the subject matter of material, motion, motion within time, and contrary motion resolving within motion and time are clearly priorities for this author, as they are within the *Physics*.[14]

This is what Philip has to say. *Materia/substancia* contains within itself motion, and within this motion, the propensity for contrary motions, motivations, or inclinations.[15] This concept of contrary motions, or contrary

---

habent proportionem ad locum multipliciter et per prius et posterius (I, 110), movetur...in aliud nunc tempore intermedio (I, 111), mensura actionis, passionis (I, 112) simul tempore (I, 113), secundum tempus quod determinetur ex motu alicuius corporis (I, 125), mensura/motus, motus/tempus (I, 130), motus cum tempore, tempus vero et motus propter mutationem de non esse in esse, motus autem ponderosi est ad medium (I, 132) illa autem non faceret vespere et mane, si esset immobilis, si vero erat mobilis (I, 144), tempus/motus, circulantus/motus (I, 145) actus (I, 161), divisionum pertinentium ad motivas, irasci motus dicunt consequentes (I, 162) de cognitiva quod primus motus est intellectus (I, 163), motivo interiori quod est intellectus (I, 164), sunt enim duo moventia appetitum intellectus et phantasia (I, 166), compositio/motus, and many more.

14    Edition index indicates Philip's quotation of Aristotle's *Praedicamenta*, *Perihermeneias*, *Analytica priora, posteriora*, *De sophisticis elenchis*, a large number of quotations from the *Physica, De caelo, De generatione et corruptione, Meteorologica, De anima* (also a large number of quotations), *De sensu et sensato, De memoria et reminiscentia, De somno et vigilia, De historia animalium, De partibus animalium, De generatione animalium, Metaphysica* (large number of quotations), *Ethica Nicomachea* (also large number of quotations) and the *Magna moralia*, showing Philip's reading of, as well as indebtedness to, Aristotle's treatises, particularly the *Physics, De anima*, the *Metaphysics*, and the *Nichomachean Ethics*; but indirect reference to, indeed, the infiltration of, these newly-available Aristotelian works is even more significant. It would appear that Philip had absorbed the major issues presented in the *Metaphysics*, exemplified in the *Physics* and the *Ethics*, and the issue, for example, of *theoria* as articulation of step-by-step process leading, even by means of contrary motions, to simultaneous resolution permeates his *Summa de bono*.

15    Edition I, p. 68 at 110. Cf. also edition, I, 4. There would seem to be an implied equivalence of seen-invisible *materia*. Both Augustine and Aristotle insist on the delimitation of the

inclinations, is a concept that can be presented within the process of intellection, as physical, directional motions, applied ultimately to life experience in terms of the juxtaposed, contrary oppositions of fear/self-interest, and lust/irascibility. In appetites that are juxtaposed in contrary opposition, there is no possibility for free will, that is *liberum arbitrium*, or, for that matter, any choice at all, as one is driven by either fear/self-preservation or lust.[16] As an example of these two opposing motions or contrary motions, intellection is, generally speaking, without fantasy, and vice versa. *Intellectus practicus* as motivational energy has a power to arbitrate, to choose, to be liberated from, the conflicting and contrary motions inherently contained within *materia/substancia*.[17]

A reconciliation of opposing intentional directions is definitely what is called for here, both by Aristotle and Philip. Of interest, however, is that this investigation of 1) truth/good(s)—2) *materia*/motion/contrary motion—3) *armonia* through time and motion in Philip's line of thought and discussion, as he attempts to explain what it is that free will does within the human spirit, produces for him a term that had, certainly, a past, but is employed in a different sense, and for a different purpose. Philip brings up, in context after context, the concept of *synderesis*—an inner spark (*scintilla*), a quiet fire, a stable flame—within the spirit, reposing within, yet leading out (*ducere/ductus*) to the desired effect. Some have named *synderesis* "conscience," "in that conscience itself does not induce, or lead to good effect, but rather, serves as remonstrance, thus achieving a certain stability within the soul."[18] The term,

---

body as a crucial differentiating factor; that is, "soulish substance" contained within the body, with the delimitation of the body, is corporeal. Containment, therefore, not *substancia* is the issue. This, I believe, is also the implication of the use of *forma*, that is, containment. See also the discussion of *silva* within Chalcidius' translation of Plato's *Timaeus* in van Deusen, "In and Out of a Latin 'Forest:' the *Timaeus latinus*, its Concept of *silva*, and Music as a Discipline in the Middle Ages," *Musica disciplina* 53 (2009) 51–70.

16    Edition, I, 94.

17    *Intellectus practicus*: id est affectus (1, 4). Concerning *phantasia* and *intelligentia practica*, cf. I.94, within lengthy discussion, *De libero arbitrio*, I, 93–99. It is important to observe that a proportional relationship exists both for Philip as well as the authors he quotes, such as Augustine and Aristotle, between invisible and visible *substantia*.

18    Edition, I, 94: angels are not subject to the contrary motions of appetite, that is, *phantasia* and *intelligentia practica*. The passage brings out all of the terms/concepts discussed above dealing with whether or not angels have "flexibile" free will: Unde ut ad predicta solvamus, dicimus quod angelus habet liberum arbitrium secundum flexibilitatem. Ideo dicit Remigius [recte Nemesius?] 'Factum est liberum arbitrium hominis vertibile sicut et angeli, divinum vero invertibile.' Ratio autem huius est quia conveniebat ipsum potius in

conscience, *conscientia*, would certainly have been available to Philip, had he required it for his discussion.[19]

What is necessary then is what Plato has also presented in the *Phaedo*, namely, the achievement of *armonia* through motion and time, through arbitration and experience, as well as through perseverance, courage, and mastery. A reconciliation of opposing intentional directions is what is called for, by Philip, Plato, and Aristotle. But Philip the Chancellor, in his attempt to explain what it is that "free will" (or a will that has been freed), does within the human spirit, comes upon and proceeds to bring up, in context after context, the concept of *synderesis*, which had a particular connotation for him, but which appears to be difficult to explain with the proper nuance, as the concept, proper to a language other than Latin, was complex.[20]

In an opening extended discussion of *synderesis*[21] Philip states that men, women, even angels, are motivated by inner driving forces by their very natures (*naturaliter*), to desire, love, and to be active—to do things. *Synderesis*, however, can be considered to be an angelic power, that is, intrinsic to angels, in that it is somehow contingent upon, yet above and beyond, reason. This, apparently, is an energy/potency (*vis*) that is neither driven by desire, nor calculating according to some perceptible rationality, but rather, a "spark" that in human beings, amazingly, did not die during and after Adam's fall. Philip then lists a catalog of opposing arguments not only of Peter the Lombard, but Augustine, giving evidence that he had been thinking about *synderesis* for a long time, was acquainted with a good deal of Latin literature on the topic, and, moreover, that he thought it was crucial. His response here: "It is a spark attached as

---

hoc statu creari quam in altero; aliter enim non mereretur. Maior autem est gloria unicuique cum confertur ei bonum ex meritis quam cum sine meritis.

19    Cf. Aristotle, *De anima*, III.5, 430ᵃ18, edition pp. 97–102, at 102: Scintilla hec conscientie otiosa esset, cum nullatenus ad effectum duceret. The spark, that is, conscience, is useless, as in no case does it lead to an effect. Conscience, therefore, is not a dynamic concept, as required by the use of *ducerent*. *Ductus* is the translation into Latin of *theoria*. Cf. van Deusen, "Figura, Ductus, Conductus: Thirteenth-Century Discussions of Rhythm in Context," in *The Harp and the Soul. Essays in Medieval Music* pp. 307–328, ibid. "Ductus, Tractus, Conductus: The Intellectual Context of a Musical Genre," in *Theology and Music at the Early University*, pp. 37–53, and ibid. "On the Usefulness of Music: Motion, Music and the Thirteenth-Century Reception of Aristotle's *Physics*," *Viator* 29 (1998), 67ff.

20    Greek: possible etymology, cf. Timothy C. Potts, "Conscience," CHLMP, pp. 687–704, at p. 690.

21    Edition, I, pp. 101–102; pp. 98–101 rely upon Chalcidius' commentary on Plato's *Timaeus*, particularly with respect to the concept of *differentia*.

much to the intellect as to the affections, and it is duplex, as in free will, in that both intellect and the affections or desires are engaged. Even diabolical powers are not able to extinguish this reciprocity between them."[22] Philip concludes here that *synderesis* is an equipoise, as conscience is also considered, in which contraries within the human spirit are brought into juxtaposition as contrary motion, and resolved; the decision-making capability is thus preserved and unified. This constitutes a harmonization of opposing impulse, namely, force and energy are reconciled with stability and immutability. Although it is doubtful that Philip himself had ever actually seen this quality operative in angels themselves, it can be imagined within the invisible, "to the naked eye," domain of angelic beings.

Philip here is invoking a distinction that he takes up in much more detail later on in his *Summa*, that between mind (*anima*) that includes cognition, thought, verbal articulation, the reception and identification of what is seen, and soul (*spiritus*), tracing a progress from principle (*lex*) to power/dynamic (*vis*); from a situation "beyond reason (*super rationem*)" to a "part containing this spark" (*pars scintilla*). What is this faculty, the mind? (*Quid est anima?*) Where, exactly, is this unseen domain where choice, resolution, harmonization of conflicting motions of intentions are simultaneously accomplished? What indeed is that "part," and where can it be found, that remains quietly directional, calmly composed, achieving then "good effect?" This "achievement" also separates, to Philip's mind, *conscientia* from *synderesis*, since *conscientia* is simply resident, so to speak, not necessarily leading to an outcome, that is, resolution of contrary motions. Invisible, yet this "part" can be exercised; it is also responsive to practice and accumulated experience.

A sentinel is on the watch; an "immobility that proceeds from grace," as Philip has it, presiding over "the diverse powers of the mind," as well as the contrary motions and properties resident by its very nature within the unseen substance or "soulish material," (*materiam spiritualem*) containing differences (*differentiae*) of potentialities of cognition and motivation. "It is manifest, first of all, that free will is not habitual, but rather potential, and that the powers of the intellect are also separate from one another, are, in fact, at contrary purposes to one another."[23] What is needed, then, is release from the contrary

---

22    Edition, I, p. 102: Respondeo. Est scintilla quantum ad intellectum et quantum ad affectum et hec est duplex: quantum ad intellectum et affectum in libero arbitrio et quantum ad intellectum et affectum in synderesi. Illa quidem que est secundum liberum arbitrium quantum ad intellectum in diabolo non extinguitur. Philip proceeds (p. 103) to discuss *De ymagine*.

23    Aristotle asks the same question, "Quid sit anima?" at the beginning of his work on that subject, *De anima*. (It is the essential opener to a disciplinary inquiry, seating what follows

purposes, one to the other; to be freed from "innate impetus" or the "appetities" inherent within *materia*, both visible and invisible. This is a medial, compositional, potential that simultaneously frees the inner substance of the "spirit" to make definitive, formative, decisions. It is the reconciliation of opposing inclinations, as well as the resolution of the verbal with the non-verbal within motion and time. Neither given over to the fantasies of sense experience or coldly calculating what might be under certain circumstances advantageous, this special sentinel, the medial "part" is able to inquire quietly, come to appropriate conclusions, all the while preserving an inner balance and harmonization of the contraries: dynamism with quietude.

"Now it came to pass in the thirtieth year, in the fourth month, in the fifth day of the month, as I was among the captives by the river of Chebar, that the heavens were opened and I saw visions of God," wrote the prophet Ezekiel, to whom Philip has recourse at this point. "And I looked and behold a whirlwind came out of the north, as a great cloud and a fire unfolding itself, and a brightness was about it, and out of the midst thereof as the color of amber, out of the midst of the fire. Also out of the midst thereof came the likeness of four living creatures. And this was their appearance; they had a likeness of a man. And every one had four faces, and every one had four wings. And their feet were straight feet, and the sole of their feet was like the sole of a calf's foot, and they sparkled like the color of burnished brass. And they had the hands of a man under their wings on their four sides; and they four had their faces and their wings. Their wings were joined one to another; they turned not when they went; they went every one straight forward. As for the likeness of their faces, they four had the face of a man and the face of a lion on the right side; and they four had the face of an ox on the left side; they four also had the face of an eagle...As for the likeness of the living creatures, their appearance was like burning coals of fire, and like the appearance of lamps; it went up and down among the living creatures and the fire was bright and out of the fire went forth lightning."[24]

---

within predictable disciplinary "space," which is also the opening question announcing a treatise within the material and measurement discipline of music, "Quid est musica?") Augustine, as Philip points out, answers the question by stating that "anima est substancia rationis" (I, p. 156). Cf. Edition I, pp. 155–165, which includes sections, *Quid sit anima, De potentiis anime, De differentiis potentiarum cognoscitivarum et motivarum,* in which "phronesis," as "interius dispositum sermonem, ex quo post aiunt provenire sermonem per linguam enarratum," is examined, *De libero arbitrio. Synderesis,* pp. 162–163 (163), is distinct from free will: Item, dividitur liberum arbitrium contra synderesim.

24    Ezekiel 1,1–14 (quoted in English translation, "King James Version"); cf. edition I, p. 158.

Ezekiel's vision was transformational. The prophet was surely never again the same. The importance of this narrative for Philip was such that he quotes it several times, in fact, returning to Ezekiel as a unifying marker by which to present and explain the simultaneous, multiplex, facets of *synderesis*. This particular context of the inward eye of the eagle deals with spiritual visions, of the resolution of oppositions such as "unspeakable words," of intellect and cognition combined with "fantasies as well as imaginations on account of spiritual visions."[25] The stable center of dynamism, the result of process or step-by-step motion to an end-goal, *synderesis* is enclosed in what this author names a "medial part" within the invisible substance of the soul.

Philip gives every indication that he is aware that he is dealing with questions that have a long history of inquiry, with many opposing answers. Many, especially Augustine, Aristotle, Jerome, Gregory the Great, Peter Lombard, had contributed to the store of comment that Philip brings to the fore. Philip here has not only presented the background to a consideration of great importance, that is, the substance of the soul, but also a progression of thought, as well as a terminology to describe and understand the inner workings of the human spirit. Taking as a given that one is speaking of the *materia/substantia* of the soul, identifying and proceeding through separate components from sense (*sensus*) to imagination (*imaginatio*) to opinion (*opinio*) to intention (*intentio*), on to intellect (*intellectus, intellectus permanens*) to application in mind/word (*figurans animam/applicatio*), he arrives at "volition with counsel" (*voluntas cum consilio*), that is counsel directed toward an end or interior perfection (*perfectio interioris*). This interior spiritual vision, seen by the unflinching eye of the eagle, is a vision of the appearance of burning coals of fire, of lamps, "and the fire was bright and out of the fire went forth lightning."[26] This "fire" is the *scintilla* of *synderesis*.

One would think that at this point Philip would have resolved the matter of *synderesis*, at least to his own satisfaction, and considered the matter closed. This, however, is by no means the case. Philip is just getting started on the topic, opening it up to the issues of complexity he was obviously convinced it deserved, and providing by means of the narrative from the Prophet Ezekiel, a template upon which to examine each issue, one by one. One hundred pages later in the edition of his work he goes after this question with an astonishing thoroughness, giving indication of the importance of the term for himself and

---

25  Edition I, pp. 158–159 at 159: phantasiam autem sive ymaginationem propter visiones spirituales...

26  Ezekiel 1,13; edition I, pp. 158–160. The four faces are also the four gospels, man-Mathew, the lion-Mark, the ox-Luke, and the eagle-John, that is, the *evangelium*.

his project—its nuances, as well as its multifarious connotations. His examination goes far beyond either the treatment, which he quotes several times, in Jerome's commentary on Ezekiel, or Peter Lombard's ensuing remarks.[27]

First, in this extended discussion of the topic, Philip goes after the topic of mind and soul, *anima* and *spiritus*.[28] A running context of scriptural passages provides both the structure and just the nuance of the attributes of *synderesis* that he is not only bringing out, but reconciling into one unified, comprehensive concept. It is worthy of note that eighty pages of the modern edition are used to bring out the following topics, one after the other. From the difficult question of what *anima* comprises, beginning his discussion with the opening question that indicates the disciplinary space of the investigation, "Quid est anima," Philip takes up the topic of simultaneity, whether what can be named "soulish substance" infused the body simultaneously, that is, *insimul create* or *simul create et infuse*. Referring to Augustine's commentary on Genesis, he states that there are eight modes of connection with the body (*Octava de modo coniunctionis ad corpus*), with a ninth existing "in place and in time" (*Nono de esse eius in loco et in tempore*). *Anima* is *substantia*, but incorporeal; and, like a spider as a sentinel residing in the middle of its web, with "little motion within its own part," so *anima* is resident *in medio* within the body, as the heart, taking hold of the motion occurring in the body, from whatever part, both extrinsically and intrinsically. This is a "beautiful example" as Philip states, from Chalcidius' commentary on Plato's *Timaeus*.[29] But this relationship of less

---

27    This discussion has also been missed by subsequent investigators, such as O. Lottiin, *Psychologie et morale aux XIIe et XIIIe siècles* II (Gembloux, 1948) as well as T. Potts, *Conscience in Medieval Philosophy* (Cambridge, UK, 1980), and in his treatment of *synderesis*, "Conscience," CHLMP, pp. 687–704, who relied almost exclusively on Lottin. Other authors, such as J.B. Korolec, "Free Will and Free Choice," CHLMP, pp. 629–641, who take up, or at least mention *synderesis*, likewise base most of their remarks on Lottin. John Marenbon, in his chapter on "Contempt, law and conscience," in *The Philosophy of Peter Abelard* (Cambridge, UK, 1997), pp. 265–281, refers to *synderesis* once in his index (p. 275, n. 30), "In fact, by treating conscience as the ability to apply general rules to particular situations, Abelard anticipated the later medieval discussion, although the later thinkers regarded the workings of conscience as a process of reasoning which could err, and they contrasted it with the infallibility of our grasp of the fundamental principles of natural law (they called this grasp *synderesis*)." It should be observed, however, that the complete edition, with its thorough documentation of both scriptural as well as commentary quotations, of Philip the Chancellor's *Summa de bono* appeared in 1985, after the publication of CHLMP (1982).

28    Discussion in the English language is impeded by the confusion and conflation of mind, spirit, soul, a confusion that is avoided in the Latin language.

29    Edition I, p. 156 at 19–24: Diffinitur autem anima a Remigio sic: "Anima est substantia incorporea regens corpus" (editor's note: *recte* Nemesius Emesenus, *De natura hominis*

motion in the middle of multiple parts maintaining, extending, infusing, directing, influence throughout the delimited, contained, "body" would also be exemplified in music composition at this time, with a stable part moving less, and multiple parts moving more, and all sounding simultaneously—a radical departure from the past, and evidence for the efficacy of music as exemplifying provocative new concepts.

As Philip brings up what can be named "soulish substance," a proportion that exists between the intrinsic and extrinsic, questions of rationality, value, intellect, and truth, *synderesis* becomes more and more important as he continues to explore the supra-rational, the distinction between *anima* and *spiritus*, articulate mind and inarticulate *spiritus*, the distinction between the "thoughts" and "intentions" of the "heart," or the core of human will, consciousness, and responsibility. He initiated the topic with Ezekiel's vision, but proceeds through a succession of relevant New Testament passages, with extensive comments from his prodigious collection of commentators. As an answer to the question of "What exactly is *synderesis*" and why does Philip think it so important, the following progression is given: first, the subject of distinction and integration of what are parts, that is, spirit, mind, and body (1 Thessalonians 5,23),[30] with the comment and subsequent quotation from the Apostle Paul, that no one knows a human spirit, except the one who possesses it. Paul also states that one must be renewed in the spirit of one's mind, showing a further distinction between "spirit" and "mind," but also a conflation of the two parts.[31] Peter Lombard comments that human beings consist of these three, spirit, mind, body—that by which we "intellect," and that by which we are visible, handle and touch, that is, our bodily presence in the world, and that often "mind" is placed or named together with "spirit, so that a duality, that is, mind

---

c. 2, eds. G. Verbeke, and J .R. Moncho [Leiden, 1975], pp. 23–50). Et ponit Calcidius pulcherrimum exemplum de aranea comparans ei animam: sicut aranea in medio tele residens sentit quantumlibet parvum motum factum in aliqua parte tele, ita anima in medio corporis residens, scilicet corde, percipit motum factum in corpore ex quacumque parte vel extrinsecus vel intrinsecus.

30    "And the very God of peace sanctify you wholly; and I pray God your whole spirit and soul and body be preserved blameless unto the coming of our Lord Jesus Christ. Faithful is he that calleth you, who also will do it." ("King James" English translation); edition, 1, p. 196: "Integer spiritus vester et anima et corpus" etc. Dicit enim Glosa quod spiritus accipitur pro ratione. Sed spiritus accipitur ibi pro synderesi, sicut habetur in glosa Gregorii super Ez. I, Dicit enim quod "hic est spiritus qui interpellat pro nobis gemitibus inenarrabilibus; nemo enim scit ea que hominis sunt nisi spiritus qui est in eo."

31    Ephesians 4, 23; Edition, 1, p. 196.

and body results."[32] But *synderesis* is not bodily, contained, presence in the world, therefore belongs to what is invisible, that is, what is intellected, thought, cogitated. Yet, Hebrews 4, 12, which Philip quotes, clearly makes a distinction, reinforced by many other scriptural passages, between *anima* (mind), and *spiritus* (intention/will), a distinction that is not always carefully maintained in the English translation. Using Hebrews 4, 12: "For the word of God is quick, and powerful, and sharper than any two-edged sword, piercing even to the dividing asunder of soul and spirit, and of the joints and marrow, and is a discerner of the thoughts and intents of the heart," Philip builds a case for the seminal distinction between what we *say* to ourselves and others, that is, "thoughts" and the articulation of these thoughts to ourselves and others; and what we in fact *are* at the most intimate core of our "intentions" (that which is known only to ourselves).[33]

Philip then brings up the parable of the woman who takes yeast to three parts of wheat until "the whole is leavened."[34] On the one hand, we are dealing with three parts, body, soul, and spirit. On the other hand, all three are

---

32    Edition I, pp. 196–197, Peter Lombard: "Nota quod tria posuit quibus homo constat, spiritum, animam, corpus, illud scilicet quo intelligimus et illud quo vivimus et illud quo visibiles et contrectabiles sumus; que rursus duo dicuntur, quia sepe anima cum spiritu nominatur." Sed synderesis non est illud quo scilicet vivimus vel quo visibiles sumus. Ergo est illud quo intelligimus.

33    Edition I, 196: Ad Hebr. IIII: Vivus est sermo Dei etc. pertingens ad divisionem anime et spiritus Glosa: "Quia cognoscit Filius Dei quomodo dividatur sensualitas a ratione et ipsa eadem a se, dum plus dedita infimis rebus inferior est vel ab hiis revocata dignior; sic etiam videt quomodo spiritus a se dividatur, dum vel in Deum inhiat de divina usia cogitans vel inferius celestia considerat vel in terra inferius de mundanis rectae agendis pertractat vel etiam quomodo spiritus, id est ratio, a sensualitate secernatur, dum quod in se, id est in ratione, est inferius superat id quod in illa est altius." Note the levels of distinction, that is, inferior/earth and superior. Further, spirit, mind, body do not disagree amongst themselves, but that "strengths of mind" (*virtutes*) are directed as one, brought together into a unity. We possess prudence in reason, anger at the odium of vice, virtue in the cupidity of desire, but, according to what Philip thought was Augustine's treatise (*De spir. et. an.* c. 4), this is not a plurality of motives and *synderesis* is not part of a duality.

34    Matthew 13, 33: Another parable spake he [Jesus] onto them; The kingdom of heaven is like unto leaven, which a woman took, and hid in three measures of meal, till the whole was leavened. Edition I, pp. 196–197: Item, tres ponuntur vires motive multotiens in Glosa, sicut in illa parabola Matth. XIII: "Simile est regnum celorum fermento quod mulier abscondit in tribus satis farine" Glosa: "Ut spiritus, anima et corpus non discrepant inter se vel ut tres anime virtutes in unum dirigantur, ut in ratione possideamus prudentiam, in ira odium vitiorum, in cupiditate desiderium virtutum." Sed cum non sint plures motive secundum Augustinum et synderesis non est altera duarum, ergo est rationabilis.

integrated. *Synderesis* is the crucial term for achieving simultaneous harmonization of apparent contradictions. *Synderesis* is not in the first instance a part of the visible body in the world, therefore, belongs to the invisible, intellecting, internal part; *synderesis*, however, is not a static presence, rather full of motion. For example, by way of answer to the question of the relationship *synderesis* has to sin, Philip states that *synderesis* "circles" around it. Further, *synderesis* is still present when "reason" has been lost.[35]

*Synderesis* then is multiplex: a "part" that is a rational, reasonable, motivational "force"; *synderesis* "serves" together with our spirit to integrate mind and body—an integration of thought and intent—that is, *synderesis* is a functional part; further, in the division between the oppositions of desire and irascibility, both of which one can say belong to the rational mind, *synderesis* is the arbitrating, judging "part," that is, the *pars rectitudinis*, and remains as such, outside of the passage of time. *Synderesis* remonstrates against sin, with the potential that is in flux, *synderesis* contemplates what is certain (*modum planum*), principle, is simple, stable, quietly, with little motion, observing "that which is."[36]

Ezekiel's eagle with its "eagle eye" reappears. *Synderesis* is part of reason, yet outside of it (*extra secundum rationem*), with a stability "not according to circumstance," unwavering, persevering (*per nobilitatem*), inflexible with respect to an "appetite for good," leading to action, productivity, and good effect. *Synderesis* is a habitual potency that is not impeded from action, and is not contingit upon the "disobedience of reason." This "habitual potency" (Philip uses the term three times in this context) is an interior face, that is not

---

35    Edition I, pp. 197–198, Job 1, 15, 16, 17, 19, the refrain "I only am escaped alone to tell thee" (of the calamities, one by one, that had befallen Job), for the question of what remains when one is still alive but "reason" (*discretio rationis*) is lost. Luke 10, 30 also occurs in this context, of "him who fell among thieves which stripped him of his raiment, wounded him, and departed leaving him half dead." What is then left, when one is "stripped" of reason?

36    Edition I, 197: Erit ergo synderesis vis illa que est ratio vel pars eius; p. 197: hoc erit synderesis quedam pars rationis et sic accipitur ibi ad Thessalonicensis in fine: "Ut spiritus vester integer servetur et anima et corpus;" pp. 197–198: tunc synderesis erit pars rectitudinis; p. 198: Quod ergo remansit synderesis dici potest. Illud enim est de se remurmurativum contra peccatum et recte contemplativum boni simpliciter et voluntarium et horum omnium est inspectrix relatione ad summum bonum ad quod principaliter se habet. Et secundum hoc non erit seiuncta potentia ab illis viribus in quantum flexibiles sunt, sed in illis existens inflexibilis eadem cum unaquaque illarum…Queritur autem quomodo dicitur synderesis remurmurare peccato? Quando enim quis peccat remurmurat illi. Ergo movetur circa illud.

impeded from action, also present for judgment in difficult situations, but it is a potency.[37] Further, *synderesis* is essential to disputation itself, leading, eventually to truth, *per modum contrapositionis propter diversas inclinationes*, Aristotle's contrary motion of diverse inclinations that must be resolved within motion and time.[38]

*Synderesis* is a "potency" and "force" that, as a motivating "part" brings together the "parts" of mind, spirit, and body into a simultaneous, integrated, unity—a unification for which the Apostle Paul prayed, a prayer that is invoked at the onset of Philip's lengthy, complicated, discussion of what was, for him, an extremely important and useful concept. The concept, in contradistinction to *conscientia* is multiplex, complex, and, in short, difficult to understand. Music, however, the exemplary discipline, proceeding from the study of particulars, to relationships, to the nature of motion, in order to apply abstract concepts to the study of invisible substance, eventually the nature of godly "substance," during and shortly after the time of Philip's writing and the inauguration of the early university, exists to make difficult concepts plain.[39]

---

37    Edition I, p. 199: Intelligentia autem vocatur illa que est cognitionis. Potentia habitualis dicitur que facilis est ad actum. Et sic synderesis dicitur potentia habitualis, quia non impeditur ab actu suo quantum in se est, sed hoc, scilicet impediri, contingit per inobedientiam rationis. Ipsa ratio dicitur potentia habitualis sed non in tantum, quia etsi impediri non possit quantum ad actum faciendi quod vult interiori facere, tamen quantum ad actum iudici in difficilibus. Cf. Edition I, pp. 201–202 for distinction between *conscientia* and *synderesis*.

38    Edition I, p. 200: Item, synderesis et fomes se habent per modum oppositionis, quia contrarias habent inclinationes, fomes ad malum, synderesis ad bonum, fomes habitus malus, synderesis potentia habitualis ad bonum. Sed peccatum est secundum fomitem. Ergo et secundum synderesim.

39    Cf. van Deusen, *Music and Theology at the Early University*, discussion of music as a "ministry discipline": in introduction to this volume, as well, especially, as the following passage, pp. XI–XII: "Music, however, for Grosseteste, was far more than relaxing entertainment in later life, after a hard day's work as Bishop of Lincoln. Music was crucial to his understanding of the most revolutionary, demanding concepts of his time. Grosseteste's great treatises, such as *De cessatione legalium*; the *Hexaemeron*, a commentary on the creation account of the book Genesis; an important, and one of the first commentaries on a portion of Aristotle's *Physica*; his short pithy work on the ten commandments, *De decem mandatis*; and his separate, succinct tracts on the liberal arts and sound, set forth underlying principles that could, in all cases, be best comprehended by using the examples provided by the quadrivial discipline of music. These principles are few in number, but utterly basic in nature. They include concepts of material, harmony, reconciliation, and the trinity, for which Grosseteste cites Augustine, as well as many other authorities from patristic and early medieval periods. He then reconciled these traditional concepts with his own external and internal intellectual milieus, coupling, for example, material with a concept of material composed of contrary, opposing parts tempered by a creator, traditional

What is "made plain" here, and how does this come about? Philip begins with visible/invisible *materia* containing motion that coalesces with the *materia* of time, contrary or oppositional motions, force, and makes a point of the concept of "parts," especially in the distinction between *anima, spiritus,* and *corpus* as *partes.* Mind (*anima*) involves articulation, the relating of one's thoughts to oneself and others, or what one might name "consciousness" today; whereas spirit (*spiritus*) is the "part" known only to oneself, bringing to the fore the distinction between what I "say" to myself as well as to others—the way in which I would like to present myself to myself and to others—and what I, in truth, am. Philip, as the Apostle Paul, writes of the dividing asunder of the "thoughts" and "intents" of the heart. The lengthy, nuanced, discussion of *synderesis* brings up not only considerations of consummate importance for the early initiation and understanding of Aristotle's *Physics,* but today in terms of consciousness, will, rationality, administrative function, mind, and dementia. The question, "What remains when reason flees," is an important one for Philip as well as for consciousness studies today. Philip's discussion provides a vocabulary that was both recognizable and infused with new significance that needed exemplification. This arsenal of words included: *medium, in medio, ductus, tenere/tenor, pars, partes, organum, de modo coniunctionis, simplex, proprietas/perfectio, modo coniunctionis* (I, p. 156), *simplex* (I, p. 157), *fundamentum,*[40] all of which found, and would continue to find, exemplification within the exemplary discipline of music. Further, this conceptual vocabulary reinforced a mental climate, with specific intellectual tools, for relating time, duration, and motion of time to the notational *figurae* of musical tones. Music notation therefore exemplifies increments of temporal duration during the early years of the formation of studies at the University of Paris.[41]

---

*armonia* with a concept of simultaneous harmony which included dissonance and consonance, a tradition of the trinity with a richer concept of unity in trinity-trinity in unity, that is, three distinct parts functioning together within a totality. In every case, a traditional concept was reconciled with a totally new, fresh way of viewing the concept." Grosseteste died ca. thirty years later than Philip, but his methodology is the same. (With respect to Philip the Chancellor's comprehensive discussion of *synderesis,* it is certainly not a coincidence that Grosseteste translated into Latin Aristotle's *Nichomachean Ethics.*) Music as a discipline is not metaphorical, rather, essentially exemplary, an important distinction that placed it at the crux of scientific methodology, explanation, and persuasion.

40 cf. *fundamentum* or ground/*grunt,* Bernard McGinn, *The Harvest of Mysticism in Medieval Germany,* 1300–1500 (The Presence of God. A History of Western Christian Mysticism IV, NY, 2005), pp. 86ff.

41 This is not an "invention" but rather firmly placed within, and exemplary of new concepts that had not only become available, but were thoroughly discussed by Philip as well as "Masters of Arts" at Paris, cf. Palémon Glorieux, *La Faculté des arts et ses maîtres au xiiie*

But, most important for this study, how did music exemplify *synderesis*? As the unflinching eye of the eagle, the spider "with little motion" in the middle of its web from this position quietly influencing its entire surrounding texture, so a section of *cantus* [42] in *medio*, in the middle of the music composition, provided the *fundamentum* as well as the organizing force (*vis motiva*) for as many as two or more upper and lower *partes*, bringing into simultaneous consonance inherent contrary motions—an exemplification of *synderesis* that could be seen in music notational *figurae* as well as heard.

Music as a material and measurement discipline uses the unseen *materia* of sound, time, and motion. It therefore makes plain invisible realities that could only with great difficulty be imagined, let alone become comprehensible. *Synderesis*, as the part, *fundamentum in medio*, with "little motion" stabilizing other "parts," both above and below, simultaneously infusing, as well, all of the parts as yeast infuses all three parts of the wheat, achieving "a good effect," was demonstrated within the illustrative discipline of music, to remain as an organizing principle for the next four hundred years.[43]

For a musical example, see Figs. 4.1–4.5: ff. 390v–392v, Firenze, Biblioteca Mediceo-Laurenziana, Pluteo 29, 1.[44] Note the following with respect to the musical example:

---

siècle (Paris, 1971), pp. 506ff., as well as van Deusen, "Change in a concept of 'mode,'" *Theology and Music at the Early University*, pp. 54–75, esp. p. 55, n. 3.

42  *cantus* commonly known as "Gregorian Chant," or "plain chant," *cantus planus*. See also van Deusen, *The Cultural Context of Medieval Music* (Santa Barbara / Denver / Oxford, 2011), introduction pp. 1–16, 185.

43  The compositional principle of *tenor* (*tenere*) imbedded within the composition, influencing the entire music composition in all of its parts can be observed in the Mass ordinary compositions, for example, of Guillaume Dufay, Josquin des Pres, Jacob Obrecht, Johannes Ockeghem, Giovanni Pierluigi da Palestrina, and many other composers of the fifteenth/sixteenth centuries.

44  This is a manuscript long associated with the cathedral of Notre Dame, but should, I believe, more properly be viewed as an anthology of *exempla* making plain the new mental tools (*organa*), and principal concepts such as the nature of substance, or pre-existent substance, process (*theoria*), *armonia*, and contrary motion culminating in *armonia*, all under discussion, as with Philip, during the first half of the thirteenth century, within the emerging interaction between the Arts and Theology faculties at the nearly-organized University of Paris. These *exempla* made these new concepts plain to the eyes on the page in terms of the *figurae* of music notation, as well as the *figurae* of the letters of the text, and made them comprehensible to the ears within the substance of sound, through time and motion. Music, thus, as the exemplary material and measurement discipline accessed the unseen substance of thought—music's irreplaceable position within the four material sciences. Rather than professional training for musicians, as is the case for the study

- *parts*, two above, with text, Veni dator previe spiritus scientie dono cuius relating both textually and musically to the *cantus*, Veni sancta spiritus reple tuorum corda fidelium et tui amoris in eius. Ignem accende, the *Alleluia* for the celebration of Pentecost, the coming of the Holy Spirit to the world.
- *Cantus*, *fundamentum*, *tenor* (also, slightly later, in the German language, *Grund, grunt*) with "little motion," here, bottom part, later, *in medio*, as the *tenor* is increasingly placed above the bass or lowest part. *Cantus* exemplified "pre-existent substance": both of the other parts are directly influenced musically as well as with respect to the meaning of the text by the *cantus*.
- The parts occur simultaneously; there is *contrary motion* both horizontally and vertically between parts. We have here a visually-perceptible exemplification of *synderesis*, in that *anima* (text) relates to and is influenced by *spiritus* (*cantus*) which is stable, moving with less motion, and constitutes the basis for the three-part composition.

---

of music particularly at North American universities today, music brought what would have been nearly incomprehensible abstractions such as "pre-existent substance," "contrary motion" to the mind in a manner that could be easily seen and grasped. Facsimile edition, 2 vols, Luther Dittmer, ed. (Institute of Mediaeval Music, NY, 1966), used by permission. *Cantus*, Alleluia for Pentecost Sunday, cf. *Liber Usualis* (Tournai, NY, 1963), 880.

f. 390ᵛ

FIGURE 4.1   *Fol. 390v, Firenze, Biblioteca Mediceo-Laurenziana, Pluteo 29, 1*

f. 391ʳ

FIGURE 4.2  *Fol. 391r, Firenze, Biblioteca Mediceo-Laurenziana, Pluteo 29, 1*

f. 391ᵛ

FIGURE 4.3   *Fol. 391v, Firenze, Biblioteca Mediceo-Laurenziana, Pluteo 29, 1*

f. 392ʳ

FIGURE 4.4  *Fol. 392r, Firenze, Biblioteca Mediceo-Laurenziana, Pluteo 29, 1*

f. 392ᵛ

FIGURE 4.5   *Fol. 392v, Firenze, Biblioteca Mediceo-Laurenziana, Pluteo 29, 1*

CHAPTER 5

# Trinity, Simultaneity, and the Music of Creation in St. Bonaventure[1]

*Peter Casarella*

Aristotle defined temporal simultaneity as "those [things] that come into being at the same time, for neither is before or after."[2] This definition became a starting for a great deal of speculation about the nature of time in the thirteenth century. For better or worse, the Aristotelian definition also accords with our everyday conception of simultaneity. Two events happen at the same time. If we see these two events as lying on distinct temporal axes, then the point of simultaneity consists of the intersection of these two axes. The perception of this intersection is arguably what it means to speak of simultaneity as a temporal moment.[3]

Theologians of the thirteenth century also inherited the notion that divine eternity could be equated with a form of simultaneity. They recognized a tradition of learning whereby Augustine had defined God's eternity as an eternal presence,[4] which is later distinguished by Boethius from the "present" of

---

1   I would like to thank Matthew McGuire for his valuable assistance in the preparation of the final redaction of this essay.

2   Aristotle, *Categories*, 13, 14$^b$24–6, *Aristoteles Latinus* I.1-5, trans. Boethius, ed. L. Minio-Paluello (Turnhout, 1961), p. 39: "Simul ergo natura esse dicuntur quaecumque conuertuntur quidem secundum essentiae consequentiam, nullo autem modo alterum alteri subsistendi causa est, et ex eodem genere quae in contrarium sibi diuiduntur; simpliciter autem simul sunt quorum generatio in eodem tempore est."

3   In our everyday understanding, we presume that simultaneity constitutes a form of unity, thus excluding (as did Aristotle on the theoretical plane) the contemporary claims made for temporal disunification. On this problem, see Rory Fox, *Time and Eternity in Mid-Thirteenth Century Thought* (Oxford, 2006), pp. 59–63.

4   Augustine, *Confessions*, XI, 14.17 *Augustinus Hipponensis—Confessionum libri tredecim*, Corpus Christianorum: Series Latina 33 (Turnhout, 1981), p. 659: "praesens autem si semper esset praesens nec in praeteritum transiret, non iam esset tempus, sed aeternitas. si ergo praesens, ut tempus sit, ideo fit, quia in praeteritum transit, quomodo et hoc esse dicimus, cui causa, ut sit, illa est, quia non erit, ut scilicet non vere dicamus tempus esse, nisi quia tendit non esse?."

© KONINKLIJKE BRILL NV, LEIDEN, 2016 | DOI 10.1163/9789004312319_007

human experience.[5] This distinction is transmitted in the scholastic tradition as the Divine "standing now" (*nunc stans*) that stands opposed to the creaturely "flowing now" (*nunc fluens*).[6] Boethius, moreover, defined the concept of eternity as a "simultaneous and total possession of interminable life."[7] Seen from this new vantage point, divine simultaneity *cannot* be a point of intersection between two temporal axes. When discussing eternity, the scholastic theologians were more likely to draw upon Aristotle's notion of simultaneity of nature than his notion of temporal simultaneity.[8] Alexander of Hales spoke to the issue in a characteristically Augustinian way when he said that the moment of creation is not with time, in time, or without time.[9] Likewise, Aquinas describes the second person of the Trinity as the *quando* ("when") of the Father.[10]

The theologians of the thirteenth century were thus faced with a dilemma: "How does one think analogically about the relation of eternity to time?" Fox rightly notes that the specific question of duration was addressed in explicitly analogical terms.[11] But there is also the broader issue of whether and how the tensed rhythms of analogical predication could transverse the span between an eternal "now" and its imprint in the temporal "nows" of the finite world. This latter issue is a critical one that did not escape the notice of one of the greatest synthesizers of the medieval doctrine of analogy of the twentieth century, namely, Erich Przywara, S.J. He opines:

---

5    Boethius, *Quomodo trinitas unus Deus ac non tres dii* (commonly, *De Trinitate*) IV, CCSL 94 (Turnhout, 1984), p. 1247: "Semper enim est, quoniam 'semper' praesentis est in eo temporis tantumque inter nostrarum rerum praesens, quod est nunc, interest ac divinarum, quod nostrum 'nunc' quasi currens tempus facit et sempiternitatem, divinum vero 'nunc' permanens neque movens sese atque consistens aeternitatem facit."

6    Cf. Albert Zimmer, *Studien zur mittelalterlichen Geistesgeschichte und ihren Quellen* (New York, 1982), p. 95 n. 80.

7    Boethius, *De Consolatione philosophie* trans. Claudio Moreschini (München, 2005) Book V, Prose VI, p. 155: "Aeternitas igitur est interminabilis vitae tota simul et perfecta possessio, quod ex collatione temporalium clarius liquet."

8    See Fox, *Time and Eternity*, pp. 287–288. The prime examples of natural simultaneity were non-causal logical examples such as half and double.

9    Alexander of Hales, *Summa fratris Alexandri*, ed. Bernardin Klumper (Quaracchi, 1924), VI.1.2.4: "Sic creatio—actio non est sine tempore—et comprehendo sub tempore nunc— quod est principium temporis—non tamen est in tempore, cum sit actio Dei prima in creaturam; nec cum tempore proprie dicitur, quia non ei associatur vel parificatur; sed non sine tempore."

10   Thomas Aquinas, *Summa theologiae*, ed. Instituti Studiorum Medievalium Ottaviensis (Ottawa, 1941), 1. 42. 2: "Relinquitur ergo quod filius fuit, quandocumque fuit pater. Et sic filius est coæternus patri, et similiter spiritus sanctus utrique."

11   See Rory Fox, *Time and Eternity*, pp. 317–319.

[As St. Thomas Aquinas states citing Boethius,[12]] "these '*nuncs*' are not parts of time; for time is not composed of indivisible '*nuncs*.'" But this negative, abstract commonality is intersected by the 'ever greater' separation between the divine '*nunc stans*' and the creaturely '*nunc fluens*,' between the "Is as now" of God and the "flux as now" of the creature.[13] Time and eternity are related to one another as "*nunc fluens*" and "*nunc stans*."[14]

Przywara certainly grasps the immensity of the challenge. It is not enough to juxtapose distinct modes of *stasis* and motion in order to elucidate eternity as a simultaneous presence. Nor was it possible to meld them. The problem of the interpenetration of the two poles had to be faced in terms of the very nature of "nowness" in the God-world relationship.

Other theologians had to face the dilemma. St. Bonaventure followed the Augustinian current of this period in thinking of the entire order of creation, including its temporal aspect, as expressive of traces (*vestigia*) of the eternal Trinity. "The created world," he writes, "is like a *book* that reflects, represents, and describes the creative trinity."[15] Moreover, he continues, the book of creation in its entirety reflects its Trinitarian author at three levels: as a trace, as an image, and as a similitude or likeness. Every created thing simply by virtue of being created reflects a trace of the trinity, even if only in a distant and unclear way. The created *image* of the Trinity is a much closer reflection and is found only in intellectual creatures. The likeness or similitude is the closest reflection and is *seen* when a rational spirit is conformed to God through grace. All three levels are "like steps of a ladder by which the human mind is designed to ascend gradually to the supreme principle who is God."[16] Regardless

---

12    Thomas Aquinas, *Expos. In* VI *Phys*. ed. P.M. Maggiolo (Torino, 1954) VI.1.11.2: "ipsa autem *nunc* non sunt partes temporis; non enim componitur tempus ex *nunc* indivisibilibus, sicut neque aliqua magnitudo componitur ex indivisibilibus." (See n. 5 above for Boethius).

13    Cf. Thomas Aquinas, *Summa theologiae*, I, q. 10, a.2, ad 1: "Sicut enim causatur in nobis apprehensio temporis, eo quod apprehendimus fluxum ipsius nunc, ita causatur in nobis apprehensio aeternitatis, inquantum apprehendimus nunc stans."

14    Erich Przywara, S.J., "Time, Space, and Eternity," in *Analogia Entis*, trans. John R. Betz and David Bentley Hart (Grand Rapids, MI, 2014), p. 587.

15    St. Bonaventure, *Breviloquium* (hereafter *Brev.*) 2.12 (S. Bonaventurae *Opera Omnia*, ed. Quaracchi V, 230): "Ex praedictis autem colligi potest, quod creatura mundi est quasi quidam liber, in quo relucet, repraesentatur et legitur fabricatrix secundum triplicem gradum expressionis, scilicet per modum vestigii, imaginis, et similitudinis."

16    *Disputed Questions on the Mystery of the Trinity* (hereafter *Disp. Questions*), trans. Zachary Hayes, O.F.M. (Saint Bonaventure, NY, 1979), p, 31. See *Brev.* 2.12 (ed. Quaracchi V, 230):

of where one stands on the ladder, one *sees* that the Triune God has in some fashion expressed himself in the order of creation.

Bonaventure thus belongs to a rich patristic and medieval heritage of reflection upon the creative Trinity (*trinitas creatrix*).[17] Like Albert the Great and Thomas Aquinas, Bonaventure sees the production of creatures as *rooted* in the procession of persons in God. This pivotal "scholastic axiom" represents a new way of perceiving the world.[18] From this vantage point, we see the life of an eternal God finds its own unique mode of expression in the natural order and in our individual, social, and ecclesial histories. Nothing, and no temporal event, in the world are untouched by God's Trinitarian irruption of beauty. Beauty thus emanates from a transtemporal source while reconfiguring itself in a temporal mode of being.

With respect to the problem of eternal simultaneity, Bonaventure asks whether trinity and eternity co-exist simultaneously (*utrum simul stent trinitas et aeternitas*).[19] In the course of offering an affirmative answer to this question, he turns to the scholastic idea of a divine production, that is, the "leading forth" of one Trinitarian person from another. This adaptation presents its own theological issues, as I will demonstrate below. His rewriting of the classical idea of divine production also allows him to introduce a new solution to the scholastic problem of time and eternity. Because eternity is based upon a relation that is productive beyond measure, Trinitarian eternity is an infinitely fecund, centrifugal emanation into the world of temporal succession.

One key implication of Bonaventure's theological creativity can be found in his hearkening to the music of creation. Bonaventure enjoins his fellow Franciscan friars not only to *see* the triune God in the created order; nor is he at all content to propose the scaling of a metaphysical or theological ladder as a solitary, private quest. Christ is "our ladder" (*nostra scala*), Bonaventure states, and Christ's death and resurrection offer the Friars an objective hope

---

"ex quibus quasi per quosdam scalares gradus intellectus humanus natus est gradatim ascendere in summum principium, quod est Deus."

17    See, above all, Gilles Emery, O.P., *La Trinité Créatrice: Trinité et création dans les commentaires aus Sentences de Thomas d'Aquin et des ses précurseurs Albert le Grand et Bonaventure* (Paris, 1995).

18    Most scholastic theologians certainly took this doctrine as axiomatic without naming it as such; however, its meaning and implications for contemporary thinking were articulated with renewed vigor by Hans Urs von Balthasar in his *Theo-Drama*, v: *The Last Act* (San Francisco, 1998), pp. 61–65.

19    *Quaestiones disputatae de mysterio Trinitatis* (hereafter *Qq. d. de trin.*), q.5, art.2 (ed. Quaracchi v, 93).

that one can possess the means to *perceive* with eyes and eyes the world created as a reflection of the creative Trinity.[20]

One highly expressive example taken from Bonaventure's preaching will
illustrate the fecundity of the creative Trinity. The *requiem* of Christ's burial
and rest during the period that spans Good Friday and Holy Saturday in the
liturgical calendar offers a penetrating insight into theological form of the
music of creation. In fact, Bonaventure is fascinated with the connection
between finding the rest offered by the Lord and the consummate rest that
Christ underwent on Holy Saturday, that is, the day in the medieval imaginary
when he went to the realm of the dead. The Latin word *requies* (literally, "rest"
or "repose") is the leitmotif of an entire sermon that he preached on the topic.
Rest is also associated by Bonaventure with the term *Sabbatum*, a word that
simultaneously encompasses the seventh day of the holy week, that is, the
interval between the crucifixion of Friday and the resurrection of Sunday (the
beginning of a new week), *and* the concept of a Sabbath rest. The verbal
connection allows Bonaventure to introduce the blessing of the seventh day of
God's creation as well as the precept to keep that day holy as the *Urtext* for
interpreting the Lord's rest. The Seraphic Doctor, therefore, paints a cosmogonic
image of divine rest as the gloss for laying Christ to rest in the tomb:

> However true it is that the Lord worked for six days and rested on the
> seventh, it is more truly the case in [his] acts of recreation, for the Lord
> worked for six *lustra* of times (i.e., a purification or census spaced out
> every five years) and rested for the seventh. From the beginning of the
> incarnation Christ was poor and engaged in work from his youth, but he
> worked the hardest during the days before the passion when he explained
> the passion and came into Bethany. On the seventh day he was placed in
> the tomb and rested.[21]

---

20   *Itinerarium Mentis in Deum*, 4.2 (*Opera Omnia*, ed. Quaracchi V, 306): "Et quoniam, ubi
     quis ceciderit, necesse habet ibidem recumbere, nisi apponat quis et adiiciat, ut resurgat;
     non potuit anima nostra pefecte ab his sensibilibus relevari ad contuitum sui et aeternae
     Veritatis in se ipsa, nisi Veritas, assumta forma humana in Christo, fieret sibi scala reparans
     priorem scalam, quae fracta fuerat in Adam."

21   *Sabbato Sancto, Sermo 1* (*Opera Omnia*, ed. Quaracchi IX, 267): "Quamquam verum sit,
     quod sex diebus operatus est Dominus et die septimo requievit, veriori tamen modo
     habet veritatem in operibus recreationis, quia per sex lustra temporum operatus est
     Dominus et septimo requievit; quia a principio incarnationis Christus pauper fuit et in
     laboribus a iuventute sua, sed potissime fuit in laboribus per dies ante passionem, quando
     se exposuit passioni et venit in Bethaniam. Septimo die positus in sepulcro et requievit."
     Translation mine.

The return to Bethany *prior* to the passion recalls the prophetic anointing by Mary in John 12, 3. She prepared him liturgically for the rest that he would undergo in the tomb. In thirteenth century Latin the word *requies* was not just "rest" since it could refer to a tomb or to a funeral Mass.[22] This artful ambiguity is also exploited by Bonaventure in his sermon on Holy Saturday. Christ's tomb is the place where he rests temporally but also the place from which he gathers the souls who will follow him to an eternal rest. The sermon continues:

> Today the most blessed soul of Christ brought it about that all holy souls are with him in the pasture of eternal delight and the repose of blessed vision. We are invited to this *requies*, and it is fitting that we find it *in life*. All labor is on account of repose (*requies*), and every motion is on account of rest (*quies*).[23]

Christ's funeral requiem thus coincides with the laborer's search for rest and the simultaneous requirement to meditate a slower pace of life by seeking the point from which all movement emanates. Thus we see in the Lord's own rest a wholly new attunement to a new symphony of daily life.

••• 

This essay lays out the foundations of St. Bonaventure's insight into the music of creation in terms of his notion of a creative trinity. After a brief overview of the main themes of his theology of the creative trinity, it draws attention to what some have called his doctrine of *expressio*. In particular, I would like to show just how Bonaventure, in line with St. Augustine, introduces divine *expressio* in a distinctly Christian fashion as a relational term. Then, the essay deals with the Trinitarian dimensions of the music of creation. Here we see that Bonaventure corrects Augustine in part because of his mentor's inability to do adequate justice to the complexity of temporal development. After offering some reflections on Bonaventure's Franciscan philosophy of the sense, space, and structure of time, this essay concludes with a synthetic evaluation of Bonaventure's contribution to a theology of simultaneity.

---

22    Charles du Fresne DuCange, *"requies,"* in *Glossarium mediæ et infimæ latinitatis* VII (Niort, 1886), p. 138a.

23    *Sabbato Sancto, Sermo 1* (ed. Quaracchi IX, 267): "Hodie beatissima anima Christi fecit, quod omnes animae sanctae secum sint in pastu aeternae iucunditatis et requie beatae visionis. Ad istam requiem invitamur, et oportet, quod inveniamus eam in vita. Omnis labor est propter requiem, et omnis motus propter quietem."

### Trinity and Creation

According to St. Bonaventure, the fact that God is trinity cannot be known by natural reason. God's trinity is very clearly a truth of faith. It is a truth of faith "that is fitting, necessary, and worthy of belief."[24] The "book of creation" offers a witness to the trinity, but the Scriptures and the "eternal testimony" of the book of life are a much more effective means to move the faithful to believe that God is one and a trinity.[25] In speaking of trinity and creation, we are therefore speaking of an illumination of Christian belief that comes from within the light of faith. No attempt is being made to *prove* the trinity solely on the basis of God's reflection in the created order.

The work of Gilles Emery, O.P., on trinity and creation in the thirteenth century remains of the utmost importance.[26] Emery offers a masterful exposition of the main themes and distinctive emphases within Bonaventure's path to the creative trinity. I will highlight some fundamental themes within Bonaventure's treatment of trinity and creation, most of which are treated exhaustively in Emery's brilliant study.

According to Emery, St. Thomas is the great systematizer of the Trinitarian thought of St. Albert and St. Bonaventure and with the following sentence, we see St. Thomas's systematic genius operating in full gear: "Ex processione personarum distinctarum causatur omnis creaturarum processio et multiplication (The procession and multiplication of every created being is caused out of the procession of the distinct [divine] persons.)"[27] In other words, the going forth of the Son and the Spirit from the Father in the trinity lies behind and is the very reason for the coming into being or creation of worldly things by God. This statement effectively summarizes the most illuminating insight of all scholastic theology into the relationship of trinity and creation. St. Thomas is *not* saying that believers can peek into the infinite mystery of the trinity by virtue of possessing knowledge of the created world. Rather, the triune God remains ever more transcendent in light of this insight into the *raison d'être* of creation. The profound epiphany of the created order demands a Trinitarian lens. The faithful Dominican Emery argues that this passage is "unparalleled in the commentaries

---

24   *Disp. Questions* q.1, art.2, resp., 128. *Qq. d. de trin.* (ed. Quaracchi V, 54): "Dicendum, quod Deum esse trinum est verum credibile tanquam congruum et debitum et dignum credi."

25   Ibid., "eternal testimony" at p. 130. (ed. Quaracchi V, 55): "ideo providit divina sapientia aeternum testimonium, quod quidem est liber vitae."

26   In addition to the dissertation cited above in n. 18, see Giles Emery, O.P., *Trinity in Aquinas* (Ypsilanti, MI, 2003).

27   *I Sent.*, d. 26, q. 2, a. 2, arg. 2, ad. 2, as cited in *Trinity in Aquinas*, p. 67 n. 105.

of Albert or Bonaventure," and on this specific point his defense of Aquinas is probably correct. I would, however, maintain that St. Thomas's systematic insight into the rootedness of creation in the Trinitarian processions is the perfect opening to St. Bonaventure's musical theology of trinity and creation.

In fact, there are distinctively Bonaventurian emphases and accents that must be considered in this context. First, there is the issue of Trinitarian primacy (*primitas*). Primacy refers to the quality of being first and can be applied to essences, causality, the three persons of the Trinity, principles, unity, or emanation. *Basically*, if something or some person of the trinity is supremely first, then it seems that the unity of three equal persons would be violated. Bonaventure's response is that the principle of primacy does *not* only *not* exclude the trinity, but actually includes it insofar as the first principle is a trinity by the very fact that it is first.[28] Trinitarian primacy is best grasped through the Father. "Primacy resides principally in the person of the Father; and for this reason, the fontal-fullness for the production of all the persons is found in Him."[29] The Father produces, as it were, supremely co-equal persons out of his supremely unbegotten fullness. This relates to the prior principle just elucidated. In the summary of Zachary Hayes: "The necessary prior condition for the production of the world, therefore, is the eternal production of another who is fully equal with God."[30] Starting with the Father's fontal fullness, the eternal production of co-equals makes Trinitarian primacy a supremely first principle. In this way, the notion of primacy is not repugnant to trinity. Supreme primacy requires a Trinitarian reading.[31]

A second distinctive feature of the Bonaventurian creative trinity is the repeated emphasis on the fecundity of the Good. The Neoplatonic maxim "*Bonum est diffusivum sui*" is practically a mantra for Bonaventure, one that he cites twenty-six times in his corpus.[32] Fecundity is also a corollary of firstness

---

28  *Disp. Questions* q. 8, concl., 263. *Qq. d. de trin.* (ed. Quaracchi V, 114): "Ad praedictorum intelligentiam est notandum, quod primitas non tantum non excludit trinitatem, verum etiam includit in tantum, ut primum principium eo ipso sit trinum, quo etiam est primum."

29  Ibid., ad. 4 (ed. Quaracchi V, 115): "sic ratio primitatis principaliter residet circa personam Patris, ratione cuius est in ipso fontalis plenitudo ad productionem omnium personarum."

30  *Disp. Questions*, 101.

31  The Bonaventurian principle of Trinitarian primacy is used by St. Thomas in the context of trinity and creation to explain how within a given genre there is a reason for the role being assigned to the first element in that genre. Emery, *Trinity in Aquinas*, p. 48, n. 45 for citation.

32  See Jacques G. Bougerol, "Bonaventure et le Pseudo-Dionysius," *Études franciscaines* 28 supplement (1968): 81ff. This essay is reprinted in Jacques Guy Bougerol, *Saint Bonaventure: Études sur les Sources de sa pensée* (Northampton, 1989).

since the very notion of supreme firstness carries with it the idea of primal fontality.[33] The goodness diffused among the three persons (what Emery calls "Trinitarian fecundity") is the condition for the possibility of goodness spreading forth in creation.[34] The supreme communication of Trinitarian fecundity is obviously a doctrine of faith, but it is also a good example of how faith aids reason to grasp the fittingness of a plurality of persons in God.[35] Some of the most original and well-crafted texts of Franciscan theology by Bonaventure, for example, the sixth book of the *Itinerarium*, are based upon the principle of Trinitarian fecundity in God and in creation: "...that diffusion in time which is seen in creation is a mere center-point in comparison to the immensity of eternal goodness. From this, it is possible to think of another greater diffusion; namely, that sort of diffusion in which the one diffusing itself communicates the whole of its substance and nature to the other."[36] The richness of language and symbolism with which he pursues a productive theology of nature (for example, as a river, as music, and so on) would be impossible to articulate without it.[37] Words for Bonaventure are fecund; they bear fruit in our minds and hearts.

Third, Bonaventure identifies in the order of creation a palpable semblance of the wholly unique intra-Trinitarian *productio*.[38] When Bonaventure treats the going forth of the Son from the Father, he underscores the irruption of an unlimited verbal act within the noun: "the one who pro-duces," he states, "possesses unrestricted actuality of power on account of its highest perfection." In the Triune God, he claims,

---

33    Cf. Emery, *Trinity in Aquinas*, pp. 46–47 and esp. G. Emery, *Trinité Créatrice*, pp. 173–184.

34    Emery, *Trinité Créatrice*, pp. 176–177.

35    *I Sent.*, d. 2, a. un., q. 2, fund. 1 (ed. Quaracchi I, 53): "Ex prima suppositione arguitur sic: si est ibi summa beatitudo; sed ubicumque est summa beatitudo, est summa bonitas, summa caritas et summa iucunditas. Sed si est summa bonitas, cum bonitatis sit summe se communicare, et hoc est maxime in producendo ex se aequalem et dando esse suum: ergo etc."

36    *Itinerarium Mentis in Deum*, 6.2 (ed. Quaracchi V, 310–311): "Nam diffusio ex tempore in creatura non est nisi centralis vel punctalis respectu immensitatis bonitatis aeternae; unde et potest aliqua diffusio cogitari maior illa, ea videlicet, in qua diffundens communicat alteri totam substantiam et naturam."

37    See, example, Sr. Damien Marie Savino, FSE, *The Contemplative River: The Confluence Between People and Place in Ecological Restoration* (Saarbrücken, 2008), an ecological theology that is based upon these principles from Bonaventure.

38    *Qq. d. de trin.* q.5, a.2 (ed. Quaracchi V, 94): "Producens enim est in omnimoda virtutis actualitate propter summam perfectionem...."

If the person who is producing is eternal, then by necessity the act of producing and the product and the mode of producing is each eternal. It follows that eternity and trinity are not only not incompatible but harmonize wonderfully and must exist simultaneously with one another.[39]

To understand this passage, we need to distinguish between human and divine production. The human production of a work of art proceeds in time. In a human work of art, the artist, the work, and the means of production are temporally independent. There is, in contrast, no temporal sequencing in a Trinitarian production. According to Bonaventure, the simultaneity of producing, product, and means is not absolute. Trinitarian creativity is eternal without being severed from the temporal sequences of the world. Trinitarian simultaneity in St. Bonaventure is such that the activity of God is not somehow frozen in time. Trinitarian eternity embraces three distinct but inseparable senses of *productio*. Rather than fusing three moments into one, Bonaventure sees an interpenetrating communion of love irrupting from within each dimension of divine production. The boundlessness and eternal going forth of the Son from the Father wholly transcends our understanding, but the creative expression of the triune God is still intelligible through temporally unfolding forms. In my conclusion, I will return to this point.

The fourth Bonaventurian theme concerns what some of Bonaventure's most lucid commentators in the twentieth century refer to as his doctrine of *expressio*.[40] A good example of Bonaventure's breakthrough is found in distinction 31 of his commentary on the First Book of Peter Lombard's *Sentences*.[41] This article contains a cogent new articulation of Augustinian Trinitarian theology as well as an intriguing and uniquely Trinitarian definition of what constitutes an image. The article presupposes that the second person of the trinity establishes an image in God. This image can be understood as either the

---

39    *Qq. de trin. q. 5, a. 2* (ed. Quaracchi V, 95): "Si igitur persona producens est aeterna, necesse est, ut aeternus sit producendi actus et ipse productus et producendi modus. Patet igitur, quod aeternitas et trinitas non solum non repugnant, sed mirabiliter consonant et necessario simul stant."

40    See Hans Urs von Balthasar, *Herrlichkeit: eine theologische Ästhetik* II: Fächer der Stile (Einsiedeln, 1962), pp. 346–349; Andreas Speer, *Triplex Veritas: Wahrheitsverständnis und philosophische Denkform Bonaventuras* (Werl / Westfalen, 1987), p. 97, and Emmanuel Falque, *Saint Bonaventure et l'entrée de Dieu en théologie: La somme théologique du Breviloquium* (Paris, 2000), pp. 151–153.

41    See in particular the conclusions of *I Sent.*, d. 31, p. 2, art. 1, q1 and q.2, and d. 31, p. 2, art. 2, q. 3, *Opera Omnia of St. Bonaventure, Commentary on the First Book of Sentences*, trans. Alexis Bugnolo (Mansfield, MA, 2014), pp. 534–536, 549.

christological "image of the invisible God" (Col 1, 15) or as the image *ad intra* of that which is expressed *ad extra* in the verse: "Let us create man in our image and likeness" (Gen 1, 26). In any case, the article presupposes, rather than argues, for the truth of a divine image. The question posed in the theological tradition was whether true image subsists in God substantially or relationally. According to Hillary of Poitiers and the Venerable Bede, image is in God according to a community of substance within the divine being. Neoplatonic reasoning likewise suggests that the exemplar from which an image actively emanates is said to be in God according to substance. If the image is more than a passive recipient of the activity of divine expression and therefore equal to God, then it stands to reason that the image too exists in God according to substance rather than relationally.

In reply to these objections, Bonaventure cites Augustine's *On The Trinity* to the effect that there is nothing more absurd than to say that something is an image of itself.[42] Bonaventure clearly recognizes the problem of dividing and multiplying divine simplicity. According to Hilary, Bonaventure writes, the image never exists by itself. Augustine had reclaimed the Aristotelian notion of a relation *pros ti*.[43] Following squarely in the footsteps of St. Augustine, Bonaventure upholds the ideas that a relation *ad aliquid* (lit. "to another") in God need not imply a relation to something exterior to God. The image of God is, therefore, an image of one divine person in another divine person. Drawing upon an obscure, ancient etymology (*"imago* is said as if *imitago"*).[44] Bonaventure concludes that wherever there is some basis for thinking of an image, one also encounters imitation. But intra-Trinitarian imaging cannot involve the imitation of something other than God. Properly speaking, image is thus said according to relation and not according to substance.

What then becomes of the relationship between the active exemplar and the passive image? An image is by Bonaventure's own definition "what expresses and imitates a second thing." Expression in an image can, however, take two different forms. Sometimes an image receives from its exemplar by virtue of a unity of nature. One example is the image of the emperor in his son. Other times an image expresses its exemplar in the diversity of nature, as when the image of the emperor appears on a coin. Whether one posits a unity or difference of nature, the relational reality of expression remains intact. All imaging

---

42   *De Trinitate* VII.1.2, *Aurelii Augustini opera*, CCSL 50, 1 (Turnhout, 1968), p. 329.
43   *Categories*, 6a, 3b–8b, trans. Jonathan Barnes (Princeton, 1984), p. 10.
44   *I Sent.*, d. 31, p. 2, art. 1, qu. 1, it. 4, *Opera Omnia of St. Bonaventure, Commentary on the First Book of Sentences* (tr. Alexis Bugnolo), 540. (ed. Quaracchi I, 540): "Imago enim dicitur quasi imitago, ergo ubi est ratio imaginis, ibi est imitatio..."

in a certain sense concerns a relationship of unity in difference. In God, the unity is that of the divine substance and the difference is that of the three persons. Created images are, therefore, not just the product of the common image of the three persons in a static sense. The divine essence as considered in the three persons (as opposed to just *simpliciter*) already contains within it a divine image and divine expression. The work of creation leaves behind traces, images, and similitudes of the eternal relation *ad aliquid* in God.

To summarize, then, the creative trinity according to St. Bonaventure is first the primal fount of all that is and is any value. Second, it is absolutely fecund, the source of all fertility in nature, man, and language. Third, it is eternally productive, a streaming forth of simultaneous goodness into the temporal orbs of history. Finally, it is infinitely expressive, viz., a communion of co-equals uniquely capable of producing likenesses in radically different forms.

### The Trinitarian Music of Creation

True to his own pluriform aesthetic principles, Bonaventure employs a variety of metaphors to describe the overflow of Trinitarian creativity into the goodness of the created order: a living book legible *both* on the exterior pages of the entire cosmos *and* on the *minor mundus* (the microcosmic world) of the human soul, an infinite sphere whose center is everywhere and whose circumference is nowhere, the outward flow from God of four rivers whose currents span the course of the globe, and finally the mellifluous harmony of musical order. I will concentrate on the Trinitarian analogy of music alone, for this focus enables us to make an important point regarding Bonaventure's potential contribution to contemporary discussion.

Bonaventure refers to the music of creation in his *Breviloquium*, a compendium of theology written for Franciscan Friars in 1257.[45] There he writes in the prologue:

> And so the entire world is therefore described by scripture as proceeding from beginning up to the end by a very well-ordered course in the mode, so to speak, of very beautifully ordered *carmen*...And just as no one can see the beauty of such a *carmen* unless his gaze is directed to the whole

---

45    What follows is a summary of my treatment of Bonaventure in Peter J. Casarella, "*Carmen Dei*: Music and Creation in Three Theologians," *Theology Today* 62/4 (January 2006), 484–500.

verse, so too no one can see the beauty of order and governance of the universe unless he observes it in its entirety.[46]

Contemplation of the world's lyricism and metrical course is neither an end in itself nor merely instrumental.[47] Bonaventure's *carmen* of creation leads one back to the radiance of Biblical wisdom. This Franciscan artistic beauty has relative permanence, and the divine beauty in creation extends beyond the life of a single man. Thus, the Holy Spirit has given us the book of Sacred Scriptures, whose "length" is equivalent in measure to the course of the governance of the universe. Bonaventure reinforces the aesthetic viewpoint and its limits using a form of reasoning found in the Bible itself:

> There is indeed great beauty in the fabric of the world (*machina mundi*) but by far more in the Church, which is adorned by the beauty of sacred charisms. But most of all [beauty lies] in Jerusalem above, superabundantly, however, in the highest and most blessed Trinity.[48]

Thus Bonaventure avoids any confusion between the intrinsic beauty of the natural world and the beauty of Biblical and ecclesial wisdom. The former is a guide to the latter, and the following ascending hierarchy of beauty obtains: (1) worldly beauty, (2) eschatological beauty and (3) Trinitarian beauty. Trinitarian beauty is the highest and most comprehensive form of beauty. With this proviso, Bonaventure leaves little room for reveling in

---

46    *Brev.*, prologus, 2,4 (ed. Quaracchi V, 204): "Sic igitur totus iste mundus ordinatissimo decursu a Scriptura describitur procedere a principio usque ad finem, ad modum cuiusdam pulcherrimi carminis ordinati, ubi potest quis speculari secundum decursum temporis varietatem, multiplicitatem et aequitatem, irdinem, rectitudinem et pulchritudinem multorum divinorum, procendentium a sapientia Dei gubernante mundum. Unde sicut nullus potest videre pulchritudinem carminis, nisi aspectus eius feratur super totum versum; sic nullus videt pulcritudinem ordinis et regiminis universi, nisi eam totam speculetur." Bonaventure also uses the phrase *carmen pulcherrimum* in 2 Sent. d 13 art. 1 qu. 2, conclusio ad 2 (*Opera Omnia*, II, 316 [ed. Quaracchi II, 315]).

47    José de Vinck renders Bonaventure's *carmen* as a "poem," and Zachary Hayes as "song." Medieval usage included poetry as well as song, and neither sense can be excluded from Bonaventure's use of the metaphor. See *The Works of Bonaventure: The Breviloquium*, trans. José de Vinck (Paterson, NJ, 1970), p. 11, and Zachary Hayes, *A Window to the Divine: A Study of Christian Creation Theology* (Quincy, IL, 1997), p. xii.

48    *Brev.*, 3, 3 (ed. Quaracchi V, 222): "Est enim pulcritudo magna in machina mundana sed longe maior in Ecclesia pulcritudine sanctorum charismatum adornata, maxima autem in Ierusalem superna, supermaxima autem in illa Trinitate summa et beatissima."

worldly beauty on its own terms, thereby reinforcing the Augustinian aversion to the enjoyment of the world for its own sake.[49] But Bonaventure underscores the activity of artistic composition as a way of grasping the divine creative act.

Bonaventure's Franciscan aesthetic is also found in his reference to the great beauty of the fabric of the world. Bonaventure refers to the world as *machina* three more times in the *Breviloquium*.[50] Here I follow de Vinck in rendering the programmatic term with an English word whose roots lie in the Latin *fabrica*, meaning a workshop or its products.[51] In the twelfth century Christian theologians operating sometimes under the influence of Arab science had begun to explore the mechanics intrinsic to nature's own operation as an image of God's artistic power (*potentia fabricatoria*).[52] Bonaventure's use of the concept *machina mundi* represents an implicit endorsement of the high medieval interest in the operations of nature. Both images, the composition of a *carmen* and the world as an organic source of new activity, portray nature as an artifact whose operations are tangible and dynamic. Natural beauty, while inferior to the beauty of Scripture, nonetheless cannot be separated from the order found in a product of human craft. Nature itself for Bonaventure is alive, for it participates in and is indwelled by a creative act. The effort issued in the human work of art is *not* the cause of its inferiority to God's art. Rather this productive side to nature itself expresses what God has wrought.[53] Without attention to the artifice inherent in nature, Bonaventure suggests, the mind cannot ascend to God.

---

49   Augustine introduces the distinction between enjoying (*fruor*) the blessed Trinity and the things in world that are to be used (*utor*) for the glory of God in *De doctrina christiana*, I, 3–5. Peter Lombard transmits this teaching in *Magistri Petri Lombardi in IV distinctae* (Grottaferrata, Roma, 1971), 1 Sent., d. 1 c. 1. Bonaventure takes up the question in *1 Sent. d. 1* (ed. Quaracchi I, 26ff.).

50   *Brev.* 2,1,1 (ed. Quaracchi V, 219): "videlicet quod universitas machinae mundialis producta est in esse ex tempore et de nihilo ab uno principio primo, solo et summo"; *Brev.* 2,3 (ed. Quaracchi V, 220): "De natura corporea quantum ad esse haec tenenda sunt, quod corporalis mundi machina tota consistit in natura caelesti et elementari,..."; and *Brev.* 4,6 (ed. Quaracchi V, 247): "...hinc est, quod quantum ad quartum modum cognoscendi cognovit Christus omnia, quae spectant ad mundanam machinam construendam, longe excellentius quam Adam."

51   *The Works of Bonaventure: The Breviloquium*, trans. José de Vinck, p. 13.

52   Ibid. See also Peter Casarella, "*Naturae Desiderium*: The Desire of Nature between History and Theology," in *Christianity and Secular Reason: Classical Themes and Modern Developments*, ed., Jeffrey Bloechl (Notre Dame, 2012), pp. 33–63.

53   In *De red.* 6, for example.

Bonaventure's comparison of the world to a *carmen pulcherrimum* could very likely be taken from Book XI of St. Augustine's *City of God*.[54] But a revealing remark about Augustine's overarching theology of music in *Itinerarium* II, 10 enables us to make an even more precise determination of Bonaventure's path to the Augustinian *carmen*.[55] Bonaventure, like most medieval theologians, knew and read book VI of Augustine's *De Musica*. In this work Augustine reflects upon the music of creation in terms of a theory of number without paying any attention to Trinitarian productivity. Augustine nonetheless identifies the beautiful composition of the world by God with a *carmen universitatis*, a metered poem that is sung like the hymns of St. Ambrose. In this *carmen* of the whole, the sum of all modulation is greater than any of the individual parts. Augustine identifies this music with six kinds of numbers: (1) sounding numbers or the numbers that are actually heard by the ear in metered verse; (2) memorized numbers or the relationships between these sounds that we retain in memory; (3) occasioning numbers (*numeri occursores*) which are the present and interior recollection of memorial numbers, (4) advancing numbers, which refer to the qualitative lengthening and shortening of musical harmonies, (5) numbers in delight (*numeri in delectatione*) which express the objective joy that derives from hearing music, and (6) judicial numbers, which are the "hidden" rational judgments made upon the entire arrangement and especially upon the overarching numerical equality present in the composition.[56]

Bonaventure cites the six kinds of numbers with great approval in his *Itinerarium*.[57] He then makes the highly elucidating remark that Augustine's judicial numbers also include *numeri artificiales* (man-made numbers).[58] Those numbers, which he classifies as a seventh kind of musical number, give prominence to the artistic conception out of which the composer creates the numerical forms that issue forth in either sounding or advancing numbers. Bonaventure's correction of Augustine on the kinds of musical numbers shows that he is much more attuned to the aesthetics of artistic creation than his Patristic teacher. Bonaventure's interest in the work of art *qua* work of art also follows from his

---

54    XI, 18, *Aurelii Augustini opera*, CCSL 48 (Turnhout, 1968), p. 337: "...tamquam pulcherrimum carmen etiam ex quibusdam quasi antithetis honestaret."
55    Ed. Quarracchi V, 302.
56    *De musica* VI, ed. Jacques-Paul Migne (PL 32 [Paris, 1845]), cols. 1161–1178.
57    *Itinerarium Mentis in Deum*, 2.10 (ed. Quaracchi V, 302).
58    Ibid.: "Ab his autem imprimuntur mentibus nostris numeri artificiales, quos tamen inter illos gradus non enumerat Augustinus, quia connexi sunt iudicialibus; et ab his manant numeri progressores, ex quibus creantur numerosae formae artificiatorum, ut a summis per media ordinatus fiat descensus ad infima."

radically Trinitarian account of the relationship of divine and human creativity. The fecundity and expressivity of the triune God are so present to the Franciscan that he cannot ignore the beauty expressed in the Franciscan *laude* and early forms of polyphony that were being composed in his very midst.[59]

## Temporal and Eternal Simultaneity

We can now return to the original question of eternal simultaneity with a greater appreciation of the underlying analogical rhythms of creative expression in Bonaventure's theology. Rory Fox makes the point that scholastic theologians were in a sense forced by the demands of an orthodox Trinitarian belief (and specifically, the co-eternity of the Son with the Father) to speculate about time and eternity.[60] So Bonaventure was by no means alone in this undertaking. In general, none of the scholastic doctors of this period were content to repeat the standard definitions of eternity without also raising the issue of the relationship of eternity to time. Looking at the development as a whole, Fox maintains, "that thirteenth century thinkers were struggling to find a language which would enable them to talk of a 'togetherness' between durations such as time and eternity."[61] In the light of Bonaventure's Trinitarian doctrine of analogical expressivity, togetherness cannot be construed in merely finite terms. Fox highlights the vocabulary of *coexistentia* ("co-existence") and *comcomitantia* ("comcomitance"), the latter also being found in the writings of St. Bonaventure. Fox also hypothesizes that these medieval doctors extended the doctrine of the analogous predication of names to an analogous predication of duration.[62] In other words, the task of defining the precise language of divine simultaneity must avoid the Scylla of univocity (that is, divine and temporal duration are the same) and the Charybdis of equivocity (that is, God and creatures share only a word "simultaneity" in common). Within the narrow but creative strait of analogical predication, one encounters a variety of viewpoints and, more importantly, the learned ignorance of speaking on the basis of knowledge

---

59   I offer this remark as a likely conjecture that still needs additional historical research. It is based above all on the fact that the most noted contemporary composer of the Franciscan order, namely, Julian of Speyr, served as choirmaster and supervisor of the readings in the refectory at the same residence as Bonaventure, namely, in the Minorite's *Couvent des Cordeliers* in Paris. See Christian Eugene, "Saint Bonaventure et le grand couvent des cordeliers de Paris," *Études franciscaines* 18 (Supplément annuel, 1968), 167–182.

60   Rory Fox, *Time and Eternity*, pp. 50–94.

61   Ibid., p. 314.

62   Ibid., pp. 314–319.

drawn from similitudes. For this very reason, Bonaventure makes the distinction between simultaneity in an instant and simultaneity *in imperceptibili tempore* ("in a time that cannot be perceived").[63]

One scientific issue that preoccupied the medieval theologians concerned the cause of temporal simultaneity or, as Fox terms it, the origin of temporal unity. In general, thinkers who favored a stricter interpretation of the Aristotelian view of time (including Aquinas) were inclined to accept a quasi-Deist position espoused by Avicenna, whereby God sets in motion the *primum mobile* as the outermost layer of the universe and this movement in turn causes the temporal motion of all life on earth.[64] Among the other explanations for the unity of time were those of Alexander of Hales and St. Bonaventure. Alexander espoused an explanation based upon exemplary causality.[65] Bonaventure undoubtedly knew of his teacher's viewpoint and even seems to allude to it. At one point in Bonaventure's *Disputed Questions on the Mystery of the Trinity*, Bonaventure notes that there is a relationship between the perpetuity of an image in the order of creation and its exemplar in the eternal Creator. Here he is referring to the fact that essence and power exist simultaneously in the human soul.[66] Bonaventure himself favors an explanation based upon the material causality of time.[67] The details of his account of material causality are quite fascinating, for they confirm the point made about regarding the centrality of expression to Bonaventure's thought. Bonaventure maintains that potentiality is the nature of matter and that all changes occur as a result of potentiality. Etienne Gilson referred to as his universal analogy because it entailed seeing the entire world (earthly and heavenly) as a unity of actuality and potentiality.[68] In brief, the mutability of matter for Bonaventure is the source of temporal change. Thus, the potentiality of matter is also the basis for thinking about what holds time together.

### Conclusion

Let us return now to the theme with which I began, namely, the creativity of the Trinity as a basis for thinking about simultaneity. Human language can

---

63  *II Sent.* d. 5 art. 2, qu. 2, ad. 2 (ed. Quaracchi II, 153): "Ad illud quod obiicitur, quod simul ceciderunt; potest dici, quod simul, non quia in eodem instanti, sed quia repente, quasi in imperceptibili tempore; et ibi bene potest esse ordo, quamvis non percipiatur."

64  Ibid., pp. 75–86.

65  Ibid., pp. 66–68.

66  *Qq. d. de trin.*, q.5, art.2 (ed. Quaracchi V, 94).

67  Rory Fox, *Time and Eternity*, pp. 69–75.

68  Etienne Gilson, *La Philosophie de Saint Bonaventure* (Paris, 1924), pp. 196–227.

express the mystery of divine simultaneity only by analogy. Bonaventure's real contribution on this score, it seems to me, lies in his reflection upon the notion of Trinitarian *productio*. He posits more than a merely abstract togetherness of divine eternity and creaturely temporality, for the *simul* in the Trinity is the source of a primal outflowing into time of unlimited fecundity. Specifically, he maintains that one can distinguish in the Trinity processions between the process of being produced (*produci*) and the product of the process (*productum esse*).[69] There is a relationship between this distinction and the mode of receptivity on the other end of the production. Here Bonaventure distinguishes between a difference between receiving (*accipere*) and having received (*accepisse*), which is plainly evident in any worldly exchange of gifts. The gift I received last Christmas, for example, is no longer being given to me. The one I would like to receive next Christmas is not yet mine. In the Trinitarian process, by contrast, there is a strict identity of receiving and having received since the Son has always received and always receives just as the Father always generates and always has generated. The divine, Trinitarian production is based upon the identity of the process and the product. The process is not ripped out of a temporal matrix so much as it establishes the reconfiguring of temporality into a new unity of producing and having been produced (or receiving existence and having received one's existence). The process discloses a new, non-temporal mode of simultaneity.

Let me conclude with three theses on Bonaventure's approach to non-temporal simultaneity.

1) All temporal processes have a beginning and an end. Non-temporal simultaneity cannot be measured in these terms. The first thesis is that the on-going process (*producere, accipere*) and the finished product (*esse productum, accepisse*) are the same in the Trinitarian production.

2) As an elucidation of this thesis, let me then offer an additional clarification. Bonaventure says that the process and the product *are* the same. The sameness (*idem*) lies in their ontological co-existence, not in specific perduring characteristics.[70] So Bonaventure's point here is not that the product at the end appears to be similar to whatever it was at the beginning of the process.

3) So far we have established that the product character of the process is simultaneous with the process character of the product. In such

---

69   What follows is based upon *Qq. d. de trin.*, q. V, a. 2, ad 10 (ed. Quaracchi V, 94).

70   As noted above, the perdurance of simultaneous characteristics in the temporal order may express a higher order of simultaneity, but this is really a distinct point.

simultaneity there is an essentially processual character to what is produced and a basic orientation to productivity in the process itself. What issues forth from an unoriginate source does not ever leave its source even while it remains eternally distinct from it.[71] The product is never finished and never incomplete. It is, to use a favored term of Bonaventure, *ars infinita*. The process, too, is open-end, but not in the sense of the utter and pointless absence of a beginning or an end, that is, what Hegel considered "a bad infinite."[72] The end of the process is the complete restoration of the temporal order to the Triune God. To view this end as *merely* temporal one is also short-sighted. Guided by the light of revelation in Luke 17, 21, Bonaventure would say that it is an eternal presence in our very midst.[73]

---

71    *Qq. de trin.* q. v, a. 2, ad 2. (ed. Quaracchi v, 95): "ideo enim sequitur causam effectus, quia ab ipsa manat, ita quod ab ipsa differt secundum essentiam; et propter hoc nec ita nobile est, sicut causa, nec aequatur ei in duratione nec in virtute aequalitate omnimoda."

72    G.W.F. Hegel, *Encyclopedia of the Philosophical Sciences in Basic Outline, Part i: Science of Logic*, eds. and trans. Klaus Brinkmann and Daniel Dahlstrom (Cambridge, UK, 2010), p. 149. "It is of great importance to grasp the concept of the true infinity properly and not merely to stop short at the bad infinity of the infinite progression. When the infinity of space and time is under discussion, it is at first the infinite progression that one tends to focus on. ...This much is indeed correct, namely that we eventually abandon proceeding further and further in such contemplation, but on account of the tediousness, not the sublimity, of the task. Engaging in the contemplation of this infinite progression is tedious because the same thing is incessantly repeated here. A limit is posited, it is surpassed, then again a limit, and so on endlessly. So there is nothing here but a superficial alternation that remains stuck in the finite."

73    St. Bonaventure, *Commentary on the Gospel of Luke*, ed. Robert Karris, O.F.M. (Saint Bonaventure, N.Y., 2004), pp. 1664–1665.

# CHAPTER 6

# Hugh of St. Cher and Thomas Aquinas: Time and the Interpretation of the Psalms

*Aaron Canty*

The psalms played a central role in the life of the Church in the Middle Ages. Whether in their function in the eucharistic liturgy and the divine office, as the basis for spiritual reading, or in the classrooms of the cathedral schools and universities, the psalms were ubiquitous texts for medieval Christians.[1] Scholastic commentaries on the psalms appropriated the exegesis of Augustine, Jerome, and Cassiodorus, among others, to shed light on the multiple layers of meaning latent in the Psalter.[2] Scholars have often noticed the move away from allegorical or tropological interpretations to the literal sense in the thirteenth century.[3] This essay provides additional evidence to that observation and also highlights the significance of an increasingly more sophisticated *divisio textus*. The development from more spiritual exegesis to more literal exegesis and from a relatively simple *divisio textus* to a more complex one can be seen in the Psalms commentaries of Dominican theologians Hugh of St. Cher and Thomas Aquinas. Moreover, their distinct modes of exegesis, emphasizing allegorical and literal respectively, reveal each mode's implicit assumptions about the ways exegesis defines spiritual presence and the spiritual use of time.

---

1   See Jacques Dubois, "Comment les moines du Moyen Age chantaient et goûtaient les Saintes Ecritures," in *Le Moyen Age et la Bible*, eds. Pierre Riché and Guy Lobrichon (Paris, 1984), pp. 261–298; Joseph Dyer, "The Bible in the medieval liturgy, *c.* 600–1300," in *The New Cambridge History of the Bible. Volume 2, From 600 to 1450*, eds. Richard Marsden and E. Ann Matter (Cambridge, UK, 2012), pp. 659–679; Theresa Gross-Diaz, "The Latin psalter," in ibid., pp. 427–445; Hans Rost, *Die Bibel im Mittelalter* (Augsburg, 1939), pp. 79–84; Pierre Salmon, *L'office divin au Moyen Age. Histoire de la formation du bréviaire du IXe au XVIe siècle* (Paris, 1967); Robert Taft, *The Liturgy of the Hours in East and West: The Origins of the Divine Office and Its Meaning for Today* (Collegeville, MN, 1986), pp. 93–213; Ceslas Spicq, *Esquisse d'une Histoire de l'Exégèse Latine au Moyen Age* (Paris, 1944), pp. 349–364; *The Place of the Psalms in the Intellectual Culture of the Middle Ages*, ed. Nancy van Deusen (New York, 1994).

2   See Jean Châtillon, "La Bible dans les Ecoles Du XIIe Siècle," in *Le Moyen Age et la Bible*, pp. 190–192; Gilbert Dahan, *L'Exégèse chrétienne de la Bible en Occident médiéval, XIIe–XIVe siècle* (Paris, 1999), pp. 146–150, 174–175, 253–255; Rost, pp. 86–87; Spicq, pp. 52–59; Jacques Verger, "L'Exégèse de l'Université," in *La Moyen Age et la Bible*, pp. 199–232 (208, 227, 232).

3   Especially Beryl Smalley in *The Study of the Bible in the Middle Ages* (Notre Dame, 1964).

© KONINKLIJKE BRILL NV, LEIDEN, 2016 | DOI 10.1163/9789004312319_008

I

Hugh of St. Cher compiled his *Postilla super Psalterium* probably in the mid- to late 1230s,[4] and his prologue to the long version emphasizes the contemplative dimension of the psalms.[5] The scriptural verse which opens the *accessus* is Song of Songs 3, 11: "Go forth, O daughters of Zion, and behold King Solomon, with the crown with which his mother crowned him on the day of his wedding, on the day of the gladness of his heart."[6] Hugh uses the opening word, "egredimini," to reflect on how spiritual life involves different kinds of *egressus* (departures, journeys). The *egressus* of evil involves moving away from God, neighbor, and oneself through pride, avarice, and selfish indulgence (*luxuriam*). The *egressus* of goodness, however, involves a series of three movements from individual selfishness to God. The first movement is from the flesh to the spirit through contrition; the second movement occurs through meditation on the divine law and is from nature to those things which are above nature; the third movement is from earth to heaven and occurs "through the dissolution of the body and soul or through contemplation."[7]

The "daughters of Zion" are clerics, theologians, and religious. Hugh is clear that these three groups of people are not sons, but daughters on account of their "exceptional fecundity," which is manifested in "true doctrine, good works, and a holy existence (*conversatione*)."[8] But why are they "of Zion"? Hugh asserts that "Zion" means speculation, and that speculation consists of three things: inquiry, prayer, and contemplation. "In the first," Hugh says, "the Bridegroom is sought, in the second, He is asked, and in the third He is found. In the first the Bridegroom speaks to the bride, in the second, the bride

---

4   Martin Morard, "Hugues de Saint-Cher, Commentateur Des Psaumes," in *Hugues de Saint-Cher (+1263): Bibliste et Théologien*, eds. Gilbert Dahan, Louis-Jacques Bataillon, and Pierre-Marie Gy (Turnhout, 2004), pp. 101–153 (141–143, 150).

5   Regarding Hugh's spiritual exegesis, see Gilbert Dahan, "L'Exégèse de Hugues. Méthode et Herméneutique," in *Hugues de Saint-Cher (+1263): Bibliste et Théologien*, pp. 65–99 (89–94). Martin Morard goes so far as to say that there is an "atrophy" of the literal sense (pp. 138–144).

6   Scriptural citations are from the RSV except where noted.

7   De mundo ad coelum, per corporis, et animae dissolutionem, seu etiam per contemplationem, *Postilla super Psalterium* (*Opera omnia in universum Vetus et Novum Testamentum* (Lyons, 1645), II, fol. 2ra).

8   Etenim non filii, sed filiae dicuntur genere foeminino, non propter sexus infirmitatem, sed propter prolis foecunditatem, quam debent habere in doctrina vera, in operatione bona, in conversatione sancta, *Postilla super Psalterium*, fol. 2ra.

speaks to the Bridegroom, and in the third, there is an embrace and an exchange of kisses."[9]

This opening reflection on spiritual ascent relates to the psalms because Hugh follows Augustine's tripartite division of the psalms. Although this division is recounted by Peter Lombard in his Psalms commentary and in the *Glossa ordinaria*, the full account can be found in Augustine's sermon on Psalm 150. Augustine provides several explanations for why the number one hundred fifty is significant, but here is Augustine's explanation for finding three divisions to the psalms:

> I do not think that it is an accident that the fiftieth psalm is concerned with repentance, the hundreth with mercy and judgment, and the one hundred and fiftieth with the praise of God offered by his saints; for this is the route by which we travel toward the blessed life of eternity. We begin by condemning our sins and continue by leading good lives, so that after repudiating our bad life and persevering in a good one, we may deserve a life that is eternal.[10]

Hugh expatiates considerably on Augustine's idea of the psalms as mirroring spiritual ascent. There are three states—penance, justice, and glory—which correspond to those beginning the spiritual life, those advancing, and those who are perfect.[11]

> The first [set of fifty] begins, the second advances, and the third brings to perfection. The first establishes a foundation, the second builds walls, and the third puts a roof over the building, and thus is the spiritual edifice perfected. One searches for King Solomon, that is Christ, in the first *egressus* and the first room, one invites him in the second, as we have said earlier, but in the third, one sees Him and embraces Him. Therefore, "Go

---

9    In primo quaeritur sponsus. In secundo rogatur. In tertio tenetur inventus. In primo loquitur sponsus sponsae. In secundo sponsa sponso. In tertio dantur oscula, et iunguntur amplexus, *Postilla super Psalterium*, fol. 2ra.

10   Non enim frustra mihi videtur quinquagesimus esse de paenitentia, centesimus de misericordia et iudicio, centesimus quinquagesimus de dei laude in sanctis eius. Sic enim ad aeternam beatamque tendimus vitam, primitus nostra peccata damnando, deinde bene vivendo, ut post vitam condemnatam malam et gestam bonam mereamur aeternam, *Enarrationes in psalmos* 150, 3 (CSEL 95/5), ed. F. Gori with the assistance of I. Spaccia, trans. Maria Boulding, O.S.B., in *Expositions of the Psalms* 6 (Hyde Park, 2004), p. 511.

11   *Postilla super Psalterium*, fol. 2ra.

forth, O daughters of Zion," in the first, second, and third *egressus*, "and
behold King Solomon, with the crown," just as He promised in Isaiah 33,
"Your eyes will see the king in his beauty." Solomon, however, is seen in
the world, and He will be seen in the Last Judgment and in his Kingdom.[12]

Using imagery found frequently in Victorine spiritual theology,[13] this text links
Augustine's emphasis on the triad of repentance, mercy and justice, and eter-
nal praise with the image of the spiritual edifice (found in the theologians of
St. Victor),[14] the image of Christ's threefold appearance in the world (from
Bernard of Clairvaux),[15] and the threefold ascent through searching, praying,
and contemplating.

This passage also returns to Song of Songs 3, 11 after a lengthy digression.
Hugh had been focusing on "Go forth, O daughters of Zion," but now he turns
his attention to the remainder of the verse. Once the daughters of Zion go
forth, the exhortation is for them to "behold King Solomon, with the crown
with which his mother crowned him on the day of his wedding, on the day of
the gladness of his heart." Seeing King Solomon means seeing with faith Christ
in the world, in judgment, and in His kingdom. Implicitly linking the vision
with the literary structure of the Psalter, Hugh says that "the first vision is of
mercy, the second is of justice, and the third is of glory. Hence Blessed Bernard
says, 'Jesus Christ was seen as small and loveable in exile, great and terrible in
judgment, and great and praiseworthy in His kingdom.'"[16] This connection

---

12   Sic ergo prima quinquagena, inchoat; secunda, promovet; tertia, consummat. Prima, ponit
      fundamentum: secunda, erigit parietes; tertia, superponit tectum, et sic consummatur
      aedificium spirituale. In primo egressu, et prima dieta, quaeritur Rex Salomon, id est,
      Christus. In secundo rogatur, ut diximus: sed in tertio, videtur et tenetur. Igitur egredimini
      filiae Sion, primo, et secundo, et tertio egressu, et videte Regum Salomonem in diademate
      suo, sicut promisit, Isa. 33.c., Regem in decore suo videbunt. Videtur autem Salomon in
      mundo, videbitur in iudicio, videbitur et in regno, *Postilla super Psalterium*, fol. 2rb.

13   Although the imagery goes back to Jerome and, even further, to Origen. See Henri de
      Lubac, *Medieval Exegesis: The Four Senses of Scripture* III, trans. E.M. Macierowski (Grand
      Rapids, 2009), pp. 211, 236, 248–250.

14   See Hugh of St. Victor, *Didascalicon de studio legendi*, ed. Charles Henry Buttimer
      (Washington, D.C., 1939), l. 6, c. 2. On the relationship between the foundation of history
      and the superstructure, see Franklin T. Harkins, *Reading and the Work of Restoration:
      History and Scripture in the Theology of Hugh of St. Victor* (Toronto, 2009), pp. 197–253.

15   Bernard of Clairvaux, *Sermones in adventu Domini* 5 (Sancti Bernardi Opera, ed. J. Leclerq,
      C.H. Talbot, and H.M. Rochais (Rome, 1957–1977), IV, pp. 188–190).

16   Prima visio est misericordiae: secunda iustitiae: tertia gloriae. Unde B. Bernardus dicit:
      Iesus Christus in exilio visus est parvus, et amabilis, in iudicio magnus, et terribilis, in
      regno videbitur magnus, et laudabilis, *Postilla super Psalterium*, fol. 2rb.

becomes explicit a little later, when Hugh exhorts, "Go forth, therefore, daughters of Zion, in the first *egressus*, see the small and loveable King David in the world, and love in the first fifty [psalms]. Go forth in the second one, see the wrathful and terrible kingdom of David in judgment, and fear in the second fifty. Go forth in the third *egressus* and see the King in his great and praiseworthy kingdom in the third fifty."[17]

Just as one can see the king in three ways, so too does the king have a threefold diadem. According to Song of Songs 3, 11, the diadem is that with which the king's "mother crowned him on the day of his wedding, on the day of the gladness of his heart." The first coronation, therefore, occurred when the Virgin Mary crowned the Son of God with flesh on the day of His espousals. The second and third coronations occurred after Christ's wedding day. The second coronation occurred when Christ's step-mother (*noverca*), the Synagogue, crowned Him with thorns during His passion. The third coronation occurred when Christ's Father crowned Him with glory in the resurrection. Only the first coronation, however, occurred on Christ's wedding day. Although Augustine is never mentioned here, the language used to describe the wedding is very similar to Augustine's description (in several places) of Christ entering into a nuptial relationship with humanity in the incarnation.[18] The nuptial union occurs in the embrace between Word and humanity in the nuptial chamber that is the Virgin's womb (one is reminded here of Bernard's meditations on the "kiss" in the opening of the Song of Songs).[19] The incarnation is Christ's wedding day and "the day of the joy of his heart."

Only after these reflections on the spiritual significance of the literary structure of the psalms, does Hugh address the author, subject matter, author's intention, and *modus agendi* (mode of exposition) of the psalms in a way reminiscent of the *Glossa ordinaria*, Gilbert of Poitiers, and Peter Lombard. Hugh's discussion of the coronation of Christ in the nuptial chamber of His mother allows for a smooth transition to a discussion of the subject matter of the psalms. David, the author, writes of the "whole Christ (*Christus integer*), that is, Head and members, Bridegroom and bride, Christ and the Church, with their

---

17   Egredimini ergo filiae Sion, primo egressu, et videte Regem David in mundo parvum amabilem, et amate in prima quinquagena. Egredimini secundo, et videte Regem David in iudicio iratum, et terribilem, et timete, et hoc in secunda quinquagena. Egredimini et tertio, et videte Regem in regno magnum, et laudabilem, in tertia quinquagena, *Postilla super Psalterium*, fol. 2rb.

18   Cf. *Enarrationes in psalmos* 17, 51; 18 (2), 6; 26 (2), 2; 30 (2), 3–4; 40, 1; 44, 1; 58 (1), 2; 61, 4; 127 (1), 3; 142, 3.

19   Cf. Bernard of Clairvaux, *Sermones super Cantica canticorum* 1, (*Sancti Bernardi Opera*, I).

respective conditions and properties."[20] The author's intention is "to show how those who became deformed in Adam were reformed in Christ."[21] And the *modus agendi* of the psalms has three dimensions. The first consists of describing the Head (in three ways, that is, according to Christ's divinity, his humanity, and *secundum transumptionem* [metonymy or metalepsis], by which what pertains to the members is attributed to the Head). The second dimension describes the members (in ten ways: apostles, martyrs, confessors, virgins, anchorites, prelates, those who are celibate, those who are married, those who are good, and those who are evil). The third dimension describes Head and members together, and Hugh elaborates no further on the *modus agendi*. All of this material is derived from twelfth-century sources, or the Early Church, with the exception of Hugh's enumeration of the ten ways that the psalms describe the members of the Church.

Hugh also says little about the title of the book of Psalms, or rather the titles. The more obvious title would be "the Book of Hymns, or of Psalms, or of Soliloquies," but Hugh, like other interpreters, also says that Psalm 1, which has no title, functions as a title for the entire collection of psalms.[22] The fact that there are two kinds of titles is analogous to the tabernacle as described in Exodus. Hugh provides an explanation for each title, and these explanations or prologues function like two of the partitions of the Israelite tabernacle. One court allows access to the threshold of the Holy of Holies, and one chamber allows entrance into the Holy of Holies. The first prologue explains the title of the book, its layout and structure, and the reason for its use in the Church, but the second prologue explains these things in light of the contents of the Psalms, thus bringing the reader from considerations of the text into the text itself.[23]

Hugh provides this second prologue in commenting on Psalm 1, which acts as a template for the entire book of Psalms. Hugh divides the text of the psalm in such a way as to illustrate its function as exemplar. Here is the text of Psalm 1:

> Blessed is the man who walks not in the counsel of the wicked, nor stands in the way of sinners, nor sits in the seat of scoffers; but his delight is in the law of the LORD, and on his law he meditates day and night. He is like

---

20    Materia libri, est Christus integer, id est, caput et membra: sponsus, et sponsa: Christus, et Ecclesia, cum suis conditionibus, sive proprietatibus, *Postilla super Psalterium*, fol. 2va.
21    Intentio Prophetae, est in Adam deformatos ostendere reformatos in Christo, *Postilla super Psalterium*, fol. 2va.
22    *Postilla super Psalterium*, fols. 2va–3ra.
23    *Postilla super Psalterium*, fol. 2va.

a tree planted by streams of water, that yields its fruit in its season, and its leaf does not wither. In all that he does, he prospers. The wicked are not so, but are like chaff which the wind drives away. Therefore the wicked will not stand in the judgment, nor sinners in the congregation of the righteous; for the LORD knows the way of the righteous, but the way of the wicked will perish.

As Hugh mentioned in the first prologue, there are three sets of fifty psalms for the states of penance, justice, and glory. The state of penance is described by the first fifty psalms, the state of justice is shown in the second fifty psalms, and the state of glory is depicted by the third set of fifty.

In the opening comments on Psalm 1, Hugh explains that the first two verses of Psalm 1 express the state of penance. Citing Gregory the Great, he says that penance consists both of lamenting one's past sins and refraining from committing sins. The first verse of the psalm, "Blessed is the man who has not walked in the counsel of the ungodly, nor stood in the way of sinners, nor sat in the chair of pestilence," shows that the just person refrains from committing sinful actions, having sinful thoughts, and saying evil things about others.[24] It also shows that one does penance by avoiding those situations that have led one to sin or could lead one to sin. In this way, Hugh believes that remorse for previous sins is implied in the opening words of the psalm. What is also necessary in the state of penance, in addition to detachment from sin, is conforming "one's will to the divine will."[25] That is what the second verse of the psalm illustrates: "But his delight is in the law of the LORD, and on his law he meditates day and night."

The state of justice or of those advancing in holiness, is described by the first part of the third verse, "He is like a tree planted by streams of water, that yields its fruit in its season." On this passage, Hugh merely comments that such a person "always grows while in the vigor of his youth."[26] The fruit to be borne will ripen in the state of glory.

---

24    *Postilla super Psalterium*, fols. 3ra–b.

25    Hoc est, quantum ad illud Gregorii, flenda non amplius committere. Sed hoc est non suf-
      ficit, si in his quae commisit remaneat, unde hoc removet subdens: *Et in via peccatorum*
      quam iam intraverat, *non stetit*, sed per poenitentiam reversus est. Hoc est illud Gregorii,
      commissa flere. Quod manifestat conformando voluntatem voluntati divinae, unde
      sequitur: *Sed in lege Domini voluntas eius, Postilla super Psalterium*, fol. 3rb.

26    Secundus status est proficientium, qui notatur, ibi: *Et erit tanquam lignum quod planta-
      tum est*, etc.; quod, scilicet, Semper crescit dum est in vigore iuventus, *Postilla super
      Psalterium*, fol. 3rb.

The third state is that of glory or of those who are perfect, and the remainder of Psalm 1 illumines various facets of this state. The remainder of verse three shows that the life of glory will not end, "and its leaf does not wither. In all that he does, he prospers." The fourth, fifth, and sixth verses contrast the truly righteous from the wicked and describe their final judgment, "The wicked are not so, but are like chaff which the wind drives away. Therefore the wicked will not stand in the judgment, nor sinners in the congregation of the righteous; for the LORD knows the way of the righteous, but the way of the wicked will perish." Since the whole purpose of the Psalms is to lead people to eternal beatitude through penance and the cultivation of virtues, it is fitting that the opening phrase of Psalm 1, a title for the whole book is "Blessed is the man..."

Of more interest to Hugh, however, than the titles of the book of Psalms is the question of why the psalms are considered prophecy when they are in the category not of prophets, but rather of hagiographers according to the Jews. Hugh explains that one is a prophet according to inspiration, office, and habit (*consuetudine*). The office of prophecy refers to those who are sent to the people of Israel to reveal something concerning the future. David was not a prophet in this sense and that is why the Jews excluded the psalms from the prophetic writings. Nonetheless, the psalms are prophecy in the sense of communicating secrets about the future as revealed by God, and they are unique among the prophetic writings because of their subject matter, the whole Christ.[27]

In a spiritual sense the psalms are a loaf of bread to satisfy the multitudes who follow Christ. By breaking the bread through interpretation (it is important to recall that Christ is recognized in the breaking of the bread in Luke's gospel), Hugh will multiply the "Davidic loaf" in order to feed the five thousand.[28] Although in the *postillae* in general Hugh will have a mystical or anagogical interpretation, in the commentary on the psalms he usually applies the literal, allegorical, and tropological senses.[29]

In the case of Psalm 1, the literal sense pertains to the Jews. It is not that the "Beatus vir" corresponds to Israel, but rather that "in consilio impiorum" refers to the Jews. More specifically, it refers to the Jews who sought help from Egypt

---

27    *Postilla super Psalterium*, fols. 2vb–3ra.

28    *Postilla super Psalterium*, fol. 2vb.

29    *Postilla super Psalterium*, fol. 3rb. The question of which senses are applied is a complicated one. In the prologue to the entire commentary on Scripture, Hugh says that there are four senses: history (*veritas historiae*), allegory, anagogy, and tropology (*Opera Omnia* I). In the *Postilla super Psalterium*, Hugh generally applies three senses: literal (*ad litteram*), allegorical, and moral. He does introduce, however, anagogical interpretations on occasion (see, for example, Hugh's interpretation of the last verses of Psalm 53, cf. ibid., fol. 138rb).

rather than from God (cf. Is. 30, 2). Allegorically it refers to Christ, who "walks not in the counsel" of wicked ministers, nor has He sat in the "seat of scoffers" ("cathedra pestilentiae" or "chair of pestilence" in Hugh's version). According to Hugh, the "chair of pestilence" is threefold, namely "by teaching error, as the heretics do; by encouraging evil actions, as false Christians do; and by offering a bad example, as evil prelates do." Christ committed none of these egregious actions and so did not sit in the "chair of pestilence."[30] The moral interpretation of Psalm 1, 1 applies to the just person in general. The righteous are tempted by three kinds of ungodly enemies who offer unwholesome counsel: the world, the flesh, and demons.[31]

There is one other phrase that merits both an allegorical and moral interpretation from Hugh, and that is verse 3: "'He is like a tree planted by streams of water, that yields its fruit in its season, and its leaf does not wither. In all that he does, he prospers." Hugh's application of this verse to the Word of God, the Second Person of the Trinity, is not obvious, but he relates it to other scriptural references to trees and their fruit. For example, Proverbs 3, 13–16 says, "Happy is the man who finds wisdom, and the man who gets understanding, for the gain from it is better than gain from silver and its profit (*fructus*) better than gold. She is more precious than jewels, and nothing you desire can compare with her. Long life is in her right hand; in her left hand are riches and honor." Hugh interprets this passage as comparing wisdom to a tree, and that wisdom ultimately is Christ. He also makes a reference to Revelation 2:7: "He who has an ear, let him hear what the Spirit says to the churches. To him who conquers I will grant to eat of the tree of life, which is in the paradise of God." Having called to mind passages that indicate the heavenly origins of wisdom, which is a *lignum vitae*, Hugh then mentions the relationship between the tree and fruit: "But before [the tree] could bear fruit, it should be planted. The Son of God, moreover, was planted when He was united to human nature."[32]

The leaf of the tree, interpreted allegorically, is the Word, which does not "fall off" from the truth, nor does it lose its power or efficacy. Everything He does "prospers," in the sense that His sufferings lead efficaciously to the salvation of the human race.[33] Only after the leaf appears does the tree bear fruit,

---

30    Et dicitur cathedra pestilentiae tribus modis. Errorem docendo, ut haeretici. Ad malum
        incitando, ut falsi Christiani. Malum exemplum praebendo, ut mali praelati, *Postilla super*
        *Psalterium*, fol. 3vb.
31    *Postilla super Psalterium*, fol. 3vb.
32    Sed antequam ferat fructum, oportet quod plantetur. Plantatus autem fuit filius Dei,
        quando unitus est humanae naturae, *Postilla super Psalterium*, fol. 3va.
33    *Postilla super Psalterium*, fol. 4ra.

and this nourishing fruit is Christ's "edible flesh" given during the Last Supper. The fruit could also be the "sacraments of the New Law," which are preceded by the leaves of the Old Testament, that is, those things pertaining to the Law, such as the sacrificial lamb, that foreshadow the sacrifice of Christ.[34]

A moral interpretation of Psalm 1, 3 links the tree, the leaf, and the fruit to the just person. The just person is planted in the running waters of graces and tears. The leaf is the person's word, which always tries to be useful and effective in some way. The person's fruit consists of the good works that naturally proceed from a humble and pious soul. It is brought forth freely and "in its season," that is, in this present life.[35]

Just as Hugh criticized "false Christians" and "evil prelates" for their actions, so, too, does he criticize theologians who fail to have both leaves (useful words) and fruit (good works) come forth from their branches. He says that evil people have neither good works nor useful words, and some, "namely theologians," have their works "fall off," but not their words. He explains further that "leaves do not fall off in summer, but in winter. Thus there are many who have good words in prosperity, but evil and less frequent words in adversity."[36]

As Hugh concludes his interpretation of Psalm 1, he notes the correspondence between the verse of the psalm and the beatitudes of Matthew 5, thus providing another lens through which to read the psalm. The poor in spirit are those who "walk not in the counsel of the wicked"; the meek are those who have not "stood in the way of sinners"; the mourners are those who have not "sat in the chair of pestilence"; those who hunger and thirst for righteousness meditate on the law of the Lord "day and night"; the merciful are "like a tree planted by streams of water"; the pure in heart "bring forth its fruit in its season"; the peacemakers do not have their leaf wither; and those who are persecuted for righteousness' sake will have whatever they do "prosper."[37]

---

34 'Quod fructum suum,' id est, carem esibilem. 'Dabit' in coena. Hunc fructum folia praecesserunt sacramenta novae legis, Isa. 4.a. 'Et fructus terrae sublimis.' Vel fructum, missionem Spiritus sancti, Ioan. 16.b. 'Nisi ego abiero,' etc. *Postilla super Psalterium*, fol. 3va.

35 'Et erit tanquam lignum quod plantatum est secus decursus aquarum,' gratiarum, lacrymarum... Sed quia crescens sic in virtutibus, debet etiam fructificare in operibus... Tempus uniuscuiusque ad operandum, est praesens vita, *Postilla super Psalterium*, fol. 3vb.

36 Et nota quod in quibusdam defluunt fructus, id est opera, et folia, id est, verba, scilicet, in malis... In quibusdam vero fructus, et non folia, scilicet in theologis. Item folia non defluunt in aestate, sed in hyeme. Sic et multi sunt qui bona verba habent in prosperitate, sed mala et defluentia in adversitate, *Postilla super Psalterium*, fol. 4ra.

37 *Postilla super Psalterium*, fol. 4ra.

II

Thomas Aquinas' incomplete commentary on the psalms, the *Postilla super Psalmos*, was begun in 1272 towards the very end of his career.[38] The commentary synthetically draws on previous patristic and medieval commentaries, but the most noticeable feature of the commentary at first glance is the *divisio textus*, which is much more complex than that of Hugh's *postilla*.[39] The scriptural verse which opens the *accessus* is not from the Song of Songs, but rather from Ecclesiasticus 47, 9: "In all that he did he gave thanks to the Holy One, the Most High, with ascriptions of glory."

Applying Aristotle's four categories of causes, Thomas divides this verse into four sections. The phrase "In all that he did" refers to God's works. These works are the matter or the material cause of the psalms, and they consist of creation, governance, reparation, and glorification. Psalm 8, 4 describes creation: "When I look at thy heavens, the work of thy fingers..." Psalm 77, 2 says "I will open my mouth in a parable," which Thomas takes to mean the stories of the Hebrew Scriptures which portray God's loving providence over creation. Psalm 3, 6 says, "I lie down, and sleep." This describes Christ's passion and death which restores humanity's relationship with God. And Psalm 149, 5 describes the "work of glorification": "Let the faithful exult in glory." Regarding the universality of these four works, Thomas says, "The material is universal, for while the particular books of the canon of Scripture contain special materials, this book has the general material of the whole of theology... And this is the reason why the Psalter is read more often in the Church, because it contains the whole of Scripture."[40] The Psalter is unique then because it is a microcosm of the entire canon of Scripture and of theology and because the matter consists of "Christ and His members."[41] This last statement removes any perceived equality among the works of creation, governance, reparation, and glorification, and indicates that the psalms portray Christ's restoration and elevation of humanity in a unique way. Thomas expresses the uniqueness of the Psalms by modifying a passage from Cassiodorus: "All the things that pertain to faith in the

---

38    Jean-Pierre Torrell, *Saint Thomas Aquinas, 1: The Person and His Work*, trans. Robert Royal (Washington, D.C., 2005), pp. 257–261.

39    See Thomas F. Ryan, *Thomas Aquinas as Reader of the Psalms* (Notre Dame, IN, 2000), pp. 52–53.

40    Materia est universalis: quia cum singuli libri canonicae Scripturae speciales materias habeant, hic liber generalem habet totius Theologiae... Et haec est ratio, quare magis frequentatur Psalterium in ecclesia, quia continet totam Scripturam, *Postilla super Psalmos* (*Opera Omnia* (Parma, 1852–1873), XIV, p. 148).

41    See Ryan, pp. 20–28.

Incarnation are so clearly set forth in this work that it almost seems like the Gospel, and not prophecy."[42]

The Psalter is also unique because of its mode or formal cause, which is expressed by the words, "he gave thanks." The books of sacred Scripture employ several modes of discourse, and Thomas enumerates four. The first is the narrative mode, and this mode is used in the historical books. The second is the admonitory, hortatory, and preceptive mode (and this is one mode), and this mode is used in the Law, the Prophets, and the books of Solomon. The third is the disputative mode, and this mode is employed in Job and in the Pauline epistles. The fourth mode is the deprecative or laudative mode. This mode is the one used in the Psalter, because, as Thomas says, "whatever is said in the other books in the modes previously mentioned, is put here by the mode of praise and prayer."[43]

It is precisely at this juncture that Thomas explains the title of the book, "The beginning of the book of hymns, or of the soliloquies of the prophet David, concerning the Christ." "A hymn is the praise of God with song; a song is the exultation of the mind dwelling on eternal things, bursting forth in one's voice. Therefore, [David] teaches how to praise God with exultation. A soliloquy is the conversation of a person with God, or within himself alone, because this is fitting for one who praises and one who prays."[44]

Hugh of St. Cher's discussion of the title occurred at the end of his prologue, after he had treated the author, matter, intention, and *modus agendi*. Thomas' placement of this discussion as pertaining to the form of the psalms and the way he interprets the meaning of the title, however, allow for a smooth transition to his discussion of the end or final cause of the psalms.

---

42  Omnia enim quae ad fidem incarnationis pertinent, sic dilucide traduntur in hoc opere, ut fere videatur evangelium, et non prophetia, *Postilla super Psalmos*, p. 148.

43  Modus seu forma in sacra Scriptura multiplex invenitur. Narrativus: Eccles. 42: *Nonne Deus fecit sanctos suos enarrare omnia mirabilia sua?* Et hoc in historialibus libris invenitur. Admonitorius et exhortatorius et praeceptivus: Ad Titum 2: *Haec loquere et exhortare. Argue cum omni imperio.* [II] Tim. 2: *Hoc commoneo, testificans coram Deo etc.* Hic modus invenitur in lege, prophetis, et libris Salomonis. Disputativus: et hoc in Job et in Apostolo: Job 13. *Disputare cum Deo cupio.* Deprecativus vel laudativus: et hoc invenitur in isto libro: quia quidquid in aliis libris praedictis modis dicitur, hic ponitur per modum laudis et orationis: *infra* Ps. 9: *Confitebor tibi, Domine, etc. narrabo etc.*, *Postilla super Psalmos*, p. 148.

44  Hymnus est laus Dei cum cantico. Canticum autem exultatio mentis de aeternis habita, prorumpens in vocem. Docet ergo laudare Deum cum exultatione. Soliloquium est collocutio hominis cum Deo singulariter, vel secum tantum, quia hoc convenit laudanti et oranti, *Postilla super Psalmos*, p. 148.

The final cause of the psalms Thomas says is prayer. Having just finished describing hymns as praise and exultation, and soliloquies as conversations with God, Thomas is prepared to talk about prayer, which he says is expressed by the words "to the Holy One, and to the Most High." Citing John of Damascus and Psalm 140, ("the lifting of my hands as an evening sacrifice") Thomas says that prayer is "the raising of the mind to God."[45] The mind can be joined to God as holy and most high in four ways. The first way is to reflect on the "loftiness of [God's] power." Thomas says that this is the "elevation of faith," expressed by Psalm 103, 24, "O LORD, how manifold are thy works!"[46] The second way is hope, that is to reflect on "the excellence of eternal happiness."[47] The third way is charity, and that is "to hold fast to the divine goodness and holiness."[48] The fourth way of raising the mind to God is to imitate divine justice.

The last clause of Ecclesiasticus 47, 9, "with ascriptions of glory,"[49] expresses the author of the psalms, the efficient cause, and this is the place where Thomas describes at length his understanding of prophecy. The Psalter contains "ascriptions of glory" in four ways. The first way is because the psalms "emanate from the glorious word of God." Thomas recalls the words of 2 Peter 1, 17–18, which describe Christ's transfiguration, "For when he received honor and glory from God the Father and the voice was borne to him by the Majestic Glory, 'This is my beloved Son, with whom I am well pleased,' we heard this voice borne from heaven, for we were with him on the holy mountain." Just as the Father's voice emanates from glory, so too do the psalms emanate from a glorious cause, namely God. The second way that the Psalter is glorious is insofar as its contents are glorious. The third way is how the Holy Spirit reveals the prophecy of the psalms, and the fourth way is the invitation of humanity to glory.

In this series of reflections on glory, the only lengthy explanation Thomas makes pertains to how the Holy Spirit reveals prophecy and how the mode of prophecy of the psalms is relatively unique in Scripture. Thomas says that prophecy is communicated in three ways. The first is through sensible objects, the second is through dreams and visions, and the third is "by the manifestation of truth itself; and such a mode of prophecy was that of Daniel, who by an interior movement of the Holy Spirit alone, without any exterior aid, uttered

---

45     Elevatio mentis in Deum, *Postilla super Psalmos*, p. 148.

46     Sed quatuor modis anima elevatur in Deum: scilicet ad admirandum celsitudinem potestatis ipsius..., *Postilla super Psalmos*, p. 148.

47     Secundo elevatur mens ad tendendum in excellentiam aeternae beatitudinis, *Postilla super Psalmos*, p. 148.

48     Tertio elevatur mens ad inhaerendum divinae bonitati et sanctitati, *Postilla super Psalmos*, p. 148.

49     Thomas' text reads "in verbo gloriae."

his prophecy."[50] This last mode of prophecy, without any exterior aid, is what allowed David to utter his prophecies.

These reflections on prophecy conclude the exposition on Ecclesiasticus 47, 9, the verse which opened the prologue. The remainder of the prologue discusses the various translations of the psalms, the interpretation of the psalms, and the literary structure and importance of the psalms. Thomas' interest in the literal sense occurs in all three of these discussions. In Hugh's prologue, there is no account of the different translations of the Psalter, but Thomas mentions the three that were currently in use (what seems to be Jerome's correction of the *Vetus Latina* used in Italy), his translation from the Greek text (used in France), and his translation from the Hebrew text (read privately "by many")).[51]

Thomas' interest in the literal sense is manifested also in his interpretation of prophecy. Hugh had articulated three senses: the literal (which is "from the Jews"), the allegorical (which pertains to Christ), and the tropological (which pertains to the just person). Thomas does not refute this understanding explicitly in his prologue (although his discussion of four senses instead of three is well known),[52] and he mentions the literal sense only in passing. He does think, however, that prophecies pertain explicitly to Christ and not to the historical events or objects that the human author might have had in mind. In fact, in several places Thomas says that a passage applies to David figuratively, while it applies to Christ explicitly. For example, Thomas says in the prologue that the first fifty psalms are about penitence, and therefore they "treat figuratively (figuraliter) those things about David's tribulations, the attacks he endured, and his liberation."[53]

Thomas explains the rationale for this method of interpretation:

> Regarding the mode of exposition, it should be known that in expounding upon the Psalter, as in other prophecies, we should avoid an error that was condemned in the Fifth Synod. Theodore of Mopsuestia said that in sacred Scripture and the prophecies nothing is said explicitly about Christ, but about certain other things; but they adapted [these texts] to

---

50 Per ipsius veritatis manifestationem. Et talis modus prophetiae convenit Danieli, qui solius Spiritus sancti instinctu sine omni exteriori adminiculo suam edidit prophetiam, *Postilla super Psalmos*, p. 149.

51 *Postilla super Psalmos*, p. 149.

52 See *Summa theologiae*, 4 vols (Turin / Rome, 1948), I, q. 1, a. 10.

53 Et ideo figuraliter tractatur in ea de tribulationibus et impugnationibus David, et liberatione ejus, *Postilla super Psalmos*, p. 150.

signify Christ. Psalm 21 is an example. "They divided my garments among them," refers literally (*ad literam*) not to Christ, but to David. This mode is condemned in that Council, and he who says that the Scriptures are to be so interpreted is a heretic. Blessed Jerome, therefore, commenting on Ezekiel, passed on to us a rule that we will observe in the psalms, namely that events are to be interpreted as prefiguring something about Christ and the Church, as indeed it says in 1 Cor. 10, "Everything happened to them as a prefiguration (*in figura*)."[54]

Thomas' point here is that even if a text refers to a past event, that event may not exhaust the meaning of the text. If a text's meaning were exhausted by a past event, that text could not be prophecy in a narrow sense. Thomas maintains in several places, however, that the primary author of Scripture, the Holy Spirit, is able to communicate not only with words but with events and objects, so that events and objects can point past themselves to a further meaning. This divine authorship, or "inspiration," is the basis for the three spiritual senses, and allows for prophetic texts to refer primarily to future events even if they also refer to past events, as well.

That is why Thomas, in his commentary on Psalm 21, repeats Theodore of Mopsuestia's condemnation and affirms that the psalm speaks literally about Christ and figuratively about David. Even if the psalm refers to David (or someone else, for that matter), the full meaning of the text is exhausted only by Christ. That is why the literal interpretation of the text pertains to a future event and the figurative interpretation (Thomas does not say allegorical or spiritual) refers to a past event.[55]

Thomas' emphasis on the psalms as prophecy also explains why they are not arranged according to the chronological order of their subject matter. They are arranged in such a way that the historical events they describe point to a meaning beyond the historical events. If they were intended simply to describe the past, there would have been a different arrangement of the psalms. Thomas

---

54    Circa modum exponendi sciendum est, quod tam in psalterio quam in aliis prophetiis exponendis evitare debemus unum errorem damnatum in quinta synodo. Theodorus enim Mopsuestenus dixit, quod in sacra Scriptura et prophetiis nihil expresse dicitur de Christo, sed de quibusdam aliis rebus, sed adaptaverunt Christo: sicut illud Psalm. 21: *Diviserunt sibi vestimenta mea etc.*, non de Christo, sed ad literam dicitur de David. Hic autem modus damnatus est in illo concilio: et qui asserit sic exponendas Scripturas, haereticus est. Beatus ergo Hieronymus super Ezech. tradidit nobis unam regulam quam servabimus in Psalmis: scilicet quod sic sunt exponendae de rebus gestis, ut figurantibus aliquid de Christo vel ecclesia. Ut enim dicitur 1 Cor. 10: *Omnia in figura contingebant illis*, *Postilla super Psalmos*, p. 149.

55    *Postilla super Psalmos*, p. 217.

does accept the division into 70 and 80, but he ultimately uses the threefold division as a basis for further dividing the psalms into groups of ten.[56] These comments about interpreting prophecy literally and figuratively (and the reference to Theodore of Mopsuestia) are absent in Hugh's prologue, as is the much more complicated *divisio textus*.

Comparing briefly Thomas' interpretation of Psalm 1 with that of Hugh, one can notice several similarities and differences. The similarities pertain to specific interpretations. For example, Thomas interprets the "running waters" that flow by the "tree" to be graces and the tree's fruits to be good works. Also, although there is almost no allegorical interpretation in Thomas' commentary on Psalm 1, he does identify the *beatus* with Christ principally and with the just secondarily. Hugh does the same, but whereas Hugh finds allusions to Christ in several places, Thomas finds allusions to Christ in only two phrases. Even in these places, however, Thomas spends almost no time discussing the psalm's relation to Christ; instead, he focuses on the moral qualities that allow the just to be configured to Christ.

Overall, Thomas' commentary is much briefer than Hugh's with a greater interest in the literal and moral senses than the allegorical or anagogical. There is no mention of the Jews or of contemporary groups of people in the Church, such as bishops and theologians. The *divisio textus* is much more complex and the comments pithier. Also, there is a more even distribution of commentary, than in Hugh's *postilla*, which lingers more frequently on certain words or phrases. Thomas inclines less to numerological schematization; for example, there is no mention of the beatitudes or of the seven reasons for the goodness of the just or the evil ways of the wicked, nor does Thomas discuss the fourfold judgment experienced at the end of time.[57] Numbers play a role primarily in distinguishing different levels of interpretation and the various strata of the *divisio textus*. These distinguishing features of Thomas' exegesis are all rooted in a greater interest in the literal structure of the psalms and their literal sense, whereas Hugh is careful to delineate more fully the allegorical and the moral senses of the psalms and to interpret them more fully using other texts from Scripture.

### Conclusion

The prologues to the psalms commentaries of Hugh and Thomas share a number of similarities. They both explain the psalms as prophetic texts, and they both are careful to describe the author, the purpose, the subject matter, and the interpretation of the psalms. In their exposition of these latter categories,

---

56    *Postilla super Psalmos*, pp. 149–150.
57    Cf. Hugh, *Postilla super Psalterium*, fols. 4va–b.

their agreement is almost complete. The Holy Spirit is the primary author of Scripture, who reveals these prophecies to shed light on the Christological depth of the economy of salvation. The readers of the psalms are invited to be transformed by the glorious message communicated in the Psalter and to be elevated into an intensifying union with God.

These prologues, however, do have some marked differences. Hugh's prologue has a more mystical and contemplative tone. The choice of a passage from the Song of Songs to open his introduction conditions the entire prologue. The psalms express a mystical ascent to God, which culminates in the words of the last psalm, "Let everything that breathes praise the LORD!" Hugh's primary interest is in the allegorical and tropological senses and how a spiritual reading of the psalms invites the reader and even propels the reader to contemplative union with God (he does say that the "daughters of Zion" are primarily religious and only secondarily theologians and clerics). Having alluded to the interpretation of the psalms as "breaking the Davidic loaf" for the multitudes, Hugh views exegesis as provision for the soul's desire for spiritual nourishment.

Thomas' prologue focuses less on the soul's love and contemplation of God and more on God's works in general, prayer, virtues, and prophecy. Even when Thomas discusses prayer, the primary categories are faith, hope, charity, and justice; he never once mentions contemplation. Also, the prophetic aspect of the psalms is decisive. In order for the psalms to be prophecy in the sense of revealing future events, they must refer literally to some future event.[58] This is the basis for Thomas' desire to emphasize the literal sense of the psalms over the figurative sense, which in his view applies to David or Solomon. Thomas also indicates his interest in the literal sense by mentioning variant readings of the text and dividing the text much more frequently than does Hugh.

Although Hugh and Thomas are both interested in textual criticism,[59] Hugh has preserved more monastic and early scholastic exegesis,[60] while Thomas has relied more on ancient authorities, more intertextuality (by weaving more scriptural passages into his commentary than Hugh), and a more sophisticated *divisio textus*. For both exegetes, however, the psalms were clear articulations, whether understood as allegorical or literal, of a future savior revealed to David, whose divinely inspired foresight allowed him to behold "King Solomon" and to give thanks to the Most High with words of glory.

---

58   See *Summa theologiae* II–II, q. 171, a. 3.

59   See Gilbert Dahan, *L'Exégèse chrétienne de la Bible en Occident médiéval, XIIe-XIVe siècle*, pp. 206–242.

60   See Morard, pp. 115–122.

# CHAPTER 7

# Walter Burley on the Time of Unknowing

*Jordan Kirk*

> Et illud nomen sic impositum non
> significabit aliquid exsistens in anima,
> quia quodlibet exsistens in anima est
> adhuc ignotum intellectui et nomen
> non imponitur nisi noto.[1]

The thirty-fourth chapter of the *Cloud of Unknowing*, a mystical treatise written in England toward the end of the fourteenth century, contains a formula of exceptional importance.[2] The formula bears on the nature and possibility of the particular devotional practice that the *Cloud* instructs in, the so-called work of unknowing.[3] "Þe abilnes to þis werk," writes the treatise's anonymous author, "is onyd to þe selue werk, wiþ-outyn departyng."[4] That is, the ability to do this work is united inextricably with the work itself. In its context, this sentence appears to do little more than extend a set of remarks counseling against taking pride in the work of devotion. The first half of the chapter in which it appears has established that whatever success you might have with the prayer is due not to any aptitude or effort on your part, but to God's prompting alone; and that God grants or withholds this prompting without regard for your merit. In one sense, the sentence in question merely reinforces this line of argumentation, by locating the ability to do the work of prayer outside of the contemplative. But its implications extend beyond the limits of a simple warning against pride. The formula establishes with precision the metaphysical status of the prayer: it is such that its possibility is inseparable from its existence. It is

---

1   Walter Burley, "Walter Burley's Middle Commentary on Aristotle's Perihermeneias," ed. Stephen Brown, *Franciscan Studies* 33 (1973), I.14.

2   The dating of the treatise remains uncertain, as does its authorship. For a summary of the considerations, cf. Annie Sutherland, "The Dating and Authorship of the Cloud Corpus: A Reassessment of the Evidence," *Medium Aevum* 71 (2002), 83.

3   An account of the word *werk* in the *Cloud* is found in Rosemary Ann Lees, *The Negative Language of the Dionysian School of Mystical Theology: An Approach to the Cloud of Unknowing* (Salzburg, 1983), p. 311. On Middle English vocabularies of labor, cf. Nicola Masciandaro, *The Voice of the Hammer: The Meaning of Work in Middle English Literature* (Notre Dame, 2006).

4   *The Cloud of Unknowing and the Book of Privy Counselling*, ed. Phyllis Hodgson, EETS 218 (Oxford, 1944), 70.3–4.

© KONINKLIJKE BRILL NV, LEIDEN, 2016 | DOI 10.1163/9789004312319_009

a prayer that can only exist when it is already underway; and, conversely, when it is not happening, it is impossible that it should ever begin. The *werk* would thus seem to name an activity that, strictly speaking, would never occur, for how can something ever take place at all, if it only becomes possible once it already exists? But the *Cloud*-author gives every indication that he is describing a prayer that he himself practices and that he believes others can practice as well. What all of this amounts to is that he endorses, in the case of the *werk* of prayer, a doctrine of existence and potentiality most famously refuted in Book IX of the *Metaphysics*, where Aristotle takes the Megarians to task for maintaining so absurd a principle as that a thing is only capable when it is acting.[5] Nonetheless, although he is not unaware of the difficulties of his position, the *Cloud*-author does not modify it; instead, he puts it only more forcefully. "Þe condicioun of þis werk is soche," he writes, "þat þe presence þerof abliþ a soule for to haue it and for to fele it. And þat abilnes may no soule haue wiþ-outyn it."[6]

One question that arises when this overlooked passage is taken seriously is *when* the work would ever actually take place. At what time is the experience of unknowing in fact possible? As it happens, this is a question that the *Cloud*-author does not neglect to address directly. His answer, which I will turn to shortly, is that the work occurs in an *athomus* of time. It is an instantaneous, momentary, entirely brief work. This temporal atomism of the *Cloud* has been discussed in two recent essays: Eleanor Johnson has discussed it in terms both of Augustinian and Boethian distinctions between time and eternity and of the *Cloud*-author's prose style; and Alastair Bennett has situated the *Cloud*'s advice within a tradition that unfolds under the auspices of the proverb *brevis oratio penetrat celum*.[7]

I would like to offer here another suggestion for understanding the momentary nature of the work of unknowing. My suggestion follows from a presupposition that, for reasons that are not entirely clear to me, does not appear to be shared by most readers of the *Cloud*. This is that the treatise makes itself understood in the light of the medieval *scientiae vocis*, the disciplines of speech that fall under the heading of the trivium: grammar, logic, and rhetoric. In this essay I wish to consider the field of logic. For it has seemed to me that the closest analogues to the *werk* of unknowing are to be found not in any of the texts

---

5    *Metaphysics, Books I–IX*, trans. Hugh Tredennick (Cambridge, MA, 1933), IX.3.

6    *Cloud*, 69.23–70.2.

7    Eleanor Johnson, "Feeling Time, Will, and Words: Vernacular Devotion in the Cloud of Unknowing," *Journal of Medieval and Early Modern Studies* 41/2 (2011), 345–368; Alastair Bennett, "*Brevis oratio penetrat celum*: Proverbs, Prayers, and Lay Understanding in Late Medieval England," *New Medieval Literatures* 14 (2012), 127–163.

grouped under the heading of "the late medieval affective tradition," nor in the auctoritates drawn on demonstrably by the *Cloud*-author (for example, Augustine, the Pseudo-Dionysius), but rather in some stray remarks in Boethius's logical commentaries and in the late medieval logical tradition that developed out of them. The particular subject of this essay is a commentary on Aristotle's *De interpretatione* produced by Walter Burley in the first decade of the fourteenth century. The improbability of the conjunction of this work with the *Cloud* does not escape me. There is absolutely no evidence that the *Cloud*-author knew it, and indeed I do not believe that he did. My aim is not Quellenforschung but something more modest: to point toward a resemblance between these works on the level of ideas. Both in Burley and in the *Cloud* certain possibilities inherent in medieval philosophy of language—the ancient accounts, for example, of *vox* and *significatio* as they were transmitted, largely by Boethius, and then elaborated by logicians and grammarians from the eleventh century through the time of the *Cloud*'s composition—become relevant as they do not elsewhere. Working with a common material, in other words, Burley and the *Cloud*-author put it into similar configurations. Whether this is the result of what I sometimes think of as an obscure kind of transmission or whether it is the result, rather, of two altogether independent itineraries, I cannot say. In any event, what I have done here is reconstruct a line of thinking gestured at by Burley, in order ultimately to see what it has to say about the problem of the time of unknowing as it arises in the *Cloud*.

• • •

Among the very first characteristics of the *werk* of unknowing that the *Cloud*-author sees fit to explain is its duration. As he writes in the fourth chapter,

> Þis werk askeþ no longe tyme er it be ones treulich done, as sum men wenen; for it is þe schortest werke of alle þat man may ymagyn. It is neiþer lenger ne schorter þen is an athomus; þe whiche athomus, by þe diffinicion of trewe philisophres in þe sciens of astronomye, is þe leest partie of tyme.[8]

The time that the *werk* takes to occur is strictly equivalent to that of a temporal "atom," the minimal element of duration. It does not last for less time than an instant—that is, it does take place in measurable time—but it does not last any longer either, so that it takes as little time as it is possible for anything to take. The *Cloud*-author's reference to the discipline of astronomy aligns his interest

---

8   *Cloud*, 17.14–18.

in the elemental form of duration with the work of the so-called "Oxford calculators," the mid-century scholars associated with Merton College (where, incidentally, Burley produced the commentary to which I will soon turn). Despite his own indications, this atomism has much more to do with the disciplines of speech than it does with those of number. For the "leest partie of tyme" in question will turn out to be precisely the duration of a *syllable*, and indeed the *Cloud*-author quickly abandons the vocabulary of "athomus" in favor of a repeated insistence on the syllable, an utterance smaller than the word but (unlike the letter) still pronounceable in itself. As Johnson has pointed out, in doing so he resumes certain indications made long before by Bede to the effect that for the discipline of grammar the syllable constitutes an atom.[9]

And in fact the *werk* of unknowing will consist in a strange practice of syllabification.[10] The technique counseled by the *Cloud*-author consists simply in the repetition ad infinitum of a one-syllable word. What word you use is of no importance, save only that it be in a language that you speak and that it be monosyllabic. What occurs in such a prayer, indifferent to the signification of its utterance, is an experience of nonsense. The entity it produces, though it is still a word, is maintained at the level of a mere syllable, and the syllable, as it was theorized both in the grammatical and in the logical traditions, is utterly without meaning. The purpose of the prayer is to reduce the word of one syllable to its status as a mere syllable and thus denude it of whatever signification would usually be attached to it. But the nonsensical quality of the prayer is perhaps so unavoidable that a reference to its syllabic nature is unnecessary. For the repetition of a word results—as everyone knows—in what is sometimes called "semantic satiation," the revelation of an uncanny underbelly of your own speech as you cease to be able to know whether your utterance is familiar or strange, even as its familiarity goes without saying.

In short, the *athomus* of time is exactly the duration of a syllable, and the syllable is a minimal unit of utterance that is in itself nonsensical. But nonsensical utterance was named and theorized not only in the discipline of grammar— that is, the study of language in terms of its formal correctness—but also in that of logic, which is concerned with the relation between language and truth. Although it has been given scant scholarly attention, the question of nonsense was a consistent preoccupation of logicians throughout the Middle Ages.

---

9      Johnson, "Feeling Time," 364, n. 15.

10    This paragraph summarizes the argument of a separate essay I am preparing on the grammatical status of the *Cloud*-author's invocation of the syllable. There I will give the detailed account of the prayer procedure or *werk* that considerations of space preclude here.

Following a gesture inherited from Boethius, logical writings from the end of the eleventh century, when new treatises began to be composed, and well into the fourteenth made reference in their introductory pages to *voces non-significativae*, or non-significative utterances. Whereas Boethius had made use, in his second commentary on the *De interpretatione*, of the Stoic examples *skindapsos* and *blityri*, as well as of his own inventions *garalus* and *hereceddy*, medieval logicians settled on their own set of exemplary nonsense words. These can already be seen in the twelfth- and early thirteenth-century treatises collected in Lambert-Marie de Rijk's *Logica modernorum*. The *Ars emmerana* adduces *blictrix* and *sindiarsis* as examples of *vox quae nichil significat*; the *Ars burana* has *blictrix* alone. In the *Introductiones parisienses* the example is *buba*. The *Logica "ut dicit"* gives *buba* and *plectrix*; the *Ars meliduna biltrix* and *buba*; and various manuscripts of the *Logica "cum sit nostra"* contain the following examples: *bon, bau*, and *beltrix*; *buba* and *bultrix*; *bon, bau*, and *bletrix*; *bou, bau*, and *beltrix*; and *bu, ba*, and *buf*.[11]

Such words appear in the same place in all of these treatises: at their very beginning, in preliminary discussions of the nature of *vox*. Their function will be seen from the use William of Sherwood makes of them in the opening chapter of his *Introductiones in logicam*, an influential logical textbook from the mid-thirteenth century that reprises the doctrines found in the treatises just cited. William invokes *buba* for the same reason as does everyone else, in the same context and in almost exactly the same language:

> Prius autem agendum est de nomine quam de verbo, quia est principalior pars quam verbum. Ideo ab eo inchoandum est. Et quia omne nomen est vox et omnis vox est sonus, ideo a sono tamquam a primo inchoandum est. Est autem sonus proprium sensibile aurium. Et dividitur sic: Sonus alius vox, alius non vox. Sonus vox est vox, ut quod fit ab ore animalis. Sonus non vox, ut strepitus pedum, fragor arborum et similia. Vox sit dividitur: Alia significativa, alia non significativa. Vox significativa est, quae aliquid significat; non significativa, quae nihil significat, ut: 'buba blictrix.'

---

11    Cf. Lambert-Marie de Rijk, *Logica Modernorum: A Contribution to the History of Early Terminist Logic* (Assen, 1967). The *Logica "cum sit nostra"* is of particular interest for my purposes, given that Walter Burley appears to have produced an adaptation of it and thus would have known it closely. It should be mentioned also that in the *Introductiones montane minores* and *Abbreviatio montana*, rather than some variation of *buba* the example given of *vox non-significativa* is the letter or syllable: insofar as it constitutes part of the word *sorex*, the syllable *rex* does not signify (that is, *king*) on its own.

The noun ought to be considered before the verb because it is a more important part than the verb, and so we must begin with it. And since every noun is an utterance and every utterance is a sound, we must begin with sound. Sound is the property to which the ears are sensitive; it is divided into vocal and nonvocal. Vocal sound is an utterance such as is made by an animal's mouth. Nonvocal sound is footsteps, the crashing of trees, and the like. Utterances are divided into significative and nonsignificative. A significative utterance is one that signifies something; a nonsignificative one signifies nothing; for example, 'buba blictrix.'[12]

The purpose of this taxonomy of sound is to distinguish the word that can serve as a term from every other kind of noise. Only the species of sound that is both vocal and significative can attain to the status of the term and thus serve as the object of logical science. Nonsense words appear in these treatises, in short, for a single purpose: so that they can be cast out of the realm of logic altogether. As is announced in the first lines of the early Oxford logic called after its incipit *Cum sit nostra*, the *materia* of the discipline of logic is *vox significativa*, and this is to say specifically that nonsense words can have no place there:

> Cum sit nostra presens intentio ad artem dialeticam, primo oportet scire quid sit materia artis dialectice. Materia artis dialectice est vox significativa, quia de voce non significativa nullus intellectus ageneratur in animo alicuius.

> Since our present concern is the discipline of logic, it is appropriate in the first place to know what is the material of the discipline of logic. The material of the discipline of logic is significative utterance, since no concept is produced in anyone's mind by non-significative utterance.[13]

*Buba* and *bufbaf* exist and circulate in the textbooks of logic solely in order that the actual subject matter of the discipline, words like *homo*, may be identified more exactly by being distinguished from them. The instances of their appearance are, so the logicians declare, exceptions: no more will ever be said about

---

12    William of Sherwood, "William of Sherwood, 'Introductiones in Logicam,' Critical Text," eds. Charles Lohr, Peter Kunze, and Bernhard Mussler, *Traditio* 39 (1983), 222–223; William of Sherwood, *Introduction to Logic*, trans. and ed. Norman Kretzmann (Minneapolis, 1966), pp. 22–23. Translation modified.

13    de Rijk, *Logica Modernorum*, p. 417. My translation.

them than that they will not appear again. Their repetition in the prefatory remarks to the arts of logic is an almost apotropaic act, meant to ward off their incursion into the treatises any further than their threshold. But for this very reason these words are decisive in the constitution of the field. The gesture by which *homo* is distinguished from *buba* is that by which logic secures the object of its research. If it failed to do so, it would not be logic at all.

The danger that logic would fail to qualify as logic at all was felt to be a real one. Logicians would not make the mistake of treating a mere sound, but they could easily fall into the trap of including a word with no meaning in a proposition. Such a proposition would be strictly indeterminate with respect to its truth or falsity: there would be no way of even beginning to decide whether *buba is on the mat* is true or false. Given that logic is the study of the truth of propositions, or of speaking truthfully (*vere loqui*), as soon as it begins dealing with a proposition containing a nonsense word it has ceased to be logic at all. In short, an encounter with *vox non-significativa* spells the end of the discipline of logic, and the ban on that encounter marks its beginning.

• • •

It is in light of logic's perennial, and constitutive, disavowal of *voces non-significativae* that I wish to consider the opening pages of Walter Burley's *Middle Commentary* on the *De interpretatione*.[14] At the beginning of the fourteenth century, Burley—who was deeply familiar with the early Oxford logical textbooks I have just surveyed[15]—began to compose new works in their tradition. Although it had not been eclipsed, as at Paris, by *grammatica speculativa*, at Oxford there had been very little new work done on the theory of terms since Roger Bacon wrote his *Summulae dialectices* in the early 1250s.[16] Burley has thus been said to represent the rejuvenation of terminism after two generations had passed without any innovation to speak of. The *Commentarius in librum aristotelis Perihermeneias*, written in the first decade of the fourteenth

---

14    The indispensable work on Burley remains Agustín Uña Juárez, *La filosofía del siglo XIV: contexto cultural de Walter Burley* (Madrid, 1978). An overview of Burley's life and works is found in Jennifer Ottman and Rega Wood, "Walter of Burley: His Life and Works," *Vivarium* 37/1 (1999), 1–23.

15    Cf. Sten Ebbesen, "OXYNAT: A Theory About the Origins of British Logic," in *The Rise of British Logic*, ed. Osmund Lewry (Toronto, 1985), pp. 2–3; de Rijk, *Logica modernorum* II.1.445–6.

16    Cf. Jan Pinborg, "The English Contribution to Logic before Ockham," *Synthese* 40 (1979), 27; Ebbesen, "OXYNAT," 4–5; William Courtenay, *Schools and Scholars in Fourteenth-Century England* (Princeton, 1987), p. 228.

century, while Burley was at Oxford's Merton College or shortly thereafter, is the third of four commentaries he devoted to Aristotle's treatise.[17] The work is a close explication of the *De interpretatione*, proceeding line by line and addressing questions that arise along the way. Like his predecessors, Burley introduces nonsense words at the beginning of the work, in a section devoted to establishing the nature of the logical term. A term, Burley writes, is a *vox*. But here he admits a doubt. It is the first *dubium* introduced in the commentary, the first matter of controversy to be entertained:

> Sed tunc est dubium: Cum eadem vox non posit bis proferri, si vox prolata esset pars nominis, videtur quod idem nomen non posit bis proferri, et sic nulla propositio prolata posset converti, et sic in syllogismo secundum eius esse in prolatione essent sex termini.

> But a doubt arises: since the same utterance cannot be pronounced twice, if the pronounced utterance is part of the noun, it would seem that the same noun cannot be pronounced twice, and thus that no pronounced proposition can be converted, and thus that insofar as the syllogism exists in pronunciation it would contain six terms.[18]

An utterance is a unique occurrence: a particular sound that comes out of someone's mouth at a particular time and place, never again to be encountered. If a logical term is an utterance, then it too would seem to be unrepeatable. But this would present a grave problem for syllogistic reasoning, which rests on the repetition of the same term now as a subject, now as a predicate. Burley's point is that logic depends for its operation on the capacity of a term to appear multiple times without ceasing to be itself. For if the term *man* is not the term *man*, nothing at all is demonstrated by the syllogism:

> Man is an animal;
> But this is a man;
> therefore, this is an animal.

Since Burley has already identified the term as a *vox*, what he must establish is that *vox* can be repeated and remain identical to itself. His reply to the doubt is that the *vox* of which a term is composed is not simply *vox*. This gesture does

---

17    On the dating of the work, cf. Brown's introductory comments to this edition and Ottman and Wood, "Walter of Burley."

18    Burley, "Middle Commentary," I.05. Translation mine here and throughout.

not at first appear unusual, as it resumes an ancient interest not in the utterance as such but in the elements into which it can be broken.[19] Boethius, for
example, had explained that Aristotle speaks, in the famous opening lines of
the *De interpretatione*, of *ea quae sunt in voce*, those things that are in the voice,
because he means to limit his discussion to certain specific forms of *vox*,
namely nouns and verbs. In the twelfth century, Peter Abelard distinguished
between *sermo* and *vox*, declaring only the former to be the object of logic.
Indeed, it is in exactly this connection that we have seen the terminist logicians introduce *buba* and *blictrix*: in order to specify that logic deals not with
*vox* as such, but only with *vox* possessed of the differentia *significativa*. But in
fact Burley is establishing something else than what these others had. He
claims that what makes up a term is not utterance itself but a resemblance
obtaining among various utterances:

> Ad istud potest dici, salvando trinitatem terminorum in syllogismo, quod
> ista vox prolata quae est pars nominis non est aliqua vox una numero sed
> est unum commune ad istam vocem prolatam et ad quamlibet vocem
> consimilem. Circa quod est intelligendum quod ista vox 'homo' prolata a
> te et illa vox 'homo' prolata a me magis conveniunt quam ista vox 'homo'
> et ista vox 'animal.'

> To this objection it can be said, preserving the trinity of terms in the syl-
> logism, that this pronounced utterance which is part of the noun is not
> some singular utterance, but is rather a single thing common to that pro-
> nounced utterance and to any other utterance similar to it. On account of
> which it should be understood that this utterance *homo* pronounced by
> you and that utterance *homo* pronounced by me accord with one another
> more than do this utterance *homo* and this utterance *animal*.[20]

The *vox* as such, as "itself," has no part in the term. The term is formed out of a
*communitas* or *conveniencia* or *consimilitudo* among multiple utterances. This
accord alone, and not the *voces* among which it obtains, is the *materia* of the
term. Nothing except what is fully iterable in the utterance matters in the constitution of the elements of a proposition. Everything else about it, all the unrepeatable characteristics of a particular vocal sound, Burley separates out from

---

19    As Daniel Heller-Roazen has shown, the ancient authorities, both grammatical and logi-
      cal, encounter *vox* only insofar as it is abstracted. Cf. his "De voce," in *Du bruit à l'oeuvre:
      Vers une esthétique du désordre*, eds. Juan Rigoli and Christopher Lucken (Geneva, 2013).
20    Burley, "Middle Commentary," I.06.

the *vox communis* that is the *materia nominis et verbis*. But even as he appears to exclude the *vox singularis* from consideration, by mentioning it here he draws attention to the variability of utterances produced by different speakers at different times and in different places.

This distinction between *vox singularis* and *vox communis* is the basis of Burley's whole theory of terms. Burley stands apart from other logicians insofar as he defines the *vox* that is pertinent to logic—that is, the kind of *vox* that can be used as a term—not as utterance that is joined with a signification but as a *consimilitudo* among multiple *voces*. And indeed he specifically distinguishes his isolation of the *commune ens* in the utterance from any question of meaningfulness, maintaining that significative and non-significative utterances alike are able to accord with one another in this resemblance. Immediately after he has declared that the utterance *homo* goes together with another utterance of *homo* more than it does with the utterance *animal*, Burley turns to nonsense words:

> Similiter, haec vox "bu" prolata a me et haec vox "bu" prolata ab alio magis convenient quam haec vox "bu" et haec vox "ba." Habent igitur aliquod commune ens quod non est commune istis, scilicet "bu" et "ba."

> Likewise, this utterance *bu* spoken by me and this utterance *bu* spoken by someone else accord with one another more than do this utterance *bu* and this utterance *ba*. Therefore they have a certain common being which is not common to those others, namely *bu* and *ba*.[21]

In citing *bu* and *ba*, familiar terminist examples of *vox non-significativa*, Burley makes explicit that he is interested in the utterance not insofar as it is meaningful but insofar as it conforms to other utterances. The vocal *commune ens* is indifferent to sense or nonsense, even as it isolates something in the utterance that is distinguishable from the mere separated noise of different vocalizations issuing at different times and in different ways.

In short, although in citing *bu* and *ba* in the first pages of his commentary Burley appears to be following the conventions of the Oxford terminist tradition in which he is steeped, in fact he uses these examples not to distinguish *vox significativa* from *vox non-significativa* but to reveal the dimension in which they cannot be told apart. The perversity of this gesture should not be overlooked: in defining the term Burley invokes, without apology, nonsense words that had been transmitted to him as nothing else than examples of what is not

---

21    Ibid.

and cannot be a term. For *bu* and *ba* not only provide him with examples of the material out of which a term might be made: they themselves take on the status of full-fledged terms. Never acknowledging that they have existed for no other reason than to exemplify what cannot appear in a proposition, Burley not only insists that *bu* and *ba* can in fact appear in a proposition, but he himself uses them in just this way. Indeed, each of the first four examples of propositions given in the *Middle Commentary* contains either *bu* or *buba*.

These four propositions containing nonsense words are adduced in order to illustrate the differences among various types of "supposition," on which some general remarks are now in order. In terminist logic, a line is drawn between a term's *significatio* and its *suppositio*, a distinction that anticipates the Fregean difference between *Sinn* and *Bedeutung*, or sense and reference.[22] It is one thing, the terminists teach, for a word to signify something: it signifies whatever it is "imposed on," that is, whatever it is assigned to as a name.[23] Thus the word *homo* signifies the mortal, rational animal: the *significatum* of a word is, effectively, its definition. But when a word comes to be used in a proposition, although its *significatio* does not vary, the word can refer to (or "supposit for") a number of discrete things. Whereas its *significatio* is what it means when it is considered in isolation, the *suppositio* of a term is the way that it stands for something in a particular proposition. As Peter of Spain writes,

> Differunt autem suppositio et significatio, quia significatio est per impositionem vocis ad rem significandam, suppositio vero est acceptio ipsius termini jam significantis rem pro aliquo.[24]

> Supposition is distinct from signification, in that signification consists in the imposition of an utterance to signify a thing, whereas supposition consists in the reference of a term that already signifies a thing to something in particular.

In the following sentences, which are the classical examples, the word *man* supposits for three distinct things and in three distinct manners, even though its signification remains unaltered:

---

22  Cf. Gottlob Frege, "Über Sinn und Bedeutung," in *Kleine Schriften*, ed. I Angelelli (Hildesheim, 1990), pp. 143–162.

23  As many commentators have pointed out, *significatio* never received a clear definition in the Middle Ages, as opposed to *suppositio*, which was minutely described. Cf. Joël Biard, *Logique et théorie du signe au XIVe siècle* (Paris, 1989), p. 15.

24  Peter of Spain, *Summule logicales*, ed. Lambert-Marie de Rijk (Assen, 1972), p. 80.

Homo currit. (A man is running.)
Homo est species. (Man is a species.)
Homo est nomen. (Man is a noun.)

For the most part, medieval logicians agreed in calling the reference of the subject terms in these propositions *simple, personal,* and *material,* respectively. Supposition theory tries to account for the fact that, while the signification of words seems to be relatively stable (words signify what they are imposed to signify), the same word will pick out different things in different contexts. The purpose of the theory is to classify and thereby control for the way that the truth of a proposition depends on the variable usage of words in contexts.

Although supposition is distinct from signification, they are not entirely separable. A word can signify without suppositing, for example if it is uttered by itself outside of a proposition, but it cannot supposit without first signifying. As this principle, common to all the early treatises, is formulated in the *Cum sit nostra,* a term supposits only when placed in a sentence, whereas it signifies whether it is placed in a sentence or outside.[25] Signification is essentially prior to supposition: in order for an utterance to pick out a particular thing, it must have first become a meaningful word. This is, effectively, a theoretical rationale for the expulsion of nonsense words from the domain of logic that has simply been asserted at the beginning of the treatises.

And yet there seems to be an exception to this rule: the third and strangest form of supposition, material.[26] *Suppositio materialis* is what takes place in a sentence on the order of *man is a noun* or *man is a monosyllable,* in which the

---

25    "Terminus supponit quando ponitur in oratione; terminus significat sive ponitur in oratione sive extra oratione." de Rijk, *Logica Modernorum* II.2.446.

26    A series of important articles on material supposition has appeared in the last decades, beginning with Elizabeth Karger, "La Supposition materielle comme supposition significative: Paul de Venise, Paul de Pergula," in *English Logic in Italy in the 14th and 15th Centuries,* ed. Alfonso Maierù (Naples, 1982), pp. 331–341. Cf. Calvin Normore, "Material Supposition and the Mental Language of Ockham's Summa Logicae," *Topoi* 16, 1 (1997); Stephen Read, "How Is Material Supposition Possible?," *Medieval Philosophy and Theology* 8, 1 (1999), 1–20; Claude Panaccio, "Tarski et la suppositio materialis," *Philosophiques* 31, 2 (2004), 295; Claude Panaccio and Ernesto Perini-Santos, "Guillaume d'Ockham et la suppositio materialis," *Vivarium* 42/2 (2004), 202–224. The indispensable account, to which my discussion is much indebted, is Irène Rosier-Catach, "La Suppositio materialis et la question de l'autonymie au Moyen Âge," in *Parler des Mots: le fait autonymique en discours,* eds. J. Authier, Marianne Doury, and Sandrine Reboul-Touré (Paris, 2003).

proposition is true only insofar as the subject term *man* supposits for itself.[27] What the doctrine of material supposition tries to explain is what is sometimes called the *autonymous* use of a term. "Autonymy," in the terminology introduced by Rudolf Carnap and adopted notably by Josette Rey-Debove, is the use of a term as the name of itself.[28] Medieval Latin did not make use of quotation marks, by which a logician in our own time might indicate the "mention" as opposed to the "use" of a term, although the Old French article *ly* was sometimes imported into Latin in order to mark the occurrence of material supposition. What is troubling about the autonymous use of a term is that in standing for itself the term appears to cease signifying its significatum. A term in material supposition does not supposit for what it signifies, inasmuch as the truth or falsity of *man is a monosyllable* has nothing to do with the nature of the rational mortal animal. Flying in the face of the principle that a word must first signify if it is going to supposit, the term in material supposition, in referring to itself, appears to break off its dependence on its signification and enter a realm in which it only supposits, and might as well be meaningless.

The non-significative nature of words in material supposition is already identified in the middle of the twelfth century by the grammarian Peter Helias. As he explains it, this variety of reference is known as *material* because what is picked out by its use is the material of the word itself. For, he says, the meaningful word can be distinguished from its material substrate. Considered in isolation, this substrate—possessed by every word—should be understood on the model of *blictrix*, one of the exemplary *voces non-significativae*:

> Dicendum est quod vocabula quandoque se ipsa nominant, ut cum dicitur, "Homo est nomen." Hic enim non de homine loquimur sed potius de hoc nomine "homo." Et hoc appellabant antiqui "materiale impositum," quod quid sit ut intelligas, materiale impositum est vox representans sepisam, id est, posita ad loquendum de seipsa et dicitur materiale impositum quia nomen, si ita contingit, representat materiam suam, id est, vocem que quasi materia preiacet ut inde fiat nomen. Ex voce namque fit nomen per

---

27    As Burley's example will show, definitions of *suppositio materialis* became increasingly complex, largely because it became clear to later writers that the phenomenon could not be explained merely as a reference of the term to itself. Nonetheless this recursive quality is inseparable from material supposition in all its various definitions. There is a parallel discussion of autonymy in sacramental theology, where certain thinkers maintained that the distinguishing feature of a sacrament is that it is a sign that signifies itself. Cf. Irène Rosier-Catach, *La Parole efficace: signe, rituel, sacré* (Paris, 2004), 41.

28    Cf. Rudolf Carnap, *Logische Syntax der Sprache* (Vienna, 1934); Josette Rey-Debove, *Le Métalangage: étude linguistique du discours sur le langage* (Paris, 1978).

impositionem. Quod inde videri potest quia "blictrix" vox est tamen non-dum nomen est, sed si alicui rei imponitur nomen erit.

It should be said that words sometimes serve as the names of themselves, as when someone says "man is a name." For we do not say this about man but about this noun "man." And this the ancients [i.e., Priscian] called *imposed materially*, which you should understand as follows: material imposition consists in an utterance representing itself, that is, imposed to speak about itself. It is called *imposed materially* because the noun, taken in this instance, represents its own material, that is, the utterance which like a material precedes the noun, which is made out of it. For the noun is made out of the utterance by means of imposition. And this can be seen in the case of *blictrix*, which is an utterance although it is not yet a noun, but it will become a name if it is imposed on some thing.[29]

Thus, in the proposition *homo est nomen*, the word *homo* is used as though it did not yet have a signification. Material supposition picks out that aspect of a word that is no more than *blictrix*, that is, its *figura vocis* or sound-shape. Now *homo* is a meaningful word being used in a way that brackets its significative function. But if the word *homo* in the proposition *homo est nomen* has been reduced to what it has in common with *blictrix*, the question arises as to why *blictrix* itself could not take its place in this proposition. In fact, *blictrix* is not a noun, so the proposition *blictrix est nomen* would be false. Still, its falsity is the sort of thing that logic can account for; and in any case, to produce a true proposition it would only be necessary to introduce a negation: *blictrix non est nomen*. For that matter, what about a proposition such as is regularly found in the treatises we have been examining, *buba est vox non-significativa*? At least at first glance, this sentence is proof that a nonsense word is fully capable of appearing within a proposition, and that the discipline that determines the truth and falsity of propositions is for its part fully capable of treating a proposition within which it appears.

In their elaboration of the doctrine of material supposition, in other words, thirteenth-century logicians found themselves unable to avoid the very words whose exclusion constituted the foundation of their discipline. Once supposition had been distinguished from signification, it became possible to conceive of supposition without signification: of propositions made up of nonsense words. As Irène Rosier-Catach has shown, Roger Bacon provided an "elegant

---

29    Petrus Helias, *Summa super Priscianum*, ed. Leo Reilly (Toronto, 1993), p. 193 (translation mine).

solution" to what should have been, by all rights, a crisis of self-definition for the discipline of logic.[30] This solution—in brief, discarding the distinction between signification and supposition and denying that any word can be used in a sentence without thereby becoming imposed, if only on its own utterance— was not, however, widely adopted. And Burley, to whom it is now possible to return, though he knew Bacon's work backwards and forwards, is not at all interested in avoiding the problem that such words present.[31]

Having identified the material of the term, Burley specifies that this *commune ens* can already take part in a proposition even before it has been assigned to name anything. A *vox non-significativa* no less than a *vox significativa* can supposit:

> Nec est solum reperire unum tale commune in vocibus significativis sed etiam in vocibus non-significativis. Unde aliquid est commune huic voci 'bu' quae profertur a me et huic voci 'bu' quae profertur a te, et pro tali communi verificatur ista 'Bu est bu.'

> Nor is such a common being only to be found in significative utterances, but also in non-significative utterances. Thus there is something common to this utterance *bu* which is pronounced by me and to this utterance *bu* which is pronounced by you, and on account of this common thing the truth of the proposition *bu is bu* can be demonstrated.[32]

*Bu est bu* is admissible and moreover true, on Burley's account, because both *bu* and *bu* naturally supposit for what is common to both of them. This resemblance is not a kind of signification. While Burley subscribes to Bacon's intentionalist account of imposition, so that the signification of a word depends on the will of a speaker, he implies here that supposition occurs apart from the

---

30   Rosier-Catach, "La Suppositio materialis," p. 51.

31   Late in his career, in the important longer version of the *De puritate artis logicae* written in about 1326, Burley would develop an unusually complex taxonomy of material supposition, and indeed he paid close attention to the phenomenon in many of his works, discussing it already at notable length in the early *De suppositionibus*, from 1302. What interests me here, however, are not his considered descriptions of the phenomenon, but his offhand remarks in the *Middle Commentary*, which lead in their own direction. For Burley's sustained treatment of the varieties of supposition, cf. "Walter Burleigh's Treatise De suppositionibus and Its Influence on William of Ockham," ed. Stephen Brown, *Franciscan Studies* 32 (1972), 15–64; *De puritate artis logicae. Tractus longior*, ed. Philotheus Boehner (St. Bonaventure, NY, 1955).

32   Burley, "Middle Commentary," I.07.

will of any speaker: there is already supposition *before* anyone utters a proposition, insofar as singular utterances in the world resemble one another and thus can stand for what is common to all of them. As he says, *voces singulares universaliter supponunt pro communibus*: singular utterances stand for what is common to them and their likes, whenever they are pronounced. The stronger resemblance between this *bu* and that *bu* than between this *bu* and that *ba* is not a matter of meaning or reference but of *conveniens*, a pattern to which disparate utterances are reducible. In the proposition *bu est bu* it is not, Burley thinks, that a speaker intends both *bu* and *bu* to name "the utterance *bu*," so that we would be confronted here with the imposition of the word, as Bacon would have argued. Rather, the proposition merely draws out the self-identity of the term insofar as it is a term, that is, something made out of what is common to both utterances. Thus, in *bu est bu*, each singular utterance supposits for a *vox communis* that is one, without the intervention of any will to signify; and the proposition is true, without prejudice to the presence of nonsense words in it.

In the final lines of Section I.07, Burley wraps up his remarks on the *commune ens* and on the possibility of supposition without signification by bringing to light a distinction that might otherwise escape notice. He has indicated that in *buba est buba* the subject term has personal supposition. In other words, when the term stands for its material as a term, that which is common to both *buba* and *buba*, the supposition that takes place is not material. Burley now makes this point explicit:

> Unde dico quod aliam suppositionem habet subiectum in ista "Buba est dissylabum" et aliam in ista "Buba est buba," quia in prima habet suppositionem materialem secundum quod est vera et in secunda suppositionem personalem.

> I maintain that the subject term has a different supposition in the proposition *buba is disyllabic* than it does in the proposition *buba is buba*, because in the first it has material supposition insofar as it is true and in the second it has personal supposition insofar as it is true.[33]

Although in both cases it appears to be *buba* itself that is referred to, the subject terms in *buba est buba* and in *buba est disyllabum* do not have the same supposition. It is here that Burley's unusual definition of the materiality of the term becomes consequential for his theory of supposition. On this account,

---

33    Ibid.

there are two ways in which a term might refer to its own utterance, because the utterance in question can be apprehended either as *vox communis* or as *vox singularis*. Material supposition occurs when a term in a proposition stands for itself as *vox singularis*. In his use of the proposition *buba est dissylabum* to exemplify material supposition, Burley follows convention, but the example is misleading to the extent that the *commune ens* for which it might also stand (in personal supposition) would itself appear to have two syllables. But this is not what he is talking about. As he presents it, *buba has two syllables* is true not because *buba*, *buba*, and *buba* can all be seen to have two syllables, but just because this *buba* does. Had he wished, he might have said that material supposition is what takes place in the propositions *buba is said loudly, buba is said with a regional accent, buba is stuttered*, and so forth.[34] Insofar as *buba est disyllabum* is true, the subject term refers only to this one unrepeatable utterance *buba*, which is not an instance of the term but a mere *vox singularis*. What this means is that *buba* does not stand for itself. It supposits for nothing else than what distinguishes *buba* from *buba*, for *buba* to the extent (and only to the extent) that it is not *buba*. Material supposition, as it is reimagined here, is the use of a term to stand for a difference. It is not the autonymous use of a term after all: for it is not the use of a term as the name of itself. It is what takes place when a term stands for itself insofar as it is not a term and can never be a term.

In a proposition that is true when its subject term is taken in material supposition, something is said about the unrepeatable noise produced by a singular human voice. Burley points toward the possibility of a propositional knowledge of the variabilities of pitch, volume, timbre, tempo, and so forth that do not go into the term itself. Burley's strange proposal is that not only can the mere *figura vocis* or *commune ens* appear in a proposition, but so can the irreproducible variability that desists from figuration. And he does not see this as a problem. For Burley, the truth of such a proposition can be known.

• • •

What does all of this have to do with the *Cloud of Unknowing*? At the most basic level, if the *Cloud*'s prayer technique consists in the production of a

---

34    Cf. the late twelfth-century Priscian commentary described by Karin Margarita Fredborg: "His example of the freedom from Latin rules in the case of French words is both humorous and extraordinary, the non-sense word '*buba*' and the pronunciation of this word—'*buuba*,' '*bubaa*,' '*búba*,' '*bùba*' are all admissible, but not '*bubá*'!" "The Priscian Commentary from the Second Half of the Twelfth Century: Ms Leiden BPL 154," *Histoire Épistémologie Langage* 12/2 (1990), 65.

vocalization that would be devoid of signification, it is fitting to examine the most concerted effort to account for such an utterance that existed in the Middle Ages. That effort was undertaken by generations of logicians wrestling with the place of the utterance in their discipline. Burley in particular is of interest here because he departs from the tradition in expecting logic to address itself to the empty noise of the bare utterance, and not simply in order to get rid of it. Proceeding on the basis of a distinction between *vox singularis* and *vox communis*, Burley believes that the unrepeatable differentiating noise of the utterance can participate in the truth of a proposition. In the *Cloud*, too—and this is its distance from the devotional works with which it is often lumped together—what is announced is that there is an access to truth possible only by way of nonsense. The repetition of the monosyllabic word to the point of senselessness produces the same encounter with the utterance as such, in its mere sounding, as does material supposition in Burley's description.

But the *Cloud*-author is, of course, primarily interested in the *effect* of his technique on the contemplative: the experience of unknowing, that is, the knowledge of God that lasts for only the briefest of moments. If I am right that Burley's commentary follows a related line of thinking to the *Cloud*'s, it too might have something to say about this matter. And thus a final question arises: what happens when someone hears a proposition containing a nonsense word in material supposition? What takes place in a human mind when it encounters *buba est disyllabum*?

In the section of his commentary devoted to the important first lines of the *De interpretatione*, Burley takes Bacon's side in the *magna altercatio* as to whether words signify concepts or things, arriving at the Franciscan's conclusion and following his argumentation to get there.[35] A word signifies a thing in the world rather than a concept in the mind, he writes, because the impositor wishes to name a thing, not a concept, and the will of the impositor is what determines the *significatio* of a word. Nevertheless, to hear a word

---

35    Cf. I.14–16. A perspicuous overview of these sections is found in Ana María Mora-Márquez, "La ontología realista de Walter Burleigh y su relación con las teorías del significado y de la suposición," in *Sobre la pureza del arte de la lógica: Tratado breve*, by Walter Burleigh, trans. Ana María Mora-Márquez (Bogota, 2009), pp. 173–227. As Mora-Marquez points out, although Burley appears to support the traditionalist position at various points, these are only moments in a larger argument that decides for the new position. A great deal has been written on the *magna altercatio*; the recent essay by Mora-Márquez on the subject is much to be recommended. Cf. her "Peri hermeneias 16a3–8: Histoire d'une rupture de la tradition interprétative dans le bas moyen âge," *Revue philosophique de la France et de l'étranger* (2011), 67–84.

spoken is to be able to gather certain things in addition to what it has been imposed to signify:

> Potest enim vox esse nota alicuius cui vox non imponitur ad significandum. Nomen enim vel verbum prolatum est nota vel signum per quod proferens habet similitudinem in suo intellectu istius rei quae significatur per nomen vel per verbum vel quod ipse aliqua passione afficitur erga eum ad quem loquitur, videlicet amore vel odio. Unde vox est signum passionis animae non quia imponitur ad significandum passionem animae sed sic est signum passionis animae sicut est signum quod iste qui loquitur est homo.

> For an utterance can be a sign of something on which it is not imposed to signify. For a noun or a verb, when it is pronounced, is a sign by which it is made known that the utterer has a similitude in his understanding of that thing which is signified by the noun or verb, or that he is affected by some feeling toward the person to whom he is speaking, for instance love or hatred. Thus utterance is a sign of a "passio animae" not because it is imposed to signify a passion in the soul, but because it is a sign of a passion in the soul, in the same way that it is a sign that the one who is speaking is a human.[36]

Even if someone uses a word with which you are unfamiliar, you can be certain that she means something by it, that she harbors feelings of one sort or another toward you and, moreover, that she is a human being. The use of a word to name something is an index of the rationality, and thus humanity, of the speaker; it is also an index of the presence of the particular concept of the named thing in the speaker's mind. If you happen to know the meaning of a word, you will know what concept its speaker has in mind; but whether you do or do not, the word will in any case indicate the mere presence of the concept without disclosing which concept it might be.

But what is crucial about Burley's adoption of this argument that a given vox will function, in Peircean terms, as both an index and a symbol is that *he restricts his discussion to the noun and verb alone*. The familiar argument that Burley is rehearsing—as a Roger Bacon, for instance, would have made it—is that *every* utterance, of whatever sort, is at least significative of the mental state of its speaker. In restricting the scope of this characterization to *voces significativae* alone, Burley effectively declares, by conspicuous omission, that nonsense words do not for their part allow their hearer to gather anything

---

36    Burley, "Middle Commentary," I.16.

about the mind of their speaker. Because *buba* has not been imposed as the name of anything, it is not the effect of a knowledge of what is named and, therefore, it does not entail, and thus signal as effect to cause, the presence of such knowledge. It does not allow you to gather whether its speaker feels this way or that other way toward you. Hearing *buba* spoken does not even allow you to ascertain whether it has emerged from the mouth of a human being. In his exclusion of *vox non-significativa* from his discussion of what is signaled by the *vox* apart from its imposition, Burley insists that there is such a thing as utterance that means nothing at all, from which nothing at all can be gathered, not even that its speaker is present, speaking, human, cognizant, and so forth. In what strikes me as an almost gratuitous fashion, Burley thus leaves hanging the question I have just posed, as to what does in fact happen when you hear a nonsense word.

If Burley does not answer this question, he does provide material for speculation. In his account of what happens when you hear a noun or a verb, the knowledge of the presence of an *intellectus* that is at once an ignorance of its identity resumes a peculiarity of the circumstances of the original imposition of the word. In section I.14, Burley has set out his views on the debate over whether utterances refer to concepts or things. Following Bacon's emphasis on the *voluntas imponentis*, Burley argues that since the meaning of a word is assigned by a person's will, as everyone would agree, there is no reason why it should not signify the thing directly if that is the will of the impositor. But Burley goes on to "prove" (*confirmo*) this argument that he has based on the freedom of the will by invoking, as in fact Bacon had done before him, a limitation that hems in that same will:

> Et istud confirmo sic: Intellectus prius intelligit rem extra quam intelligit aliquid exsistens in eo, quia intellectus non intelligit aliquid exsistens in eo nisi per reflecionem; nunc intellectio directa praecedit intellectionem reflexivam; igitur in illo priori in quo intellectus intelligit rem extra potest imponere nomen ad significandum rem extra. Et illud nomen sic impositum non significabit aliquid existens in anima, quia quodlibet exsistens in anima est adhuc ignotum intellectui et nomen non imponitur nisi noto; igitur vox potest significare immediate rem extra et non oportet quod primo significet passionem animae.

> And I prove this as follows: the understanding understands an exterior thing before it understands anything that exists within itself, because the understanding does not understand anything existing within itself except by reflection; but direct understanding precedes reflexive understanding;

and therefore, in order to signify the exterior thing, it can impose a name on that prior thing in which the understanding understands the exterior thing. And this name, imposed in this way, will not signify anything existing in the mind, because anything existing in the mind is still unknown to the understanding and a name is imposed only on what is known; therefore the utterance can signify an exterior thing immediately, and it does not have to first signify an impression in the mind.[37]

Whereas Burley has just demonstrated the possibility that an utterance will signify a thing directly by recurring to the freedom of the will of the impositor, he now demonstrates the *necessity* of its signifying a thing rather than a concept by placing a constriction on the will. As it turns out, it is not possible for a name to be assigned directly to a concept. Burley's argument rests on a dictum that he ascribes to Boethius, although the latter never said anything of the sort: *vox non imponitur nisi noto*, a name is not given to something unless that something is already known.[38] Since the *intellectus* can only know its own operations in a secondary, reflexive moment, the thing known in the moment of imposition has to be a *res extra*. Proceeding now on the basis of the distinction between *id quo* and *id quod* made famous by Thomas Aquinas in his discussions of *species intelligibilis*, Burley says that the concept is not that which is itself known but that by which something else is known.[39] What this means is that the mind (*intellectus*) is always necessarily unaware of its cognitions (*intellectus*) while they are occurring. While it is certainly possible for the mind to think and even to name one of its own cognitions, it can only do so by the mediation of another cognition which will remain for its part unthought for as long as it is in operation. The only way for the *intellectus* to know itself is through an infinite speculative regression of thinking but unthought cognitions. If the impositor wishes to name a concept in his own mind, he can do so only by forming another concept of it by which to know it, which is to say by treating it as a thing. In short, although the impositor appears to be free to name anything, which is to say either concepts or things, he must know something in order to name it; and because he cannot in the first instance know a

---

37    Ibid., I.14.

38    Burley makes use of this precept, derived from Averroes, throughout his career; cf. Paul Spade's note to Burley's ascription of an associated dictum to Boethius in the *De puritate*: "The quotation has not been found." Burley, *On the Purity of the Art of Logic*, 88, n. 37.

39    On *species intelligibilis*, cf. Leen Spruit, *Species Intelligibilis. 1. Classical Roots and Medieval Discussions* (Leiden: Brill, 1994); Robert Pasnau, *Theories of Cognition in the Later Middle Ages* (Cambridge, 1997).

concept, in fact if he names anything it will be a thing. The prior knowledge of the named thing on which every imposition depends is immediately the failure of that knowledge itself to be known.

It is this perpetual postponement of knowledge on the part of the speaker of a word that is signaled to its hearer when the utterance of that word entails the presence of a concept without disclosing the concept itself. But something different happens with the nonsense word. Rather than being the *effect* of a *prior* knowledge, the pronunciation of a nonsense word is the *cause* of a *subsequent* failure of knowledge. Such an utterance is the production of an inability on the part of the mind to cognize any object. This prompting of a lapse in the *intellectus* is the reason that Boethius had banned nonsense words from the field of logic in the first place:

> Si quis vero huiusmodi vocem ceperit, quae nihil omnino designet, animus eius nulla significatione neque intelligentia roboratus errat ac vertitur nec ullis designationis finibus conquiescit.

> If someone hears an utterance of this sort, which signifies absolutely nothing, his mind—bolstered neither by signification nor understanding—wanders around, turning upon itself, and does not come to rest at any such limit as would be provided by a signification.[40]

But Burley does not think that this effect is a reason to exclude the *vox quae nihil omnino designat*.[41] His innovation consists in suggesting that this predicament of mental breakdown can itself be known. In *buba est disyllabum*, the unimposed word standing for itself in its unimposability produces a failure of knowledge. The propositional knowledge of the *vox singularis* in material supposition is thus not, after all, just an apprehension of the physical characteristics of a vocal entity. It is the experience of the failure of knowledge when the mind is deprived of any object. As sheer differentiation, the *vox singularis* cannot be known. But the truth of the proposition in which it appears is a demonstrable and knowable truth.

What I am suggesting is that the use of a nonsense word in material supposition should be understood as the direct inverse of the scene of imposition. If to assign an utterance as the name of something is to *presuppose* (and thus signal) a knowledge that is itself unknown, to use an utterance non-onomastically to

---

40    Boethius, *Commentarii in librum Aristotelis Peri hermeneias, pars posterior, secundam editionem continens*, ed. Carl Meiser (Leipzig, 1880), p. 74. My translation.

41    On Burley's relations to Boethius, cf. Uña Juárez, *La filosofía del siglo XIV*, pp. 345–347.

stand for its singularity is to *produce* an absence of knowledge that is itself known. Rather than the unknowability of the mind to itself, what is produced in material supposition is a knowledge of the unknowable voice object. While the mind cannot ever hope to know its own operations without mediation, Burley implies that it is possible for it to know its failure to cognize the utterance as such. Now this is just what the *Cloud*-author claims: that the failure of knowledge amounts to a knowledge of that failure, and that both the knowledge and the failure are produced in the reduction, by repetition, of a word to its status as a bare utterance. As he writes toward the end of the treatise, although a person cannot "bi þe werk of his vnderstondyng com to þe knowyng of an vnmaad goostly þing, þe whiche is nouȝt bot God," this does not mean that knowledge of God is impossible. For he continues: "Bot by þe failyng it may; for whi þat þing þat it failiþ in is noþing else bot only God" (125).

• • •

It is at last possible to answer the question with which I began. The *Cloud's* *werk* of unknowing, which I take to be the production of the same sort of truth that inheres in a proposition containing a subject term in material supposition, takes place in an *athomus* of time. What corresponds to this *athomus*, in Burley's schema, is the minimal delay that interposes itself between the *intellectus* as cognizing and the *intellectus* as cognized. The irreducible lag that prevents the mind from ever being entirely present to itself, forcing it to know itself only across a duration, be it ever so slight, is the time of an unknowing at the basis of every operation of naming. In the de-naming or un-imposition that occurs in material supposition, the mind is forced to while away in this duration. And the temporal non-coincidence that is a failure of self-understanding is that same lag that opens up, sometimes, in all its brevity into the frozen time of a word repeated to the point of senselessness.

The truth is that this notion of Burley's fizzled out. He did not maintain it in his mature works on supposition, his contemporaries did not take it up, and his successors developed theories about the material basis and the properties of terms that do not invite such speculation. Indeed, the Oxford logic of the following generations—namely, as it was taught during the time of the composition of the *Cloud*—as it is summed up for example by a Paul of Venice, would never allow for such notions to emerge at all. It has sometimes seemed to me that once the nonsense word—an entity that had persisted as a problem at the margins of the discipline of logic since its inception—finds, in Burley, a champion, logic flares up in a kind of immune response and manages at last to rid itself of this irritant. But just when the success of this operation becomes inevitable, the irritant reemerges elsewhere, in an improbable place: a vernacular treatise on prayer. However that may be, what the constellation of the various appearances of *vox non-significativa* I have brought together here allows to be discerned is the opening of a minimal duration, in the time it takes the bare utterance to sound, of a mental failure that attains to truth.

# CHAPTER 8

# Theological and Social Time: The Case of the Beguines

*Vera von der Osten-Sacken*

Despite many similarities and close contact with the Cistercians, the Beguines were, as far as we can tell from their *Vitae*, not part of an emerging religious movement that encouraged enthusiasm for monastic life. Rather the Beguines worked in the world, caring for the terminally ill and the dying, especially for lepers, trying to face the human suffering in the world by imitating Christ's sufferings. Moreover, scholars working on early Beguine piety, using Jacques de Vitry's *Vita Mariae Oigniacensis* (1215), the earliest exemple of Beguine *Vitae*,[1] as well as other *Vitae*, see the characteristics of the *mulieres religiosae*[2]—their good works—as not totally accounted for by referring

---

1  The earliest example of such a *Vita* is Jacques de Vitry, *Vita beatae Mariae Oigniacensis* (VMO), ed. Daniel von Papebroch, in *ActaSS Junii* IV (1707), pp. 630–666, *Iacobvs de Vitriaco, Vita Marie de Oegnies*, ed. Robert Burchard Constantijn Huygens (Turnhout, 2012 [CCCM 252]). The *Vita* of Mary and other examples are available online and can be read in the *Acta Sanctorum database*: http://acta.chadwyck.co.uk/. For an English translation see *Jacques de Vitry, Thomas de Cantimpré, Two lives of Marie d'Oignies*, trans. Margot H. King, Hugh Feiss OSB (Toronto, 2003). Daniel von Papebroch undertook the classifying into *capitula* that correspond to the thirteenth-century manuscripts, even though the wording occasionally varies. His titles can probably be traced back to Jacques de Vitry. Papebroch also divided the text into larger sections and added consecutive numbering of paragraphs. His numbering is used here to make the quotations clear. Papebroch's counting, which uses chapter 11 twice, for the concluding chapter of the *Vita*'s prologue and for the first chapter of book I, is adopted to avoid confusion. I have used quotations from the *Vita* of Mary indicating the prologue, book I, or book II as sources. Joseph Greven provides evidence that the women described in the *Vita*'s prologue are all identical with those for whom the term "Beguines" is used. See Joseph Greven, *Der Ursprung des Beginenwesens nach den Legenden, der geschichtlichen Literatur und den Aussagen der Quellen* (Münster i.W., 1911), p. 46f, and *Die Anfänge der Beginen* (Münster i. W., 1912 [VRF 8]), p. 46. Greven works with the story of a witness concerning a Cistercian from Aulne (VMO, prologue, cap. 4) that Jacques de Vitry uses later in one of his sermons. See Jacques de Vitry, *Sermo ad viduas et continentes*, ed. Jean-Baptiste Pitra, *Analecta novissima Spicilegii Solesmensis, Altera Continuatio* II (Paris, 1885–1888).

2  Jacques de Vitry used this term. The word "Begine" was originally a dirty word that subsequently was adopted by the women themselves. For this term and its meaning, see Vera von der Osten-Sacken, "Dangerous Heretics or Silly Fools? The Name 'Beguine' as a Label for Lay

to their pursuit of poverty and virginity,[3] that is, by their harsh ascetic practices, their dedicated Eucharistic devotion, and the mystical excesses by
which they imitated the suffering of Christ. Jacques de Vitry, Thomas de
Cantimpré and other theologians who wrote the *Vitae* of the Brabantian
women, were introducing them to representatives of the traditional clergy as
new models for female sanctity.[4]

Biographers of the first Beguines often split the women's lives into two periods, but not necessarily chronological ones. The *vita exterior* and *vita interior*
of Marie d'Oignies, for example, illustrate two different aspects of her entire
life: first, her works of active charity, associated in many cases with the biblical
figure of Martha,[5] reported as practical ministries, for the first Beguines were
engaged in activities such as tending to the diseased or the terminally ill; second, her devotion to Mary in contemplating both the incarnated and the risen
Christ as her mystical bridegroom, in preparing for physical death in order to
reunite with Christ. The domains of Martha (temporal) and Mary (eternal)
were thus brought together in the life of one person.[6]

Indeed, in many *Vitae* women seem to exist between time and eternity. An
especially dramatic example of this is Christina Mirabilis whose *Vita* is narrated *post mortem*. Thomas de Cantimpré thus begins the description of her
life with her death.[7] She returns to the physical world when asked whether
she wants to enjoy eternal blessedness with the risen Christ or to fight alongside the incarnated Christ against the devil and against sin. She says she
wants to rescue as many souls as possible from purgatory. Her *Vita* is the

---

Religious Women of Early Thirteenth-Century Brabant," in *Labels and Libels. Naming
Beguines in Nothern Medieval Europe*, eds. Letha Böhringer, Jennifer Deane and Hildo van
Engen (Turnout, 2015), pp. 99–116.

3 These virtuous practices are understood as the striving for monastic life. See Brigitte Degler-
Spengler, "Die religiöse Frauenbewegung des Mittelalters, Konversen—Nonnen—Beginen,
Albert Bruckner zum 13. Juli 1984," *Rottenburger Jahrbuch für Kirchengeschichte* 3 (1984),
pp. 75–88.

4 Maria Grazia Calzà, *Dem Weiblichen ist das Verstehen des Göttlichen "auf den Leib" geschrieben.
Die Begine Maria von Oignies (†1213) in der hagiographischen Darstellung Jakobs von Vitry
(†1240)* (Würzburg, 2000), p. 45, considers the life of Mary as evidence for a new specifically
female perception for which she established the German term "Somatophonie."

5 See Luke 16, 20–31.

6 See Martina Wehrli-Johns, "Maria und Marta in der religiösen Frauenbewegung," in
*Abendländische Mystik im Mittelalter*, ed. Kurt Ruh (Stuttgart, 1986), pp. 354–367.

7 See Thomas de Cantimpré, *S. Christina Mirabili Virgine Vita*, in *ActaSS, Julii* v (1727), pp. 637–
660, cap. 5: "Et factum est post haec, ut ex interno contemplationis exercitio virtute corporis
infirmata, vita excederet."

post-mortem life of a woman who has returned to life, but from the perspective of eternity. Her *Vita* is in this way the life of woman in the temporal world who perceives suffering as an opportunity, given by grace, to do penitence[8] for herself and others, to intervene for souls in this world, and to fight the devil and his demons as a member of the *militia Christi*. Indeed, with the help of the *fides oculata*,[9] the first Beguines and other pious people were supposedly able to see invisible beings such as angels and demons and to be able to understand doctrinal truth.[10] They were *milites Christi* who, in the secular world, fought actively against the devil and demons to snatch dying souls from hell.[11] Such fights were frequently depicted as inner female acts to help other humans by means of prayers or ascetic intercessions. Many *Vitae* give vivid descriptions of still invisible demons—the women concerned remained visible.

The Beguines were, of course, connected to this world in such domains as liturgy,[12] pastoral care, and church politics. Grundmann's influential thesis does not take into consideration that the first beguines openly voiced criticism of the church and the behavior of the clergy.[13] The Beguines supported internal

---

8    See Vera von der Osten-Sacken, *Jakob von Vitrys Vita Mariae Oigniacensis. Zu Herkunft und Eigenart der ersten Beginen* (Göttingen, 2010 [VIEG 223]), pp. 123–145.

9    See VMO, prologue, cap. 2 and lib. II, cap. 71: "Domino revelante invisibilia, quasi visibiliter fide oculata percipiebat."

10   See for example VMO, lib. II, cap. 72, 83.

11   See VMO, lib. II, cap. 50: "Tunc illa, consuetae gravitatis & innatae verecundiae quasi oblita, ad lectum aegrotantis cucurrit, & immundis spiritibus se opponens, non solum precibus pugnabat, sed etiam pallio suo tamquam muscas abigebat." The motif is considerably older. Jacques de Vitry was familiar with the *Vita* of the desert father Antonius, ascetic, who survived fights with demons. See Athanase d'Alexandrie, *Vie d'Antoine*, ed. G.J.M. Bartelink (Paris, 1994 [Sources Chrétiennes 400]), chap. 5. For an English translation see Athanasius of Alexandria, *The Life of Antony. The Coptic Life and the Greek Life*, trans. by T. Vivian and A.N. Athanassakis (Kalamazoo, 2003 [Cistercian Studies Series 202]).

12   The most famous example is Juliana of Cornillon and her commitment that resulted in the introduction of the Feast of Corpus Christi. For Juliana of Cornillon, see Ernest W. McDonnell, *The Beguines and Beghards in Medieval Culture, with Special Emphasis on the Belgian Scene* (New Brunswick, 1954), pp. 299–310; Barbara R. Walters, *The Feast of Corpus Christi* (University Park, 2006), pp. 3–35.

13   Herbert Grundmann, *Religiöse Bewegungen im Mittelalter. Untersuchung über die geschichtlichen Zusammenhänge zwischen der Ketzerei, den Bettelorden und der religiösen Frauenbewegung im 12. und 13. Jahrhundert und über die geschichtlichen Grundlagen der deutschen Mystik* (Berlin, 1935), p. 338. English translation: Herbert Grundmann, *Religious Movements in the Middle Ages: The Historical Links Between Heresy, the Mendicant Orders,*

church reform and ecclesiastical efforts to improve the clergy's preaching and pastoral care, efforts that arose as a result of reforms that Parisian theologians, especially in the Petrus Cantor circle, voiced.[14] Evidence suggests that the Beguines were inspired by supporters of the Chanter who cared for early Beguine communities in the Liège area at the beginning of the thirteenth century. Jacques de Vitry had contact with Petrus Cantor, as did Jean de Nivelles who, according to de Vitry, founded a Beguine community in Nivelles and, like many other Cantor supporters, was engaged in preaching campaigns to recruit for the crusades and to condemn usury.[15]

• • •

The first *mulieres religiosae* thus seem to fit into the context of lay confraternities, as Kaspar Elm has argued. According to Elm, the distinctive features of their religious lives led to their being called "semireligious." Moreover, Elm argues that the first Beguines made direct use of scripture, which correlates with the women's affinity for preaching. As far as we know, the Beguines not only attended the sermons of Cantor supporters, but they read the Bible and developed sermons themselves. Jacques de Vitry mentions in a letter he sent back to his friends and fellow Augustinians in Oignies, after he had visited the papal court at Perugia in 1216, that Honorius III would allow the Beguines to live as a community and to sermonize, but on moral themes and for domestic purposes only.[16]

As we know, the *mulieres religiosae* found their main activity in the urban environment by caring for the sick and the despised members of society. For example, around the turn of the thirteenth century, citizens of the towns gradually took over the hospitals, strengthened them financially, improving

---

and the Womens' Religious Movement in the Twelfth and Thirteenth Century, with the Historical Foundations of German Mysticism, trans. Steven Rowan with an intro. by Robert E. Lerner (Notre Dame, 1995). Grundmann is incorrect in having separated the *mulieres religiosae* from so-called heretic groups. The Beguines' unique characteristics were strongly influenced by their contact with wandering preachers recruiting for the Crusades, leading penitence campaigns condemning usury, and by their own urban and wealthy origin.

14   For Petrus Cantor and his disciples, see John W. Baldwin, *Masters, Princes and Merchants, The Social Views of Peter the Chanter and His Circle* (Princeton, 1970).

15   For the interaction between the women and the Parisian theologians, see von der Osten-Sacken, *Vita*, pp. 12–37, 93–133.

16   See Robert Burchard Constantijn Huygens, *Lettres de Jacques de Vitry* (Leiden, 1960), p. 74 (rev. ed. [Turnhout, 2000 = CCCM 171], pp. 491–652). Also the "catholic poor" were allowed to assemble for mutual admonition and edification, see Grundmann, *Bewegungen*, p. 111.

their staff and gradually incorporating these church institutions into public ones so that a municipal hospital system arose. In this process many *mulieres religiosae* worked in prominent positions to care for the sick and to found hospitals. One of the most famous examples is Ivetta,[17] who reorganized the hospital of Huy. According to her biographer, Hugo of Floreffe, Ivetta's desire to care for lepers grew out of her desire to remedy the lack of nursing as well as pastoral care; scarcely anybody was willing to care for the physical needs of lepers and masses were rarely said for them.[18] Hugo describes Ivetta's quotidian activities admiringly: men and women apparently came from great distances to follow Ivetta's example, to watch her laying the table for the sick, washing and bedding them and doing their laundry. She did all this in such a reverential way "ut in omnibus esse Christum crederet, & Christum reuereri videretur in singulis."[19]

*Mulieres religiosae* incurred general disapproval, however, because they begged for alms for lepers. Nonetheless, in his *Historia Occidentalis*, Jacques de Vitry assured men and women caring for the sick that the less they were valued in terrestial life, the more they would be honored in heaven. According to him, those caring for the sick underwent the severest penance by living for them, enduring squaler and fetid circumstances: "Tantas autem plerumque pro Christo sustinent infirmorum immunditias et fetorum molestias pene intolerabiles, sibimet violentiam inferentes, quod nullum aliud penitentie genus huic sancto et pretioso in conspectu dei martyrio posse arbitrer comparari."[20] According to Jacques de Vitry, Marie d'Oignies gave herself over to Isaiah's "Suffering Servant" whom Jerome identified with the Suffering Messiah:[21]

---

17  For Ivetta of Huy, see Anneke B. Mulder-Bakker, *Lives of the Anchoresses. The Rise of the Urban Recluse in Medieval Europe* (Philadelphia, 2005), pp. 118–147; *Verloren Vrouwen. Kluizenaressen in de middeleeuwse stad* (Hilversum, 2007), pp. 116–129.

18  See Hugo von Floreffe, *Vita de B. Ivetta, sive Iutta, vidua reclusa, Hui in Belgio, auctore Hugone Floreffiensi, ActaSS Januar* I (1743), 863–887 (chap. 10, 33).

19  See Hugo von Floreffe, *Vita*, chap. 10, 33: "Videbant siquidem eam ponentem infirmis mensam, fundentem aquam manibus, insuper & pedes manusque leprosorum lauantem, cum ipsimet eos lauare non possent; vestimenta eorum lauantem & excutientem, leprosos decumbentes lectulo ponentem, de lectulo leuantem, & cuncta sancta obsequia fideliter exhibentem, & indefesse cum tanto studio, & reuerentia, vt in omnibus esse Christum crederet, & Christum reuereri videretur in singulis."

20  See John Frederick Hinnebusch O.P., *The Historia Occidentalis of Jacques de Vitry, A Critical Edition* (Fribourg, 1972 [SpicFri 17]), p. 148.

21  See *Commentaires de Jérôme sur le prophète Isaïe*, texte établi par R. Gryson et C. Gabriel, Livres XII–XV, avec la collaboration de H. Bourgois et V. Leclercq (Freiburg i. Brusgon, 1998), p. 22, lines 8–10, 68–78.

"vere languores nostros ipse tulit et dolores nostros ipse portavit et nos putavi-
mus eum quasi leprosum et percussum a Deo et humiliatum."[22] The three
attributes—*leprosus, percussus* and *humiliatus*—thus exemplify the way many
of the first *mulieres religiosae* sought to follow the ideal of the suffering Christ.
Bernhard of Clairvaux added that using the term *leprosus* allegorized that
Christ not only adapted human flesh, but the flesh of a sinner.[23] The disease
brought together the concept of *ecclesia militans* with the future: the timeless
and eternal world of the *ecclesia triumphans*. Because they lived *as if* they had
already died, the sick stood no longer in the *saeculum*, but at the borderline
between time and eternity. Jacques de Vitry judged leprosy to be an anticipated
purgatory that, like martyrdom, kept the sick from all punishment after
death.[24] Perhaps this is the reason why Jacques de Vitry quotes Marie d'Oignies
as saying that every sick person should welcome his or her disease.[25] Indeed,
many of the *mulieres religiosae* not only strove to help their fellow human
beings in pain as a service to Christ, but also wanted to achieve martyrdom by
allowing themselves, for example, to be infected by lepers and then to die from
the lingering illness.

We should add that nearly 20 years after the death of Marie d'Oignies,
Thomas of Celano, in his first hagiography of Saint Francis of Assisi, completed
in 1228, mentioned that as a young man Francis, who anxiously avoided accidentally encountering a leper, made, a few years later, nursing lepers his highest priority. The very climax of Francis's *conversio* is the moment when he
overcame the fear and abhorrence he felt at nursing and turned to a poor sick
man in the street to kiss him.[26]

---

22  See Isaiah 53, 4.

23  See Bernhard of Clairvaux, *Pro Dominica I. Novembris, Sermo I* (*De verbis Isaiae, VI, 1, Vidi
    Dominum sedentem*), PL 183, col. 345b: "Multifarie ergo multisque modis non solum locu-
    tus in prophetis, sed et visus est a prophetis. Agnovit eum David minoratum ab angelis;
    Jeremias etiam vidit eum in terris cum hominibus conversantem; Isaias modo super
    solium excelsum, modo non solum *infra* angelos aut inter homines, sed tanquam lepro-
    sum se vidisse testatur, id est non in carne tantum, sed in similitudine carnis peccati."

24  Jacques de Vitry, *Sermo II ad leprosis*, eds. Nicole Bériou and François-Olivier Touati,
    p. 123, lines 152–155: "Lepra enim corporis est uobis purgatorium in hac uita quam si pati-
    enter sustineatis pro martyrio uobis reputabitur et aliud purgatorium non habebitis."

25  See VMO, lib. II, cap. 74. "quod numquam infirmum vidisset [sc. Marie d'Oignies], quin
    ejus, quantacumque esset, infirmitatem desiderasset."

26  Thomas of Celano, *Vita Prima sancti Francisci*, in *Fontes Franciscani*, eds. Enrico
    Menestò and Stefano Brufani, (Porziuncola, 1995) p. 259–424 (292). The motif of the
    kiss is older, however; it is already in Venantius Fortunatus, *Vita* of Radegundis, *ActaSS*,
    *Augusti* III (1737), 67–74, dating from the sixth century, where a saint kisses lepers.

As for the status of lepers, the *Sachsenspiegel*, written as a record of law in Germany during the first half of the thirteenth century, stated that lepers were considered dead and "gone." They were not allowed, for example, to hold a fief or to own a title.[27] The *Coutumes de Beauvaisis*, which Philippe de Beaumanoir wrote ca. 1280, refers to the lepers as having drifted out of the living world: "il est mort quant au siècle."[28] Once they were dismissed from the *saeculum* using a ritual phrase like "sis mortuus mundo, vivens iterum Deo,"[29] lepers were often very strictly limited to their dwellings or were not even allowed to go out without wearing special clothing or giving audible warning of their presence to avoid contact with healthy citizens. This kind of complete seclusion must have been attractive to the *mulier religiosa*, Marie d'Oignies. Poor, suffering, and despised by the world, the sick reminded her of the passion Jesus underwent. The life of the lepers made the *imitatio Christi* possible.

• • •

Cf. Venantius Fortunatus, *Vita Sanctae Radegundis*, chap. 15: "Ipsa exinde mulieres variis leprae perfusas maculis comprehendens in amplexu, osculabatur in Deo, eas toto diligens animo. Deinde posita mensa, ferens aquam calidam, facies lavabat, manus, ungues & ulcera, & rursus administrabat, ipsa pascens singulas. Recedentibus praebebat auri vel vestimenti solatium, vix una teste munifica. Ministra tamen praesumebat eam blandimentis sic appellare; Sanctissima Domina, quis te osculetur; quae sic leprosas amplecteris? Cui respondebat benevole: Vere si me non osculeris, hinc mihi nulla cura est."

27 See Peter Browe, *Die häufige Kommunion im Mittelalter* (Münster, 1938), p. 102 with notes 19 and 21. To the legal status of lepers, see Friedrich Merzbacher, "Die Leprosen im alten kanonischen Recht," *ZSRG.K* 53 (1967), 27–45 and Antje Schelberg, *Leprosen in der mittelalterlichen Gesellschaft. Physische Idoneität und sozialer Status von Kranken im Spannungsfeld säkularer und christlicher Wirklichkeitsdeutungen* (Göttingen, 2000), electronically published at GOEDOC Dokumentenserver der Georg-August-Universität Göttingen (SUB Göttingen), http://webdoc.sub.gwdg.de/diss/2003/schelberg/schelberg.pdf (release: 1 May 2010).

28 See Philippe de Beaumanoir, *Coutumes de Beauvaisis*, ed. Amédée Salomon (Paris, 1899–1900), II n. 1617.

29 Kay Peter Jankrift, *Leprose als Streiter Gottes, Institutionalisierung und Organisation des Ordens vom Heiligen Lazarus zu Jerusalem von seinen Anfängen bis zum Jahre 1350* (Münster, 1996), p. 68, refers to the phrase "sis mortuus mundo, vivens iterum Deo" in the context of the thirteenth and fourteenth centuries. Siegfried Reicke, *Das deutsche Spital und sein Recht im Mittelalter* I (Amsterdam, 1932) p. 278 with n. 4, dates the mode "ejiciendi seu separandi leprosos a sanis in diocesi Trevirensi" for the diocese of Trier to the end of the 15th century. To further suspensions of rites see Edmund Martène, *De antiquis ecclesiae ritibus* (Antwerpen, 736; rpt. Graz, 1967), II, lib. 3, kap. 10.

With respect to their insular life, zeal for chastity,[30] perceived as a virtue not as a gift or *charisma*, was a prominent feature of Beguine religious life.[31] The *mulieres religiosae* refused earthly marriages, practicing celibacy in already existing marriages. The first Beguines not only demanded sexual abstinence (*cœlibem, & vere Angelicam vitam ducentes*),[32] but they also expected, according to contemporaries, extreme asceticism so that their bodies would not be an obstacle to a blessed and eternal union with Christ.[33] They wanted to receive Christ mystically, thereby making their bodies vessels of his eternity. They took upon themselves the suffering that Christ had endured: "stigmata Domini nostri Jesu Christi in…corpore [earum]."[34] Through mystical excesses the *mulieres religiosae* imitated the crucified Christ's suffering. They contemplated the Incarnate Christ as the Man of Sorrows in visions and mystical experiences.[35] They drew images of affection for the suffering Christ from the *Song of Solomon* and from the exegesis of these texts by monastic theologians such as Bernhard of Clairvaux and Rupert of Deutz.[36] In his *Sermones super Cantica Canticorum*,

---

30    Virginity is the only vow that was certainly required of future Beguines. Cf. *Klosterfrauen, Beginen, Ketzerinnen*, eds. Amalie Fössel and Anette Hettinger (Idstein, 2000), pp. 139–147 (Gregory IX, 1235 (139f), Fritzlar 1244 (141f), cardinal Hugo to provost Heinrich 1251 (140f), Vienne 1311 (146f)). Also Jacques de Vitry in his *Sermo ad virgines* mentions virginity as the outstanding characteristic of female piety. This is astonishing in its clearness and reveals much about what Jacques de Vitry draws from the ancient church. See Jacques de Vitry, *Sermo III ad Religiosas* (*Sermo ad moniales albas cisterciensis ordinis*), ed. Jean Longère, in *Quatre Sermons ad Religiosas de Jacques de Vitry* in Michel Parisse, ed., *Les Religieuses en France au XIIIème siècle* (Nancy, 1985), pp. 215–300 (275), and *Sermo IV ad Religiosas* (also a sermon to Cistercian nuns), ed. Jean Longère, p. 292f.

31    Already in the twelfth century the virtue of virginity was intensified and exaggerated by the idea of mystical marriage with Christ. See Ludwig Hödl, *Jungfräulichkeit*, LMA V (1999), col. 808f.

32    See VMO, prologue, cap. 3.

33    See VMO, Lib. II, cap. 65.

34    Galatians 6, 17. See VMO, lib. I, cap. 12: "licet stigmata Domini nostri Jesu Christi in nostro corpore ferre debeamus; scimus tamen quod honor Regis judicium diligit, nec placet Domino sacrificium de rapina pauperis."

35    This phenomenon suggests the idea of an imitation of Mary that was mainly cultivated in a Cistercian environment. For the Beguines' imitation of Mary, see, for exemple, Martina Wehrli-Johns, "Haushälterin Gottes. Zur Mariennachfolge der Beginen," in *Maria. Abbild oder Vorbild? Zur Sozialgeschichte der mittelalterlichen Marienverehrung*, ed. Hedwig Röckelein (Tübingen, 1990), pp. 147–167 and Joanna E. Ziegler, *Sculpture of compassion: the Pietà and the Beguines in the Southern Low Countries: c. 1300–c. 1600* (Rome, 1992).

36    For the term "monastical theology" in contrast to "scholastic theology," see Ulrich Köpf, "Wurzeln reformatorischen Denkens in der monastischen Theologie Bernhards von

Bernhard connected mystical experience to penance and the grace of charity, allegorizing the biblical role of Mary and Martha and their brother Lazarus whose name, especially in the Brabantian area ca. 1200, was also associated with the "poor" in the Beatitudes.[37]

In a gloss that was most likely added to the *Vita* of Norbert of Sempringham by an unknown Norbertinian, the lepers are also named after Lazarus: "Lazari etiam nunc in Belgio nominantur leprosi: quia horum contubernia plura, ac nominatim apud Hierusalem. Quin etiam Romae extra urbem teste a Lapide, habentur fundata sub invocatione S. Lazari, evangelici pauperis, cujus ulcera canes lingebant."[38] Since Lazarus was not apparently at home when Jesus came to visit his sisters, Bernhard assigns to him the part of the penitent, "cleaning his soul" as a house made ready for quests by the active charity of Martha and filled by the contemplative *amor Dei* of Mary.[39] Rupert of Deutz related his whole exegesis of the *Song of Solomon* to the incarnation of Christ, entitling it *De incarnatione Domini*,[40] interpreting the "bride" as the Virgin

---

Clairvaux," in *Reformation und Mönchtum. Aspekte eines Verhältnisses über Luther hinaus*, ed. Athina Lexutt (Tübingen, 2008), pp. 29–56 (31).

37    Luke 16, 20–31. Already in John's version of the Maria-Marta-pericope, Lazarus has a more prominent position than in Luke's Gospel. Lazarus moves closer towards Jesus, for example, he sits next to Jesus at a table and is to some extent integrated into the inner circle of his disciples. According to John 11, 45–57, his revival motivated many people to convert—people flocked to Bethan not only for Jesus, but especially for Lazarus. Accordingly, Lazarus was ordered to be killed.

38    See PL 170, col. 1307d with n. 82.

39    See Bernhard of Clairvaux, *Super Cantica*, Sermo 57, IV, 10, in *S. Bernardi Opera*, ed. Jean Leclercq (Rome, 1957–1977), II, p. 125. See also Giles Constable, *Three Studies in Medieval Religious and Social Thought* (Cambridge, 1995), p. 6 n. 13.

40    Rupert appears to have required an interpretation of the written word as well as of the spoken word as it was discussed with regard to Waldensians, and also Humiliates, 80 years later. After the year 1117, Rupert wrote to Archbishop Friedrich of Cologne: "Das weite Ackerland der heiligen Schrift ist allen Bekennern Christi gemeinsam zu eigen, und niemandem kann das Recht, es zu bearbeiten, abgesprochen werden, wenn er nur mit unversehrtem Glauben seine Meinung sagt oder schreibt. (The broad field of the Holy Scripture is owned by all confessors of Christ, and nobody can be denied the right to deal with it, when he states his opinion with a sound faith)," from Friedrich Ohly, *Hohelied-Studien. Grundzüge einer Geschichte der Hoheliedauslegung des Abendlandes bis um 1200* (Wiesbaden, 1958), p. 123 with n. 1. Rupert of Deutz detached himself from the patristic tradition. On the basis of a direct and mystical divine experience and legitimized by his conscience and mercy, he replaced references to the patriarchs' sayings in his texts with reports of visions. Such reports that have legitimacy were absolutely a matter of course for female *Vitae* a hundred years later.

Mary[41] who was connected to all three persons of Trinity in different ways.[42] As the "church," she could not remain in motionless contemplation. According to Rupert, it was her part to not keep her secret knowledge private, but to share it by teaching and example (*verbo et exemplo*).[43] That is exactly what Jacques de Vitry and his fellow preachers had heard from Petrus Cantor and what they were about to put into action when they met the first Beguines.

• • •

As for their relationships with other religious communities, many of the first Beguines, as Simone Roisin points out, lived in close contact with female Cistercians or spent their remaining years among them. Jacques de Vitry depicts this extensive interaction between nuns and Beguines. This is perfectly in line with the large percentage of Cistercian copies of existing manuscripts of the *Vita* I have found during my own research. However, many of the first Beguines started their life as laypersons and did not aspire to affiliate themselves with a monastic community. Herbert Grundmann[44] assumed that they were connected with a female religious movement that arose in the twelfth century. By arguing this, he responded to the traditional assumption that the occurrence of numerous women, who predominantly aspired to personal poverty and a celibate life in the twelfth and thirteenth century, was largely a result of propaganda made by Premonstratensian and Cistercian communities. According to Grundmann the women were acting of their own accord when they first turned their attention to the Premonstratensian and then to the Cistercian communities and after both communities restricted the affiliation of female members, produced the Beguine religious life as a reaction.[45]

---

41   Like Origenes, Rupert identifies in the *Song of Songs* the seventh verse of the Old Testament, dealing with love, through which God's son became human. Rupert legitimizes the aim of his works with two visions of his own and with the vision of another who has him sitting next to the saints around Christ's throne holding the *Song of Songs* in his hands, while discussing the aspect of Trinity with the Virgin Mary and listening to the following: "Femina mente Deum concepit, corpore Christum: Integra fudit eum nil operante viro" (see Ohly, *Hohelied-Studien*, p. 125).

42   The father as *vera sponsa*, the son as *sponsa et mater* and the spirit as love of both is *templum proprium caritatis*. Rupert understands Mary as a teacher of the apostles, as witness, and as co-author of the Gospel.

43   See Ohly, *Hohelied-Studien*, p. 132.

44   See Grundmann, *Bewegungen*, p. 320.

45   This view on the first Beguines striving for monastic life and accordingly on the first Beguinages as nunneries manqué is still influencing the current debate on character and

Of course, a significant secondary aim that especially Jacques de Vitry, but also Thomas of Cantimpré, most likely had in mind while writing the women's *Vitae* as an account of saintliness was to win the Cistercians' favour for the *mulieres religiosae*. But this cannot be the only explanation. In their works, Jacques and Thomas described attributes of actual women with whom they had come into contact. Accordingly, for the *mulieres religiosae* no direct line of continuity can be traced to an existing evangelical movement pursuing the imitation of Christ since the eleventh century, but the *mulieres religiosae* were strongly influenced by the spirit of their time. Their emergence, especially their caring of the sick, constitutes a new direction—and mandate—enabling a *vita religiosa* and a *vita activa in saeculo*, giving the perception that they were at the borderline between time and eternity. According to their biographers, the Beguines were aware of the option to live in visible and invisible spheres of society and also in the eternal world. Their intensive transchronological and antichronological mysticism went beyond the borderline between time and timelessness, for biographers describe how the mystics receive access to the afterlife, where their souls could temporarily stay, while their bodies remained in the here and now.

---

origin of Beguine piety. Jocelyn Wogan-Browne and Marie-Élisabeth Henneau, "The Medieval 'Woman Question,' and the Question of Medieval Women," in *New Trends in Feminine Spirituality. The Holy Women of Liège and Their Impact*, eds. Juliette Dor, Lesley Johnson, and Jocelyn Wogan-Browne (Turnhout, 1999), p. 5: "They [the Beguinages in the late twelfth century] were largely formed by women from upper social classes who sought to follow monastic ideals (such as voluntary poverty) and for whom there was inadequate institutional provision."

CHAPTER 9

# Medieval Mechanical Clocks

*Gerhard Jaritz*

On December 28, 1372, in the small Lower Austrian town of Tulln on the Danube, about 30 miles west of Vienna, a man named Niclas Swaelbl, burgher of Breslau (Wrocław) in today's Poland, swore an expiatory oath in front of the judge and the members of Tulln's town council.[1] He came before the court, because he had killed Konrad, the local town clerk. The judge and the council decided that, for the "improvement of his soul" (*ze pezzrung seiner sel*), he should construct an *arloy*, that is, a clock, for St. Stephen's parish church in Tulln, *das sich selber slach an welher glokken man im zaigt* (a clock with striking mechanism). Moreover, Niclas was ordered to make a pilgrimage to Rome. All this was to be done between then (December 28, 1372) and Saint Michael's day of the following year (September 29, 1373), that is, during the next nine months. Then he would again be granted freedom of movement in Tulln and Austria. He should be liable with all his goods in Breslau and anywhere else.

Niclas Swaelbl was a clockmaker, one of these specialist artisans who, particularly in the second half of the fourteenth century and first half of the fifteenth century, were internationally acknowledged and needed. He apparently did the job imposed in Tulln, although there are no more surviving sources about this particular novelty of a striking clock in the small Lower Austrian town at a time when even the nearby capital of Vienna did not yet possess one.[2] This is not the only case that shows the attempts of small communities to outdo larger and wealthier cities by possessing a mechanical clock.[3] We should add that the local court's decision to "improve" Swaelbl's "soul" may also be compared with the well-known late medieval donations for the support, building, and repair of other common and public necessities of communities,

---

1 St. Pölten (Lower Austria), Niederösterreichisches Landesarchiv, charter Tulln n. 36; Anton Kerschbaumer, *Geschichte der Stadt Tulln* (Krems, 1874), pp. 374–375. See also Ernst Englisch and Gerhard Jaritz, *Das Leben im spätmittelalterlichen Niederösterreich* (St. Pölten / Vienna, 1976), p. 13; Gerhard Dohrn-van Rossum, *History of the Hour: Clocks and Modern Temporal Orders*, trans. Thomas Dunlap (Chicago / London, 1996), p. 158.

2 Concerning the earliest clocks in Vienna, see Ferdinand Opll, *Leben im mittelalterlichen Wien* (Vienna / Cologne / Weimar, 1998), pp. 12–13.

3 Dohrn-van Rossum, *History of the Hour*, p. 141, mentions, for example, the small Silesian town of Schweidnitz near Breslau. As early as 1370 the town representatives ordered "a clock equal to the one in Breslau or better."

like roads, paths and bridges. These were also seen as pious contributions meant for the salvation of the donors' souls.[4]

• • •

Some years ago, the German historian Gerhard Dohrn-van Rossum conducted an extensive study of the development, function and perception of mechanical clocks in the late Middle Ages.[5] My contribution is meant to offer some additional observations concerning the variety of values that were connected with clocks, their function, use, and public presentation.

The first European boom of the new invention of the mechanical clock was between 1370 and 1380, as the story about Niclas Swaelbl indicates.[6] The earliest evidence of public clocks originates from northern Italy in the first half of the fourteenth century. Around the mid-fourteenth century, an "internationalization" can already be demonstrated, with clocks in urban centres and residences of England, Sweden, France, and Germany.[7] Some of the clockmaker specialists traveled through all of Europe, even to today's Ukraine and further (for instance, Moscow, 1404[8]), constructing clocks in a large number of mainly urban communities that had enough funds to afford the new mechanical devices to measure time, prestigious objects to be seen and heard in public.

With regard to surviving late medieval specimens, one of the most famous examples is the mechanical clock and astronomical dial on the southern wall of the Old Town City Hall in Prague (Fig. 9.1), dating back to 1410 and constructed by a local clockmaker and a professor of mathematics and astronomy at Charles University.[9] This *Prague orloj* was one of the most prestigious pieces among the complex mechanical and astronomical clocks designed and constructed during the fourteenth and fifteenth centuries.

---

4   Concerning pious donations for building bridges, see Franz Falk, "Die Kirche und der Brückenbau im Mittelalter," *Historisch-politische Blätter für das katholische Deutschland* 87 (1881), pp. 87–110, 184–194, 245–259; Erich Maschke, "Die Brücke im Mittelalter," *Historische Zeitschrift* 224 (1977), pp. 265–292. Regarding indulgences for building bridges and roads, see also Nikolaus Paulus, *Geschichte des Ablasses am Ausgang des Mittelalters* (Paderborn, 1923; rpt. Darmstadt, 2000), pp. 370–374.

5   Dohrn-van Rossum, *History of the Hour*.

6   Concerning this boom, see ibid., pp. 157–159; with regard to the invention of the mechanical clock and its dissemination during the Middle Ages and the early modern period, generally, see ibid., pp. 45–215.

7   Ibid., pp. 129–133.

8   Ibid., p. 161.

9   See Jakub Malina, *The Prague Horologe—A Guide to the History and Esoteric Concept of the Astronomical Clock in Prague* (Prague, 2005).

Mechanical clocks could be used to determine when work was to start and when to end, when church services and other events were to begin and finish, meaning that they had a practical value ("Gebrauchswert").[10] But they also were valuable and prestigious objects; thus they possessed a supplementary value ("Zusatzwert") as well.[11] Quite frequently, the supplementary value seems to have been more important than the practical value. Clocks signified the order and welfare of the community and, at the same time, they represented wealth and progress. They became "legal and symbolic expressions of communal autonomy," as Gerhard Dohrn-van Rossum has put it.[12]

In this way, clocks were integrated into the specific late medieval group of other special objects and devices of public exposure that played an important role in the representation of the ideal that urban space was in good order. These objects also marked the success of the communities and their inhabitants, for instance, red roof tiles,[13] glass windows,[14] or paved roads.[15] In this context, the phenomenon of the mechanical clock's innovation played a particularly important role.[16] All of these objects jointly contributed to the beauty of an urban community.[17]

---

10    Concerning this practical value of clocks for late medieval urban merchant and artisan society, see Jacques Le Goff, "Merchant's Time and Church's Time in the Middle Ages," in Le Goff, *Time, Work, & Culture in the Middle Ages*, trans. Arthur Goldhammer (Chicago / London, 1980), pp. 36–37; Le Goff, "Labor Time in the "Crisis" of the Fourteenth Century: From Medieval Time to Modern Time," in ibid., pp. 48–52.

11    Concerning this "Gebrauchswert" and "Zusatzwert" of clocks, see Gerhard Dohrn-van Rossum, "Uhrenluxus—Luxusuhren. Zur Geschichte der ambivalenten Bewertung von Gebrauchsgegenständen," in *"Luxus und Konsum"—eine historische Annäherung*, eds. Reinhold Reith and Torsten Meyer (Cologne, 2003), p. 97.

12    Dohrn-van Rossum, *History of the Clock*, p. 139.

13    See, for example, Harry Kühnel, "Normen und Sanktionen," in *Alltag im Spätmittelalter*, ed. Harry Kühnel, 3rd ed. (Graz / Vienna / Cologne, 1986), pp. 24–25; Gerhard Jaritz, "Alltag in der Stadt des 15. und 16. Jahrhunderts," in *Alltagserfahrungen in der Geschichte Österreichs*, ed. Ernst Bruckmüller (Vienna, 1998), pp. 51–52.

14    See, for example, ibid., pp. 58–59; Helmut Hundsbichler, "Wohnen," in *Alltag im Spätmittelalter*, ed. Kühnel, p. 263.

15    See, for example, Jaritz, "Alltag in der Stadt," pp. 52–55.

16    Concerning the general role and function of innovations and new objects in the material culture of the late Middle Ages, see Gerhard Jaritz, "Das 'Neue' im 'Alltag' des Spätmittelalters: Annahme—Zurückweisung—Förderung," in *Alltag und Fortschritt im Mittelalter*, Sitzungsberichte der Österreichischen Akademie der Wissenschaften, philosophisch-historische Klasse 470 (Vienna, 1986), pp. 83–87.

17    Concerning this 'beauty' of late medieval urban communities, see Helmut Hundsbichler, "Stadtbegriff, Stadtbild und Stadtleben des 15. Jahrhunderts nach ausländischen

In studying fifteenth-century visual representations of towns, churches or monasteries, one comes across the situation that, there, clocks were already seen as an integral and necessary part of public space and its visualization.[18] This is also true for depictions where the clocks had to be portrayed as such small details so that an ordinary beholder certainly was not able to see or recognize them. But they simply had to be there: like in a panel painting donated in 1462 by a cleric from the small Upper Austrian town of Freistadt (Fig. 9.2) showing a view of the town in the background (Fig. 9.3) that includes a tower with the public clock on it (Fig. 9.4). Or when one looks closely at a medallion from around 1490 depicting the twelfth-century saintly Austrian Margrave Leopold III (1095–1136), with two Lower Austrian Benedictine monasteries in the background, Melk (right) and Klein-Mariazell (left) (Fig. 9.5), one again finds a clock, this time above the entrance to the church of the religious house of Klein-Mariazell (Fig. 9.6).

Other visual representations of clocks may be seen as even more marginal but, nevertheless, they still show the importance of the object and its role in society, as, for instance, a simple sketch done by a town clerk from the Estonian city of Reval (today's Tallinn) in an account book dating from the first half of the sixteenth century. There, the rather schematic drawing of the clock by the scribe (Fig. 9.7)[19] denotes entries about the costs of the new clock on the Holy Ghost church.

Soon, the sought-after constructors of mechanical clocks and time-measuring devices were included among the artisans who were illustrated in the depictions of the Planet Children of Mercury (Fig. 9.8).[20] Again, their special position played an important role in this kind of visual representation, making mechanical clocks and their relevance still more familiar to fifteenth-century beholders.

---

     Berichterstattern über Österreich," in *Das Leben in der Stadt des Spätmittelalters*, Sitzungsberichte der Österreichischen Akademie der Wissenschaften, philosophisch-historische Klasse 325, 2nd ed. (Vienna, 1980), p. 123.

18    See the examples, mainly from book illuminations, presented by Dohrn-van Rossum, *History of the Hour*, pp. 120, 132, and 144–154.

19    Juhan Kreem, *Sketches of a Clerk. Pen-and-Ink Drawings in the Margins of the Medieval Account Books of Reval (Tallinn)*, Medium Aevum Quotidianum, Sonderband 18 (Krems, 2006), p. 45, ill. 35.

20    See, for example, the clockmaker with mechanical clock and sundial, detail from the planet image "Children of Mercury," pen-and-ink drawing, Hausbuch, 1460–1480, castle of Wolfegg; Christoph Graf zu Waldburg Wolfegg, *Venus und Mars. Das Mittelalterliche Hausbuch aus der Sammlung der Fürsten zu Waldburg Wolfegg* (Munich / New York, 1997), pp. 38–39. Photo: detail out of ibid., p. 39.

The development towards the use of mechanical clocks can also be found in small urban communities as well as in rural space. In some cases, as seen above, this was as early as the fifteenth and sixteenth centuries,[21] but often still later, with the introduction of public clocks on church towers or other communal buildings only in the eighteenth or nineteenth century. The *torre dell'orologio* from the small Apulian community of Uggiano Montefusco (Fig. 9.9) is one of the many Mediterranean examples of such late prestigious developments. There, "time" was certainly again in the centre of town, on a very public square, "a politically highly sensitive site."[22] This tower on the *piazza* was still used in the twentieth century for the important publication of the names of the members of the community who had died in the World Wars.

Important and prestigious material public objects of diverse communities regularly led to the putting up of a variety of imitations. For the late Middle Ages and the Early Modern period, for instance, red roof tiles were imitated by shingles or wooden boards painted red,[23] glass windows were imitated by false windows painted on walls (Fig. 9.10),[24] as were marble, brocade, and so on. Similar actions were also applied to clocks, although surviving examples are rare. In the rural areas near Bozen (Bolzano) in South Tyrol, for instance, one finds clocks on church towers of small villages constructed no earlier than the eighteenth century, like the one on the church tower of St. Verena in Rotwand (Fig. 9.11). Nearby, another church, St. Nicholas in Mittelberg[25] (Fig. 9.12), today still shows a false clock, that is, a cheaper imitation as a wall painting on the church tower. The indication of time was less important than giving the impression of a small rural community possessing a prestigious mechanical clock. It may be assumed that from the fifteenth to the nineteenth century a number of similar imitations were constructed.

The clock's role as a special object also led to its utilization as part of the fifteenth-century *artes memorandi* figures. The *loci* of these figures, mainly used for the memorization of sermon-texts, were not supposed to contain commonplace objects. Based on ancient mnemonic rules, they were to be

---

21    Concerning such early examples of mechanical clocks in villages, see Dohrn-van Rossum, *History of the Hour*, pp. 152–155.

22    Dohrn-van Rossum, *History of the Hour*, p. 139, where he emphasizes this position of the clocks at a "politically highly sensitive site" already for the earliest specimens.

23    See n. 13 above.

24    False window, Teutonic Order commandery of Lengmoos (South Tyrol), seventeenth century. See Josef Weingartner, *Die Kunstdenkmäler Südtirols* 2, 7th ed. (Bozen / Innsbruck / Vienna, 1991), p. 145.

25    See Weingartner, *Kunstdenkmäler Südtirols* 2, pp. 142–143.

images of extraordinary things.[26] As the explanatory texts say, they should be "new or rare or comical or unheard, enormous, wonderful or delightful," which could be memorized more easily than everyday things.[27] The *turris cum horologio* (tower with a clock) (Fig. 9.13 [detail of Fig. 9.14]) seems to have fitted into this collection of extraordinary devices around the *balneator nudus cum pileolo* (naked surgeon barber with a cap), the *duo naute in navicula super aqua* (two rowers in a boat on the water), the *fenix conburens se in igne* (a phoenix burning itself in the fire), and so on (Fig. 9.14).[28] The role of the clock on a high, strong, and well defended tower may have been to point the preacher to the sermon's motif of a certain plan of time and a given order to be observed or may have been seen as an allegory of the virtue of *temperantia*.[29]

• • •

An advertisement for *Rolex* watches in *Time* magazine from February 19, 2007, concentrated on the "supplementary value" of the product, ignored more or less the watch's practical value but emphasized it as a species of personal improvement:

> In its search for the perfect lasting colour, Rolex created an exclusive alloy, everose gold. The colour of regular pink gold tends to fade over time as it is exposed to environment, particularly chlorine found in the sea and swimming pools. 18 ct everose gold contains platinum, the most noble of all metals. The platinum protects the alloy from the elements and locks in the colour. The exclusive colour of 18 ct everose gold marries perfectly with the steel of the rolesor models. Tests specifically developed for this new alloy all prove the same thing: that your Rolex will remain as beautiful as on the day you first set eyes on it.

---

26    See Sarah Khan, *Diversa diversis. Mittelalterliche Standespredigten und ihre Visualisierung* (Cologne / Vienna / Weimar, 2007), p. 22.

27    See Susanne Rischpler, "Die Ordnung der Gedächtnisfiguren: der bebilderte Mnemotechnik-Traktat im Cod. 5393 der Österreichischen Nationalbibliothek," *Codices Manuscripti* 48/49 (2004), p. 77.

28    *Ars memorandi*, Austrian, 1465. Vienna, Austrian National Library, cod. 5393, fol. 333r: *locus* of the surgeon barber. Photo: Institut für Realienkunde, Austrian Academy of Sciences, Krems.

29    Concerning this *locus* and attempts of its interpretation see Khan, *Diversa diversis*, pp. 213–232, 411–414.

From the first occurrence of the mechanical clock in the fourteenth century until today's personal mechanical clocks, instruments for measuring time have always been objects that surpassed the practical function of telling time. This is particularly true with respect to their value as objects of wealth, prestige, and competition worthy of public exposure. The latter phenomenon was particularly important with regard to the clocks of late medieval and early modern communities. With the construction of smaller clocks (for instance, Fig. 9.15) and, later, of watches, wealth value, beauty, and competition moved from the communal sphere of clock-owners and aspects of collective representation to individual social elite owners of clocks or watches. In this way, the instruments to measure time kept their special and prestigious, sometimes luxury, supplementary functions.[30] This situation can be seen connected with the reverse development from the necessary practical value of clocks to measure exact time for a rather small group of people in the late Middle Ages to the collective value essential for "everybody" today.

Clocks and watches have always been special objects and thereby also time itself, less the experienced but more the measured and publicly exposed aspects of time, which played an important role in the medieval period as well as in post-medieval society, up until today. Time, its representations and the instruments to measure it, have not only meant or stood for units, points and space, for order, autonomy, organization, and accuracy, but also for other, material values, qualities, and prestige that often have proved much more important.

---

30    See Dohrn-van Rossum, "Uhrenluxus—Luxusuhren," pp. 97–116.

FIGURE 9.1 *The mechanical clock and astronomical dial of the Old Town City Hall, Prague, 1410*
PHOTO: AUTHOR

FIGURE 9.2   *The Virgin and Child, St. Catherine, St. James the Greater, Saint Nicholas (?) and*
             *St. Sebastian with the donor Caspar Hornberger, panel painting, Austrian (?), 1462.*
             *Vienna Diocesan Museum*
             PHOTO: INSTITUT FÜR REALIENKUNDE, UNIVERSITY OF SALZBURG,
             KREMS (AUSTRIA)

FIGURE 9.3
*View of Freistadt in the back-*
*ground of the panel painting*
PHOTO: INSTITUT FÜR
REALIENKUNDE, UNIVERSITY
OF SALZBURG, KREMS
(AUSTRIA)

FIGURE 9.4  *The tower with the public clock as part of the town view of Freistadt*
PHOTO: INSTITUT FÜR REALIENKUNDE, UNIVERSITY OF SALZBURG,
KREMS (AUSTRIA)

FIGURE 9.5   *Margrave Saint Leopold, medallion of the Babenberg family tree, panel painting,*
*c. 1490. Klosterneuburg (Lower Austria), Gallery of the Austin Canon House*
PHOTO: INSTITUT FÜR REALIENKUNDE, UNIVERSITY OF SALZBURG,
KREMS (AUSTRIA)

FIGURE 9.6 *The clock above the entrance to the church of Klein-Mariazell*
PHOTO: INSTITUT FÜR REALIENKUNDE, UNIVERSITY OF SALZBURG,
KREMS (AUSTRIA)

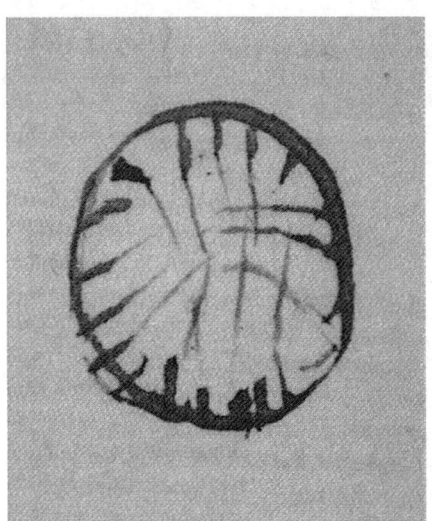

FIGURE 9.7
*The clock of the Holy Ghost church in Reval,
depicted in the city's account book, beginning
of the sixteenth century*
PHOTO: JUHAN KREEM, TALLINN

FIGURE 9.8 *The clockmaker with a sundial and a mechanical clock as one of the children of
Mercury, pen-and-ink drawing, Hausbuch, 1460–1480, castle of Wolfegg*
PHOTO: DETAIL OUT OF CHRISTOPH GRAF ZU WALDBURG WOLFEGG, *VENUS
UND MARS. DAS MITTELALTERLICHE HAUSBUCH AUS DER SAMMLUNG DER
FÜRSTEN ZU WALDBURG WOLFEGG* (MUNICH/NEW YORK, 1997), P. 39.

FIGURE 9.9  Torre dell'orologio, *Uggiano Montefusco* (*Apulia*)
        PHOTO: AUTHOR

FIGURE 9.10    *False window, wall painting, 17th century, Teutonic Order commandery of
Lengmoos (South Tyrol)*
PHOTO: AUTHOR

FIGURE 9.11    *Church tower with clock, 18th century, St. Verena in Rotwand (South Tyrol)*
               PHOTO: AUTHOR

FIGURE 9.12
*Church tower with false clock (wall painting),*
*St. Nicholas in Mittelberg (South Tyrol)*
PHOTO: AUTHOR

FIGURE 9.13
*Image of the* turris cum horologio
*in the* locus *of an Austrian* ars
memorandi *manuscript (detail out*
*of Figure 9.14)*
PHOTO: INSTITUT FÜR
REALIENKUNDE, UNIVERSITY OF
SALZBURG, KREMS (AUSTRIA)

FIGURE 9.14    *The* locus *of the surgeon barber* (balneator) *holding the* turris cum horologio,
*illustration in an Austrian* ars memorandi *manuscript, 1465. Vienna, Austrian
National Library, cod. 5393, fol. 333r*
PHOTO: INSTITUT FÜR REALIENKUNDE, UNIVERSITY OF SALZBURG,
KREMS (AUSTRIA)

FIGURE 9.15    *Traveling clock, end of the sixteenth century. Budapest, Museum of Applied Arts,*
*inv. 62.1419*
PHOTO: INSTITUT FÜR REALIENKUNDE, UNIVERSITY OF SALZBURG,
KREMS (AUSTRIA)

# CHAPTER 10

# Providence, Temporal Authority, and the Illustrious Vernacular in Dante's Political Philosophy

*Jason Aleksander*

One of the striking features of the metaphysics underpinning the dramatic elements of Dante's *Divine Comedy* may be seen by attending to the difference between the way in which the pilgrim and his interlocutors inhabit and perceive time. For instance, near the midpoint of *Paradiso* the pilgrim himself attests to this difference in a question posed to his ancestor Cacciaguida:

> O cara piota mia, che sì t'insusi
> che, come veggion le terrene menti
> non capere in trïangol due ottusi,
> così vedit le cose contingenti
> anzi che sieno in sé, mirando il punto
> a cui tutti li tempi son presenti:
> mentre ch' io era a Virgilio congiunto
> su per lo monte che l'anime cura
> e discendendo nel mondo defunto,
> dette mi fuor di mia vita futura
> parole gravi, avvegna ch' io mi senta
> ben tetragono ai colpi di ventura;
> per che la voglia mia saria contenta
> d'intender qual fortuna mi s'appressa,
> ché saetta previsa vien più lenta.

> Oh my dear root, who raises yourself so high that, as earthly minds see that two obtuse angles cannot fit in a triangle, so you see contingent things before they come to be, gazing at the point to which all times are present: while I was with Virgil upon the mountain that restores the souls and descending into the dead world, grave words were said to me about my future life, although I feel prepared well foursquare against the blows of events; so my will would be contented to understand what fortune approaches me, for the arrow foreseen comes more slowly. [17.13–27][1]

---

1  Italian references to the *La Divina Commedia* follow the Edizione Nazionale sponsored by the Società Dantesca Italiana, ed. Giorgio Petrocchi (Milan, 1966–1967); this edition can be found

© KONINKLIJKE BRILL NV, LEIDEN, 2016 | DOI 10.1163/9789004312319_012

In short, the pilgrim's interlocutors in Paradise see all contingent things in the temporal world's past, present, and future by gazing upon them *sub specie aeternitatis*. Dante's representations of the damned, too, are permitted a glimpse of the temporal world under the aspect of eternity, though for them the knowledge of either the temporal present in which the pilgrim's journey takes place or the eternal present of the divine mind is crucially absent. Thus, as John Freccero has put it in a recent essay devoted to canto 10 of Dante's *Inferno*, the damned only know a present that is granted to them by the presence of the pilgrim who, like the drink offered by Odysseus in Book XI of the Odyssey, "brings them to 'life' momentarily when their static existence intersects his human time."[2]

In one sense, then, in his interactions with those in Paradise or Hell, only the pilgrim has knowledge of the temporal present, and he is in motion in such a way that his motion marks time—a time that corresponds allegorically (as is generally recognized in the secondary literature) to the temporal death and resurrection of Christ. But while the pilgrim's cognition of time is that of a mortal human, the very fact that his conversations are with those for whom the phenomenology of time is quite different suggests that the metaphysics underlying the drama of the *Commedia* is a neo-Platonic one in which time is but the image of eternity. Indeed, the very geography of Paradise and Hell in the *Commedia* suggest why this must be the case. The damned, for their part, inhabit the concentric circles of the underworld. Their motions, if they are permitted any motion at all, are a mockery of the motions of the celestial spheres in that they are eternal but fruitless. The blessed, by contrast, do not even reside in the temporal world but, instead, inhabit the Empyrean beyond the boundaries of even the eternal motions of the celestial spheres in which the pilgrim first encounters them. Dante even draws explicit attention to this detail when, in *Paradiso* 4.28–48, he has Beatrice resolve the pilgrim's confusion regarding the reflected *appearance* of these blessed souls within the moving celestial spheres. According to her explanation, although all of the blessed truly inhabit the Empyrean (beyond time), they are reflected in specific celestial spheres according to an allegorical logic that signals the nature and extent of their spiritual exemplarity. In fact, perhaps the only souls that the Pilgrim encounters in his journey through the afterlife whose cognition of time resembles his are those purging themselves of sin on Mount Purgatory. The physical geography of this region, after all, differs from that of the other regions in two crucial respects:

---

online at both the Dartmouth and Princeton Dante Project websites. English translations are my own.

2   "Epitaph for Guido: 'Inferno' X," *Religion and Literature* 39/3 (2007), 1–29 (12).

(1) it is fundamentally an earthly geography in which the pilgrim's and the penitents' cognition of time is reinforced, for instance, by the apparent motion of the sun; and (2) the penitent motions of its inhabitants are fundamentally the mixed motions of a spiral (combining both circular and linear motion) as they travel both around the mountain and up to the summit.

Moreover, to the extent that one reads the *Commedia* as assuming the possibility of a human experience of happiness in the state described in *Paradiso*, the tension between the phenomenology of earthly time and its metaphysically imitative relationship to the reality of the eternal presence of the Divine—an eternal presence that is both in the temporal world and beyond it—fundamentally undergirds the allegorical purpose of the *Commedia*. On this reading, time is a feature of human fallenness. Consequently, it is only through the temporal striving for the Good that a human being becomes capable of receiving, through the grace of God, the eternal blessedness that would repair that fallen condition. Nevertheless, because the fallen world remains the image in the likeness of the eternal and unchanging model on which it is created, the cautious interpreter can read each specific fate of the damned, the penitent, and the blessed as indicative of the same providential ordering of human history; that is, each moment of every fate is a sign of the eternal presence of the divine in the world. Indeed, in *Purgatorio* 22, Dante, through his representation of the Roman poet Statius, suggests that, because of this providential ordering of human history, even a pagan poet like Virgil can guide others to Christianity:

> Facesti come quei che va di notte,
> che porta il lume dietro e sé non giova,
> ma dopo sé fa le persone dotte,
> quando dicesti: "Secol si rinova;
> torna giustizia e primo tempo umano,
> e progenïe scende da ciel nova."

> You did as one who goes at night, who carries the lamp behind and does not benefit from it, but teaches the people who follow when you said: "The age turns new; justice returns and the first human time, and a new progeny descends from heaven." [67–72][3]

---

3   Indeed, *Purgatorio* 21–22 suggest that Virgil's poetry has a greater capacity to guide others to Christianity than that of Statius, for even though Virgil is a pagan and lacks the faith that would be necessary for salvation, he was in life, unlike Statius, free of any moral defects (see *Inferno* 4.34–42). I discuss some of the philosophical implications of Dante's depiction of the

That Virgil is able to do this even though he cannot himself attain spiritual salvation should not surprise the attentive reader since, for Dante, the greater one's intellectual capacity, the greater is one's ability to see in each and every creature and event the providential structure of history—thus representations of the souls of the damned and saved alike are, in the *Commedia*, signs of "quella fede che vince ogne errore" (that faith that vanquishes every error [*Inferno* 4.48]). It should not be surprising, therefore, that a poet of Virgil's ability and wisdom is able to grasp and signify a greater part of the truth than even that of which he himself is aware. Indeed, this conclusion is also suggested by Virgil's own comment that the merits of Homer, Horace, Ovid, and Lucan win them some (limited) degree of grace even in Hell (see *Inferno* 4.78).

In light of these brief glimpses of some of the *Commedia*'s metaphysical presuppositions, this essay intends to explore how Dante's ethico-political philosophy operates within the crucial tension between the phenomenology of time as the condition for the possibility of human moral development and also as, metaphysically speaking, the privation and imitation of eternity. Rather than focus on the *Commedia* itself—a text which is structured by this tension in both its philosophical assumptions as well as the narrative and dramatic logic through which its fiction operates—this essay confines itself in what follows to a discussion of how Dante's understanding of time structures his ethico-political philosophy articulated in his three major philosophical treatises: *De vulgari eloquentia*, *Convivio*, and *Monarchia*. I will begin with a discussion of Dante's definition of the illustrious vernacular in *De vulgari eloquentia*: I will argue that Dante's understanding of the poetic and rhetorical function of the illustrious vernacular is tied to his political philosophy in a way that depends upon a rich but ultimately unresolved tension between (a) the demand that only an atemporal, unchanging vernacular would be suitable for the tasks of universal monarchy and (b) the recognition that only a temporal, localized, and changing illustrious vernacular could possibly bring about the existence of the universal monarchy.

In the second half of the essay, I will turn to Dante's treatment of the providential grounding for the independence of spiritual and temporal authority in *Convivio* and *Monarchia*: I will argue that Dante's understanding of divine providence provides common justification for the temporal and spiritual authorities whose independence he otherwise insists upon. Then, drawing on the letter to Cangrande della Scala (the authorship of which is disputed),

---

relationship between Virgil and Statius in "The Aporetic Ground of Revelation's Authority in the *Divine Comedy* and Dante's Demarcation and Defense of Philosophical Authority," *Essays in Medieval Studies* 26 (2010), 1–14.

I will discuss how, for Dante, the providential ground for the legitimacy of temporal authority can only be discerned through the allegorical interpretation of history itself. In light of my discussion of these themes in Dante's political philosophy and its dependence on his understanding of divine providence, I will conclude with a brief reflection on how Dante's understanding of divine providence might help us better appreciate important aspects of the neglected legacy of Renaissance humanism in the history of early modern philosophy.

## Political Authority and the Illustrious Vernacular

In *Renaissance Humanism: Studies in Philosophy and Poetics*, Ernesto Grassi points out that Dante's understanding of the nature of the illustrious vernacular harbors an indissoluble tension. On the one hand, Dante's political philosophy endorses the rule of a universal monarch who, "unlike the individual kings, rules the whole world, and whose task consists of preventing any alteration in the eternal order established and politically institutionalized by way of intellectual insight."[4] Accordingly, Dante's philosophy of language requires for such a political state the possibility of a universal language that "in its universality and abstraction, avoids all local and temporal variations" (p. 7). Dante, however, also ascribes to the vernacular-poetic function of language the role of a tool necessary for concretely disclosing the "historical range of his nation" (p. 9) and for temporally bringing about the political unity of the universal monarch. In this section I will discuss why this tension arises in Dante's philosophy of language and how it depends upon Dante's understanding of the providentially guided unfolding of human history.

In *De vulgari eloquentia*, a work which was probably begun around 1302 and abandoned sometime around 1305, Dante defines vernacular language as

> eam qua infantes assuefiunt ab assistentibus, cum primitus distinguere voces incipiunt; vel quod brevius dici potest, vulgarem locutionem asserimus, quam sine omni regula, nutricem imitantes, accipimus.

> that which infants acquire from those around them when they first begin to distinguish sounds; or, to put it more succinctly, I declare that vernacular

---

4   Medieval & Renaissance Texts and Studies 51, trans. Walter F. Veit (Binghamton, 1988), pp. 6–7.

language is that which we learn without any formal instruction, by imitating our nurses. [1.1.2][5]

Thus, vernacular language is distinguished from grammar, which is a "locutio
secundaria" (secondary kind of language) possessed by the Romans and Greeks
and others, "sed non omnes" (but not by all [1.1.3]).[6] Unlike vernacular language, these grammars are a product of art and can only be acquired through
the cultivation of a *habitus* in a dedicated study of its rules and doctrines (1.1.3).
And also, as he explains in 1.9, unlike vernacular language, grammar

> nichil aliud est quam quedam inalterabilis locutionis ydemptitas diversi-
> bus temporibus atque locis...nulli singulari arbitrio videtur obnoxia, et
> per consequens nec variabilis esse potest.

> is nothing less than a certain immutable identity of language in different
> times and places...[and] subject to no individual will; and, as a result, it
> cannot change. [1.9.11]

In relation to this distinction, Dante argues in 1.1 for the superiority of the vernacular language—and here it will also help to note that Dante treats the
modes of the vernacular as if they belong to a single species of language—for
three reasons:

> Tum quia prima fuit humano generi usitata; tum quia totus orbis ipsa
> perfruitur, licet in diversas prolationes et vocabula sit divisa; tum quia
> naturalis est nobis, cum illa potius artificialis existat.

> First, because it was the language originally used by the human race;
> second because the whole world employs it, though with different

---

5    Latin references are to *De vulgari eloquentia*, ed. Pier Vincezo Mengaldo (Padua, 1968).
     Translations are those of Steven Botterill (Cambridge, UK, 2005).

6    The term *secundaria* may imply in this context that it is a "second-rate" or "inferior" language
     as well as a kind of language that can only develop *after* one acquires a vernacular. One of the
     most common mistakes in interpretations of *De vulgari eloquentia* is to suppose that Dante
     believed that Romans and Greeks had *only* this secondary kind of language—that there were
     never Latin or Greek vernaculars. It seems clear to me, however, that Dante must recognize
     that there were Latin and Greek vernaculars before there were developed Latin and Greek
     grammars. In any case, as a matter of the logical claim here, Dante is saying that not all cul-
     tures have grammars, not that the Greeks and Romans *only* had grammars.

pronunciations and using different words; and third, because it is natural to us, while the other is, in contrast, artificial. [1.1.4]

For Dante, despite the superiority of the vernacular over grammar, the very diversity of modes of the vernacular indicates its deficiency. Specifically, Dante argues that no existing mode of the vernacular is a language that, in any of its specific, municipal varieties—varieties that differ by convention—is perfectly equipped to produce or express virtues consistent with any of the ends of human nature. On behalf of this argument, Dante first appears to proceed empirically by means of a hunt for a perfect mode of the vernacular among roughly fourteen existing municipal forms of the Italian mode of the vernacular. Although I'll indicate below why we must regard this supposedly empirical hunt as a fantasy, it is worth noting a few of the deficiencies that he says mark the failures of existing municipal forms of the Italian mode of the vernacular. For instance, in 1.11, Dante judges the "tristiloquium" (vile jargon) of the contemporary inhabitants of Rome to be the "turpissimum" (ugliest) of all species of Italian expression and not even worthy of being called a vernacular. And, he adds, "nec mirum, cum etiam morum habituumque deformitate pre cunctis videantur fetere" (this should come as no surprise, for they also stand out among all Italians for the ugliness of their manners and their outward appearance [1.11.2]). Other local vernaculars fare little better. For instance, Dante writes that, when it comes to the Tuscans, it is evident that that their "turpiloquium" (foul jargon [1.13.3]) is little better than that of the Romans.

Dante's discussion of these Italian modes of the vernacular indicates two significant features of his philosophy of language. First, that Dante associates the vileness of custom with a corresponding deformity of the vernacular suggests that he also maintains that one's idiomatic language and one's ethico-political dispositions are highly correlated. Accordingly, for Dante the task of cultivating a virtuous person or a just state corresponds to the use of, or cultivation of, a virtuous mode of the vernacular. Second, what is significant about Dante's judgments regarding these modes of the vernacular (as well as the rules he provides in book two for the composition of poetry in the illustrious vernacular) is that he speaks on behalf of a universal authority, that is, as if he possesses an *a priori* knowledge of the underlying eternal form of the illustrious vernacular—when evaluating the aesthetic expression of the diverse existing modes of the vernacular. For instance:

Si quis autem quod de Tuscis asserimus de Ianuensibus asserendum non putet, hoc solum in mente premat, quod si per oblivionem Ianuenses amitterent *z* licteram, vel mutire totaliter eos, vel novam reparare oporteret loquelam. Est enim *z* maxima pars eorum locutionis: que quidem lictera non sine multa rigiditate profertur.

If there is anyone who thinks that what I have just said about the Tuscans could not be applied to the Genoese, let him consider only that if, through forgetfulness, the people of Genoa lost the use of the letter *z*, they would either have to fall silent for ever or invent a new language for themselves. For *z* [constitutes] the greater part of their vernacular, and it is, of course, a letter that cannot be pronounced without considerable harshness. [1.13, translation altered.]

There is probably no reason to dispute Dante's characterization of the phonetic or ethical quality of the letter *z*, but the passage suggests two interesting questions that correspond with my observations above. First, why would a phonetic feature of a particular mode of the vernacular have anything to do with ethical considerations? Second, on what authority is Dante entitled to the judgment that an illustrious mode of the vernacular would not frequently include this phoneme? At the level of the question of the grounds for his competence as an authority[7] capable of judging particular vernaculars Dante's argument can only be that the reason why any particular feature of a mode of the vernacular fails to produce or express the virtues pertinent to human ends is that it is a corrupted imitation of a timeless illustrious vernacular. However, as we shall see, it also turns out that the only non-corrupted imitation of this illustrious vernacular must itself be subject to local variation in response to the present needs of a particular political community.

In *De vulgari eloquentia*, Dante never explicitly makes such an argument. Consequently, to understand why this interpretation of Dante's philosophy of language is warranted requires beginning with a consideration of how he carries out his argument that the variability of municipal modes of the vernacular indicates their imperfection through a phylogenetic investigation of the origins of the modes of the vernacular language. In *De vulgari eloquentia* 1.3,

---

7   Albert Russell Ascoli's recent *Dante and the Making of a Modern Author* (Cambridge, UK, 2009) provides a masterful analysis of Dante's appeals to and transformations of a variety of traditional understandings of "authority." My own claims about Dante's conception of authority in this context are much more limited, and I am ultimately more concerned with Dante's philosophical understanding of the grounds of authority in aesthetic/political judgment than with the complicated rhetorical and dialectical processes through which Dante marshals and deploys this authority.

Dante maintains that humans are defined by having been endowed with vernacular language since, unlike other animals and also unlike angels,

> Oportuit ergo genus humanum ad comunicandas inter se conceptiones
> suas aliquod rationale signum et sensuale habere; quia, cum de ratione
> accipere habeat et in rationem portare, rationale esse oportuit; cumque
> de una ratione in aliam nichil deferri possit nisi per medium sensuale,
> sensuale esse oportuit.

> So it was necessary that the human race, in order for its members to com-
> municate their conceptions among themselves, should have some [sign]
> based on reason and perception. Since this [sign] needed to receive its
> content from reason and convey it back, it had to be rational; but since
> nothing can be conveyed from one reasoning mind to another except by
> means...[of] the senses, it had also to be based on [the senses]. [1.3.2,
> translation altered]

The endowment of the vernacular is thus interpreted as a divine gift since,
"divinitus in nobis esse credendum est quod in actu nostrorum affectuum ordi-
nato letamur" (we may believe that our joy in the ordered activity of our affects
is of divine origin [1.5.2, translation mine]). And, according to Dante, this
accords with the account of Genesis since the original human language, the
language of Adam, was given to mankind by God so that human beings would
be capable of glorifying "ipse qui gratis dotaverat" (He who had freely given so
great a gift [1.5.2]).

This original vernacular, Dante claims, was a "certam formam locutionis a
Deo cum anima prima concreatam fuisse" (certain form of language created
by God along with the first soul [1.6.4]). It was spoken by Adam and by
"omnes posteri eius usque ad edificationem turris Babel" (all his descendents until the building of the Tower of Babel [1.6.5]). And, had it not been
for human presumption, "omnis lingua loquentium uteretur" (would have
continued to be used by all speakers [1.6.4]). After the confusion, it was used
only by the Hebrews, "ut Redemptor noster, qui ex illis oriturus erat secundum humanitatem, non lingua confusionis, sed gratie, frueretur" (so that
our redeemer, who was to descend from them, insofar as He was human,
should not speak the language of confusion, but that of grace [1.6.6]).
According to this view, the cause of the imperfection and diversity of human
language would not be the original fall from grace of Adam and Eve, but only
the later hubris that had compelled Nimrod and his followers to construct
the Tower of Babel.

Specific aspects of the *De vulgari eloquentia*'s discussion of language are later altered in Dante's treatment of the subject in the *Commedia*. For instance, in *Paradiso* 26, Adam himself explains that:

> La lingua ch'io parlai fu tutta spenta
> *innanzi* che a l'ovra inconsummabile
> fosse la gente di Nembròt attenta.

> The language that I spoke was entirely extinct *before* the people of Nimrod turned their attention to that unattainable work. [124–126, italics mine]

In light of this discrepancy, it may be that Dante simply "changed his mind," as Umberto Eco puts it,[8] about the origins of the diversity of vernacular modes. It is, in any case, clear that the later view is that "born of humanity's natural disposition towards speech, languages may split, grow and change through human intervention. According to Adam, the Hebrew spoken before the building of the tower...was not the same as the Hebrew spoken in the earthly paradise" (Eco, p. 47).

But even though Eco is correct that *Paradiso* provides a different conclusion about the origins of the differentiation of the vernacular modes from each other, such a view is not at odds with the metaphysical presuppositions of *De vulgari eloquentia*. In the earlier work, as Eco is aware, Dante explains that the *forma locutionis* of the original vernacular is a form in respect to "quantum ad rerum vocabula et quantum ad vocabulorum constructionem et quantum ad constructionis prolationem" (words used for things, and to the construction of words, and to the arrangement of the construction [1.6.4]). Eco points out that this statement allows the inference that, "by *forma locutionis*, Dante refers to a lexicon and a morphology and, consequently, a determined language" (p. 42). But while the *De vulgari eloquentia* seems to allow the interpretation that this "determined language" was spoken, as Hebrew, not only until the building of the Tower of Babel, but even afterwards, Eco insists that the text requires a more nuanced interpretation. According to Eco, when Dante writes "hac forma locutionis locutus est Adam" (In this form of language Adam spoke [1.6.5]), one must interpret Dante as suggesting that the activity of speaking "constructs" the Hebrew language not as a "determined language," but as a language that is temporally changing and so only contextually determinate as regards the

---

8   "The Perfect Language of Dante," in *The Search for a Perfect Language*, trans. James Fentress (Oxford, 1995), pp. 34–52.

relationships between words and things.[9] And, according to Eco, this interpretation is to be favored because, later in this same passage, Dante uses *lingua* rather than *forma locutionis* when speaking of the Hebrew spoken by Jesus. Moreover, such an interpretation is warranted by the final sentence of 1.6, where Dante writes that "Hebraicum ydioma illud quod primi loquentis labia fabricarunt" (the Hebrew language was that which the lips of the first speaker [fashioned]).[10] In short, for Eco, the *forma locutionis* of the *De vulgari eloquentia* "was neither the Hebrew language nor the general faculty of language, but a particular gift from God to Adam that was lost after Babel. It is the lost gift that Dante sought to recover through his theory of an illustrious vernacular" (p. 43). Consequently, even though *Paradiso* offers what appears to be a different account of the origins of the municipal modes of the vernacular, both the *Divine Comedy* and the *De vulgari eloquentia* allow similar conclusions about the relationship of these modes of vernacular to the original *forma lucutionis* of the human vernacular.

Similarly, although I will confine my remaining remarks about the illustrious vernacular to a discussion of the *De vulgari eloquentia*, both it and the *Divine Comedy* allow the same conclusions about Dante's understanding of the role that an illustrious mode of the vernacular would play in political life. That is, both texts support the conclusion that, regardless of when human vernaculars became species differentiated by convention from the original, prepolitical *forma locutionis*, an "illustre, cardinale, aulicum et curiale" (illustrious, cardinal, aulic, and curial [1.16.6]) vernacular would be a mode of the vernacular that, in its lived, temporally changing modes of expression, approaches the perfection that pertains to the original *forma locutionis*. As Dante insists in *De vulgari eloquentia* 2.1–2, the illustrious vernacular ought not be used by all

---

9   Both "constructs" and "determined language," as terms employed in Fentress' translation of Eco's text, connote a modern understanding that may not be fully consistent with Dante's meaning. I have left these terms unaltered here and below when citing Eco's text. In relation to "determined language" it seems clear to me that Eco is speaking of a language in which there is an unchanging, universally determined relationship between word (*verbum*) and thing (*res*). In a temporal mode of the vernacular, however, the relationship between word and thing could not, according to Dante's argument, be determined universally/atemporally: because this relationship for temporal modes of the vernacular is structured by convention, it would be determinate only contextually.

10  Translation altered. In Fentress' translation of Eco's text, this sentence is rendered as follows: "It was thus the Hebrew *tongue* that was constructed by the first being *endowed with speech*" (p. 43). This translation is also consistent with the Latin sentence, and it correctly emphasizes the fact that Dante is not speaking of Hebrew as a *forma locutionis* but as a lived, temporally changing language. Here *fabricarunt* is translated as "constructed."

writers, but only by those who possess both *scientia et ingenium*,[11] that is, by those who possess both knowledge and genius (DVE 2.1.8). In the mouths of these rightful users, the illustrious vernacular would be a language that would express and produce the virtues that correspond to the tripartite aspects of a human soul: prowess in arms, which pertains to the vegetal part of the soul's determination to seek that which is useful; love, which pertains to a rectitude of the appetitive or animal part of the soul; and rectitude of will, which pertains to the rational part of the soul. That such a language can produce these virtues is precisely what Dante means in 1.17 when he calls such a language "illustrious" since such a language, by giving off light or reflecting the light it receives from elsewhere, is exalted by mastery and power, that is, the illustrious vernacular is equipped to rule because it is capable of changing human hearts so as to "nolentem volentem et volentem nolentem faciat" (make the unwilling willing and the willing unwilling [1.17.4]). And, in 1.18, he explains that such a language is *cardinalis* (cardinal or "pivotal") because it "platant" (plants) what is best and peculiar to the local vernaculars and "extirpat" (removes) what is disreputable. It is "*aulicus*" (aulic or courtly) because "est ut omnibus sit comune nec proprium ulli" (it is common to all yet owned by none [1.18.2]). And it is "*curialis*" (curial) because it has been weighed by the scales of justice in a court unified either under the rule of a singular monarch (e.g., judged in a unified court such as that of the King of Germany) or, as is the case for the Italians, who lack a unified court in the former sense, according to a court unified "gratioso lumine rationis" (by the gracious light of reason [1.18.5]). This last criterion seems also to require that such a vernacular express what is necessary, that is, "what cannot be expressed otherwise" (Eco, p. 46), since: "quia curialitas nil aliud est quam librata regula eorum que peragenda sunt" (the essence of being curial is no more than providing a balanced assessment of whatever had to be dealt with [1.18.4]). To this extent, it would share with Adamic language the capacity to express a primordial affinity between word and thing.

It is in light, then, of Dante's discussion of the relationship between the illustrious vernacular and Adamic language that we can see in what sense Grassi was right to emphasize the tension between Dante's insistence that the authority of such a mode of the vernacular derives from its resemblance to an

---

11    Dante's use of the subjunctive in this context seems to me more than a bit misleading. When he asks whether all poets should be permitted to use this language, the question seems to imply that it would be possible for them to do so. But if the account that I am offering here is correct, it would in fact only be possible for those who ought to use such a language actually to do so.

unchanging *forma locutionis* and, at the same time, his insistence that such a language function temporally as a political tool for bringing about the stable rule of a universal monarch. This tension between the universal, atemporal, transhistorical authority of an illustrious vernacular and its intrinsically localized context of signification[12] constrains how Dante must address the question raised earlier concerning the grounds of his authority to judge the diversity of Italian municipal vernaculars. In the first place, "cardinality," for instance, would seem to be a criterion that establishes what is most just with respect to a particular context, that is, according to this interpretation, the cardinality of an illustrious vernacular would pertain not to its universal authority but to its suitability to produce justice in particular political contexts. Indeed, it is telling that in Dante's discussion of cardinality, the Bolognese receive the greatest admiration. He judges this municipal vernacular to be closest to an illustrious vernacular because it judges other Italian forms and appropriates what is best in them and dispenses with what is not. However, as will become clear, even in this instance, Dante implicitly maintains that such pivotal judgments about what to fetch and what to remove are possible only by reference to an underlying principle of universal authority.

I will discuss Dante's understanding of political authority in greater detail in the second half of this essay. However, even at this point in the discussion of his philosophy of language, it should be clear that his understanding of justice in its relation to the authority of a universal monarch departs significantly from Aristotle's understanding of justice as a virtue that can pertain only to a particular political constitution (it is worth noting that, for Aristotle, there are both just and unjust versions of different political constitutions). At the same time, however, Dante does acknowledge explicitly in *Monarchia* 1.14.5 that there must be variations of customs even within a universal monarchy. Thus, even where his political theory differs so markedly from Aristotle's, he nevertheless recognizes that even an illustrious vernacular must function in local contexts of signification that differ with respect to time and place.

A more revealing key to Dante's attitude about what grounds the authority to judge vernacular languages may be found in his selection of examples of those who write in a vernacular that most closely approaches the illustrious. All of the authors that Dante cites in a positive regard depart from the municipal vernacular that is native for them. More importantly, Dante explicitly draws

---

12    This tension is also discussed—to different ends—in Albert Russell Ascoli, "'Neminem ante nos': Historicity and Authority in the *De vulgari eloquentia*," *Annali d'Italianistica* 8 (1991), 186–231; and Zygmunt G. Barański, "'Significar *per verba*': Notes on Dante and Plurilingualism," *The Italianist* 6 (1986), 5–18.

attention to the significance of his own exile as a feature pertinent to his ability to judge in the name of a universal authority:

> Quicunque tam obscene rationis est ut locum sue nationis delitiosissimum credat esse sub sole, hic etiam pre cunctis proprium vulgare licetur, idest maternam locutionem, et per consequens credit ipsum fuisse illud quod fuit Ade. Nos autem, cui mundus est patria velut piscibus equor, quanquam Sarnum biberimus ante dentes et Florentiam adeo diligamus ut, quia dileximus, exilium patiamur iniuste, rationi magis quam sensui spatulas nostri iudicii podiamus.

> Whoever is so misguided as to think that the place of his birth is the most delightful spot under the sun may also believe that his own language— his mother tongue, that is—is pre-eminent among all others; and, as a result he may believe that his language was also Adam's. To me, however, the world is a homeland, like the sea to fish—though I drank from the Arno before cutting my teeth, and love Florence so much that, because I loved her, I suffer exile unjustly—and I will weight the balance of my judgment more with reason than with sentiment. [1.6.2–3]

In relation to the question of the universal grounds of authority, this comment suggests that the very condition for the possibility of evaluating diverse modes of the vernacular is "exile"[13] or estrangement from any particular mode of the vernacular. Moreover, such an exile is also necessary for acquiring an illustrious mode of the vernacular since it would be impossible to acquire a perfect language within any particular political community already deformed by local, imperfect conventions. Thus, Dante's unjust exile from the imperfect community of Florence suggests a reversal of Adam's exile from Eden, an interpretation that is also justified by the fact that in this same section Dante proceeds from this comment on his own suitability to judge Italian vernaculars immediately to a discussion of the *forma locutionis* of Adamic language.[14] From this, it

---

13    The significance of "exile" in Dante's philosophy of language receives extensive treatment in Marianne Shapiro's *De vulgari eloquentia: Dante's Book of Exile* (Lincoln, 1990).

14    Moreover, that Dante, out of love, suffers this exile from the imperfect (fallen) community also perhaps suggests a Christological feature of his thought on this point. If so, Dante's understanding of the possibility of reclaiming a perfect language is also clearly messianic (in the sense of Christian messianism): such a language would be a language appropriate only for a redeemed humanity and would require the unification of the ethical and spiritual ends of human existence.

becomes clear that illustriousness, cardinality, courtliness, and curiality apply to the illustrious vernacular universally and independently of the conventions of a local political context. Or, put differently, the illustrious vernacular is a mode of the vernacular language that must function within a local political context in order to produce virtue in citizens through particular political conventions and habits, but its own authority to do so depends on or relates to that of universal political authority. In short, the illustrious vernacular could only be a language appropriate for a universal monarch who rules all local communities with the same authority, that is, the mode of the vernacular that corresponds most fully to the original, timeless and unchanging *forma locutionis*. Thus, beyond the aesthetic evaluations of specific defective features of the Italian modes of the vernacular noted earlier, the underlying principle according to which Dante judges all local modes of the vernacular as unequipped to produce or express the virtues pertinent even to earthly ends of human existence is that the illustrious vernacular must, in principle, serve the authority of the universal Monarch, and so, in each specific contemporary context, it must be whatever municipal mode of the vernacular that is most appropriate for the task of guiding a people to natural human felicity insofar as it is possible.

A last comment on the universality of this illustrious vernacular is in order, however. As I pointed out at the beginning of this section, Dante distinguishes between all vernaculars and all grammars in part on the basis of the mutability of all species of vernacular and the immutability of all species of grammar. That is, although the purpose and authority that underwrite the employment of an illustrious mode of the vernacular is itself unchanging—and so, in this sense, constitutes its authority over all human beings—the purpose of this authority to affect the will(s) of any of its subjects requires that the extrinsic manifestations of this vernacular be temporally subject to change. First, this illustrious vernacular must be local in its range of signification so that it is meaningful to those who hear it. Second, the actual signs employed by such a vernacular must also be mutable in relationship to the specific temporal needs of specific political communities, even of the specific temporal needs in a universal political community. In other words, the illustrious vernacular is whatever vernacular is used by one who is endowed with the ability to lead human beings from a condition of ethical and/or spiritual failure to a condition of ethical felicity and/or spiritual blessedness. In truth, then, the illustrious vernacular is an unchanging language only in the sense that, when it is employed by one capable of using it, the purpose and authority of linguistic expression is properly attentive to the universal needs of human nature. That is, the illustrious vernacular is the language used by one for whom the events of mundane history are recognized as the *de jure* unfolding of divine providence ordering all things in time toward the same universal end. Thus, this vernacular, like the manifestations of the souls of the blessed in the celestial spheres to which Beatrice refers in *Paradiso* 4, is accommodated to the specific needs of its audience in a particular time and place, but the authority of a speaker of such a language to deploy this vernacular is grounded in the speaker's understanding of the necessity of the unfolding of contingent events from an eternal, providential authority. In light of this recognition, then, I will discuss Dante's understanding of political authority in greater detail in the second half of this essay.

### Divine Providence as Grounds for Temporal Authority

There are two fundamental principles that ground Dante's political philosophy: one is psychological, the other cosmological. The psychological principle, articulated succinctly in the last section of *Monarchia*, is that the human being, "solus inter omnia entia in duo ultima ordinetur, quorum alterum sit finis eius prout corruptibilis est, alterum vero prout incorruptibilis" (alone among all beings is ordered to two ultimate goals, one of them being his goal as a corruptible being, the other his goal as an incorruptible being [*Monarchia* 3.16.6]).[15] This principle is sufficient for Dante's demonstrating the necessity of the independence of temporal and spiritual authority. However, for Dante the very fact that the human being is ordained to these two goals by its creator implies that his cosmological conception of divine justice is the more fundamental of the principles of his political philosophy. From this more fundamental principle, Dante derives the conclusion that temporal authority no less than spiritual authority derives from the providential unfolding through human history of divine justice in the world:

> Cumque dispositio mundi huius dispositionem inherentem celorum circulationi sequatur, necesse est ad hoc ut utilia documenta libertatis et pacis commode locis et temporibus applicentur, de curatore isto dispensari ab Illo qui totalem celorum dispositionem presentialiter intuetur. Hic autem est solus ille qui hanc preordinavit, ut per ipsam ipse providens suis ordinibus queque connecteret. Quod si ita est, solus eligit Deus, solus ipse confirmat, cum superiorem non habeat.... Sic ergo patet quod auctoritas temporalis Monarche sine ullo medio in ipsum de Fonte universalis auctoritatis descendit.

---

15      All references to the Latin are from Prue Shaw's edition of the Cambridge Medieval Classics series (Cambridge, UK, 1995); translations are also Shaw's.

And since the disposition of this world is a result of the disposition inherent in the circling of the heavens, in order that useful teachings concerning freedom and peace can be applied appropriately to times and places,
it is necessary for provision for this protector to be made by Him who
takes in at a glance the whole disposition of the heavens. For he alone is
the one who preordained this disposition, making provision through it to
bind all things in due order. If this is so, then God alone chooses, he alone
confirms, since he has none above him. Thus it is evident then that the
authority of the temporal monarch flows down into him without any
intermediary from the Fountainhead of universal authority. [3.16.12–15]

In *Convivio*, an unfinished treatise composed circa 1304–1307[16] in which he had
also discussed these issues, Dante also makes clear why it is that he maintains
that "non da forza fu principalmente preso per la romana gente, ma da divina
provedenza, che è sopra ogni ragione" (this office, then, was obtained by the
Roman people not principally by means of force, but by divine providence,
which is the ultimate ground of all reason [4.4.11]).[17] Dante advances this argument as a response to the hypothetical objection that the Roman emperors—
the paradigm for Dante's Emperor—obtained rule not by reason or right but
by force. Dante's response, simply stated, is to admit that force may have been
the instrumental cause that allowed for the establishment of Roman imperial
authority, but that the moving or efficient cause was divine reason itself just as,
by analogy,

---

16   *De vulgari eloquentia* was probably composed in 1303–1305, but was abandoned before its
     completion. There was probably some overlap between the periods in which Dante was
     working on both the *De vulgari eloquentia* and *Convivio*. In fact, there are some similar
     expressions between the two texts—e.g., the tripartite conception of the human soul as
     vegetal, animal, and rational (cf. *De vulgari eloquentia* 2.2 and *Convivio* 3.2). However,
     there are certainly important inconsistencies between the texts in certain specific areas.
     For a discussion of these, see Pier Vincenzo Mengaldo, Introduction and Notes to *De vul-
     gari eloquentia* in *Opere minori, tomo II*, ed. P.V. Mengaldo et alia (Milan, 1979), pp. 6–7;
     Ileana Pagani, *La teoria linguistici di Dante* (Naples, 1982), pp. 152–54; and Stefano Rizzo, "Il
     *De vulgari eloquentia* e l'unità del pensiero linguistic di Dante," *Dante Studies* 87 (1969),
     69–88. Nevertheless, both at a fundamental level and as pertaining to this essay, I believe
     that the *Convivio* and *De vulgari eloquentia* depend upon fundamentally compatible eth-
     ico-political philosophies and so permit a synthetic treatment of his philosophy of lan-
     guage and political philosophy.

17   Italian references to *Convivio* follow the Edizione Nazionale sponsored by the Società
     Dantesca Italiana, ed. Franca Brambilla Ageno (Florence, 1995); this edition can be found
     online at the Princeton Dante Project website. Translations are those of Christopher Ryan
     (Stanford, 1989).

li colpi del martello cagione [instrumentale] del coltello, e l'anima del fabro è cagione efficiente e movente; e così non forza, ma ragione, [e ragione] ancora divina, [conviene] essere stata principio dello romano imperio.

the blows of the hammer are a cause of the knife, but the mind of the smith is the efficient or moving cause. So, too, reason not force, and indeed divine reason, must be regarded as the source of Roman rule or empire. [4.4.12][18]

As evidence for this claim, Dante provides two arguments to show that this single providential structure is the common basis for the legitimacy of both temporal and spiritual authority. First, he argues that, after the fall of man, when God wanted "l'umana creatura a sé riconformare" (to refashion the human creature into a likeness of Himself once again 4.5.3), the earth had to become perfectly disposed for the coming of Christ, and this required two things: a political community, namely Rome, that would bring about this disposition and

una progenie santissima, della quale dopo molti meriti nascesse una femmina ottima di tutte l'altre, la quale fosse camera del Figliuolo di Dio: e questa progenie fu quella di David, del qual discese la baldezza e l'onore dell'umana generazione, cioè Maria.

a family-line of the highest sanctity, into which, after it had boasted many members of great virtue, would be born a woman surpassing all others, who would be the resting place of the Son of God. This was the line of David, from whom was born the joy and glory of the human race, Mary. [4.5.5]

Accordingly, human reason is able to discern in the coincidence of the birth of David and the arrival of Aeneas in Italy the fact of "la divina elezione del

---

18   In *Monarchia*, Dante claims that he, like the hypothetical interlocutor of the *Convivio*, once considered the matter only superficially and regarded the Roman people as having obtained their authority to rule over all not *de jure* but merely by force. "But," he states, "medullitus oculos mentis infixi et per efficacissima signa divinam providentiam hoc effecisse cognovi" (I penetrated with my mind's eye to the heart of the matter and understood through unmistakable signs that this was the work of divine providence [2.1.3]).

romano imperio" (that the Roman empire was willed by God [4.5.6]). Second, claims Dante,

> da Romolo cominciando, che fu di quella primo padre, infino alla sua perfettissima etade, cioè al tempo del predetto suo imperadore, non pur per umane ma per divine operazioni andò lo suo processo.

> from the time of its founding father, Romulus, until the age of its highest perfection, under the above-mentioned emperor [Ceasar Augustus[19]], was the result of divine, not merely human, activity. [4.5.10]

After all, claims Dante, without God's direct involvement, it would be impossible to explain the moral uprightness of such an array of civically minded leaders (and Dante cites a number of examples that he believes justify this claim), as well as the fact that, for instance, the mere cackling of a goose could at one point save Rome "quando li Franceschi, tutta Roma presa, prendeano di furto Campidoglio di notte" (when the Franks, after capturing the whole of Rome, were on the point of capturing the Capitol by stealth under cover of darkness [4.5.18]).

In *Monarchia*, which was written sometime after 1314,[20] Dante reiterates these latter arguments: that the Romans placed public good above personal gain is discussed with similar examples in *Monarchia* 2.5; that a series of portentous events is evidence of divine providence is reiterated in *Monarchia* 2.4. But in *Monarchia* 2.6–11, Dante offers additional lines of argumentation. First, Dante argues that, since the judgment of God can be disclosed through contests between many striving for a single prize as well as through contests of

---

19    For Dante, an explicit consequence of this argument is that not only was Rome chosen by God, but, since the world could *only* have been best prepared to receive Christ at the specific time of his coming, Rome must also have, at that time, secured the greatest degree of peace the world will know until, arguably, after the Last Judgment (see *Convivio* 4.5.7–8). Dante also reinforces this point in the *Commedia* where, in Justinian's account of the history of Rome, the following terzina refers to Augustus: "Con costui corse infino al lito rubro; / con costui puose il mondo in tanta pace, / che fu serrate a Giano il suo delubro" (With him it [Rome] raced to the Red Sea; with him it brought the world such peace that the temple of Janus was locked [*Paradiso* 6.79–81]). Commentators are unanimous in noting that, for Dante, Augustus shutting the gates of the temple of Janus for only the third time in the history of Rome is coincident with the birth of Christ.

20    For a summary of scholarly debates about the dating of the *Monarchia* see pp. xxxvii–xli of Prue Shaw's introduction to her Cambridge Medieval Classics edition of the *Monarchia*.

strength between two champions, Rome's universal authority is made manifest by its having won both an athletic contest among various rivals for "pal-
mam monarchie" (the palm of monarchy [2.8.8, my translation]) as well as
having won a duel between two champions. In the case of athletic competi-
tions, Dante argues in 2.8 that while many (most notably Alexander of
Macedonia) had sought to conquer the world, all failed but Rome. In the case
of contests of strength between two champions, Dante argues in 2.9 that
"Romanus populus per duellum acquisivit Imperium" (the Roman people
acquired the empire through trial by combat [2.9.12]), and he cites many examples of such contests between champions decided in Rome's favor (for example, between Aeneas and Turnus, between the Romans and the Albans,
between Scipio's forces and those of Hannibal, and so on). From both of these
lines of argumentation, Dante concludes that Rome acquired rule not merely
*de facto* but *de jure* and even states at the end of the section on duels that this
"est principale propositum in libro presenti" (is our main thesis in this present
book [2.9.21]). Also in both of these cases, Dante claims that he argues not from
articles of Christian faith but from what he regards as "rationalibus principiis"
(rational principles [2.10.1]) accepted by all people. Thus, in interpreting God's
judgment, Christians as well as non-Christians are entitled to use the same
evidence: Aeneas's victory over Turnus is of the same variety as David's over
Goliath and Hercules' over Antaeus. The *rationalia principia* here is that in
such contests, properly undertaken—for example, in a duel, both parties have
to be seeking the same end, namely justice, and in an athletic contest both
parties must abide by the same regulations—God's judgment, by whatever
name it is recognized,[21] becomes manifest.

    While Dante claims that this first line of argumentation establishes Rome's
*de jure* authority by rational principles, he presents additional lines of argument in *Monarchia*, 2.10–11, to prove Rome's authority "ex principiis fidei
cristiane." In the first of these, Dante emphasizes especially that he seeks to
undermine the arguments of those—by whom he means especially the
popes and those who have supported them in the usurpation of secular
authority[22]—who regard themselves as "zelatores fidei cristiane" in their

---

21    "Pirrus 'Heram' vocabat fortunam, quam causam melius et rectius nos 'divinam providen-
      tiam' appellamus" (Pyrrhus called Fortune "Hera," that very cause we name, more cor-
      rectly and righteously, "divine providence" [2.9.8]).

22    To some extent the argument may also have been targeted at a traditional, Augustinian
      understanding of the nature of just authority. Anthony Cassell suggests that "Dante's major
      concern in book 2, although he never names Augustine outright in this connection, is to
      mollify the anti-Roman negativity of that saint's treatise *The City of God*, that had narrowed

having raged and meditated inane things against the "romanum Principatum" (2.10.1). The basic argument of these sections is that Christ himself sanctioned Rome's authority over all men. The first argument on behalf of this claim is that

> Sub edicto romane auctoritatis nasci voluit de Virgine Matre, ut in illa singulari generis humani descriptione filius Dei, homo factus, homo conscriberetur: quod fuit illud prosequi.

> Christ chose to be born of his Virgin Mother under an edict emanating from Roman authority, so that the Son of God made man might be enrolled as a man in that unique census of the human race; this means that he acknowledged the validity of that edict. [2.10.6][23]

The second argument on behalf of the conclusion that Christ sanctioned the authority of Rome is slightly more complicated, but it would also perhaps have been more compelling for Dante's intended audience. In the final section of *Monarchia* 2, Dante argues that for Christ to have suffered for the sins of all mankind means that he had to be punished by an authority with jurisdiction over all mankind:

> Et *supra* totum humanum genus Tyberius Cesar, cuius vicarius erat Pilatus, iurisdictionem non habuisset, nisi romanum Imperium de iure fuisset.

---

and transformed the primitive Christian acquiescence in the 'powers that be,' preached by both Christ and the Apostle Paul, into antagonism" (*The Monarchia Controversy* [Washington, D.C., 2004], p. 66). Put differently, Augustine does recognize an actual existence of two structures of authority in the world, but he denigrates the temporal authority as illegitimate whereas Dante argues for the providential grounds of temporal authority's legitimacy. But it should also be noted that Dante's disagreement with Augustine is not over the issue of providence, for as Cassell recognizes, Augustine, though he may have "derided the ancients for the idolatry that led them to a proud and obsessive pursuit of worldly glory through war and gore," also "proclaimed in *City of God* 5:21 that 'the Roman rule [*Romanum regnum*] was established by God, from whom all power comes, and by whose providence all things are ruled'" (p. 66, bracketed supplement in Cassell).

23     Dante pushes the argument even further in 2.10.7: "Et forte sanctius est arbitrari divinitus iliud exivisse per Cesarem, ut qui tanta tempora fuerat expectatus in sotietate mortalium, cum mortalibus ipse se consignaret" (And perhaps it is more holy to believe that the edict came by divine inspiration through Caesar, so that he who had been so long awaited in the society of men might himself be enrolled among mortals).

And Tiberius Caesar, whose representative Pilate was, would not have had jurisdiction over the whole of mankind unless the Roman empire had existed by right. [2.11.5]

Anthony Cassell has addressed a number of these arguments in detail in *The Monarchia Controversy*. According to him, Dante's arguments concerning the *de jure* authority of Rome are, for the most part, validly constructed even if based on faulty or arbitrary premises. However, Cassell gives greater attention to the arguments from principles of faith regarding Christ's sanctioning the Roman Empire both by his birth and by his death. According to Cassell, these arguments rely on rhetorically persuasive enthymemes that suppress antecedents such as "the birth of Christ signified approval" and "whatever Christ suffered as punishment he approved" (Cassell, p. 79). Yet, even excepting these latter arguments of *Monarchia*, one has the sense that Cassell's basic attitude toward all of Dante's arguments concerning the nature of divine providence might best be summed up when he writes: "It is hard today not to see the poet's instances as preposterously quaint even while we inevitably admire the intensity of his personal persuasion. His arguments have, after all, nothing to do with claims of using the light of human reason or even of adducing credible authority" (p. 70).[24]

While I do not wish to imply that I regard Dante's arguments as compelling (especially regarding the truth of the premises from which his demonstrations proceed), it is more important, for my purposes, to understand why Dante himself believed the arguments to be sound, especially in the cases of those arguments that Dante insists are constructed from "rational principles" rather than those he acknowledges to have been demonstrated from articles of Christian faith. Doing so, I believe, requires attending to the understanding of history subtending Dante's arguments.[25]

---

24    The comment is made specifically in relation to Dante's argument that portentous events signify the hand of God at work in the growth of Rome; however, I think the tone of the comment basically captures Cassell's attitude toward the majority of Dante's arguments about divine providence.

25    An alternative way to account for Dante's understanding of the soundness of his method of argumentation (which I shall not discuss in this essay) would involve examining, for instance, the discussion between the pilgrim and Saint Peter in *Paradiso* 24. In this canto, Dante is "tested" on the fundamentals of the nature of faith and its role in theological argumentation. He responds to Peter's question about why "faith" is regarded as a substance: "E io appresso: 'Le profonde cose / che mi largiscon qui la lor parvenza, / a li occhi di là giù son sì ascose, / che l'esser loro v'è in sola credenza, / sopra la qual si fonda l'alta spene; / e però di sustanza prende intenza. / E da questa credenza ci convene / silogizzar,

As Jacob Klein has pointed out in his essay "History and the Liberal Arts," "the primary liberal disciplines listed by Dante in *Convivio* and linked to the ten heavens of the world...are Grammar, Logic, Rhetoric, Arithmetic, Geometry, Music, Astronomy, Physics and Metaphysics, Ethics, Theology. History is not one of them."[26] Accordingly, Dante's understanding of the subject matter of history is certainly a pre-modern one. However, it would be a mistake to think that Dante has no distinct view about the subject matter of history. Rather, Dante's view of history resembles Aristotle's since, as Klein explains,

> History in this sense is founded on completely 'unhistorical' points of view. That is why this kind of history writing does not constitute a specific domain like physics or even poetry. Note that Aristotle, the great systematizer of human knowledge, in the face of such history—the only one he knew—did not treat it as a *pragmateia*, a discipline in its own right. (p. 130)

According to such an understanding, Dante interprets the events of mundane or temporal history as symbols of the moral and spiritual development of humankind. Thus, for instance, "Troy and its destruction are symbols of man's pride and man's fall" (Klein, p. 132).[27] And, to this extent, like Aristotle's understanding of history, Dante's history is, as Klein puts it, akin to history of the "pragmatic and genealogical kinds" (pp. 132–133), which are, in the case of pragmatic history, attempts to "measure the significance and importance of events" "by the desire to derive a lesson for the future either from mistakes and failures or from exemplary actions in the past" or, in the case of genealogical history, attempts to measure the significance and importance of events "by the consideration of the present state of affairs, the salient features of which want

---

sanz' avere altra vista: / però intenza d'argomento tene'" (And I responded: "The profound things that appear to me here, are so concealed from the eyes below that they exist in belief alone, and on that belief is founded the high hope; and therefore this belief refers to substance. And from this belief, without having another view, we must assent to syllogisms; therefore, it refers to argument" [70–78]). In short, Dante claims that faith, because it intends substance, may be utilized as a first principle in demonstrative arguments. Cf. Thomas Aquinas, *Summa theologica* II–II, q4, a1.

26   "History and the Liberal Arts," in *Lectures and Essays*, eds. Robert B. Williamson and Elliott Zuckerman (Annapolis, 1985), pp. 127–138 (133). In *Convivio*, see 2.13.

27   Although dramatic elements in the *Divine Comedy* support this interpretation insofar as speakers describe historical examples, an obvious confirmation of Klein's interpretation would seem to be the divinely-carved *exempla* of historical virtues and vices in *Purgatorio* 10 and 12.

to be traced back to their origins" (p. 129). To the extent that history is employed in either of these ways, history is, according to Klein, understood to provide the *pragmata* relevant to narrating or symbolizing matters under consideration within another discipline (*pragmateia*), and this employment is especially pertinent with respect to the highest of the disciplines which are, by their nature, most distant from immediate experience.

Of these two modes of historical investigation, pragmatic history is especially pertinent to Dante's thought since, to quote Klein once again, "as far as pragmatic history is concerned, the selection is based on our sense of moral virtues or our understanding of practical maxims of conduct" (p. 130). History, in this sense, serves a practical end, and, in the disciplines of ethics and theology, it serves as the signs with which Dante, like his own guide Virgil, guides his readers to a state of ethical and spiritual perfection. In short, Dante's "history" is a tool of dialectical inquiry or rhetorical or poetic production,[28] and it operates through allegorical interpretation. To the extent that Dante's view would permit history to function pragmatically not merely for ethics (which Aristotle allows) but also for theology (which Aristotle to some extent rejects—see, for instance, the last paragraph of *Metaphysics* Λ.8), his view would, of course,

---

28 For Dante, dialectical inquiry supplies first principles of a science since, following Aristotle, he maintains that "verum, quia omnis veritas que non est principium ex veritate alicuius principii fit manifesta, necesse est in qualibet inquistione habere notitiam de principio, in quod analectice recurratur pro certituine omnium propositionum que inferius assummuntur. Et quia presens tractatus est inquisitio quedam, ante omnia de principio scruptandum esse videntur in cuius virtute inferiora consistant" (since every truth which is not itself a first [that is, indemonstrable] principle must be demonstrated with reference to the truth of some first principle, it is necessary in any inquiry to know the first principle to which we refer back in the course of strict deductive argument in order to ascertain the truth of all the propositions which are advanced later. And since this present treatise is a kind of inquiry, we must at the outset investigate the principle whose truth provides a firm foundation for later propositions [*Monarchia* 1.2.4]). Thus, although Dante does not explicitly state that his inquiry is dialectical, his method is rooted in the rubric for this kind of inquiry laid out in the *Topics*, where Aristotle plainly states: "dialectic, being exploratory, is the path to the principles of every inquiry" (in *Aristotle: Selected Works*, 3rd edition, trans. and ed. Hippocrates Apostle and Lloyd Gerson [Grinnell, Iowa, 1991], A.2, 101*b*4). In discussing the letter to Cangrande della Scala and *De vulgari eloquentia* below, I will explain Dante's understanding of how a principle, once grasped, may subsequently guide productive (e.g., poetic or rhetorical) and practical (e.g., ethical, or political) activities. For now it suffices to note that, for Dante, when ruled by intellect (specifically, by the possession of a principle), poetic language and rhetoric may help produce desirable political ends such as a universal monarchy. Concerning Dante's familiarity with Aristotle's *Organon*, see Cassell, pp. 27–33.

depart from Aristotle's. However, the more significant departure from Aristotle is not in mobilizing historical allegoresis on behalf of theology, but in the underlying understanding of the very activity of historical allegoresis.

Irrespective of its authenticity, the famous letter to Cangrande della Scala[29] provides an important insight into Dante's understanding of the proper allegorical employment of history. In section 7 of this letter, the letter's author explains that, with respect to its service to practical ends, "history" must be understood in more than one way, for although various kinds of "sensus mystici variis appellentur nominibus, generaliter omnes dici possunt allegorici, *cum sint a litterali sive historiali diversi*" (mystical meanings are called by various names, they may one and all in a general sense be termed allegorical, *inasmuch as they are different from the literal or historical*, § 22, emphasis mine).[30] In short, Dante's notion of history is, like his own *Divine Comedy*, "polysemos" (polysemous) in that its "primus sensus est qui habetur per litteram, alius est qui habetur per significata per litteram" (first meaning is that which is conveyed by the letter, and the next is that which is conveyed by what the letter signifies, § 20). Accordingly, to the extent that the literal sense of the *Divine Comedy* is "status animarum post mortem simpliciter sumptus" (the state of the souls after death, pure and simple, § 24), the *Divine Comedy's* "history" records the placement of the souls of those who had died prior to approximately 1300, though it must be accepted that Dante regards such a literal sense, even if "historical," as fictive as well as poetic.[31] But the allegorical "subject" of the *Divine Comedy* "est homo

---

29   The question of this letter's authenticity continues to be hotly debated. For recent discussions, see Robert Hollander's *Dante's Epistle to Cangrande* (Ann Arbor, 1993) and *Seminario Dantesco Internazionale. International Dante Seminar 1*, Proceedings of the First Congress at the Chauncey Conference Center, Princeton, October 21–23, 1994, ed. Z.G. Baranski (Florence, 1997). In any case, insofar as the letter is relevant to this essay, its arguments are roughly consistent with (though more detailed than) those offered in the first book of *Convivio*, so there is little reason for me to attempt any adjudication of the matter here.

30   Latin references and section numbers follow the Testo critico della Società Dantesca Italiana, ed. Ermenegildo Pistelli (Florence, 1960). This edition can be found online at the Princeton Dante Project website. Unless otherwise noted, translations are from *Dantis Alagherii Epistolae*, 2nd ed., ed. and trans. Paget Toynbee (Oxford, 1966).

31   On the various aspects of the allegorical form of the *Commedia*, see § 27 of this letter. I see no reason to believe that Dante actually made a journey into the afterlife (nor, for that matter, to believe that Dante himself believed he had made such a journey, nor even that he believed he had received an epiphanic mystical vision). On the contrary, his "historical" narration must be regarded as fictive even if supporting a treatment that is also "definitive, analytical, probative, refutative, and exemplificative" (diffinitivus, divisivus, probativus, improbativus, et exemplorum positivus). In fact, this explains why Dante relates "encounters" with non-human or quasi-human souls who are known only through

prout merendo et demerendo per arbitrii libertatem iustitie premiandi et puniendi obnoxius est" (is man as he is deserving of reward or punishment by justice according to merits or demerits in the exercise of his free will, § 25). Thus, as the letter's author explains in section 39, "finis totius et partis est removere viventes in hac vita de statu miserie et perducere ad statum felicitates" (the aim of the whole and of the part is to remove those living in this life from a state of misery and to bring them to a state of happiness). And, as he explains in the following section, consistent with this *finis*, the branch of philosophy to which the *Divine Comedy*'s history (in its allegorical employment) belongs is that of "morale negotium, sive ethica" (morals or ethics).[32]

In relation to this allegorical practice, the distance between Dante and Aristotle is therefore plain if we attend to a feature of Dante's argument that draws its force initially from a reference to Aristotle. In *Convivio* 4.5, as I have already noted, Dante explains that one of the best pieces of evidence that Rome was intended to rule by divine providence is that such a state of affairs would have most perfectly prepared mankind to receive Christ. Here Dante draws upon an argument he had made in *Convivio* 4.2.5 to the effect that, in order to understand matters properly, "quel tempo in tutte le nostre operazioni si dee attendere" (we should in all our actions wait for the right time). For Dante, this conclusion is explicitly warranted by the fact that Aristotle himself defines time in *Physics* Δ as "'Numero di movimento secondo prima e poi,' e 'numero di movimento celestiale'" (number of motion with respect to before and after, and number of celestial movement [my translation]), but which, according to Dante in the very same sentence, means that time "dispone le cose di qua giù diversamente a ricevere alcuna informazione" (disposes things here below to receive information in diverse ways [4.2.6, translation mine]).

But despite the ostensive references to Aristotle's *Physics*, there is a distinctly non-Aristotelian metaphysical understanding of human reason at work here. First, at the very beginning of *Monarchia* 2.6, the section in which Dante begins

---

myths (e.g., the "giants" of *Inferno* 31) as well as those who are known through historical or quasi-historical accounts and those whom Dante knew personally in his own lifetime.

32   Toynbee translates *morale negotium* as "morals" in this section and *moralis neogtii* as "ethics" in § 10 (§ 3 in Toynbee's divisions). Given Klein's discussion of *pragmateia* above, note that, at the earlier instance of the term *negotium*, Toynbee explains: "*negotium* is the rendering in the *Antiqua Translatio* of the *Ethics* of the term πραγματεία" (n. 5, pp. 169–170). It should therefore be kept in mind that the connotations of both *morales negotium* and *ethica* are both different and somewhat broader in meaning than the English terms "morals" and "ethics" and that the author of the letter to Can Grande likely has in mind a philosophical "discipline" that involves the sorts of topics that are discussed in the *Nicomachean Ethics*.

to lay out his "historical" arguments for the independence of Roman authority from the spiritual authority of the Church, Dante states:

> Et illud quod natura ordinavit, de iure servatur: natura enim in providendo non deficit ab hominis providentia, quia si deficeret, effectus superaret causam in bonitate: quod est inpossibile.

> Besides it is right to preserve what nature has ordained, for nature in the measures it takes is no less provident than man; if it were so, the effect would surpass its cause in goodness, which is impossible. [2.6.1]

Second, in *Paradiso* 27, Dante also makes clear that all motions (both temporal and eternal) are related to a single common measure in the movements of the Primum Mobile:

> E questo cielo non ha altro dove
> che la mente divina, in che s'accende
> l'amor che 'l volge e la virtù ch'ei piove.
> Luce e amor d'un cerchio lui comprende,
> sì come questo li altri; e quel precinto
> colui che 'l cinge solamente intende.
> Non è suo moto per altro distinto,
> ma li altri son mensurati da questo,
> sì come diece da mezzo e da quinto.

> And this heaven has no other where than the divine mind in which are enkindled the love that turns it and the power it rains down. Light and love encompass it in a circle just as it does the others; and that enclosure only the one who encloses understands. Its motion is not marked by another's, but from it the others are measured just as ten is from a half and a fifth. [109–117][33]

Moreover, in *Convivio's* discussion of the Primum Mobile, Dante claims that without this common measure for the motion of all the other spheres,

> Non sarebbe quaggiù generazione né vita d'animale o di pianta; notte non sarebbe né die, né settimana né mese né anno, ma tutto l'universo sarebbe disordinato, e lo movimento delli altri sarebbe indarno.

---

33   This understanding of the "movement" of the Primum Mobile most resembles what Aristotle seems to have in mind when he discusses eternal motion.

There would be no generation here below, either of animal or of plant
life; there would be no night or day, or week or month or year, but rather
all the universe would be disordered, and the movement of the other
heavens would be in vain. [2.14.17]

Now, it would be more surprising if Dante, firmly in the tradition of Christian
neo-Platonism, were not to associate the ends of Nature with the Good, however un-Aristotelian this may be. But what is relatively new[34] in this formulation of history is a secularization of understanding of divine providence
according to which history becomes a subject of allegorical interpretation in
light of distinctly human ends. Put differently, for Dante, the order of nature
and God's will are coextensive, and the mundane world is ordered in such a
way as to prepare mankind for its moral and spiritual perfection. But the order
of nature is understood by humans only in relation to their own distinct temporal and spiritual ends. Accordingly, Dante's understanding of history, by
stressing the context through which all mundane events may be interpreted,
requires a metaphysical grounding for the conclusion that what occurs naturally is "preserved" *de jure* and knowable through "rational principles."

It is not the practice of allegoresis that is new in the Renaissance, let alone
in Dante's thought; rather, what is new in Dante's philosophy is that allegoresis,
as applied to both the temporal and spiritual human ends, has turned from the
objective representation of historical events to the subjective interpretation of
the context of history. It is possible for Dante to reconstruct and interpret the
past in such a way as to lead others to natural and supernatural perfection only
to the extent that nature itself has prepared him in his own "proper moment"
to understand its innermost causes. To one who is unprepared to understand
the innermost causes of things, the events of mundane history might appear to
be merely a *de facto* procession of chance. But to one who is properly disposed
by history itself to understand the innermost causes of things, the events of
mundane history are recognized as the *de jure* unfolding of divine providence
ordering all things toward the same universal end.

It should now be clear why the enthymemes that Cassell regards as merely
rhetorical or eristic in Dante's arguments regarding the evidence for divine
providence have already, for Dante, been grounded by this peculiar understanding of history as a context of interpretation in which both the natural and
the supernatural human ends can come to be unified and understood most

---

34    According to Walter Ullmann, Dante's predecessors in this regard would include Otto of
      Freising and William of Malmesbury (see *Medieval Foundations of Renaissance Humanism*
      [Ithaca, 1977], pp. 64–67).

fully. So as long as one accepts the arguments that the significant and unique human ends are temporal and spiritual perfection, the historical evidence, however arbitrary it may seem, is ordered *de jure* in light of these ends. As a consequence, Cassell's judgment that Dante cannot reasonably expect his readers to accept an enthymeme involving a suppressed antecedent that Christ sanctioned his own crucifixion cannot be correct. To the contrary, it is all too clear that Dante does not regard, nor does he expect his readers to regard, such a premise as dubitable in any way. With respect to both temporal and spiritual ends, it is *necessary* that Christ did condone his own suffering and that the event is an expression of justice; otherwise one would have to deny what Dante assumes to be the inevitable truth that divine reason caused these events in order to make human perfection possible through its own nature. It is only when such events are abstracted from their precise order in nature, that is, abstracted from time itself, that they appear to be contingent and unjust.

### Conclusion

In short, then, both Dante's political philosophy and his philosophy of language are tied to a common understanding of the nature of divine providence. Dante's political philosophy does not merely hold that divine providence is the common ground of both temporal and spiritual authority. Rather, Dante also maintains that the very grounds for the legitimacy of a sovereign can only be measured by one who attends to that sovereign's specific and local significance in the providential ordering of human history. Moreover, this understanding of political legitimacy is also embedded in Dante's philosophy of language since Dante's conception of the illustrious vernacular is constrained by the competing demands that it both help produce justice in a local environment and that it derive its authority from its universal suitability.

In light of these comments, then, it is clear that Dante's understanding of divine providence continually opens onto the aporetic relationship between the temporal function of the illustrious vernacular and its atemporally determined authority to perform this function. On the one hand, Dante explicitly maintains that the significance of singular historical events or the value of particular modes of expression can only be interpreted or judged in relation to their unique historico-political contexts. Moreover, he seems to acknowledge that conditions that might produce justice in one context are local, unique, and non-transferrable to another context. On the other hand, Dante persistently seeks a universal ground for the authority to judge and interpret these events and modes of expression. The principle that establishes such a jurisdiction is cosmological in that it understands history as the unfolding of particular con-texts of signification in relation to divine providence. Thus the legacy of Christo-Platonism in Dante's thought is clearly expressed in the statement quoted above that the ultimate grounds for the legitimacy of human judgment "flows down...without any intermediary from the Fountainhead of universal authority" (*Monarchia* 3.16.15).

What is new, though, in Dante's understanding is the way in which this providential ground for secular authority operates in relation not only to the spiritual ends of the human being but also in relation to the earthly ends of human nature. Or, in other words, Dante seems to embrace a tension between two major modes in the prior history of philosophy. On the one hand, he embraces elements of an Aristotelian ethico-political philosophy that empha-sizes the importance of the temporal ends of human activity and that to some extent undermines the significance of divine providence in legitimizing those activities. On the other hand, he is also deeply committed to a Christo-Platonic psychology and metaphysics in which a predominant conception of the role of divine providence makes very few concessions to the legitimacy of historical and cultural differentiation.

In light of this tension within his thinking, I would like to close by suggest-ing that, with respect to later philosophies of history that might bear some resemblance to Dante's peculiar understanding, what may be considered dis-tinctly "modern" is: (a) that Dante understands history as a realm of inquiry in which the significance of mundane events is interpreted in relation to the total context of history in which they are embedded and (b) that, for Dante, such an interpretation is made possible only by virtue of a universal metaphysical prin-ciple that secures a common measure in relation to which these events may be judged. For Dante, this common measure is itself always linked to the particu-lar point of view of the individual for whom it is appropriate to interpret his-tory allegorically or, what may be the same thing, to produce virtue in others through poetic expression. It may therefore be appropriate to think that Dante's understanding of the proper grounds for individual authority might also be viewed as a prototype of the modern "subject" since the principles by which such an individual authority judges are authoritative for that subject only to the extent that they are, in principle, common to all by nature.

CHAPTER 11

# Time, Myth and the Quarrel between the Ancients and the Moderns: Racine and Fontenelle[1]

*Sara E. Melzer*

> The quality of a barbarian...would be natural to men unless a good education corrected for it.
>
> FURETIÈRE, *Dictionnaire universel*

> Mon mal vient de plus loin.
>
> RACINE, *Phèdre*

The issue of human time is central to the *Quarrel between the Ancients and Moderns*, a series of debates in early modern France that, broadly speaking, began in the early sixteenth century and extended to the mid eighteenth-century.[2] Did the imagined course of human time travel downwards towards decline? Did each successive era degenerate in relation to a more perfect civilization in the past when giants and geniuses supposedly walked the earth? The dominant view of the ancient Greeks and Romans issued a resounding yes. Hesiod articulated this perspective in his *Works and Days*, describing the deterioration of the human race over time. "A golden race of mortal men" marked its high point, which then degenerated into humans made of increasingly baser metals: silver in a subsequent era, then followed by bronze and

---

1 I would like to thank Sylvie Romanowski and Eric Gans for their insightful readings of this essay.

2 Some scholars situate the *Quarrel* within a narrower frame, depending upon the issue they highlight. For example, see Joan de Jean, *Ancients against Moderns: Culture Wars and the Making of a Fin de Siècle* (Chicago, 1997). She dates the Quarrel from 1687–1715. Larry Norman, in *Shock of the Ancient. Literature and History in Early Modern France* (Chicago, 2011), also frames the Quarrel roughly within this time frame, but acknowledges it can be expanded in both directions by a few decades. Other scholars interpret it more broadly, see Terence Cave, "Ancients and Moderns: France," in *The Cambridge History of Literary Criticism* III: *The Renaissance*, ed. Glyn P. Norton (Cambridge, 1999); Gilbert Highet, *The Classical Tradition: Greek and Roman Influences on Western Literature* (London, 1967); Hubert Gillot, *La Querelle des anciens et des modernes en France* (Geneva, 1968); Hélène Merlin-Kajman, *Public et littérature en France au XVIIe siècle* (Paris, 1994).

then iron, which were "unlike the golden one in thought or looks."[3] Such a view meant that those peoples who were born at a later historical moment were inherently inferior to those who had arrived first on the scene of human (that is, European) history.

Although the French were late-comers in history, their cultured elite, comprised mainly of men and women of letters, imitated the Greek world of thought, including its underlying assumptions about the nature of time, with its supposed degenerative effects on the human soul, mind, and character. This educated elite who most ardently favored imitating the Greeks and Romans were called the "ancients" in the Quarrel. (I use the term "ancients" with a small "a" placed in quotes to differentiate the early modern French partisans of Greek and Roman thought from the Greeks and Romans themselves, which I indicate with the capitalized term Ancients without quotes.) The "moderns," however, challenged the imperative to imitate the Ancients and rejected their view of time. They sought to reverse the imagined path of time, to slant it upwards so that the future could conceivably improve upon the past. Such a reversal would make the concept of progress imaginable, thus ushering in a foundational notion of modernity.

This debate about the slope of time shaped every arena of human thought: science, philosophy, economics. It also had profound implications for France's elite world of letters, which was dominated by Humanist thought.[4] Imitating Greco-Roman thought created a serious dilemma for the French cultured elite, trapping them in a circular bind. To civilize themselves, the French cultured elite imitated the Greeks as the highest embodiment of civilization, with the Romans a cut below them. But what they were imitating were thought structures that assumed time and history were slanted downwards towards decline. This meant that no matter how well the French imitated Greco-Roman thought structures and forms of expression, they would never come close to their exalted models because the playing field was not level. No matter how strongly they believed in the myth of the *translatio studii*, in which the Greeks and Romans supposedly chose them as heirs and kin to their learning, the educated

---

3  Hesiod, *Works and Days*, trans. Apostolos N. Athanassakis (Baltimore, 1983), pp. 109–120.

4  Humanism, as it is conventionally defined in literary studies, was an intellectual movement that began in Renaissance Italy and then made its way into France from 1400–1600. Its central belief was that the ancient Greeks and Romans were at the pinnacle of intellectual and human achievement, and Europeans should look to them as their models. Humanism had a notion of time built into it: the height of human civilization existed in the past, with the present representing a fall away from it; the goal of the era's contemporary writers was to revive this lost past through imitation.

elite could never measure up to this imagined promise. The degenerative slope of time meant that the French imitators would fall far short of the ideal whose perfection lay in a more perfect past. Perceiving that they were inferior, the French elite imitated their civilized models yet more assiduously. But the underlying thought structures they were imitating only aggravated their fears of their own inadequacy. Both the "ancients" and the "moderns" fell into a vicious circle of a civilizing process, but each proposed different ways to escape this circular bind, as I have developed elsewhere.[5]

This essay explores one of the most important thought structures that the French literary elite imitated: myth. My goal is to open up Racine's use of myth in *Phèdre* to show how it was a site of conflict between two warring notions of time, reflecting the fact that this dramatist was both an "ancient" and a "modern," torn between the two. I couple my reading of *Phèdre* with that of Fontenelle, an unequivocal "modern," because his exploration of myth illuminates how Racine employed mythical references in much broader and more complex ways than most scholars have realized and in ways that reflected the fundamental tension of the Quarrel between the Ancients and the Moderns.

Racine used myth in two opposite senses. In a first sense, which was the dominant understanding of the era, *mythos* was quite different from how we use it today. It was a true story that unveiled the true origin of the world and the nature of the human soul. It issued primarily from Greek and Roman culture, conveying its deepest truths that could not be conveyed otherwise. As such, myth existed outside of human time, linked to notions of a cultural sacred. Yet the Greeks and Romans also inserted myth in a historic time, with Hesiod and others believing in the myth of Golden Age that subsequently deteriorated. The Greeks supposedly possessed golden souls, inhabiting a golden era, and thus they were closer to, and better able to capture the universal truth that was eternally present.[6] The French cultured elite thus borrowed Greco-Roman myths to recapture this transcendent truth and to endow their language and stories with the prestige and cachet of these Ancient civilizations.

I argue that Racine harbored a conflicting understanding of myth, one that Fontenelle clearly articulated, making explicit what was implicit in Racine. Fontenelle presented an alternative view of myth that existed in incipient form in the seventeenth century but emerged more fully in the eighteenth.

---

5   Sara E. Melzer, *Colonizer or Colonized: The Hidden Stories of Early Modern France* (Philadelphia, 2012).

6   For a general discussion of myth, see Mircea Eliade, *Myth and Reality*, trans. Willard R. Trask (New York, 1963).

Fontenelle situated myth within a historical time frame and questioned its sacred status. Far from existing outside of time, myth was deeply embedded in it, marking the primitive beginning of an evolutionary stage that reflected the earliest phase of human development. Conjuring up an evolutionary continuum, myth was an expression of a key historical moment—a barbaric state before humans evolved to a more advanced phase of civilization. Ideally, historical time would progress from the barbaric, mythical stage to the more rational and ordered structures of civilization.

This progression, however, was difficult if, as the "ancients" believed, historical time was slanted downward towards decline, since it meant all late-comers were of a lesser status, more vulnerable to slipping back into a primitive stage. But if, as the "moderns" believed, this historical time could be reversed and oriented upwards towards a future that improved upon the past, a progression from the barbaric stage of myth to civilization was possible. But in either case, myth was inserted in a historical notion of time that was human and linear, be it regressive or progressive. As a "modern," Racine wanted to progress towards civilization, which meant eliminating the mythic to create something new and independent, thus helping his nation and its world of letters in their journey. And yet as an "ancient," he was haunted by a mythic and a degenerative view of time that shaped his writing of *Phèdre* (1677), both thematically and in its dramatic structure, as I will show. The degenerative slant of time made the progression towards civilization a struggle much akin to Sisyphus' attempt to roll the ball upwards on a downwards slope. This reflects Racine's tragic vision. Trapped in the "ancients'" view of time, *Phèdre* dramatized the regressive implications of Ancient thought structures for the French elite and their pursuit of a progressive development towards civilization.

As a "modern," Fontenelle, like Racine, felt the suffocating effects of a degenerative concept of time. In his essays, *Digression on the Ancients and Moderns* (1688) and in *On the Origin of Fables* (1724), Fontenelle sought a way out of the circular bind. He did so by targeting the phenomenon of myth and asked what it represented. Did it embody the height of civilization? Was it the mark of a high style, as the Humanist tradition had suggested, conveying the highest and most important universal truths that were ahistorical? Or was it situated inside of time, pointing to the lowest stage in an evolutionary development, representing a remote past that was barbaric? Fontenelle used myth to open up a new concept of time to reverse the slope of history by recasting the Greeks as primitive and their myths as the most archaic mode of thought that needed to be transcended. Fontenelle's understanding of myth will help us see how the mythic structures in Racine reflected this dramatist's deep-seated fears of a "barbarian within." This fear was the effect of the "ancients'" view of time: the

French, as late-comers in history, were made of a baser metal than the Ancient Greeks and Romans. The French elite's struggles to civilize themselves were constantly at risk for regressing back into the most primitive state of the barbaric.[7] And yet Racine expressed this dilemma in a form that could be called "modern."

In discussing myth, I want to differentiate between two overlapping terms: "mythic" and "archaic." The term mythic most often connotes a mode of thought that is outside of human time, possessing no beginning or end, and existing prior to the emergence of a linear or historical development. The term "archaic" frequently implies a beginning of human, historical time, indicating a primitive stage in a linear and evolutionary continuum preceding a more advanced stage of civilization. This distinction is clear in Fontenelle who, as a "modern," sees the Greeks as "archaic" and denies to them the more elevated meaning of the "mythic." Racine, however, because he is partly attached to the "ancients," though not completely, experienced the overlap between the "mythic" and the "archaic" as a source of tension, embracing both understandings in a conflicted mix.

## On Racine's Use of Myth

Racine's use of myth in *Phèdre* is troubling, even scandalous.[8] The particular myths he chose reflect the most degraded and barbaric form of the human, centering on highly taboo subjects: incest and bestiality.[9] These taboos are pivotal in the journey away from the barbaric, since they must be eliminated in order for a civilized order to take place. The outlawing of incest, according to Claude Lévi-Strauss, "has been the minimum and ever-present condition for the passage of any community of men from a state of nature to one of culture."

---

7   It may seem strange to suggest that the French cultured elite feared that they were barbarians given the official rhetoric that expressed confidence in their greatness. For a development of this argument, see Melzer, *Colonizer or Colonized*. See also Rémi Brague, *La voie romaine* (Paris, 1992).

8   For further discussion of the scandalous nature of myth in this play, see Francesco Orlando, *Toward a Freudian Theory of Literature, with an analysis of Racine's Phèdre*, trans. Charmaine Lee (Baltimore, 1978). See also Mitchell Greenberg, *From Ancient Myth to Tragic Modernity* (Minneapolis, 2010); Sylvaine Guyot, *Racine et le corps tragique* (Paris, 2014). Most scholars, however, tend to rationalize it.

9   For further discussion of incest and bestiality, see Sara Melzer, "Incest and the Minotaur in *Phèdre*: The Monsters of France's Assimilationist Politics," in *Biblio* 17/131 (2001), 431–445; Sara Melzer, "Myths of Mixture in *Phèdre* and the Sun King's Assimilationist Policy in the New World," in *Racine for the New Millenium*, ed. Harriet Stone, *L'Esprit créateur* 35/2 (Summer, 1998), 72–81.

Its prohibition is the "very basis of what we call civilization."[10] Similarly, bestiality is subject to the same interdiction. And yet both are at the center of Racine's drama.

In the play's incest plot, Phèdre was obsessed by her love for her step-son Hippolyte and tried to kill herself to avoid revealing this awful truth. But she received news that her husband, Thésée, was dead. Believing her love legitimate, she was persuaded to live and confessed her love to her step-son, Hippolyte, even though her love for a step-son still qualified as incest, according to seventeenth-century canon law. After exposing this secret truth, Phèdre discovered her husband was not dead. When he returned, he received a false report accusing his son of being the seducer. Outraged, Thésée called upon Neptune to avenge him; Hippolyte was then punished by a particularly gruesome death. Upon learning this news, Phèdre finally confessed the truth before dying from the poison that she had just taken.

Racine framed this incest plot within a mythical story about bestiality. Phèdre's mother, Pasiphae, fell in love with a bull. Not content to worship this bull from afar, she had sex with this animal and thereby produced the Minotaur, a monster who was half-human and half-bull. Although this monster was contained within a labyrinth, it nevertheless terrorized the city. Thésée became the founder of civilization after he slew the monster in the labyrinth, thus becoming the first king of Epire.

A third scandalous myth was at the root of both the mother's bestial lust and the daughter's incestuous passion. It involved Venus who, in Racine's version (and following both his Euripidian and Senecan models), exhibited her most barbaric side: the need for an excessively cruel vengeance.[11] Venus, the wife of Vulcan, had been deceiving her husband by engaging in dalliances with Mars. Her husband thus set a trap to catch her in *flagrant délit* and expose her infidelity. Apollo was his accomplice; this sun god used his rays to illuminate Venus' guilt, exposing her shame for all to see. Venus retaliated, directing her ire not simply against Apollo, but also against his descendents: Pasiphae and Phèdre, causing both to harbor and act on two of the most uncivilized of passions.

These myths are puzzling because of their extreme barbaric and immoral character. But they are also troubling because they were so contrary to the conventional use of myth during Racine's era. Mythic references had once flourished, especially in the Renaissance, constituting an evident sign of learning,

---

10    Cited in Orlando, *Freudian Theory of Literature*, p. 16.

11    *Le Mythe de Phèdre. Les Hippolytes français du dix-septième siècle*, ed. Allen Wood (Paris, 1996.)

lending prestige and nobility to all forms of artistic expression. However, the status of myth began to wane in seventeenth-century theater, discredited due to Cartesian rationalism, on the one hand, and to the strengthening of Christianity, on the other, as Christian Delmas has shown. By the last quarter of the seventeenth century, myth had virtually disappeared from theater, although it was still used in opera.[12]

Several of Racine's contemporaries were particularly scandalized by his use of incest. After his *Phèdre* was performed, an anonymous author lambasted him in *Sur les Tragédies de Phèdre et d'Hippolyte* for selecting incest as his subject.[13] Euripides and Seneca, this critic argued, could present such a subject with impunity because they did not know any better. After all, they were pagans and were not enlightened by Christian thought; they did not truly understand the horror of what they were doing. But Racine should have known better. For him to present such an immoral subject in an enlightened Christian era was outrageous. This critic further attacked Racine's play for his portrait of Thésée and his descent into the underworld, arguing that it was completely unbelievable and irrational.

Phèdre's love for Hippolyte was the subject of four other plays in the seventeenth century, in addition to that of Racine. But in contrast to Racine, these playwrights changed the story, taming its scandalous nature to conform to the French Academy's rules of *les bienséances*. Their plots did not include incest or bestiality.[14] For example, Phèdre was merely Thésée's fiancée, thus minimizing the incestuous dimension. In addition, these plays provided logical motivations for the key events that in Racine's version were explained by mythological rationales. However, these dramatists did not completely expunge the mythical elements because they were part of a long literary tradition that gave prestige to poetic discourse and stature to dramatic characters. But they did reduce the mythical to an ornamental status to conform to the conventions of the time that emptied them of their significance and function, as Delmas has described.[15]

Racine's use of myth in *Phèdre*, however, was far from ornamental. He wove the myths so finely into the heart of the play that it is difficult to situate the action firmly in the dramatic present or in an imagined, archaic past. The main characters, Thésée, Phèdre, and Hippolyte, were real historical figures, situated

---

12    Christian Delmas, *Mythologie et mythe dans le théâtre français (1650–1676)* (Genève, 1985), p. 7. See also Delmas, *Mythe et histoire dans le théâtre classique* (Paris, 2002).

13    Anonyme, *Sur les Tragédies de Phèdre et d'Hippolyte*, in Racine, *Oeuvres complètes*, ed. Georges Forrestier (Paris, 1999), pp. 877–904. Some scholars have attributed this text to Sieur de Subligny.

14    See Wood, *Le Mythe de Phèdre*; Orlando, *Freudian Theory of Literature*.

15    Delmas, *Mythologie et mythe*, pp. 7–10.

in a current drama of political succession created by the (false) news that
Thésée, the king of Epire, was dead. Yet all three figures were also mythological
figures pulled back into a remote, primitive past, with their stories revolving
around Thésée's act of killing a mythological monster, the Minotaur who had
been lodged in a labyrinth. Killing this monster established him as the founder
of civilization and also the first king. In sum, the boundaries between these
two temporal frames were confused; the main characters slipped between the
play's present moment and a hazy, mythological past.

Racine's dramatic structure is based on this slippage. On the formal level,
the play's dramatic action moves forward in real time of 24 hours, following a
rational, chronological sequence. However, the action is propelled forward in
real dramatic time through mythical events or memories that pull the
characters back into an archaic past of primitive barbarism. The characters
cannot stabilize themselves in the rational order of the present, but continually slide back into a violent, irrational past. In other words, the play's structure
moves forward into a dramatic future by taking us back into increasingly
remote layers of a mythological past. In sum, the mythic, then, is essential to
the plot structure.

How then to understand this puzzling use of myth? And of a concomitantly
strange temporality? I must begin my answer to this question by repeating the
question I asked earlier: What did myth represent for this dramatist? Most
scholars who have studied it have understood Racine's use of myth as representing that which exists outside of time and history. Because myth is
grounded in a civilized Greco-Roman heritage that conferred an important
cultural weight upon its use, scholars have frequently recuperated mythological elements by seeing them as a key feature of civilization and poetic language. This perspective has encouraged some scholars to rationalize and
normalize Racine's use of the mythic, by seeing it as an expression of repressed
archetypical psychological desires or patterns that communicate transcendent truth.[16]

Other scholars, such as Timothy Reiss, have seen the mythic as the mark of
what prevents the evolution towards a rational, civilized world; hence, the
mythic is what needs to be eliminated. While this certainly is correct, Reiss
still views myth as representing that which exists outside of time, for he
opposes myth, with its "fictions of eternity," as he describes it, to history, which
is firmly rooted in time. For him, the play dramatizes the "passage from myth

---

16    For example, see Muriel Bourgeois, "Des invraisemblances de la vraisemblance classique,"
      in *Le Vrai et le vraisemblable. Rhétorique et poétique, Revue des sciences humaines* 280/4
      (2005), 49–65. See also Christian Delmas, *Mythologie et mythe.*

to history."[17] This progression is what makes Racine modern for Reiss. Both scholarly approaches take the edge off of myth, taming its scandalous nature, by viewing it as a transcended stage on the way to achieving a higher, rational order. Such a view, I argue, is partial because it dulls the edge of Racine's scandalous use of myth by rationalizing it. While myths can refer to truths that are outside of time and history, they also are situated within them. The mythic in *Phèdre* conjures up the most primitive mode of thought, a pre-rational stage in the evolution of the human before the founding of civilization. Racine's use of myth, I will show, expressed his fear of backsliding on an evolutionary continuum, a fear that was built into the "ancients'" view of time as degenerative. Such a view makes the institution of a transcendent, stable, civilized order highly problematic.

To illuminate Racine's use of myth, I would first like to examine how Fontenelle, his contemporary, used myth, enabling us to read *Phèdre* backwards via Fontenelle's understanding of this complex phenomenon. Through this retrospective lens, we can see how Racine had a conflicted understanding of myth and how it was a site of struggle in the Quarrel between the Ancients and the Moderns and in their respective views towards time and history.

### On Fontenelle

For Fontenelle, myth became a key instrument in the battle of the "moderns" against the "ancients'" notion of the degenerative slant of time. Fontenelle brought out its negative implications by parodying the Ancients/"ancients" view of time in his *Digression on the Ancients and Moderns*, insisting, for example, that the trees of antiquity were superior to those of France, larger and more beautiful because nature was more generous towards those trees that had come first. By the seventeenth century, poor mother nature was exhausted, having expended all her energy on the Greeks; subsequently, trees grew smaller over time, with the French versions approaching the equivalent of a bonsai. Similarly, in the world of letters, as the "ancient" view would have it, French thinkers and writers resembled mental and creative pygmies by force of this degenerative slant. This was because Plato, Demosthenes, and Homer were formed "from a finer clay and were better prepared than our philosophes."[18]

---

17    Timothy Reiss, "From *Phèdre* to History: The Truths of Time and the Fictions of Eternity," in *Tragedy and Truth* (New Haven, 1980), p. 287.

18    "...d'une argile plus fine [et] mieux preparée que nos philosophes." Bernard de Fontenelle, "*Digression sur les anciens et les modernes*," *Rêveries diverses: Opuscules littéraires et philosophiques* (Paris, 1994), pp. 31–32. All translations are my own.

The Greeks and Romans "had more spirit than us" because "the brains of their time were better ordered, formed of more delicate or firmer fibers, full of more animal spirits." It followed that "one only has spirit as much as one admires [the Ancients who] cannot be equaled in these last centuries."[19] With the slope of history slanting downwards, the French would always remain a few cuts below their models. The implications of this downward slope of history are that the French elite would always be doomed to feel inadequate and barbaric in relation to their supposedly more civilized models. No matter how faithfully the elite imitated their Greco-Roman models, they would still be inferior.

To challenge the status of the Greeks, Fontenelle, in *On the Origin of Fables*, targeted their myths, emptying them of the elevated position they had in the Renaissance and early seventeenth century, making them merely archaic, and thereby denying them any claims they might have to a transcendent truth. Examining the Greek myths, Fontenelle argued that the Greeks lived in "barbarousness so extreme that we can't even represent it."[20] Their gods, he wrote, prized "the force of the body"[21] and "are almost as brutal as [the Greeks themselves], only a bit more powerful."[22] The Greeks projected their highest values onto their gods whose values were "physical force...not wisdom." Their gods were "cruel, bizarre, unjust, ignorant," all qualities that reflected the barbarism of the Greeks themselves.[23] He continually called the Greeks, "these poor savages" and described how the Greek gods used their brute force to satisfy their most base and animalistic desires. Far from protecting humans, the gods used innocent individuals as pawns, trapping them in petty and nefarious squabbles. The myths about their feuds normalized and legitimated a primitive violence. These myths were told so frequently that, by dint of repetition, they acquired an aura of truth; no one dared question them and these stories became increasingly sacred, obscuring the fact that they were basically immoral and irrational, a "heap of chimeras, reveries and absurdities."[24]

19    "...avaient plus d'esprit que nous," "les cerveaux de ce temps-là étaient mieux disposés, formés de fibres plus fermes ou plus délicates, remplis de plus d'esprits animaux," "l'on n'a d'esprit qu'autant qu'on admire [the Ancients who] ne peuvent être égalés dans ces derniers siècles," ibid., p. 31.

20    Fontenelle, "De l'origine des fables," in *Œuvres complètes* 3, ed. Alain Niderst (Paris, 1989) pp. 187–202 ("La barbarie [qui] dûrent être à un excès que nous ne sommes presque plus en état de nous représenter," p. 187).

21    "La force du corps," ibid., p. 191.

22    Ibid., pp. 187–197.

23    "Cruels, bizarres, injustes, ignorans," ibid., p.197.

24    "Un amas de chimères, de rêveries et d'absurdités," ibid., p. 187.

Perceiving the Greek myths as an archaic, primitive mode of thought, Fontenelle developed the notion of an evolutionary history in which the Greeks represented its earliest and most crude beginning. This concept emerged out of his readings of the best-selling reports about the New World "savages." These reports, known as *les relations de voyage*, were written by the French explorers, missionaries, and merchants who had been in contact with these peoples since the beginning of the sixteenth century.[25] The descriptions of primitive peoples in the New World provided Fontenelle with a new lens to conjure up an image of what the Greeks were once like at their beginning stages. "Imagine the Caffirs, the Lapps or the Iroquois," wrote Fontenelle, asking his readers to use these peoples as mirrors to look back in history at the Greeks who lived in a "barbarousness so extreme that we can't even represent it."[26] In fact, the Greeks were even more primitive, since the Iroquois, Lapps, and Caffirs had "certainly arrived at some degree of knowledge and politeness that the first men [the Greeks] did not have."[27] For Fontenelle, the Greeks did not have the sacred status as an origin existing outside of time; rather, he inserted them inside of time, in an earlier but historical human time at its humble beginnings, like the primitive Amerindians of the current era.

The phenomenon of mythic thought is what stimulated Fontenelle to compare the Greeks and Amerindians. As he wrote in *On the Origin of Fables*, there was "an astonishing similarity between the myths of the Americans and those of the Greeks."[28] Their similarity "shows that the Greeks were once savages just as the Americans are now, and that they were taken out of the barbaric stage by the same means."[29] Because the Greeks had once used myths as a major mode of thought, the ultimate mark of primitivism, this awareness highlights that they once inhabited a barbaric stage. "Put a new people on this earth, and their first stories will be fables," he continued.[30] Myth is fundamentally irrational, reflecting a pre-civilized era. When the Greeks were "a new People, they did not think any more rationally than did

---

25    Cf. Melzer, *Colonizer or Colonized*, chap. 3.

26    Fontenelle, "De l'origine des fables" ("Figurons nous les Cafres, les Lappons ou les Iroquois" and "La barbarie [qui] durent être à un excès que nous ne sommes presque plus en état de nous représenter," p. 187).

27    "Ont dû parvenir à quelque degré de connoissance et politesse que les premiers hommes n'avoient pas," ibid., p. 187.

28    "Une conformité étonnante entre les Fables des Américains et celles des Grecs, " ibid., p. 197.

29    "Ce qui montre que les Grecs furent pendant un temps des *Sauvages* aussi bien que les Américains, et qu'ils furent tirés de la barbarie par les mêmes moyens," ibid.

30    "Mettez un Peuple nouveau sous le pôle, ses premières histoire seront des Fables," ibid.

the barbarians of America, who were also, according to all appearances, a rather new People when they were discovered by the Spanish."[31] Like the Amerindians, the Greek world of myth was full of giants and magicians; logic had little power over their minds. Myths were an explanation of the unknown, a result of complete ignorance. Fontenelle described how these "first men" invented false divinities and absurd stories to explain what exceeded their reason.

In challenging the prestige of Greek myths, Fontenelle sought to turn history on its head by deflating the value of *firstness*. Several analogous thought structures in early modern France were also predicated on an imagined superiority of that which came first, for example, the law of primogeniture, stipulating that the firstborn son should inherit the entire family property, established for complex economic, social, and political reasons. Analogously, the *noblesse d'épée* claimed an elevated status based on firstness. As the old feudal, military elite, they claimed superiority over the more recently arrived young upstarts, the *noblesse de robe*, who had bought their way in.

To challenge the Greeks' elevated status, Fontenelle situated them on an evolutionary continuum, reframing them as "ces pauvres sauvages qui ont les premiers habité le monde."[32] Repeatedly using the term "the first men," as opposed to the neutral, denotative term, "Greeks," Fontenelle emphasized that their firstness in history meant the most primitive stage, a beginning point rather than its highest peak. To attribute greatness to the Greeks because would be "as if one boasted about having been the first to drink the water of our rivers, and insulted us for not having drunk as much as others."[33] These "first men" were not necessarily wise; they did what all inexperienced, ignorant people do on their first try. They bumbled their way through many mistakes, trying as best they could to make sense of the world. But their first stabs were crude, inventing fantastical and irrational mythical stories. After much trial and error, they finally got it right and produced a rich, sophisticated literature and civilization.

Fontenelle anticipated the obvious objection that the Greeks did in fact transcend their crude first stage to develop a highly advanced civilization. He responded that their cultural and intellectual advances were grounded in the

---

31    "Un Peuple nouveau, [ils] ne pensèrent point plus raisonnablement que les Barbares de l'Amérique, qui étoient, selon toutes les apparences, un Peuple assez nouveau lorsqu'ils furent découverts par les Espagnols," ibid., p. 198.

32    Fontenelle, ibid., p. 189.

33    *Digression*, "J'aimerais autant qu'on les vantât sur ce qu'ils ont bu les premiers l'eau de nos rivières, et que l'on nous insultât sur ce que nous ne buvons plus que leurs restes," ibid., pp. 34–35.

primitive thinking of their earlier stage since "we explain the unknown in nature by what we have in front of our eyes,"[34] Fontenelle argued that Greek rationality was built on the earlier, irrational foundation of Greek fables, rendering the thought structures built upon them shaky and unsuitable as models for the French elite.

His re-conceptualized concept of human time was meant to help the nation achieve dignity and self-confidence and to escape the vicious circle of a degenerative temporality. Refusing to look at Greek and Roman culture as existing either in a timeless vacuum, or at the height of a historical human time, according to the "ancients'" view, he inserted them in the flow of a new kind of time, in a new linear, historical human time that we might call evolutionary or anthropological. Turning history on its head, he saw the Greeks not as a sacred origin or as representing a *telos*, where the end existed in a more perfect past. Rather, he framed their antiquity as archaic, akin to "savages," like those of the New World. Their primitiveness provided a model for imagining what the Greeks had once looked like.

To reverse the slope of time and counteract the Greek theory of degeneration, Fontenelle included the New World on the same linear trajectory of human history as France and the Ancient World. The Greeks were barbaric first men, cultural infants in the "childhood of the world."[35] France was in "the age of virility where [France] reasons with greater force and more enlightenment than ever."[36] Fontenelle pushed that evolutionary frame yet further to suggest that "the *beaux esprits* of a future time could be the Americans."[37] Science and knowledge might shift so that in the future "one can hope to see great Lapp or Negro authors."[38] The primitive Americans would one day be more civilized than the Ancients. Nature could be cultivated to develop a future that not only differed from, but also improved upon, the past. This reversal opened up a new concept of time, with a future that allowed for the notion of progress, thus escaping the stranglehold the past had on the French imagination.

Fontenelle's notion of an evolutionary continuum was not quite as radical or aberrant for his era as it might perhaps seem. Such a concept was implicit in how the French church and state imagined the nation's contact with "savages"

---

34   *Fables*, "Nous expliquons les choses inconnues de la nature par celles que nous avons devant les yeux," ibid., pp. 189.

35   "L'enfance du monde." *Digression*, p. 35.

36   "L'âge de virilité où [la France] raisonne avec plus de force, et a plus de lumières que jamais," ibid., p. 43.

37   "Les beaux esprits de ces temps-là, ...pourront être des Américains," ibid., p. 34.

38   "L'on peut espérer de voir de grands auteurs lapons ou nègres," ibid., p. 33.

when they sought to civilize and evangelize the Amerindians, justifying their attempts to assimilate them into an expanded version of France. By looking at these primitive inhabitants of the New World more closely, it became possible to get a sense of the passage from the "primitive" to "civilization." Thus Fontenelle's argument is a logical outgrowth of an evolutionary discourse that was foundational to the nation's colonial stance towards the New World "savages," as it was portrayed in the *relations de voyage*.[39] These texts—best-sellers during much of the seventeenth century—were propaganda arms of the nation's colonizing ambition, and its evangelizing project was part of the Catholic Reformation.[40] These *relations* justified assimilating the New World *sauvages* by viewing them, not as creatures of a fundamentally different order of being, but simply as humans at the beginning stage of a continuum that led from barbarism to civilization. Such a continuum was built into the etymology of the term *sauvage*. The Latin *salvaticus* (meaning, "of the woods") implied that the concept of the "sauvage" was rooted in the world of agricultural growth. The "sauvage" simply designated a wild and primitive plant or a being that was neglected. This seedling or bud could grow, however, to flower and reach its highest potential if it were cultivated properly. In this colonial and evolutionary discourse, the "sauvage" was thus placed on the same continuum as the French and the Greeks, but at a beginning point.

In sum, myth for Fontenelle signified a mode of thought that was archaic, inside of time and history, at the very beginning of a linear, evolutionary continuum that progressed from barbarism to civilization. Since "the Greeks... were taken out of the barbaric stage by the same means" as the Amerindians, Fontenelle suggested that the path towards civilization meant rejecting the mythic.[41] Using myth to reverse the slant of time, seeing the Greeks as archaic, Fontenelle suggested that the French borrowing of their barbaric mythic forms signaled a regressive move back into a primitive stage. If the slope of time were not reversed, the French elite would be doomed to an endless circular bind, seeking to civilize themselves by imitating a past that forced them into a position of perpetual inferiority and perceived barbarism.

---

39  Melzer, *Colonizer or Colonized*, chap. 3.

40  I use the term "Catholic Reformation" instead of "Counter-Reformation" because the former is broader, implying that the Catholic Church sought to counter not simply the Protestant Reformation but also to spread its influence by targeting the Muslims and New World Amerindians. See Louis Châtellier, *La religion des pauvres: les missions rurales en Europe et la formation du catholicisme moderne au XVIe–XIXe siècles* (Paris, 1993).

41  "Ce qui montre que les Grecs furent pendant un temps des *Sauvages* aussi bien que les Américains, et qu'ils furent tirés de la barbarie par les mêmes moyens," *Fables*, p.197.

### On Racine's *Phèdre*

The dramatic structure of *Phèdre* conjures up the kind of evolutionary continuum that Fontenelle had described in his essay on fables. Presumably Racine's characters had already progressed from barbarism to civilization since they lived under the rule of law, possessed behavioral norms for polite society, were clothed in civilized dress and spoke a highly civilized language of poetic alexandrine verse. And yet the play's main characters were continually haunted by their remote past, regressing back towards the barbaric. In fact, the play's dramatic structure is predicated on that backsliding, slipping between two different temporal frames. As stated above, the play's dramatic action moves forward in real time of 24 hours, following a rational, chronological sequence. However, that rational structure and action are propelled forward in real time, and contrast with the memories of mythical events that pull the characters back into an archaic past.

The opening lines of *Phèdre* place us in a logical, linear action, anchored in a historical and political reality. Thésée, the king, is mysteriously missing. After having struggled to institute a political order, the king is not triumphantly seated on his throne, taking care of business. Where is he? Hippolyte wants to find him, but his confidant, Théramène, seeks to dissuade him, recounting that he had already searched for him in vain. The king play soon shifts abruptly to a mythical level that is not guided by reason since his confidant reveals he had searched in the underworld: "J'ai demandé Thésée aux peuples de ces bords/ Où l'on voit l'Achéron se perdre chez les morts" (lines 11–12).[42] We learn later that Théramène's instincts were correct since that is precisely where he was— in the underworld. Not only that: Thésée went there voluntarily. And he returned alive to tell the tale, describing how he had descended into "des cavernes sombres" near "l'empire des ombres" (965–967). Trapped there for six months, Thésée finally escaped and subsequently returned to the present world of the play.

Thésée's absence was structurally necessary for the play.[43] However, the motivation for that absence did not need to be conveyed through a myth. Some of his contemporaries criticized his play for its recourse to mythological explanations, which offended the ideal of *le vraisemblable* and the era's

---

42    Henceforth, all references to the play give its line numbers.

43    Thésée's supposed death is structurally necessary because it permits two characters to reveal their transgressive passions: Phèdre expresses her forbidden love to Hippolyte and in a parallel move, Hippolyte confesses his forbidden love to Aricie. These confessions set the plot in motion.

rationality.[44] To highlight this point, it is revealing to note that Racine's contemporary, Gabriel Gilbert, who wrote a version of the Phèdre story in 1647, changed his plot so that Thésée's absence was explained in rational terms: he was away in a war. And in Mathieu Bidar's 1675 version of this story Thésée never left his throne or the city.

Racine's use of the underworld, a far-fetched mythological explanation, is more significant than most scholars have realized. It should not be rationalized away through a poetic license, accepting the poet's claim that he was simply imitating his source in Plutarch;[45] nor should it be normalized as a form of *vraisemblance*, as some scholars, such as Delmas, have attempted to do. Thésée's descent into a mythologized world had a sharp edge that cuts to the quick of the play. This king's decision to leave city walls, *the polis*, to travel to Hades takes on a greater weight once we remember that the Greek world was a highly polarized space, starkly opposing the civilized order of the *polis* to the wildness of the outside. As Anthony Pagden has described it, to be truly human and civilized was only possible inside the city, since the forces of the city were what helped civilize the wilder side of the human.[46] The *polis* was the sole place of human flourishing. Outside its walls was a space of the subhuman, a less evolved species, the barbarian who is prone to animal-like behavior and passions. The Greek world had highly polarized inside/outside boundaries; but Racine's representation of boundaries made that opposition yet even more rigid. Most characters in Racine's tragic universe were not free to cross the boundary lines between the inside and outside world, as Roland Barthes has noted.[47] They were enclosed within a claustrophobic, insular space. Thésée's leave-taking and return are striking exceptions because he had founded the city as a place of civilization and left it for its antithesis: the primal chaos of an underworld.

By opening his play with a reference to Thésée's mythological descent into the underworld, Racine is establishing a frame in which his characters regress

---

44   Recall that the anonymous critic of the play, *Sur les Tragédies de Phèdre et d'Hippolyte*, referred to earlier, was offended by Thésée's mythological descent into hell. For an extended discussion of Racine's *Phèdre* in relation to the other contemporary versions of the play, see Allen Wood, *Le Mythe de Phèdre*. See also Orlando, *A Freudian Interpretation of Literature*.

45   In his preface, Racine, recognizing that Thésée's descent into the underworld was unbelievable, justified his choice by stating that he simply followed the story as Plutarch had presented it.

46   Anthony Pagden, *Lords of All The World: Ideologies of Empire in Spain, Britain and France c. 1500–1800* (New Haven, 1995), pp. 17–24.

47   Roland Barthes, *Sur Racine* (Seuil, 1967).

to an archaic world, showing that even the founder of civilization is prey to a backsliding on the evolutionary continuum. Thésée's motives for leaving the *polis* were neither political nor even noble and they conjured up another strange mythological story. Thésée left to serve as a willing accomplice to his friend's "imprudente flamme," helping his friend "ravir la femme" of the "tyran d'Epire" (957–958). The verb "ravir" belongs to the polite language that the French Academy legislated, so it is difficult to know what action is really designated. Scholars have not focused on this issue, at least not to my knowledge, so it is difficult to know how they would weigh in on this question. But when I informally suggested to several French scholars that it could refer to a rape, I met with a great deal of resistance and repugnance. They prefer to keep the ambiguity of the word so that it never falls into the cruelty of such a brutal sexual violation. However, "ravir" does include the meaning of rape, as Furetière's seventeenth-century *Dictionnaire universel* defined it, viewing the words "ravisseurs" and "violateurs" as synonymous:

> VIOLATEUR: qui viole. Les ravisseurs et violateurs des femmes ou filles, ceux qui en abusent par force, sont punis de mort.

Furetière defined "ravir" as "emportement de quelque chose violemment. Ce jeune homme a ravi l'honneur à cette fille." It refers particularly to "des personnes qu'on enlève pour les captiver, ou en abuser." To interpret "ravir" as rape is consonant with the portrait that the play's other characters paint of Thésée. For example, when Théramène mentioned Thésée's absence, he assumed that this king was in hiding, "cachant de nouvelles amours,/Ce héros n'attend point qu'une amante abusée..." (20–21). Théramène provided more details of Thésée's abuse of women:

> Sa foi partout offerte et reçue en cent lieux;
> Hélène à ses parents dans Sparte dérobée;
> Salamine témoin des pleurs de Péribée;
> Tant d'autres, dont les noms lui sont même échappées.
> Trop crédules esprits que sa flamme a trompés:
> Ariane aux rochers contant ses injustices,
> Phèdre enlevée sous de meilleurs auspices (84–90).

This confidant interpreted the mystery surrounding Thésée's absence as an indication that this king, as founder of civilization, has himself regressed into the primitive disorder he had previously fought against by instituting the social and political order.

Furthermore, Racine's mythic story highlights Thésée's backsliding into the primitive, archaic past by conjuring up another form of barbaric violence in the underworld: cannibalism. When Thésée returned from the underworld, he described the activities he encountered there: "Le tyran m'a surpris sans défense et sans armes./J'ai vu Pirithous, triste, objet de mes larmes,/Livré par ce barbare à des monstres cruels./*Qu'il nourissait du sang* des malheureux mortels" (962–4, emphasis mine). Detailing the tyran's atrocities in this underworld, Thésée then observed: "D'un perfide ennemi j'ai purgé la nature;/*A ses monstres lui-même a servi de pâture;*" (969–970, emphasis mine). Clearly, the underworld that this new king chose over the civilized *polis* was a place where human bodies were feasted upon. In short, the wild and scandalous irrationality of Thésée's descent into a mythological world underlines his dual nature. His soul was not ultimately so removed from the Minotaur monster he had killed. Placed at a different stage on the same continuum, this founder of civilization could slip back down to the bottom to join—or become—this monster, if he were not restrained.

Like Thésée, Phèdre lives in the ordered, political world of Ancient Greece and is its Queen. And like Thésée, Phèdre is also pulled back on an evolutionary continuum, descending into the archaic wildness of her past, a mythological underworld that is psychic rather than physical. Her descent into a mythological underworld occurs at two key moments that push the dramatic action forward in time. Those key moments are two confession scenes when Phèdre reveals her incestuous love first to her nurse/confidante OEnone and then directly to Hippolyte himself.[48] This pull backwards is what moves her story forward in dramatic clock time.

The first plot device that advances her story occurs when we first meet her and she wants to die in order to keep secret her incestuous love for her step-son Hippolyte. But she reveals her secret to her nurse OEnone when she lets slip a small segment of her archaic past: "O haine de Vénus! O fatale colère!/ Dans quels égarements l'amour jeta ma mère!" (249–250). These words are highly packed, yet OEnone and the reader can easily unpack them since they connect Phèdre's archaic past to the mythic story of Venus who, as discussed earlier, sought vengeance upon Apollo, by punishing Phèdre's mother, Pasiphae, with lust for a bull. Venus' hatred haunts the play, dramatizing that the gods themselves were primitive, barbaric creatures, just as Fontenelle had described them.[49] This pull back into Phèdre's mythological past is the first major point

---

48   My analysis of Phèdre's confession scenes owes much to Orlando's analysis in *Toward a Freudian Theory of Literature*, and to Eric L. Gans' *Le Paradoxe de Phèdre* (Paris, 1975).

49   Melzer, *Myths of Mixture*.

in the plot that pushes the action forward. When she hears Phèdre's rambling words, OEnone deduces that Phèdre, like her mother, harbored an illicit love. But more importantly, OEnone then subsequently sets the play's action in motion by suggesting to Phèdre, after learning the false news of Thésée's death, that her love is now legitimate. OEnone plants the seed of doom in this eponymous heroine by encouraging her to reveal her passion directly to Hippolyte himself. Taking OEnone's advice, Phèdre is drawn yet further back in her mythological past, pushing the action even further forward when she confesses her love to Hippolyte himself.

Phèdre frames her confession by retelling another part of the same mythological story in a hallucinatory moment. She begins by recounting the standard version of the narrative about the founding act of civilization in which Thésée slew the Minotaur by entering the labyrinth in Crete where this monster was lodged. Ariane played a key role by providing a thread to guide his way out of this maze. Killing this monster was a foundational moment marking the progression from barbarism to civilization.

Phèdre retells this same story; however, her version is about a regression from civilization to barbarism. In so doing, she conveys her love to Hippolyte. Her confession begins with the narrative's socially sanctioned frame. She expresses love for her husband, emphasizing his heroism; he is "tel qu'on dépeint nos Dieux," these words evoking the story of the monster-slaying and establishing the civilizing values of the community.[50] Her regressive version is structured by a hallucinatory series of substitutions: first she replaces her husband with Hippolyte, imagining that her step-son played the same role as Thésée did in that same story (645–649). In her perverse telling of the story, Hippolyte embodies the values that had founded the civilized order. Then, in a second substitution, she displaces her sister Ariadne; Phèdre herself then holds the thread outside the labyrinth to help Hippolyte find the exit. Then the thread suddenly seems insufficient; her own person becomes the thread itself (657). Then she reverses the direction of their journey: she seeks not the exit but the maze's center. The labyrinth's dark, hidden recesses are the imagined place for their rendez-vous. In an unstated, substitutive move, her incestuous love replaces the monster in the maze. In short, Phèdre rewrites the heroic mythic story of civilization's founding as a regression away from civilization. In so doing, she pushes the dramatic action forward by revealing her forbidden passion for her step-son.

Phèdre's regression on the evolutionary continuum echoes that of her mother, Pasiphae, who had crossed the boundaries of the civilized world to

---

50    Eric Gans has a very similar analysis of this passage in *Le Paradoxe de Phèdre*.

become sexually intimate with an animal. The mythic stories of Phèdre, Pasiphae, and Thésée, then, are all similar in that they each slip back to a more primitive stage of existence. In all cases, the mythological versions of the characters, as opposed to their historical counterparts, are the ones who conjure up that backsliding.

An archaic, mythological past also surfaces at other key plot junctures to push the action forward, creating a strange circularity. One of the most decisive moments occurs when Thésée, confused and upset by the false accusation that Hippolyte tried to seduce his wife, seeks vengeance upon his son. Rather than provide a logical, rational way to accomplish his goal, he calls upon a mythological god, Neptune, who causes a mythological sea monster to arise, transmuting itself out of the ocean's foam to form a bull (and harking back to the bull who was father to the Minotaur). This monster then frightens Hippolytus' horses, causing him to die in a particularly violent way. Racine's invocation of Neptune and other mythological figures in this play are not ornamental but provide structural mechanisms to push the action forward in real dramatic time. But this structure is circular since the action is propelled forward to the future by circling back to the past via myth.

•••

How then to understand Racine's puzzling use of myth? Some scholars such as Mitchel Greenberg have explained it as a representation of the psychological notion of an unconscious.[51] Other scholars have provided religious or social paradigms to account for this oddity. However, any single explanation is partial since we need a fuller understanding. I am suggesting an additional paradigm in which Racine was caught in between warring concepts of time's trajectory, which were at the root of the Quarrel between the Ancients and the Moderns.

Some scholars situate Racine within a modern camp, based on his concept of time. Timothy Reiss, for example, attributes the modernity of *Phèdre*, in part, to the play's ending in which a completely new and more rational order emerges. Aricie represents that new order, one that negates the past and transcends it by installing a more rational social and political structure. The play's last two lines highlight a sudden change: Aricie, who had once been shunned as an enemy of the state, is adopted as Thésée's daughter, incorporated into the center of a new order. Phèdre and Hippolyte, both of whom were attached to the past, die, whereas Aricie lives and represents the future.

---

51    Mitchell Greenberg, *From Ancient Myth to Tragic Modernity* (Minneapolis, 2010); Orlando, *Towards a Freudian Theory of Literature*.

The underlying assumption is that time is slanted so that progress is possible, with the future representing a possible advance over the past.

Reiss links Aricie to the modern, moreover, because Racine liberates her from the reign of myth. Reiss sees Phèdre, Hippolyte and Thésée as "representatives of myth"; they are all well-known figures of Greek mythology, tied to stories that were told so often that their future was known before the story begins. By contrast, Aricie is free of this legacy. She has no mythological heritage and is thus emancipated from the weight of this kind of past.[52] As a purely invented character, Aricie is liberated from any backstory in Euripides or Seneca. Furthermore, Racine separates Aricie yet further from the realm of myth by highlighting that mythological explanations for real world events have no sway over her. Ruled by reason, she rejects as absurd the reports that Thésée had returned from the underworld; she is astonished that other people could possibly believe in such explanations (380–91). For Reiss, then, Aricie's triumph at the end, which "deliberately negate[s] *all* the 'past' of (and in) the play" is a triumph of a new modern order in which reason replaces myth.[53] In sum, Reiss links Racine's modernity to a new view of both time and myth: the play progresses towards a view of time that allows for the concept of progress in which two of the characters who are "representatives of myth" die and the one character who is free of myth triumphs in the end.

I do not wish to dispute this interpretation of Racine's modernity, but rather to question its weight. I contend that it has to be counterbalanced with an opposing view: Racine's understanding of time and myth are also consonant with the "ancients." This opposing view should be given equal weight since both are partially true and partially false, since Racine was trapped between both. In deciding what weight to attribute to the play's end, it is important to note that Aricie is a very minor figure, with the least amount of stage presence and given by far the fewest lines. It is too facile to assume that the play's last two lines erase the power of all that happened in the preceding 1650 lines and also transcend it. The fact that the new order is indicated via the two final lines is not sufficient to counteract the weight of the entire play, which is mired in an archaic order.

I am proposing then that Racine's use of myth is a site of conflict between two warring notions of time. As modern readers ourselves, we understand best the modern concept of time and focus on that since we are more accustomed to it. However, the dominant understanding of Racine's era was more closely linked to the "ancients'" view, which had the notion of a backsliding built into

---

52    Reiss, "From *Phèdre* to History," pp. 272–280.
53    Ibid., p. 264 (his emphasis).

it since the concept of progress had not yet been established. But more to the point, the entire play, with its striking use of myth, dramatizes the negative effects of being trapped in a degenerative notion of time, in which history was slanted downwards towards decline, making the notion of "progress" difficult to fully conceive. Yes, the play's end does negate this, but how seriously should we take it? How fully installed is this modern view in relation to the "ancient" view that dominates throughout the entire play? It would seem that both exist in a continuing tension.

The virtue of reading Racine in tandem with Fontenelle is that it illuminates how myth, for this playwright, is a site of conflict, revealing the tension between the Ancients and the Moderns. Like Fontenelle, Racine inserted myth within a historical frame that included a primitive, barbaric stage on an evolutionary continuum leading to civilization. However Racine understood that continuum quite differently since Fontenelle was a "modern," whereas Racine harbored a conflicted stance. For Fontenelle, the evolutionary continuum represented a positive, liberating frame of thought; it enabled him to see Greek myths as archaic, analogous to the myths that the primitive peoples of the New World were inventing. He thus rejected the frame of thought that attached their myths to an elevated notion of civilization. By reframing the mythic as archaic, and casting the Greeks as cultural infants, Fontenelle saw the French as adolescents, representing an advance over the past. In this way, he was able to break decisively with the Ancients by reversing the slope of time and history to suggest that the future could improve upon the past, enabling the French elite not only to equal the Ancients but even surpass them.

Racine, however, was both an "ancient" and a "modern" and thus saw myth as a curious mixture of the "high" and the "low," existing on both ends of the evolutionary continuum. On the high end of the continuum, myth reveals an important component of Greco-Roman thought, creating an elevated style and carrying great prestige because of its connection to the venerated traditions of Ancient Greece and Rome. Viewed in this light, myth can be interpreted as existing outside of time, able to communicate transcendent truth, an important element of civilization itself. These truths can be interpreted to support Racine as either an "ancient" or a "modern," depending on how one reads them.

On the lower end of the continuum, myth was archaic and represented a threat to civilization for Racine. He found this perspective troubling because the "ancient" belief in a degenerative notion of time meant that the French could never equal the Greco-Romans. But as troubling as it was, Racine had internalized this paradigm so fully that his characters were trapped within its mentality, despite their struggles to escape it. His characters, although they act and speak in a highly civilized manner on stage through perfectly crafted

alexandrine verse, all remain vulnerable to a backsliding, falling prey to the
barbarian monster within that comes out of their past. Just as Thésée had left
his throne at the center of civilization and descended back into the underworld,
all the play's characters were continually prey to similar regressive movements.
The "ancients'" degenerative notion of time is what made Racine's characters
all the more vulnerable to this regressive pull back in time. Racine's use of
myth thus illuminates his fears about being trapped in degraded temporal
scheme in which the past weighed so heavily on the present that all escape was
virtually impossible.

As a "modern," Racine wanted to believe that the future could break from
the past; he wanted to reject the "ancients'" view of time as degenerative.
Certainly Aricie in *Phèdre* gestures in this direction, as does Astynax in
*Andromaque*.[54] But these are simply gestures and do not have great solidity,
given that in the rest of the play, the characters are powerless to break with
their past. Racine, his drama, and the belief systems were thus torn between the
two competing views of time that were at the heart of the Quarrel between the
Ancients and the Moderns. Fortunately for France, Fontenelle and the "moderns" won the battle and reversed the slant of time. Fontenelle and others
understood that the Ancients/"ancients" view of time was itself a myth that
they sought to debunk.

---

54     Leo Bersani, *A Future for Astyanax: Character and Desire in Literature* (Boston, 1976).

# CHAPTER 12

# Time and Space as Manipulated Materials in Rameau's *Les Cyclopes*

*Mark Howard*

Time and space are integral to the experience of listening to music. In tangible sound, time as intangible substance is expressed in the forms of measure, rhythm, duration, and repetition. Space is also unseen substance and is defined through blocks or *chunks* of time with prescribed limits. Thus these two properties are materials that the composer may shape and manipulate in several, creative ways. Time and space ultimately produce *modulation* or movement over the course of a piece of music. They govern the listener's experience in elemental ways and cooperate with musical components such as the fundamental bass, a compositional tool Jean-Philippe Rameau (1683–1764) devised in order to understand harmonic particularities, relationships, and eventually modulation or movement. This order of harmonic concepts was established in his earliest published treatise *Traité de l'harmonie* (1722) and again used in the late stages of his writing career in *Code de musique pratique* (1760). It is an order rooted in the medieval educational system, which continued through the eighteenth century.

Rameau published his second book of harpsichord pieces two years after *Traité de l'harmonie* in 1724. This collection featured *Les Cyclopes*, a violent, virtuosic piece whose figures, conceptual design, and emotional substance were largely unprecedented. The subject of the work is centered on the race of one-eyed giants as portrayed in Homer's *Odyssey*. Rameau himself expressly cites *Les Cyclopes* as a piece of characterization in a letter written to Houdar de la Motte (1672–1731), a librettist at the *Académie française*, on October 25, 1727. Rameau writes: "If you will but come and hear how I have characterized the song and dance of the *Sauvages* who appeared in the Italian Theater one or two years ago and how I carried out the following titles, *les Soupirs*, *les Tendres Plaintes*, *les Cyclopes*, *les Tourbillons*...you would then see that I am not a novice in the art and that above all it does not seem that I show a great display of science in my productions, where I endeavor to conceal art within itself...."[1] Rameau's distaste for showmanship and art for art's sake may be gleaned from

---

1  Quoted in Michaela Maria Keane, *The Theoretical Writings of Jean-Philippe Rameau* (Ph.D. dissertation, The Catholic University of America, 1961), p. 76.

© KONINKLIJKE BRILL NV, LEIDEN, 2016 | DOI 10.1163/9789004312319_014

these statements. Earlier in the letter, he trivializes composers who so obsess themselves with mere "combinations of notes" that they sacrifice "common-sense, emotion, feeling, reasoning."[2] Rameau's ultimate goal in characterizing the Cyclops is that emotional aspects of the Cyclops and features of the story are eventually represented and reflected in his "combination of notes."

The original myth in Homer's *Odyssey* tells of an episode during Ulysses' journey in which he and his seafaring crew encounter a race of one-eyed giants called the Cyclops. Polyphemus, a brash, brutish, irascible, and violent, monster, kills some of Ulysses' crew, eats them, and imprisons Ulysses within his cave. Polyphemus asks Ulysses for his name, which he states, in Greek, as "Nobody." Polyphemus tells Ulysses that he plans to eat him as well soon. Over the course of the evening, Polyphemus drinks himself to sleep, during which Ulysses grabs a spear and impales Polyphemus' single eye. Polyphemus' compatriots investigate the source of Polyphemus' screaming rage, asking, "Who has hurt you?" to which Polyphemus replies, "Nobody has hurt me." During all of the commotion, Ulysses manages to evade and escape his captor. If Ulysses tried to take on Polyphemus in a physical fight, he would surely have quickly and embarrassingly lost, and perished like his crew. Ulysses, however, uses something far superior to brute strength to beat the Cyclops: his wisdom, his ability to strategize. These were qualities that Polyphemus undervalued. He also disrespected lawfulness and the gods, particularly Zeus, who Ulysses attempts to use as a reason to be set free. The Cyclops is a monster with which one cannot reason; wisdom and rationality were useless. It is a story of David versus Goliath.

Musicologists have failed to note the origin of the subject of this work. Girdlestone simply states that he knows of no other work with this title, while mentioning that *Persée* (1682), a *tragédie mise en musique* by Jean-Baptiste Lully (1632–1687), which contains a Cyclops character wholly unlike that portrayed in Rameau's work, was revived around the time that Rameau published this book of pieces.[3] The attitude that musical works could only be inspired by other musical works does Rameau's piece a grave disservice by suppressing its meaning and inspiration, a meaning that held considerable weight in eighteenth-century French culture and that, due to its Homeric origins, relates to profound themes found in both art and literature.

The seen and unseen materials of time and space, with which Rameau works, are inextricably woven with the subject and meaning of *Les Cyclopes*. This essay examines how these two properties aid Rameau in realizing them

---

2   Ibid., p. 75.

3   Cuthbert Girdlestone, *Jean-Philippe Rameau: His Life and Work* (New York, 1969), pp. 25, 597.

and even creating meaning from them on macro and micro levels throughout the composition. We begin by examining how Rameau appropriates and repurposes the past for his present, analyzing the *rondeau* as cyclical time and space, and finish with the manipulation of musical time; all in order to demonstrate how greatly the materials of time and space impact the composition and its meaning.

## Rameau Translates the Past into His Present

One of the most observable and useful aspects of ancient writing is how authors/narrators, such as Homer and Plato, use stories to illustrate important points about their subjects, stories that encourage the reader to reflect. The parables of Jesus Christ are one such example of this practice. Rather than directly answer someone seeking his advice, Jesus prefers to tell stories that elucidate his points. The story in question presumably occurred in the past, a crucial component whose essence is only significant to the teller as it relates to the present and then the future. A transaction takes place, *modus* or movement, that moves forward from teller to listener/reader, from past to present. Understanding of the point on the part of the listener is also a way of moving or *modus*. In this sense, the stories, as they relate to past, present, and future, are temporal substance, material with which the author/teller may work in order to make his/her points clear. The past as related in stories also brings with it an assumed authenticity on the part of the listener/reader, an authenticity that is strengthened through a relationship with the point that the author is trying to make.

Jean-Philippe Rameau utilizes the past as temporal substance in both his musical compositions and theoretical writings. For example, in his treatises the *Code de musique pratique* and *Nouvelles réflexions sur le principe sonore* (1760), there are several references to past theorists, artists, and mythological and biblical figures who are used to validate specific points.[4] Some of these figures include the seventeenth-century playwright Molière (1622–1673), the sixteenth-century music theorist Gioseffo Zarlino (1517–1590), the mythological King Midas, and even Adam, the first man God created according to the Old Testament book of Genesis. These figures collectively become one, grounded on the same principles within the context of Rameau's writing. They individually translate the past to Rameau's present or the time of his reader;

---

4   See Mark Howard, "Rameau's *Code de musique pratique* and *Nouvelles réflexions sur le principe sonore* (1760): A Translation with Commentary" (Ph.D. dissertation, Claremont Graduate University, 2011) and Mark Howard, *Decoding Rameau: Music as the Sovereign Science* (Lucca, 2016).

that time, now 250 years later reveals how Rameau's examples take on a time-less quality. The *Code de musique pratique* as a "Code" or "body of laws," as some early, eighteenth-century reviews deemed it, sought to communicate this identity through argumentative structures and carefully selected words exhibiting its judicial inspiration. The particularities of the language used, ideas, and modes of thought specific to eighteenth-century French culture, and in particular to Rameau were the channels or *figures* that best repurposed past material into useable material for their present.

Thirty-eight years before the *Code*, Rameau incorporated the past, for example, using quotations from Zarlino's *Le istitutioni harmoniche* (1558) in the first chapter of Book Two of *Traité de l'harmonie*, to deepen the significance of the role of the bass in music. Rameau's first major contribution to music theory was what he deemed the fundamental bass that, simply put, is a series of bass tones as a foundation for chords, resulting in a harmonic progression. Rameau writes:

"As the part containing the fundamental sound is always the lowest and deepest, we call it the *bass*. Here is what Zarlino says on this subject:

> Just as the earth is the foundation for the other elements, so does the bass have the property of sustaining, establishing, and strengthening the other parts. It is thus taken as the basis and foundation of harmony and is called the bass—the basis and support, so to speak.

After imagining how, if the earth were to disappear, all the beautiful order of nature would fall into ruin, Zarlino says:

> In the same way, if the bass were to disappear, the whole piece of music would be filled with dissonance and confusion. [...] Thus, when composing a bass, the composer should make it proceed by movements which are rather slower and more separated, i.e. more spread out, than those of the other parts. In this way the other parts can proceed in conjunct motion, especially the treble, whose property it is to move in this manner, etc.

But if we contrast this clear and accurate definition of the fundamental part of harmony with the rules and examples given by this author, we find everywhere contradictions which leave us in doubt and uncertainty."[5]

---

5  Jean-Philippe Rameau, *Treatise on Harmony*, trans. Philip Gossett (New York, 1971), p. 59.

As ironic as Rameau eventually makes it seem, Zarlino's ideas about the bass are as chaotic and confusing as a nature without order. Nonetheless, Rameau first uses Zarlino as a historically recognized authority to shape a revered attitude of the bass in his reader's mind. Over the course of the first five chapters or so, Rameau, however, attempts to outline what he perceived were Zarlino's shortcomings. Whether writing approvingly or critically about Zarlino, Rameau used the past, that is, Zarlino's writings and reputation, to clarify and strengthen his own points.

As an alternate case in point, and one whose subject matter is more specific to the present discussion, 124 years after Rameau published *Code de musique pratique*, Mark Twain (1835–1910) published his American classic, *The Adventures of Huckleberry Finn* (1884). Chapters 5 through 7 are a masterful retelling of the Cyclops myth. Huck's father is an overbearing, pushy, abusive drunk, who has a disdain for school and religion; all qualities that the Cyclops Polyphemus possesses in Homer's *Odyssey* and that Rameau's harpsichord piece *Les Cyclopes* illustrates and actually disparages through its figures. Huck's father, like Polyphemus, threatens to kill Finn in the morning after a night of heavy drinking. The following day, Huck is forced to strategize, just like Ulysses, and hatch an elaborate plan of faking his own death in order to escape imprisonment from a secluded cabin in the forest where Huck's father tries to shield him from society and its government, religion, and education. The Cyclops myth is translated from the past to Twain's present through the setting and language of the American South, a context alien to Rameau's culture. Twain and Rameau, however, share how the past, that is, myth, is used as temporal material in order to illustrate that wisdom (or education) and the ability to use it to overcome problems far outweigh limitations in physical strength.

Just as in *Traité de l'harmonie*, *Code de musique pratique*, and *Huckleberry Finn*, musical compositions may translate meaning from the past as unseen substance via sound, manifested in various ways that become relevant to its listener. For example, unlike its use by Mark Twain, which was set in nineteenth-century America, Rameau uses *Les Cyclopes*, recognizable as Homer's figure, with its main character not being a human appearance of everything the Cyclops represents but rather an actual Cyclops. Myths from both Homer and Plato were embedded in eighteenth-century French culture and inspired countless works of poetry, music, and visual art. *Les Cyclopes* was also composed for harpsichord, one of the primary keyboard instruments of Rameau's time, and includes *batteries* and ornaments, which are melodic figures idiomatic to that instrument. Ultimately the translation of the past into the present provides Rameau the opportunity to manipulate time on a local level and even stop it as a meaningful gesture in its own right.

## Rameau's *Rondeau* in *Les Cyclopes* as Cyclical Space/Time

The *rondeau* was used in several different genres of eighteenth-century music and signified an opening section that alternated with contrasting sections. The *rondeau* played a similar role in poetry, where a section of verse was repeated throughout recitation. Its use as a device in Baroque poetry is perhaps older than in music. Seventeenth- and eighteenth-century dictionaries in particular pay strong focus on its textual content. Sébastian de Brossard (1655–1730) does not include a definition of *rondeau* in the third edition of his *Dictionnaire de musique* (1708). However, *rondeau* received a musical definition in the *Dictionnaire de l'Académie française* (1762): "*Rondeau* is also called a piece of instrumental Music, in which the first section is repeated after each of the others. *A gavotte, a sarabande in rondeau.*"[6] A musical definition is absent from the 1694 edition of the *Dictionnaire de l'Académie française*. The *rondeau* certainly existed around the time of Brossard's dictionary, since Rameau used the *rondeau* pattern in the *gavote* from his 1706 collection of harpsichord pieces. From a simple survey of eighteenth-century harpsichord music, the *rondeau* was used as a structural pattern quite frequently, especially in descriptive pieces from the second quarter of the century following the publication of Rameau's *Les Cyclopes* in 1724. The *rondeau* pattern's cyclical nature is its most distinctive feature, which, in the case of music, gives the composer a malleable state of space and time. According to Karol Berger, time prior to approximately the middle-to-late eighteenth century was viewed to progress cyclically as opposed to teleologically.[7] Berger uses the "da Capo" aria in works such as Johann Sebastian Bach's *St. Matthew Passion* (1729) as well as additional examples, to show how returns to individual sections reflected this view of time and in particular its religious connotations.

The *rondeau* as cyclical pattern presents a wide array of possibilities with respect to the various content and priorities of eighteenth-century composers. The sections of a *rondeau* varied in length, though many pieces, also by Rameau, are divided equally. *Les Cyclopes* returns to its original *rondeau* twice and to an abbreviated version once between the first and second *reprises*. The space and time within a *rondeau* section become fixed when limited to a certain length and only repeated as such. The only variance in that state results from how contrasting sections affect the nature of repeated material. *Les*

---

6   Robert Morrissey, ed. s.v. "rondeau," *Dictionnaires d'autrefois*, ARTFL *Project*, http://artflsrv02
    .uchicago.edu/cgi-bin/dicos/pubdico1look.pl?strippedhw=rondeau&headword=&docyear
    =ALL&dicoid=ALL (November 12, 2013).

7   Karol Berger, *Bach's Cycle, Mozart's Arrow* (Berkeley, 2007).

*Cyclopes* is wholly different in this regard. Rameau disrupts time and space by abbreviating the middle return of the *rondeau*. In this way, the violent content of the opening of the piece is greatly diminished by content that is ephemeral, transitory. From a practical standpoint, the second *reprise* is much more akin in character and material to the opening *rondeau* than the first *reprise*. As expressed in his treatises, especially *Code de musique pratique*, achieving variety in composition was a lifelong priority for Rameau; and repeating all the material from the *rondeau* followed with similar content might have proved too monotonous.[8]

Of wider significance is how Rameau's abridgment of the first repeat relates to the first *reprise* with respect to the Cyclops myth. As discussed below, Rameau manipulates time by stopping it altogether with a full measure of rest. This moment simulates the hesitation Ulysses experiences before resolutely piercing the Cyclops' lone eye, defying all the reprehensible and dangerous qualities that the Cyclops possesses with the single downward thrust of a spear. Fig. 12.1, taken from the 1724 edition, shows the end of the first *reprise* and return to the *rondeau*.[9]

Compared to its counterpart, the abridged version of the *rondeau* is weakened by the material representing humanity's pursuit of justice and religious order, which are qualities that the Cyclops, and his Twainian equivalent, lack. The important point here is that Rameau reshaped the temporal and spatial properties of the cyclical *rondeau* in order to achieve this desired meaning. The material that follows the first *reprise*, though certainly related to the *rondeau*, receives new significance because of its relationship to what precedes it. This is an example of what in medieval writings is referred to as *figurae in modis* or figures within motion. The appearance of new figures throughout a composition completely changes its complexion and gives it new direction. In the case of Rameau's *rondeau* in *Les Cyclopes*, the contrasting first *reprise* colors the first repeat of the *rondeau*. The Cyclops' violent nature only gradually regains its full momentum with the return of the original *rondeau*. Rameau continues to make the manipulation of temporal and spatial substance the primary mode for expressing his content. It is a process only realized in time.

A truly meaningful manifestation of Rameau's use of the cyclical properties of the *rondeau* is realized in an unseen *figura* distinctive to the subject of this piece, the eye. Ironically, it is through sight (that is, the eye) that one obtains knowledge or unseen substance, as Aristotle stated in the opening to his

---

8  Variety as a concept is also introduced in *Génération harmonique* (1737) but discussed to a much less degree than in *Code de musique pratique*.

9  Jean-Philippe Rameau, *Pièces de clavessin*, Performers' Facsimiles 156 (New York, [s.d.]), p. 31.

*Metaphysics.*[10] Described in the entry on "eye" in the *Dictionnaire de l'Académie française* is a person with "good eyes," that is, someone who "has the ability to see into (*penetrate*) affairs and not be mistaken..."[11] This definition, among others in the entry, concentrates on the unseen substance behind visual observation. In *Les Cyclopes*, Rameau beckons the listener to find this *figura* through the process of time and space itself. She/he must become part of the movement or *modus* that occurs from the act of listening and gain insight into the emotional substance so visceral to the subject of the Cyclops. Fig. 12.2 shows how *Les Cyclopes* progresses in order to delineate the figure of a single eye within the mind's eye of the listener. The opening *rondeau* represents the white of the eye (mm. 1–51). Beginning at the top of the iris, the first *reprise* (mm. 52–84) only proceeds a half-circle, due to its short duration, in order to reach the abridged *rondeau* at measure 85. Since the middle *rondeau* acts as a proxy for the original *rondeau*, it likewise travels full circle to reach its starting point, from which the second *reprise* (mm. 97–123) follows half-circle. We now connect to the beginning of the original *rondeau*, which also goes full circle for a final time, outlining the figure of an eye. Through this temporal/spatial process, Rameau cleverly imitates how an eye physically moves when focusing on, or peering into, something, seeking to reveal unseen or intangible substances (for example, emotions, knowledge) to its beholder.

### Rameau's *Batteries* and the Partitioning of Musical Time

Rameau considered *Les Cyclopes* a crowning achievement in the composition of his harpsichord music. In the preface that accompanies his *Pièces de clavecin en concerts* (1741), Rameau instructs the reader that, should she/he need to learn how any of the ornaments should be executed, just consult the book with the Cyclops.[12] This seemingly innocuous statement reflects the widespread reputation of this very unique, and downright ferocious, piece of harpsichord music during the first quarter of the eighteenth century. With *Les Cyclopes*, Rameau was able to take a culturally significant and recognizable story and recast it in such a way that its musical representation became synonymous with that story itself.

---

10    Aristotle, *The Metaphysics*, trans. John McMahon (Amherst, NY, 1991), p. 11.

11    Robert Morrissey, ed. s.v. "œil," *Dictionnaires d'autrefois*, ARTFL Project, http://artflsrv02
      .uchicago.edu/cgi-bin/dicos/pubdico1look.pl?strippedhw=oeil&headword=&docyear=A
      LL&dicoid=ALL (November 13, 2013).

12    Jean-Philippe Rameau, *Pièces de clavecin en concerts* (Paris, 1741), preface, "*Avis pour le clavecin.*"

*Batteries*, which Rameau claims are completely new and of his own devising in the *Méthode pour la méchanique des doigts* that prefaces the collection containing *Les Cyclopes*, present the composer with an opportunity to work with new musical material substance or delineatory *figures* that enhance his subject (that is, the inspiration behind the *dessein* or design) and reshape how music is articulated. Design is neatly defined in Book Three, chapter forty-four of *Traité de l'harmonie* as a "general term encompassing everything we put forth, that is: movement, key and mode, melody, and harmony suitable to the subject, all of which a skillful musician will envisage from the start."[13] Therefore, the measure (that is, time) becomes a material substance that may be divided and manipulated in several ways, all for the purpose of a composer's *dessein*. The French word *temps* is defined in the *Dictionnaire de l'Académie française* (1694) as "a measurement of movement."[14] The use of "measurement" in this context reflects its purpose as useable material, something inherent to music itself. Rameau began his compositional career manipulating musical space and time in the unmeasured *Prélude* from his first book of harpsichord pieces.[15] Once Rameau conceived *Les Cyclopes*, its *batteries* became a new mode of rearticulating rhythm and dividing musical time for his contemporaries, who proceeded to imitate him in this dialogue with time as useable material.

Rameau describes two types of *batteries* in his harpsichord primer. The first, as Rameau writes, imitates the sound of two drumsticks beating a drum. Fig. 12.3, drawn from *Les Cyclopes*, depicts this first type with its quick, repetitive striking of the same note divided between the performer's left and right hands. The second type, shown in Fig. 12.4, occurs when the left hand crosses over the right in order to play notes above where the right hand is playing.

Generically speaking, the two *batteries* are simply identified as two distinct types. Their practical application, however, reveals that Rameau was able to employ them constantly in a variety of ways by altering register, note placement, figure, key, harmony, and hand movement. His division of the measure into eight eighth-notes remained virtually the same throughout the piece.

---

13    Rameau, *Treatise*, p. 348.

14    Robert Morrissey, ed. s.v. "temps," *Dictionnaires d'autrefois*, ARTFL *Project*, http://artflsrv02
      .uchicago.edu/cgi-bin/dicos/pubdico1look.pl?strippedhw=temps (November 8, 2013).

15    This piece features two sections, unmeasured and measured, respectively. The former
      type of section is mostly uncontrolled by the limits of time, while the second is so much
      so that it is almost entirely based on a single rhythmic idea. The unmeasured prelude had
      its roots in the mid-seventeenth century. Louis Couperin, one of the earliest composers of
      unmeasured preludes, indicated the metered section of his G-minor prelude with the
      words "change du movement," which signified a fundamental change in how motion or
      time progressed (that is, from un-metrically to metrically).

Their realization in sound, however, is different from section to section and from measure to measure. Fig. 12.3 shows immediately how Rameau alters the effect of the two drumsticks by placing the second of each two-note repetition on the strong beat of the measure. Should the listener focus attention on two-note groups consisting of the same note, the effect is vastly different than if one simply focuses on the rhythm as Rameau published it. This balanced unevenness is resolved when the *batterie* reaches its high point (m. 7). The performer may introduce this effect by slightly holding the first quarter note that commences the *batterie* and allowing the eighth-notes in the right hand to be audibly paired with their counterparts in the left hand.

Fig. 12.4, similar to Fig. 12.3, demonstrates how Rameau uses hand crossing to punctuate the lowest notes in the passage (*F-G-E-F-D-E-C♯*) all while maintaining an eighth-note division of the beat. The remaining notes are accompanimental. Variances in notes and register aid Rameau in keeping the passage musically interesting.

In addition to the examples illustrated in Fig. 12.3 and Fig. 12.4, Fig. 12.5 and Fig. 12.6 show yet other *batteries*. The *batterie* is located in the left hand of Fig. 12.5, where the index finger crosses over the thumb in order to play the third note in each group of four eighth-notes. Just as in Fig. 12.4, Rameau divides the measure into eight eighth-notes, yet he achieves a different result because the *batterie* only occurs in one hand. Over the course of this passage, it should also be noted that the bass in the left hand changes against a repeat of the melody in the right hand, coloring the second statement of the melody with a different harmony. The texture of Fig. 12.4 is certainly thinner than that of Fig. 12.5, in which the melody is placed in the right hand against the accompaniment of the left. This latter *batterie* continues for much of the remainder of the *rondeau* (m. 51).

The *batteries* in Fig. 12.6 are similar to those in Fig. 12.4. The left hand crosses over the right to either play a note higher than the one in the right hand or the same note. Notably different is the manner by which Rameau punctuates only the first eighth-note of each measure rather than the first and fifth. Like all its counterparts, the division of the measure is the same, yet this accent on the first note yields a passage distinct from its predecessor. This divisional commonality among all four examples exemplifies how Rameau was able to maintain continuous movement from a single division of the measure, while varying musical elements such as the texture, articulation, pitch, register, and harmony, giving the work as a whole an erratic, chaotic, quality and testing the technical capabilities of the performer.

The Cyclops as described by Homer is violent, irascible, unlawful, arrogant, and blasphemous. The fast, perpetual, movement and fluctuating manipulations of time and texture shape a composition that perfectly characterizes the Cyclops, who personifies everything eighteenth-century audiences disparaged.

As we shall see in the following section, the continuous eighth-note division of the measure reinforces the most dramatic moment in the story of the Cyclops, the stabbing of Polyphemus' eye, a moment where Rameau stops time as a symbolic confrontation with the reprehensible qualities of the Cyclops.

### Rameau Stops Time

As much as Rameau's *batteries* express the Cyclops' violent nature, lawlessness, and religious scorn, of greater significance is Rameau's expression for their opposites, a gesture that is perfectly communicated by stopping time altogether. The definitive, courageous, moment in the story of Homer's Cyclops is Ulysses' penetration of the Cyclops' lone eye; a symbolization of brains over brawn, and order (that is, lawfulness and religion) over chaos (that is, lawlessness and contempt for religion, or the gods). These distasteful qualities were used by Madame Anne Dacier (d. 1720) to describe the Cyclops in her French translation of the Odyssey in 1716.[16] Ulysses, the Cyclops' captive, pleads his release by citing the gods. In Dacier's version, the Cyclops responds:

> Foreigner, you lack good sense, or you come from good care, you who exhorts me to respect the gods and to have humanity. Knowing that the Cyclops do not care about Jupiter nor the other gods; because we are stronger and more powerful than them...I would have compassion for you and your Companions, if my heart itself felt pity.[17]

The Cyclops' contempt for the gods is clearly evident in this passage, and Ulysses' spear symbolizes everything this monster is not, that is, human, pious, and lawful. In order to depict this scene musically, Rameau brings his *batteries* high up the keyboard before making a dramatic leap down on to a chord on C in measure 79 and then stopping time with a full measure of rest in measure 80 (Fig. 12.7).

This suspension of time thus signifies a moment of silent eternity, the opposite of time as measurement of movement, which is defined as such in the *Dictionnaire de l'Académie française* of 1694.[18] Modulation (that is, appropriate movement) or *modus* ceases. Fundamental bass movement—in this case C to

---

16   Madame Anne Dacier, *L'Odyssée d'Homere, traduite en françois, avec rémarques* (Paris, 1716). This translation was reprinted in 1731 and 1756.

17   Homer, *L'Odyssée*, trans. Dacier, 2: 280.

18   Robert Morrissey, ed. s.v. "temps," *Dictionnaires d'autrefois*, ARTFL *Project*, http://artflsrvo2 .uchicago.edu/cgi-bin/dicos/pubdico1look.pl?strippedhw=temps (November 9, 2013).

*F*—is momentarily suspended, interrupted. The listener experiences what she/he believes could be a spiritual connection with God. Music's disciplinal role as making plain relationships in which the nature of God was understood was systematized within the medieval educational system, still in place during the eighteenth century.[19] "Eternité" or "eternity" as a concept was associated with God and the afterlife. Definitions of "eternité" included "God is for all eternity" and "a happy or unhappy state, where souls eternally exist after death."[20] Time is immaterial to that state.

Time ironically was, however, material in music, and thus ultimately limited by its own figures or *figurae*. The whole-rest could only last as long as indicated, barring a fermata, and its relationship with the surrounding music (that is, the music in measured time) defines it as time that is essentially timeless.[21] Therefore, measure 80 is literally measured but figuratively unmeasured from the perspective of the listener. This type of transaction, in which visible *figurae* represented non-tangible concepts or experiences, exemplified the allegorical mode of learning in the medieval education system.[22] Ulysses's encounter with the Cyclops Polyphemus was an ideal story for Rameau to illustrate musically because of how this monstrous figure represented the ugliness of humanity. As Charles Dill has shown, eighteenth-century music critics viewed monstrosity, due to its irrationality, as a quality completely alien to nature.[23] Nature, or *natura*, was held as the supreme example of order and rationality. Rameau sought, as Christensen writes, to explain the natural, mechanistic, operations behind music's processes.[24] Rameau's *batteries*, the major type of *figurae* driving *Les Cyclopes* forward, thus become colored as monstrous themselves through their representations of Cyclopean qualities. Ironic, however, is that

---

19  Nancy van Deusen, *The Cultural Context of Medieval Music* (Santa Barbara, 2011).

20  Robert Morrissey, ed. s.v. "eternité," *Dictionnaires d'autrefois*, ARTFL Project, http://artflsrvo2.uchicago.edu/cgi-bin/dicos/pubdico1look.pl?strippedhw=eternite (November 9, 2013).

21  Rameau also effectively uses the stopping of time in his *tragédie en musique*, *Hippolyte et Aricie*, Act II, Scene 4 (*"Puisque Pluton est inflexible..."*), where Theseus, Hippolyte's father, beckons Neptune, the god of the sea, to help him convince Pluto to let him return to earth from the underworld. The music is dark and ethereal, moving as Theseus sings each line and stopping with dead silence at each line's end, as if Theseus is waiting for Neptune's responses. Time seems to stand still at these moments, increasing the scene's drama with every passing moment.

22  Nancy van Deusen, *The Cultural Context*, pp. 10–11.

23  Charles Dill, *Monstrous Opera: Rameau and the Tragic Tradition* (Princeton, 1998), p. 13.

24  Thomas Christensen, *Rameau and Musical Thought in the Enlightenment* (Cambridge, UK, 1993), p. 1.

all of the components of a musical work (for example, melody, harmony, figure, fundamental bass, key, mode, and so on) that Rameau uses to depict irrationality were stated by Rameau in his writings to originate fundamentally
from within the rational order of nature itself.

### Conclusion

We have seen how Rameau shaped and manipulated the material of time (and
space) on macro and micro levels throughout his harpsichord piece, *Les Cyclopes*.
The significance of the Cyclops myth for eighteenth-century audiences and its
culture was profoundly transformed through the composition's temporal elements, which included the myth itself, the *rondeau, batteries*, and time's inter-
ruption. Of deeper significance is how these features of the work interacted with
other musical components, such as the fundamental bass, key, harmony, register,
instrument, and ornament. For example, time and motion are inherent to the
movement of the fundamental bass, a fundamental bass whose harmonic progression is dictated and ultimately realized by the rhythm of its notes or motion.
Stopping time with absolute silence not only directly affected the piece's forward
momentum but also consequently halted the fundamental bass. The drive to create meaning through figures and gestures is manifested in limitless ways. Rameau
so powerfully and uniquely represents the Cyclops that, against the backdrop of
eighteenth-century harpsichord pieces, in which composers sought to imitate
their subjects as deeply as Rameau had done, the popularity and significance of
its musical manifestation within *Les Cyclopes* cannot be overstated.

FIGURE 12.1     Les Cyclopes (*1724*)

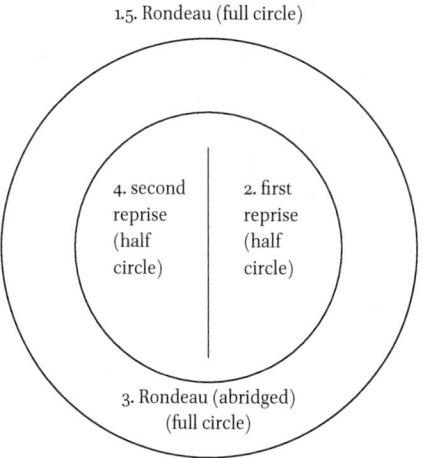

1.5. Rondeau (full circle)

4. second reprise (half circle)

2. first reprise (half circle)

3. Rondeau (abridged) (full circle)

FIGURE 12.2
*Eye* figura *as shaped by the time and space of the* rondeau

FIGURE 12.3    Les Cyclopes, *mm. 5–8*

FIGURE 12.4    Les Cyclopes, *mm. 16–20*

FIGURE 12.5    Les Cyclopes, *mm. 29–32*

FIGURE 12.6      Les Cyclopes, *mm. 97–103*

FIGURE 12.7      Les Cyclopes, *mm. 74–81*

CHAPTER 13

# Resisting Death's Finality: Jules Dalou's *Blanqui* Tomb and the Dialectics of Memorialization

*Andrew Eschelbacher*

In the ninety-first division of Paris's Père Lachaise cemetery, the tomb of Auguste Blanqui shivers with an undercurrent of angst-riddled vigor. A bronze funerary shroud covers the sculpted body of the controversial revolutionary, responding with folds and wrinkles to the angles and undulations of Blanqui's pained frame. His lean right arm, vitalized by its throbbing musculature, emerges from the cloth, extending beyond the space defined by the sculpture's simple rectangular plinth. Blanqui's head falls jarringly backward and to the right. An engorged vein swells in the center of the revolutionary's forehead. His seemingly animated body and psyche complicate his presumably posthumous state.

Upon seeing this tomb monument at the Salon of 1885, the critic Gustave Geffroy proclaimed that the bronze effigy represented "what art should be, a commentary on History that speaks above man and events."[1] For Geffroy, the tomb, sculpted by Jules Dalou, transcends narrative details to embody the spirit of the century—France's continuous revolutionary pursuit of an ideal republic. In the next decade, Geffroy would draw inspiration from the image of struggle that he found coursing through Dalou's sculpture. It served as a source for his *L'Enfermé*, a biography of Blanqui that became the most influential text in popularizing the transhistorical portrait of the revolutionary as a Promethean hero.

Both at Père Lachaise and in the Salon, the *Blanqui* tomb participated in fin-de-siècle France's culture of memorialization. The nation mediated its visions of the present and the future through objects and physical spaces that teemed with revisionist, or at least reductive, historical value. Much has been made about the phenomenon of *Statuemania* during this period, an era between 1870 and 1914 when private and government committees erected over 150 public statues in Paris—more than six-times the number of monuments the city saw installed during the Reformation (1815–1830), the July Monarchy (1830–1848), the Second Republic (1848–1851), and the Second Empire (1851–1870) combined.[2] The period's didactic sculptural interventions presented social

---

1   Gustave Geffroy, *La Justice*, 9 May 1885.
2   June Hargrove, *Statues of Paris: An Open Air Pantheon* (Antwerp, 1989), p. 105.

© KONINKLIJKE BRILL NV, LEIDEN, 2016 | DOI 10.1163/9789004312319_015

and political visions of France's history as a set of finished experiences, embodying fixed values that espoused monolithic, though often contested, national narratives. Tomb sculptures contributed significantly to this commemorative culture, as they had in previous generations, playing an important role in establishing martyrs whose deaths could be exemplars of political and moral virtue as well as serving as loci for rites and rituals of national commemoration.[3]

Dalou's *Blanqui* tomb engages this monumental culture in the muddled spaces where memorialization, as an active, culturally shared, phenomenon, inhibits historicization.[4] Similar conflicts often arose when edifying memorials of competing ideologies engaged one another in the commemorative landscape, or when historicizing sculpture obfuscated contentious experiences embedded within the physical geography of an installation site.[5] But in funerary monuments of the nineteenth century, and the vast majority of the period's public sculpture more broadly, Dalou's *Blanqui* is remarkable in that it actively embraces the unsettled memories of the past. Rejecting the temporal fixity and boundedness of its funerary and ideological contexts, the Blanqui engages the ambiguities within French experience, challenging linear constructions of national histories as well as the roles of time and permanence in memorial culture.

• • •

The impetus for the *Blanqui* tomb surfaced just weeks after the revolutionary's death when his followers began plans to erect a memorial that could be a site of future pilgrimage.[6] Dalou received the commission for the monument, which was to be funded by national subscription, a year later. When fundraising efforts failed to raise enough money to fully finance the sculpture, Dalou

---

3  Antoinette le Normand-Romain, "En hommage aux opposants politiques. Monument funéraire ou public?" *Revue de l'Art* 94/94 (1991), 74–80; Suzanne Glover Lindsay, *Funerary Arts and Tomb Cult—Living with the Dead in France, 1750–1870* (Farnham, UK / Burlington, vt, 2012).

4  Pierre Nora, "Between Memory and History: *Les Lieux de Memoire*," *Representations* 26, Special Issue: Memory and Counter-Memory (Spring 1989), 8–9.

5  The consideration of memory sites has dominated many scholarly writing on public monuments in the last twenty years. Beyond June Hargrove's essay in Nora's original *Les Lieux de Memoire*, *La République*, scholars such as Neil McWilliam, Kirk Savage, and Janice Best have expanded on Nora's theories and enriched contemporary understandings of how monuments function as part of an interconnected and historically rich urban landscape.

6  Brunet letter published in *Ni Dieu Ni Maître* (January 23, 1881).

offered his labors to the committee without hope of proper remuneration.[7] By 1883 the sculptor had finished a model for his statue—not a bust as originally planned, but a full-length recumbent portrait.[8] Dalou, and sympathetic contemporaries, presented this change as the generosity of an artist who felt duty-bound to participate in the social battles of his era.[9]

Questions about the relationship between the nation's history and its future plagued the early-1880s climate in which Dalou worked. The structural and ideological shifts of the 1860s and 1870s produced a French culture where past experiences no longer served as an appropriate guide to predict the future.[10] The Paris Commune of 1871 had not initiated this condition, but as its revolutionary pressures weighed on shifting social, geographical, industrial, and class systems, it accelerated its advance. The ensuing culture of confusion, where the underlying factors of French social instability became indecipherable, engendered profound fear about the causes and repercussions of the massive civil bloodshed.[11]

In the Commune's wake, the national governments, first the archconservative Government of Moral Order and later the moderate, republican Opportunist regime, as well as France's bourgeoisie created a discourse that allowed the nation to cope with the horror and instability the revolt produced. Marginalizing the political concerns and divergent attitudes of Communards, the dominant milieu vilified the uprising as a product of boorish and wanton lawlessness perpetrated by a working class replete with criminals and devoid of morals.[12] These attitudes, and the legislative action that emerged from them, advanced an existing social dichotomy rooted in a split between the bourgeoisie and the proletariat, and separated the 1871 uprising from the revered revolutionary tradition that stemmed from 1789.[13] At the same time, socialist, anarchist, and

---

7   Le Normand-Romain, "En hommage aux opposants politiques," 78.

8   Winant, *Le Républicain Socialiste de Centre*, (October 14, 1883).

9   Geffroy, "Dalou," *La vie artistique* (Paris, 1903), p. 321.

10  On the disconnect between "spaces of experience" and "horizons of expectations," see Reinhart Koselleck, "'Space of Experience' and 'Horizon of Expectation': Two Historical Categories," in *Futures Past: On the Semantics of Historical Time*, trans. Keith Tribe (New York, 2004), 255–275.

11  Peter Starr, *Commemorating Trauma: The Paris Commune and its Cultural Aftermath* (New York, 2006), p. 3.

12  Paul Lidsky, *Ecrivains conte la Commune*, (Paris, 2010), pp. 97–115; Richard D.E. Burton, *Blood in the City: Violence and Revelation in Paris 1789–1945* (Ithaca / London, 2001), p. 143.

13  Emile Carron, "Rapport Supplémentaire fait au nom de la commission chargée d'examiner la proposition de loi de M. de Pressensé," *Annexe au process-verbal de la séance du 20 mai 1874. Assemblé Nationale, Année 1874* N° 2392; "The Reactionaries: The Commune as Social

communist commentators structured an oppositional narrative, a counter-discourse, that complimented the bourgeois view that the Commune was a departure from France's traditions of revolution. Such accounts celebrated the uprising, and what some considered to be a brief "dictatorship of the proletariat," as a new prism through which to consider modern social and political engagement.[14]

When the Opportunist regime took power at the end of the 1870s, it attempted to actively reclaim and reframe the revolutionary tradition in order to help legitimize its new power. These efforts included the establishment of July 14—the anniversary of the storming of the Bastille (1789) and the first *fête de la fédération* (1790)—as the *fête nationale*. The moderate republicans also advanced the cult of Voltaire, a symbol of the universal republic, over that of the people's hero Jean-Jacques Rousseau, whom some associated with the *sans-culottes* and the Terror as forerunners of the Commune.[15]

Similar projects took visual form in the Parisian landscape. In 1883 the capital's Municipal Council, as part of an effort to reestablish links with the national government, installed a monument at Place de la République that included a parade of reliefs retelling a moderate, bourgeois version of revolutionary history. At the base of the formally staid ensemble, capped by an antiquated image of Marianne, the series of reliefs begins with the Tennis Court Oath of June 20, 1789 and culminates with the July 14, 1880 establishment of the Fête nationale. Along the way, the sculptors Leopold and Charles Morice highlighted a number of specific dates from a conservative narrative, ignoring potentially controversial instances of revolt that could complicate the path towards the Opportunist hegemony. For instance, the popular beginnings and street barricades of the 1830 revolution are not pictured, but the subsequent *Orleanist* ascension of those July days is prominent; the revolutionary events of 1848 are conspicuously absent, but the abolition of slavery in March of that year earns a choice spot, and naturally, the Commune is nowhere to be found.

Plague," in *The Paris Commune in French Politics, 1871–1880: The History of the Amnesty of 1880, I The Partial Amnesty*, ed. Jeane T. Joughin (Baltimore, 1955), pp. 67–114; Martin R. Waldman, "The Revolutionary as Criminal in 19th Century France: A Study of the Communards and 'Deportes,'" *Science & Society* 37/1 (Spring 1973), 31–55.

14   Karl Marx, "Adresse du Conseil général de l'Association internationale des travailleurs (30 mai 1871)," in *La Guerre civile en France 1871* (Paris, 1963), pp. 41–89; Frederick Engels, "Introduction," *Der Bürgerkrieg in Frankreich* (Gloke, 1891), np.

15   Michel Delon, "1878: un centenaire ou deux ?," *Annales historiques de la Révolution française*, 234 (1978), 643–647; Georges Benrekassa, *et al.*, "Le premier centenaire de la mort de Rousseau et de Voltaire: significations d'une commémoration," *Revue d'Histoire littéraire de la France*, 79:2/3, Voltaire, Rousseau, 1778–1978 (Mar.–Jun. 1979), 265–295.

Such revisionist and reductive narratives fit comfortably with neither Dalou's personal biography nor Blanqui's evolving posthumous legacy. At the age of nine, Dalou is reputed to have climbed the workers' barricades of the 1848 revolution.[16] Twenty-three years later, he was an active participant in the Commune, serving as an adjunct curator of the Louvre museum and as a member of the Artists' Federation. Following the revolt's bloody end, he fled to London where he remained in exile for the duration of the decade. Yet Dalou's vision of the Commune was far from the version that reactionary authors and politicians espoused. The sculptor consistently described his actions with an echo of 1848, claiming that he fought to establish an order that would "give bread, concord, and work to all, while beginning a new era of faith and love, joy and enthusiasm."[17] The contemporary author Paul Cornu concurred with Dalou's self-appraisal, writing that the sculptor was never a "communard honnteux," in a text that celebrated Dalou's unrepentant attitude and republican convictions.[18]

In the 1880s, Blanqui's reputation and posthumous position also slipped into the fissures in national narratives. In the first half of the century, the romantic-socialist revolutionary had risen to prominence as a leader of a series of insurgencies aimed at overthrowing governments and shepherding in a republic. Always willing to consider violent strategies to accelerate revolutionary progress, Blanqui made himself a constant target of government prosecution and incarceration, spending more than thirty-three years in a series of French prisons. By the 1860s, a new generation of insurgents revered him as *Le Vieux* and venerated him for his devotion to the cause.[19]

On March 17, 1871 Adolphe Thiers's provisional government jailed Blanqui for his participation in a Parisian uprising of late October 1870. The timing of this imprisonment, just a day before the unofficial beginning of the Commune, meant that Blanqui could not actively participate in the spring revolt. Nevertheless, Communards continued to cite him as an example of revolutionary will.[20] They believed he was so crucial to their cause that they offered to trade all of their hostages, including the Archbishop of Paris, in return for his release. The Versailles government concurred about Blanqui's

16    Henry Roujon "Preface," in Maurice Dreyfous, *Dalou: Sa vie et son œuvre* (Paris, 1903), ii.

17    Dalou to Cornaglia, 25 February 1879, Louvre Manuscrits, 321.

18    Paul Cornu, "Jules Dalou," *Portraits d'hier* (Paris, 1909), p. 228.

19    Patrick Hutton, *The Cult of the Revolutionary Tradition: The Blanquists in French Politics, 1864–1893* (Berkeley / Los Angeles / London, 1981), p. 22; Maurice Paz, *Un révolutionnaire professionnel: Auguste Blanqui* (Paris, 1984), p. 133.

20    Hutton, *The Cult of the Revolutionary Tradition,* p. 86.

importance and, fearing that his organizational skills and ability to direct the insurgents would greatly boost their military prospects, denied the exchange.[21]

A year after the Commune, a new government commission sentenced Blanqui to life in prison for inciting civil war.[22] Despite the intended length of this punishment, by the end of the decade a number of socialist groups began to place Blanqui on election ballots for a spot in the Chamber of Deputies. Their hope was that in electing Blanqui to national office they could force the government's hand and lead it to release *Le Vieux* from prison.[23] The power of his name made such goals too modest. Inside the first campaign, one for a legislative seat from Marseilles, the issues at stake magnified and engaged existential questions about the Republic's character. In this contentious period of the late 1870s, *Le Vieux* found himself at the center of the national movement for Communard amnesty.[24]

Competing 1880s efforts to create politically tinged French histories complicated Blanqui's posthumous legacy. While some maintained that he represented the most sinister element of France's revolutionary cult, after his death journals sympathetic to the government, no longer concerned that he could undermine the regime, praised his character and resolve in an effort to appropriate his revolutionary fortitude.[25] Certain prominent radicals of the 1870s and 1880s believed that had Blanqui participated in the Commune, he would have been a moderate among the Parisian revolutionaries. Henri Rochefort, for one, suggested that Blanqui's influence would have prevented many of the Communards' acts of atrocity.[26] At the same time, as Blanqui's followers splintered into diverse republican and radical groups, new sects of extremists invoked *Le Vieux*'s name to pursue violent class struggle without basing their social agitation in Blanqui's political ideologies.[27]

Amnestied, Dalou returned to France in 1880. In the years immediately following his repatriation, he sought to structure a new national paradigm in sculptures such as *Fraternité* (1879–1883) and *Triumph of the Republic* (1879–1899) that would reconsider the Commune and its connection to the revolutionary tradition. He began the *Blanqui* project with similar aims, in the

---

21    Max Nomad, *Apostles of Revolution* (London, 1939), p. 68.

22    Joughin, *The Paris Commune*, p. 161.

23    The efforts to force amnesty through the republican value of universal suffrage dovetails with Georges Clemenceau's rhetoric of 1879.

24    Joughin, *The Paris Commune*, pp. 161–169; *L'égalité* 19 Mai 1878.

25    Maurice Dommanget, *Auguste Blanqui au début de la IIIᵉ République (1871–1880): Dernière prison et ultimes combats* (Paris, 1971), p. 143.

26    Henri Rochefort and Ernest W. Smith, *The Adventures of My Life* (London, 1896), p. 433.

27    Samuel Bernstein, *Auguste Blanqui and the Art of Insurrection* (London, 1971), p. 356.

moment when the revolutionary's legacy was in flux and intersecting with constructions of French history and its relation to Commune memory. John Hunisak suggests that the importance with which the sculptor regarded the commemoration of Blanqui and his historical position led Dalou to show such generosity in modeling a monument that far exceeded the budgetary restrictions and committee plans.[28]

• • •

Dalou's sculpture calls on the republican conventions of modern French funerary traditions, specifically in its recumbent composition, to position Blanqui as a revolutionary martyr. The artist surely found inspiration in Alphonse Dumilatre's tomb of the balloonists Joseph Eustache Crocé-Spinelli and Théodore Sivel.[29] This well-known group, which commemorates aeronauts who in 1875 plummeted to their deaths in pursuit of scientific achievement, won a first class medal at the Salon of 1878. Despite the clear similarities between Dalou's monument and this prize-winning ensemble, critics largely ignored the formal connections between them. Instead they almost universally stressed the relationship between the *Blanqui* and François Rude's renowned 1847 tomb sculpture of mid-century journalist and leading republican reformer Godefroy Cavaignac.[30]

Rude, who in this instance also donated his efforts to the republican cause, presented an inert Cavaignac on top of an elevated and modestly ornamented rectangular plinth with an effect, as Linda Nochlin argues, that conveys "the finality of death as an ultimate fact."[31] Rude partially covered the journalist with a death cloth but stimulated the fabric's folds, exposing Cavaignac's naked chest and enlivening the effigy with a romantic verve that expressed the tomb's vitality. The calmness of the journalist's face proves a contrast from the active drapery, and its idyllic qualities, with limited indications of age and veins, reveal a historicized departure from the blemishes and bags that mark Cavaignac's death mask.[32] Critics pointed to Rude's treatment of the author's

---

28    Special thanks again to John Hunisak for sharing with me his unpublished manuscript on
       Dalou. In his third chapter he describes the Blanqui monument and commission.
29    John Hunisak, *The Sculptor Jules Dalou: Studies in His Style and Imagery* (New York /
       London, 1977), p. 127.
30    On Cavaignac's biography, see Lindsay, *Funerary Arts*, pp. 178–181.
31    Linda Nochlin, *Realism* (London, 1990), p. 65.
32    Lindsay, *Funerary Arts*, pp. 195–196.

visage an expression of Cavaignac's ever-present thoughtfulness.[33] At the journalist's side, a pen and a sword comprise the allegorical attributes that contextualize his modern crusade and, though he died of tuberculosis, define
Cavaignac as a revolutionary martyr of the republican cult.

Rude's sculpture recalled the tradition of French *gisant* monuments common to royal mausoleums at Saint Denis. In updating the mode of commemoration, Rude presented Cavaignac free from architectonic trappings and
auxiliary figures that create physical and spiritual distance between the viewer
and the deceased. Suzanne Lindsey notes that this composition startled viewers who were not accustomed to experiencing funerary sculptures with such
visual immediacy. She argues that the site of installation at the cemetery of
Montmartre, not in a burial chapel, only amplified this effect, helping the
monument take its place as the founding stone of a democratic martyrium.[34]

Aimé Millet's 1872 tomb of Alphonse Baudin builds from Rude's precedent,
adding journalistic qualities to the new democratic *gisant* tradition. In
December 1851 troops loyal to Napoleon III shot Baudin, a representative in
the National Assembly, as he stood on a people's barricade in Paris. Two
decades later, Millet represented the fallen republican clothed as he had been
when he died on the capital's streets. His head turns slightly to the right, with
a bullet hole marking his brow. Despite the evidence of violence, the counte-
nance and bodily composition are serene. Baudin's lifeless right arm drapes
over a stele inscribed with "La Loi." Additional inscriptions provide further edifying context, clarifying the credentials for the assemblyman's martyrdom.

When Dalou showed the *Blanqui* in the Salon of 1885 critics were almost
universal in indentifying Dalou's manipulation of the modern *gisant* tradition, specifically pointing out similarities between the *Blanqui* and the
*Cavaignac*. The right-wing author Olivier Merson was among several scribes
who censured Dalou for a lack of originality in drawing on such a clearly
established precedent.[35] Other critics took a different view, understanding the
formal quotation to elevate Blanqui's tomb and its sculptor to the heights of
the *Cavaignac* and Rude, respectively.[36]

---

33   Philibert Audebrand, "Godefroy Cavaignac, romancier," *La Nouvelle Revue* 1 (October
     1882), 656.
34   S. Lindsay, "Rude's Cavaignac Tomb: The Symbolism of Cast Bronze in a Modern
     Democratic Martyrium," *La Sculpture au XIX^e Siècle, Mélanges pour Anne Pingeot*, (Paris,
     2008), pp. 116–7. For a more thorough analysis see the introduction and chap. 6 in Lindsay's
     *Funerary arts and tomb cult.*
35   Olivier Merson, "Salon de 1885," *Le Monde Illustré*, July 11, 1885.
36   Judith Gautier, "Salon de 1885," *Le Rappel*, May 5, 1885.

While the *Blanqui* manifests intentional compositional echoes of the *Cavaignac* and *Baudin*, its straining muscles and psychological angst offer a far different construction of death than its predecessors. Whereas the faces of the *Cavaignac* and the *Baudin* are calm and their musculatures are languid, the *Blanqui*'s visage is agitated, its muscles strained and bulging. The mannered drapery of the *Cavaignac* adds vitality to the portrait of the journalist, but in the *Blanqui* the contours of the revolutionary's body animate the shroud's folds and creases. In opposition to Baudin's arm, which falls limply on the stele, Blanqui's right hand reaches out, and in modern times often holds a rose or a carnation, attempting to grasp at life, or perhaps a viewer's hand.

With its active right arm and the signs of flowing blood, the *Blanqui* has much in common with Auguste Bartholdi's 1872 tomb of National Guardsmen in the cemetery of Colmar, a work outside the *gisant* tradition. In the monument commemorating the fortitude of French troops who strive against death, hoping in vain to pick up their swords and wage one more fight, Bartholdi imaged a vigorous right arm emerging between two stone sheets that cap the grave. The arm's muscles clench and its fingers dig into the surface. Blood swells in the veins from the effort of resisting the finality of the tomb.

Bartholdi's monument employs the anonymous roused arm to communicate ideas of national valor, accelerating a *Revanchist* ideology about the continuation of the fight. The *Blanqui*'s vibrancy similarly expresses the spirit of unfinished struggle, but because the arm and blood belong to a specific person, the tomb complicates ideas of individual death and temporal continuity. With its vigor and strain, the *Blanqui* attests to a type of spiritual and physiognomic struggle that still courses through the revolutionary. This palpable vibrancy subverts the expectation that the image of a dead man should create an inert sculptural environment that would promote reflection and contemplation. Instead, the appearance that Blanqui refuses his deceased state produces sublime exchanges where viewers negotiate their position *vis-à-vis* the tomb, both physically and psychologically, as they contend with the unbounded temporal representation that the sculptural object presents.

Since the end of the nineteenth century scholars have associated a similar set of Modernist sculptural concerns with Auguste Rodin.[37] Alex Potts notes that a number of Rodin's contemporaries believed that the artist's work "embod[ied] a distinctively modern psychological sense of becoming that dissolves fixity and

---

37    Alex Potts, *The Sculptural Imagination: Figurative, Modernist, Minimalist* (New Haven / London, 2000), pp. 60–101.

boundedness."[38] Rodin achieved these ends often through a manipulation of form that created a *musculature vivante* where the formal animation of muscle and sinew communicated psychological states. This is precisely how Dalou treated the body of Blanqui, but whereas Rodin's sculptures often took the modern condition and psychology as ends to themselves, Dalou's treatment of Blanqui engaged fractures within the construction of the past.

Reinhart Koselleck proves instructive in understanding how Dalou's unbounded and unfixed intervention into the memorial landscape functions as only a partial historical representation. He argues that historical representation depends on a combination of structures and events. He defines a structure as something that may be described to "illuminate long-term duration, stability and change." Events can be experienced or captured as a "discernable unity capable of narration."[39] Koselleck insists that presenting structures and events require different modes of communication, descriptive and narrative respectively, but it is impossible to achieve a complete historical representation without a melding of both components.

For the *Blanqui* tomb, the visual references to the traditions of the *Cavaignac* and *Baudin* mark its participation in the conventional French mode of describing martyrdom within the republican structure. At the same time, the effigy's refusal of death's finality resists seamless entry into this metanarrative. The disruption of expectations encourages a reconsideration of the idea of the republican-revolutionary tradition itself, an issue pertinent to Dalou's intervention into 1880s cultural discourse. In addition, the vitality of the portrait limits Dalou's historical representation because it leaves Blanqui's death the necessary "event," seemingly incomplete.

Dalou compounded such representational instabilities through his restrained allegorical and iconographic program. A single thorny crown rests at Blanqui's feet. Despite its ostensible symbolic power, the wreath exists in the blurred space between allegory and realism. The prominence of the brambles clearly alludes to a martyr's crown, compounding the Christly insinuations of the deposition pose and shroud. Dalou's Naturalist inclusion of precisely rendered leaves, however, fogs the singular interpretation by recalling the actual circumstances of Blanqui's 1881 funeral when wreathes overwhelmed his cortege on its way to Père Lachaise Cemetery. The monument's inscription also refuses to elucidate a specific commemorative message. It notes only that public subscription funded the project, providing a vague reference to the tomb's democratic origins.

---

38    Ibid., p. 76.
39    Koselleck, *Futures Past*, pp. 105–108.

The limited narrative details—no wounds, no edifying inscriptions, no precise allegorical messages—render Blanqui a martyr with an imprecise meaning, not simply as an icon of permanent struggle. This of lack clarity leaves Blanqui available for appropriation by almost any republican or radical ideology. Geffroy's *L'Enfermé* offers one reaction—the Promethean model of republican fortitude—but the author's view was far from universal in the 1880s when the tomb helped Blanqui achieve only a vague status as an imprecise symbol of the revolutionary tradition.[40] Nevertheless, Geffroy's reaction is instructive, offering an example of the active mode of viewing that the unfinished monument mandated.

As viewers engage the effigy, they become responsible for constructing the complete representation—the fusion of structure and event—by fixing a historical narrative onto the psychologically open tomb and structure. To do so, they must negotiate the monument's formal program with an individual matrix of memoires about the nation's past, the revolutionary tradition, as well as Blanqui and Dalou themselves. The shifting nature of Blanqui's legacy and the varied constructions of France's past ensured that interactions between the monument and its audience would never produce consistent meaning. While Monarchists, Bonapartists, and others entrenched on the French right would likely dismiss Blanqui as nothing more than a vile revolutionary, within the leftist milieu a broad spectrum of reactions was possible. Such subjectivity of experience created an environment where Blanqui could not be historicized, and instead the tomb environment unlocked a commemorative space that elicited a continuous memorial dialogue.[41]

• • •

The effigy's destined inauguration site, the trauma-laden terrain of Père Lachaise Cemetery, added a problematic layer of meaning to this active process of memorialization. For nearly the first four score of the nineteenth century, the cemetery existed outside of the socio-geographical bounds of France's dominant sphere. This cultural position was not simply dependent on the cemetery's location in the working-class sectors at the extreme of eastern Paris, but also due to the cemetery's function as what Danielle Tartakowsky calls "un pantheon de substitution" for nineteenth-century liberals.[42] The events of 1871

---

40    Hutton, *Cult of the Revolutionary Tradition*, pp. 24, 161.

41    On the difference between memory and history see Nora, "Between Memory and History."

42    Danielle Tartakowsky, *Nous irons chanter sur vos tombes: Le Père-Lachaise, XIXᵉ–XXᵉ siècle* (Paris, 1999), p. 15.

amplified the division between the dominant sphere and the cemetery. It gained widespread infamy—or notoriety depending on the perspective—as a site of violent conflict in the last days of the Commune's Bloody Week, and its eastern wall, where government soldiers summarily executed nearly 150 Communards, became a site where Commune memory solidified.

In the period immediately following the revolt, the Government of Moral Order accentuated the space as a site of contestation, erecting the double tomb of Claude-Martin Lecomte and Jacques-Léonard Clément-Thomas on the cemetery's ceremonial axis. The two generals, the Versailles government's first martyrs of the Commune, had been interred originally at the cemetery of Saint-Vincent à Montmartre. A year later the government translated their remains to the more contentious burial site when the reactionary National Assembly committed to installing a monument in their memory. Though the sculptor Léon Cugnot and his architect Coquart had originally planned a contemplative memorial, under the regime's direction the definitive tomb became an aggressive pronouncement of the victory of the forces of order over Commune anarchy.[43]

The political transition at the end of the decade, which saw the Opportunist Republic ascend to power, precipitated a deft political gesture that altered the nature of the cemetery's memorial environment. Beginning in 1880, the government unofficially sanctioned Père Lachaise as a site for the commemoration of the Commune.[44] It allowed the red flag to fly on certain days and even tolerated a certain amount of pilgrimage to the so-called Wall of the Federates. In relinquishing Père Lachaise to those who would celebrate the Commune and its violence—largely socialists, anarchists, and rebels—the government concentrated the memory of the uprising around an eastern Paris site that highlighted 1871 as a moment of bloodshed, radicalism, and failed revolt.[45] For anti-Communards, such a narrative, emanating from Eastern Paris, reinforced their vision of the insurgency as an expression of an uncivilized proletarian nature. But pilgrims used this memory as a reminder of their martyrs of 1871 and as a call for future revolts. These efforts teemed with the unspoken undertones, and actual manifestations, of renewed class violence, seemingly confirming

---

43    Le Normand-Romain identifies a sketch in the Maciet albums of the Bibliothèque de L'Union centrale des Arts décoratifs as Cugnot and Coquart's initial plan. See A. Le Normand-Romain, *Mémoire en Marbre. La sculpture funéraire en France 1804–1914* (Paris, 1995), p. 98 (fn 82).

44    Madeleine Rebérioux, "Le Mur des Fédérés: *Rouge 'sang craché*,'" in *Les Lieux de mémoire*, dir. Pière Nora, 1 La République (Paris, 1986), p. 626.

45    Betrand Tillier, *La Commune de Paris, Révolution sans images?* (Seyessel, 2004), p. 422.

the criminality and barbarism that the dominant sphere associated with 1871 and its followers.

In 1881 Blanqui's internment intersected with the development of such Commune narratives. The massive public outpouring that lined his funerary procession from the Boulevard des Italiens to Père Lachaise proved that Blanqui still enjoyed an immense popularity with Paris's working class. Despite his move towards a more moderate program in his last years, the day of his funeral saw the red flag of the Commune return to the streets of the capital for the first time since 1871.[46] In the cemetery itself a huge crowd massed to listen to former Communards such as Louise Michel eulogize the deceased.[47] Cries of *Vive la Commune* and *Vive la Révolution* created further unease as right wing journals suggested that the "most sinister survivors of the Commune" led the crowd.[48]

According to Blanqui historian Patrick Hutton, Dalou insisted on delaying the official installation of the monument at the cemetery in order to place it in the Salon of 1885.[49] It is impossible to determine the sculptor's exact motivation for exhibiting the work in this space of cultural meaning produced by the dominant, and largely anti-Communard, bourgeoisie who embodied the Opportunist Republic's ideals. That said, it is likely that the Dalou, always feeling unstable in his social position, was thinking of the audience that would see the sculpture and how that would impact his own renown. Implicit in this theory is that Dalou understood that the nation's cultural elite were less likely to venture to the proletarian quarters of eastern Paris to view the effigy in the oppressive memorial environment.

The intervention of the *Blanqui* into the Salon sphere is a marked departure from the tone of the 1885 exhibition. For instance, Henri Gervex, whom J-K Huysman called a leader of the "Bourgeois Modernist" style that dominated the Salon, exhibited *Une séance du jury de peinture au Salon des Artistes français*.[50] This painting showcases the Salon jury, a group of artists in the bourgeois uniform of frockcoat, top hat, and walking sticks, reviewing a Naturalist nude and a landscape. Gervex's canvas includes three workers in beige jumpsuits, jotting notes and moving canvases. Their marginality in the scene insists on their position as functionaries doing the bidding of the bourgeois gentlemen

---

46    Dommanget, *Blanqui: au début de la IIIe Republique*, p. 145.

47    Hutton, *Cult of the Revolutionary Tradition*, p. 122.

48    Dommanget, *Blanqui: au début de la IIIe Republique*, p. 151.

49    Dommanget, *Blanqui: au début de la IIIe Republique*, 146; Hutton, *Cult of the Revolutionary Tradition*, p. 123.

50    J-K Husymans, "Le Salon de 1885," *L'Evolution Sociale*, May 23, 1885.

whose decisions codify official taste and republican cultural expression. Like Père Lachaise, they remain removed from the dominant milieu, serving as exemplars of the pliant laborers whom the bourgeoisie celebrated as the antithesis of the Communards.

Despite the Salon's distance from the cemetery, both geographically and socially, viewers could not have separated the *Blanqui* effigy from its future space of installation. The Salon catalog noted that the monument was "Destined for his [Blanqui's] tomb at Père Lachaise cemetery," and reviewers continuously insisted on the monument's future location.[51] If these details were not enough to spark connections, three weeks into the national exhibition of 1885 a particularly violent standoff between government forces and Commune pilgrims at the cemetery covered newspaper front pages and made the active memory of the site more profound.[52]

Because of the tomb's embedded connection with Père Lachaise, the effigy within the Salon environment transfers the site of Commune memory from the city's margins to its cultural center. It does so, however, through a sculptural vessel that recalls republican visual structures and refuses overtly radical narratives. These collisions offer the possibility of reconsidering the Commune outside of the marginalizing memories active at the eastern Paris burial ground, pushing viewers to contend with the revolt's messy link to a broader national history.

A similar disruption of memorial expectations occurred when Dalou finally installed the effigy at Père Lachaise in Auguste 1885. The sculpture's ambiguities and association with Rude's *Cavaignac* negate obvious narrative links between the tomb and the increasingly militant and extremist tone of the memorial environment. Moreover, because the work had been vetted, and often celebrated, within the republican Salon, the effigy could rekindle the cemetery's traditional role as a pantheon for liberal heroes. It existed in the blurred space between Commune discourse and counter-discourse, forcing a reconsideration of these structural ideas and their relationship to broader French experience.

• • •

Traditionally in Western culture, monuments exist as vessels into which memory is divested, fixing history as a finished entity.[53] Because of this sense of

---

51    Joseph Noulens, *Artistes français et étrangers au Salon, 1885–1886* (Paris, 1885–7), p. 210; Alfred Lanson, "Salon de 1885," *L'Artiste* 1 (1885), 324.

52    *L'univers illustré*, (January 15, 1881), 37; Rebérioux, "Le Mur des Fédérés," 626.

53    James E. Young, *The Texture of Memory: Holocaust Memorials and Meaning* (New Haven / London, 1993), pp. 1–7; Nora, "History and Memory," pp. 7–9.

permanence, monuments express a quality of timelessness and cultural stability that often obfuscates the shifting socio-historic dynamics that active memorialization reveals. As many of these projects exist outside the viewer's temporal environment—as indeed the funerary tombs of *Cavaignac* and *Baudin* and the Morices' group at République function—a number of critics from Nietzsche on have questioned the viability of the monumental mode for a modern era that Charles Baudelaire defined as ephemeral, fugitive, and contingent.[54]

The *Blanqui* operates somewhat apart from this tradition. Its insistence on an unfixed sculptural environment actively cultivates a sense of inconsistent meanings through memorial negotiation, renewal, and dialogue. The sculpture's unboundedness and engagement with its physical environments, and their connections to the traditions of funerary rites, Commune pilgrimages, and revisionist narratives, creates an environment where the past is undefined and refuses petrification. Viewers interact with the tomb at the intersection of multiple conflicting spaces of physical and social meaning, all crucial to contemporary notions of French identity.

In the context of the 1880s, Blanqui's bronze shiver encourages a reconsideration of the historical structures and narratives through which the dominant and counter-discourses had defined the Commune and the revolutionary tradition. It exposes the contested spaces where Frenchmen navigated their varied identities and pasts. Beyond these specifics, however, the intentional creation of a sculptural space where history is contested, and not where memory is divested, offers a new mode of monumentality rooted in an overtly dialectic process of memorialization. Here the monument does not exist as an object fixing time and history, but instead embodies the temporal and narrative ambiguities that public commemorations had traditionally sought to deny.

---

54    Friedrich Nietzsche, *The Use and Abuse of History*, trans. Adrian Collins (New York, 1957), pp. 14–17; Lewis Mumford, *The Culture of Cities* (New York, 1938), p. 438; Charles Baudelaire, *The Painter of Modern Life and Other Essays*, ed. and trans. Jonathan Mayne (London, 2006), p. 12.

CHAPTER 14

# End of Story: Closed Form and Open Time

*Hans J. Rindisbacher*

> What then is time? If no one asks me, I know: if I wish to explain it to one
> that asketh, I know not.
>
> AUGUSTINE, *Confessions*[1]

> 'To be' means to know one 'has been.' 'Not to be' implies the only 'new'
> kind of (sham) time: the future. I dismiss it. Life, love, libraries, have no
> future.
>
> VLADIMIR NABOKOV, *Ada*[2]

In narrative, authors necessarily deal with both time and space. Time is *of the
essence* for the task of telling a story. In fact, it appears doubled as both narrating time (the moment and duration of the storytelling itself) and narrated time,
the (usually past) stretch of time in which the plot unfolds. Yet there is a hint of
derivativeness in time, at least partly accounting for Augustine's—and everybody else's—difficulty in discussing it: it is easy to talk about space without
reference to time and to envision space as timeless, but it is almost impossible
to talk about time without referring to space, certainly without using spatial
metaphors, often container or distance and extension metaphors.

The contrarian, Vladimir Nabokov, claims the opposite: "I cannot imagine
Space without Time but I can very well imagine Time without Space" (*Ada*,
p. 411)—imagine, perhaps, but not conceptualize, for the vocabulary of time is
thoroughly spatialized. Accepting George Lakoff and Mark Johnson's claim
that metaphor is deeply conceptual rather than merely rhetorical decoration,
we can understand discussions of time by analogy to the discussions of certain
modes of sensory perception transferred into the vocabulary of other modes
for increased conceptual clarity.[3] Our sense of time is so thoroughly spatialized
that spatial concepts have come to dominate even narrative forms and genre

---

1   *Augustine's Confessions*, trans. E.B. Pusey, bk. 11.14.17. Online at <http://faculty.georgetown
    .edu/jod/augustine/Pusey/book11> (accessed April 22, 2015).

2   Vladimir Nabokov, *Ada or Ardor: A Family Chronicle* (New York, 1969), p. 425. The whole of
    Part 4 of *Ada* is a literary essay on time.

3   George Lakoff and Mark Johnson, *Philosophy in the Flesh: The Embodied Mind and Its
    Challenge to Western Thought* (New York, 1999).

© KONINKLIJKE BRILL NV, LEIDEN, 2016 | DOI 10.1163/9789004312319_016

concepts: we talk about "open" and "closed" forms, assume that the "past is behind," the "future is in front," and use terms such as "foreshadowing," "narrative span," and so on. The lyrical poem is the domain of the noun, hence less temporally inflected than the play—it is the realm of action and the verb, whereas the classic epic narrative is the leisurely world with room for the adjective. Nabokov connects time, the senses, and conceptual thinking in yet another way: "Physiologically the sense of Time is a sense of continuous becoming, and if 'becoming' has a voice, the latter might be, not unnaturally, a steady vibration. …Philosophically, by contrast, Time is but memory in the making" (*Ada*, p. 425). "Becoming," as we will see, is also a key concept for Mikhail Bakhtin and Gary Saul Morson.

The tension between the necessarily closed narrative form and open time, which goes on as a continuous stream of events, is a fundamental fact of narrative. The former is the realm of the story, the latter of history. These two English-language terms characterize a difference that in German is marked through the distinct use of the same term, *Geschichte*, in either the singular or plural. *Geschichte* covers the functions both of story and history, but only the singular means, as it were, *History* with a capital-H; the plural *Geschichten*, in contrast, denominates the unlimited number of stories that can be told, hinting perhaps at the concept of history as an imagined sum total, the supreme incorporating framework of all stories.[4] The conceptual difference between history and story is at the center of the present inquiry, the difference between the ongoing, broad, and chaotic mass of narratable events, on the one hand, and the singled-out, individual, and ordered actual narrative on the other: the potentially infinite and open grand narrative of history, and the finite, closed literary art form of the story. Both structure time differently. Gary Morson's study[5] is probably the most

---

4   I am predominantly focusing on fictional, but realistic stories (technically *realist* because the term includes certain formal conventions) of an old-fashioned, everyday kind with a beginning, a middle, and an end. Such stories are often more or less realistic accounts of realistic characters, that is, accounts of a biographical kind, with biographical time as their primary clock. They have as their ultimate limits the birth and death of their characters, the biological life-span of individual human beings. Of course, they may be shorter or about many characters in sequence but as stories they will focus on individuals and single out specifics. As history and historical fiction, in contrast, narratives often summarize and totalize and thus foreground the essential fact that in individual lives *potentialities* always exceed *actualities* and that there are narrative choices to be made.

5   Gary Saul Morson, *Narrative and Freedom: The Shadows of Time* (New Haven, 1994). Morson worked closely with Michael Bernstein whose own book on narrative theory, *Foregone Conclusions: Against Apocalyptic History* (Berkeley, 1994) appeared practically simultaneously.

thorough account to date of narrative openness/closedness and the underlying temporal concepts and their ethical implications. His book stands at the
outset of this inquiry, serving as its theoretical backbone, followed by three
literary examples that illustrate the general points and broaden the theoretical
spectrum. The first is the *Communist Manifesto* by Karl Marx and Friedrich
Engels, read here for its narrative-temporal strategies rather than its political
implications; the second is Max Frisch's novel *A Wilderness of Mirrors*; and the
last is Anna Seghers's short story "Ausflug der toten Mädchen." These three
texts represent different genres (manifesto, novel, short story) as well as three
different modes of opening up time frames: through a historical long-term perspective; through formal structural devices; and through psycho-pathological
temporal distortions.

• • •

Before turning to my literary examples, however, let me look first briefly at two
examples of the crucial distinction between the literary aesthetic need for narrative closure and the resistance of the world of events to such closure and
then at Gary Morson's account of narrative openness/closedness and his concept of sideshadowing.

My first example of the need for narrative closure and the world's resistance
to such closure is the iconic moment of the American President George W.
Bush's photo op in May 2003, on the aircraft carrier *USS Abraham Lincoln*. Only
two months after the start of the war in Iraq, Bush landed dramatically in full
aviator gear on the deck of the warship with the banner "Mission accomplished" displayed prominently behind him.[6] This controversial media event is
a classical example of a historically contextualized attempt to "close the story,"
while the plot, the war itself, was still ongoing. The narrative closure, sought in
this case for political gain rather than aesthetics, failed, and the intended narratable story form has been kept open by the unstoppable flow of history. The
lesson to be drawn is that it is never safe, indeed often embarrassing, to attempt
to "close the book" too early on a (historical) plot still in progress. Accordingly,
in July 2003, the President was forced to reopen his account, as it were. He did
so by means of another controversial statement, the cocky "bring 'em on!" in
reference to the elusive enemy. The narrative has since moved increasingly

---

6   See <http://www.cnn.com/2003/ALLPOLITICS/10/28/mission.accomplished/> (accessed April
    22, 2015). This CNN site shows the banner behind the president at the rostrum and discusses
    the politics of this event.

into the present tense in expressions such as "we are winning," or "we have Al Qaeda on the run," and so on.

The second example of the distinction between the need for closure and resistance to it is literary and explicitly plays on narrative structure: Scheherazade's story *in Thousand and One Nights*. The plot device that provides the cohesion for this famous collection of tales consists in Scheherazade's efforts to keep her future husband, King Shahryar, from killing her by entertaining him with a series of stories. The king, jealous of his (former) wife, had her executed and, for good measure, decided to take ongoing revenge on all womankind. Night after night, therefore, he consorts with some beautiful girl, only to have her beheaded the next morning. But when he meets Scheherazade, the beautiful and clever daughter of his vizier (whose sad duty it is to carry out the executions of the women), the king takes her into his company every evening. She had promised her distressed father, who trusts her storytelling abilities, that she would engage the king in her tales so that he would be unwilling to execute her for fear of missing the end of the story. In this way she would eventually persuade him to stop his murderous practice. The plan works: piqued by curiosity to know the outcome of each story, the king stays the hand of the executioner and, after a thousand and one nights, marries his storyteller. Taking Scheherazade's reality as the historical world in which stories are told, her playing with narrative form at crucial moments thus becomes more than a mere ploy: manipulating narrative form saves her life.

The empirically and philologically minded nineteenth century raised the issue of narratability and narrative form not just in literature but in history, too. In the German context, Leopold von Ranke's famous demand for historiography to tell the past "as it actually happened" stands at the outset, in 1824, of the then emerging discipline of modern historiography,[7] a discipline initially preoccupied with issues of a philological kind, but over time increasingly reflecting on its own narratologies, assessing not only what is told, but increasingly also how and in what context. This development reached its apogee in the New Historicism in the 1980s. Narratological issues, narrative forms and strategies, perspectives, narrators, and so on, issues traditionally associated

---

7   Leopold von Ranke (1795–1886), "wie es eigentlich gewesen" ("as it actually was"). "History has had assigned to it the office of judging the past and of instructing the present for the benefit of future ages. To such high offices the present work does not presume; it seeks only to show the past as it really was." *Geschichte der romanischen und germanischen Völker von 1494 bis 1514 (History of the Latin and Teutonic Nations from 1494 to 1514)* 1824. See Die Deutsche Geschichte in Dokumenten und Bildern (DGDB) <http://germanhistorydocs.ghi-dc.org/sub_document.cfm?document_id=358&language=german> (accessed December 28, 2015).

with and theorized in literature and literary criticism, became recognized territory where history and fictional narrative overlap. In light of the insight that narrative is never a representation of what happened, but always its (re)construction, Ranke's desideratum to tell the past "as it actually happened" came to be understood as an impossible dream. Any story is always only one version of the events that constitute its contents. Many others are possible; history writing is an infinite task.

For the present inquiry to remain manageable, the grand difference between literary and historical narrative must remain focused on the narrative devices, specifically those relating to the shaping of time and temporal structures. Every good story has a beginning, a middle, and an end, but capital-H History as the narrative account of the never ending stream of events has only a middle, as it were; it is always *in medias res* and in order to be narratable, it too must choose beginnings and endings, generally based on topic or specific events, epoch, geography, person, and so on. The perceived need at the outset of history books to describe and justify their limits is thus the constitutive mechanism to turn capital-H History into story in order to make it narratable. This is a commonality of historiography and literature. The difference is this: the historiographer brackets *out* a surplus of material that is practically unlimited and open in time; the fiction writer brackets *in*—namely all that is needed for a complete, rounded, aesthetically pleasing totality—but also not more: stories are closed.

This raises a simple but important question: how can fictional narrative retain the openness of real-life eventness and yet come to a formally satisfying conclusion as a well-rounded story? How can the messiness of real life and human decision-making be brought into an aesthetically satisfying closed form? How can the impression be avoided—and this applies to the writing and reading of both history and fiction—that everything is told from a narrator's fixed position outside, that everything is all a foregone conclusion, a fixed game? What are the narrative strategies or devices, therefore, that an author might use to convey a sense that his or her characters might be making actual decisions, have options, and tell their (life) story not from the end, but from its unfolding? The last question is important because it raises not merely a formal aesthetic consideration, but an ethical one within a broader understanding of literature as a field that discusses and mediates human experience, an understanding I would like to suggest and conceptualize through Gary Morson's study.

• • •

Morson addresses the issue of an unfolding story within the broad historical framework of "narrative and freedom" in communist societies. As concepts of time are shaping our outlook on life and the world, the temporal structure of narratives (about acting in the world and, hence, the parameters of such action) are placed at the center of his inquiry. A belief in either determinate or indeterminate time, "fate or choice, unfolding or becoming, gradual or sudden change, and temporal closedness or openness" may well "affect how people conduct their daily lives" (p. 3). As a Slavist, Morson reads the great Russian narrators of the nineteenth century against the Soviet reality of the twentieth whose grand narrative, the Marxist sociopolitical teleology, knew its own ending and whose political practice was, accordingly, structured from that final perspective. Aligning himself with Bakhtin, Morson notices how "dominant cultural models close down time by thinking away its processual nature" and "leave no real place for creativity or choice. Life comes to resemble a finished product, in which everything has already been fixed" (p. 21).[8] In Dostoevsky, Tolstoy and others, by contrast, Morson recognizes true narrative openness and finds characters whose decision making process the reader can follow in detail, and who live their lives not from the perspective of a predetermined end but as "becoming." Morson sees literature as both a cultural treasure trove as well as a field of experimentation and modeling for lives-in-time and temporal structures of behavior. His analysis is a critique of communist or, more broadly, teleologically oriented literary practices in the medium of narrative, manifested above all in temporal structure. Among the three fundamental modes of narration he analyzes—one characterized by "foreshadowing," the other by "backshadowing," and the third by what he calls "sideshadowing"—he prefers the third as the mode closest to actual life in allowing for, and showing, characters in situations where they have real alternatives, make real choices, and hence come to bear responsibility. Foreshadowing, and even more clearly backshadowing, are elements of a deterministic, closed form of narration.

How, then, is this desirable narrative openness achieved? While "foreshadowing robs a moment of its presentness" (p. 117) by depicting it as merely the shadow of a future event that obviously already has happened in the author's mind, backshadowing characterizes the past as "having contained signs pointing to what happened later." Backshadowing may lead the observer to "a kind of temporal egotism...to endow our own actual present with special privilege." The backshadowing observer (who may or may not be a common type among historians) may even exhibit a "tone of superiority" vis-à-vis those who failed to read the signs in due time (p. 234). Both fore- and backshadowing are inimical

---

8 It might be worthwhile to consider TV commercials from this perspective. They are the quintessential modern mini narratives constructed from the end, often providing answers to questions or problems viewers have not even realized they might have.

to openness, eventness, and the conveying of the fact that in life possibilities always outnumber actualities. Underlying Morson's analysis of fictional temporality and the representation of past, present, and future is the philosophical debate between a "tensed view" of the world and a "tenseless view."[9] Morson inclines to the former which, briefly, holds that "there are genuine distinctions between past, present and future." The present moment may "be privileged in the radical sense that, as it is sometimes put, 'only the present is real'" (Cockburn, p. 4). The tenseless view holds that "while there are tensed judgments there are no tensed facts" (Cockburn, p. 18). But Morson also warns us not to forget that even realistic fiction rests on conventions that are different from real life: essentially we agree that *what is still unknown will be treated as equivalent to what is still undetermined*," and he adds that "Ignorance is not the same thing as indeterminacy, any more than the suspension of disbelief is the same as belief" (p. 175, italics his).[10]

Temporal openness, and thus the possibility of real decision-making, privileges the present moment. When Morson discusses what he calls "four diseases of presentness," he invokes these fundamental philosophical positions. He diagnoses a "desiccated present" where we "lose the sense that the present has special importance" (p. 188), and an opposite "isolated present" that feels "as if it were an island cut off from tradition and from consequences." There is also "hypothetical time," where it seems that "every actuality were a mere possibility" and "multiple time," a special case of hypothetical time, where "the temporal sequence in which we live loses its significance because it is not the only sequence" (p. 189). These temporal diseases—particularly the last one—correlate with certain psychic states and can be matched up with the textual examples in the second part of this essay.

Sideshadowing, which is Morson's favorite literary temporal structure, may be viewed as the antithesis to foreshadowing (p. 117). "While we see what did happen, we also see the image of what else could have happened. ...two or more alternative presents, the actual and the possible, are made simultaneously visible. This is a simultaneity not *in* time but *of* times. ...Sideshadowing, therefore, counters our tendency to view current events as the inevitable products of the past" (p. 118). Sideshadowing is presentist and emphasizes the here and now as the moment of decisions. When in 1989 Francis Fukuyama mused on the "end of history," he did not mean, to be sure, the end of time nor the end of the stream of human-engendered events that constitute the raw material for

---

9    For a useful introduction to this debate see David Cockburn, *Other Times. Philosophical Perspectives of Past, Present, and Future* (Cambridge, UK, 1997), p. 4.

10   This is precisely the difference between "realistic" and "realist." See n. 4 above.

history-*qua*-narrative.[11] Rather, he reacted to a development, the impending end of communism which had accompanied a large part of twentieth-century history as but a shadow of a possibility. Now the grand historical narrative that had been conceived as the struggle between communism and capitalism was ending with the demise of the former. Capital-H History was turning out to be only small-h, a mere story. But the end of a story can only be pronounced from a position outside of it in the "happy ever after" that every concluded story implies and fairy tales explicitly invoke in this formula. Fukuyama's variant of the ever-after is the *post-histoire* that dissolves History, dramatically conceived as struggle or battle, into petty stories, and reduces it to prosaics.[12] It is our muddled present. This was, however, only the Western perspective. For what also ended was, for those nations that had lived under Marxism (communism), the *end of history* itself! As Marxism had operated with the assumption of a historical teleology that would inevitably lead to its universal triumph, in other words, the assumption that Marxism itself constituted—with some fine-tuning still to be done, no doubt—the end stage of history, its very end opened up history again for millions of people. This is a point Morson makes in the introduction to his book. In fact it is the point that motivates his deeply anti-grand/closed-narrative stance and his inquiry into the possibility of conveying the openness of historical realistic narrative. Only this will restore the "eventness" of real life to people.

In contrast to Bakhtin's *polyphony*, however, Morson's idea of sideshadowing is one of potentiality more than actuality. The multiplicity of possibilities is not actually present as is the multiplicity of voices in a polyphonic text; instead sideshadowing, even if only dimly, alerts the reader to roads not taken. The linguistic surface structures associated with sideshadowing often contain subjunctives or even past subjunctives, images of longing and desire, musings and imaginings. The mental attitude of irony and the emotional state of jealousy are further literary modes with rich sideshadowing potentials.[13] The

---

11    Francis Fukuyama, "The End of History?" *The National Interest 16* (Summer 1989). The full text is available at <https://ps321.community.uaf.edu/files/2012/10/Fukuyama-End-of-history-article.pdf> (accessed April 22, 2015).

12    Gary Saul Morson and Caryl Emerson, *Bakhtin: Creation of a Prosaics* (Stanford, 1990). As an aside that cannot be pursued further here: one of the invaluable analytical advantages the idea of prosaics offers, that is, acting in small, seemingly insignificant steps, is the conception of *agency* that transcends the traditional human protagonist, a kind of plot-moving not necessarily tied to individual action. It is thus eminently suited to historical narrative where plot outcomes accumulate over time and as a process, not a specific single action.

13    Max Frisch was a great writer on jealousy, and *A Wilderness of Mirrors* discussed here is an excellent example.

English-language expression "two-timing" is an entirely fitting way of depicting this state of affairs as it were. Further textual microstructures include comparisons and similes that momentarily evoke a parallel universe, as well as
reminiscences in (auto)biographical narratives. As a genre, the detective novel
is an interesting case. It offers the reader a large number of alternatives and
keeps them open for as long as possible. But ultimately it is also the genre of
closure; solving the case eliminates every story but one. In contrast, irony, jealousy, the subjunctive mode are rarely the devices of the historian. Few things
make historians more uncomfortable than the subjunctive "what if" question?
What if Hitler had or had not done this or that? And what if Stalin had? They
did what they did—and that is the history to be analyzed.

• • •

My first example of the problem of narrative openness and closedness is the
*Communist Manifesto*, a strange mixture of history, high literariness, and *urtext*
of Marxism itself.[14] The *Manifesto* lends itself to a discussion both of closed
and open narrative form and issues of story and history.[15] A manifesto is a text
meant to be actualized. It stands in a different temporal relation to the real
world than either history or realist fiction, both always post factum, the latter
an alternative, a sideshadow of a reality that constitutes itself in each reader's
mind in the very act of reading. A manifesto, in contrast, provides ground for
action. It is presentist, emphasizing the here and now as the moment of
decision with regard to shaping a specific future.

The *Communist Manifesto* is a tripartite document in three different temporal modes. Its first part is a closed history of the bourgeoisie and its achieve-
ments (we might also say "capitalism") in a breathtaking, apodictic sketch.
Sections II–IV taken together form a present-tense dialogic exchange, a kind of
Q&A on what communists are and what they want; a survey of earlier types of
cooperative social arrangements and existing literature and debate; and a
statement on the communists and their relationship to other contemporary
political forces. The *Manifesto* ends with the exhortation to the workers of the

---

14    The text used here is *Karl Marx, The Communist Manifesto*, ed. Frederic L. Bender (New
      York, 1988). This English translation is a slightly amended version of Samuel Moore's 1888
      translation from the German original, generally considered the English urtext because
      Engels himself had chosen Moore as the translator and participated in the preparation
      of that text.

15    For a discussion of the *Manifesto* with strong literary inclinations, see Marshall Berman,
      *All that is Solid Melts into Air: the Experience of Modernity* (New York, 1982).

world to unite. Our focus is primarily on Part I. In content, it is a history, in style literature, in form a good story—and overall a great read. Marx sketches a large-scale historical development, the shaping of the bourgeoisie, which is still ongoing. Eleven breathless paragraphs open with a phrase structured on the model "The bourgeoisie has achieved ..." (pp. 57–59)—a present-perfect form that inevitably leads to the present, raising the question: "what now?" The *Manifesto* is not after all meant to be a praise of the bourgeoisie—although Marx is clearly taken in by its secret charms. He faces the task of finding a way to introduce the new actor, his real hero: the proletariat.

Marx sets up the change of focus carefully, shifting the action to the new agent and then presenting that agent as yet another, although perverse, achievement of the bourgeoisie: "But not only has the bourgeoisie forged the weapons that bring death to itself; it has also called into existence the men who are to wield those weapons—the modern working class—the proletarians" (p. 61). The scene is set: "The proletariat goes through various stages of development. ...At this stage the labourers still form an incoherent mass.... Thereupon the workers begin to form combinations" (pp. 62–63) and the bourgeoisie itself continues to "furnish...the proletariat with weapons for fighting the bourgeoisie" (p. 63) until finally—this the last sentence of Part I, "the bourgeoisie...produces, above all, ...its own grave-diggers. Its fall and the victory of the proletariat are equally inevitable" (p. 66). The last word is interesting. A negative-prefix adjective opens the narrative to a presumably open future that it immediately narrows down to one possible outcome. It makes a promise of contested openness, but assures a specific, inevitable outcome.[16] As experienced readers we are thus prepared for the obstacles every protagonist in every good fight must overcome, but we can also look forward to ultimate victory. Yet we are disappointed. As Marx pulls back from his masterful account of the bourgeois success story to the much drier matter of communist politics, past and present, and the account of various predecessors, he loses his narrative drive. As he switches from the sweeping historical outline (mostly in the present-perfect) that sustains him to the present moment, he becomes bogged down in the detail of both present-time discussion and present-tense discourse, the messy and open world of events, the vagaries of contingency. Only in the very last sentence of the whole document, the famous "Working men of all countries, unite!" (p. 86), grammatically an imperative and as such necessarily future-oriented, does he revert to the action mode of the first part— there in the past, here for the future. Yet in doing so he concedes that the

---

16    The German original is "unvermeidlich," an analogously built adjective; the findings are thus not an effect of the translation.

opponent of the bourgeoisie is not yet ready for the fight. What was "inevitable" is now considerably toned down because of the obstacles that might lurk (and historically did lurk) behind the imperative to unite. Nonetheless, as the shadow of a doubt creeps in at the end of the *Manifesto* so does the possibility of openness through the very form of the imperative.

• • •

In Max Frisch's *A Wilderness of Mirrors*, a first-person narrator is looking for stories that would fit his life experience.[17] The author (1911–91) was preoccupied throughout his life and career with existential issues whose autobiographical roots he often left visible. He was notably interested in the (non) congruence of narrative forms and the underlying life experience, the very fact that infinitely ramifying experience and memory, in order to be communicable, must be pressed into finite narrative form. *Wilderness of Mirrors* is a novel that uses this tension as its plot-driving device. The narrator, an anonymous man in his fifties, is trying on stories for size, a formula that appears early on in the text. Frisch's problem is familiar to us: stories have a beginning, a middle, an end. Given that his narrative urge is (auto)biographical and his stories therefore are life-stories about relationships, jobs, social situations and interactions, and so on, by the middle he knows how it all ends. Often he already fears the beginning, the end does not interest him any more because the recording of experience inevitably straight-jackets and thus falsifies it. For Frisch the danger is real that stories determine and cast in stone experience and identity. Life stories in particular come only in certain models, as it were, and the difficulty is always in fitting them onto the person. "Boy-meets-girl," "married-with-children" "the immigrant-tale," or "rags-to-riches" are just a few popularly recognized fixed tales.

   In contrast to Morson's examples from the realist tradition of the nineteenth century, where alternative possibilities and potentials show up only as *sideshadows*, Frisch's sought-after openness is manifested in the very structure of the narrative, more robust than sideshadowing, but also more disruptive. The basic premise of the novel is simple: "A man has been through an experience,

---

17    Max Frisch, *A Wilderness of Mirrors*, trans. Michael Bullock (New York, 1966). The German
      original is *Mein Name sei Gantenbein* (Frankfurt am Main, 1964). The original German
      title, *Mein Name sei Gantenbein*, iterally means "let my name be Gantenbein" or "may I be
      called Gantenbein." Gantenbein is an actual, although not a very common Swiss surname,
      and the subjunctive construction implies the search for and final acceptance of an iden-
      tity. The English title connects to this theme differently.

now he is looking for the story to go with it—you can't live with an experience that remains without a story, and I often used to imagine that someone else had exactly the story to fit my experience..." (p. 11). The narrator then sets out to find the story. His tool is imagination and the phrase "I imagine" appears more than two dozen times over the length of the novel. At one point early on he puts it differently: "I try on stories like clothes" (p. 21), thereby emphasizing the narrative device. The result is a post-modern experimental text that lacks the coherent surface structure of more traditional realist prose and is broken open as a form, with the many individual stories accumulating around the narrator who creates them in the first place, evaluates them, rejects them. In the course of the book, Frisch drifts toward one character, namely the figure that appears in the German title, Mr. Gantenbein, and his stories, returning to him, developing him, and finally accepting him. Narrative alternatives in Frisch appear as fully developed plotlines, not merely as the possible switching points that Morson identifies in the great Russians whose emphasis is on presentness as the crucial point of moral and political decision. Frisch is more focused on past (individual) experience and how to cast it narratively. His specific temporal inflection centers on the inevitability of repetition, the problem, say, of how to *work through*, which inevitably means how to *narrate*, the experience of a series of love relationships over the course of a lifetime.

Yet even the Gantenbein figure that the narrator finally accepts as a possible design for himself is slippery: a seeing man who pretends to be blind! This opens up the world of possible stories in unexpected ways. Gantenbein is able to do a double take on everything, observing the real events and situations that unfold with disarming openness in front of him, while enjoying—or suffering from—the stories that his interlocutors tell him and want him to believe. That this is a prefect setup for the literary treatment of jealousy—the not uncommon result when someone "two-times" his or her partner—hardly needs mentioning. But it offers other possibilities too: in one episode, the narrator imagines Gantenbein as a blind tour guide who shows a group of visitors the Acropolis. As he cannot possibly see what he describes, his group begins really to look and to see in order to inform their poor guide. They learn more about the Acropolis than they had ever thought possible. In another story, Gantenbein is a house-husband with a wife, Lila, who is an actress and thus often absent. How can he possibly keep the house clean and do chores? After some trial and error, he finds the right balance: not too much, which would likely give him away; but not too little either so that Lila, who is a delicate and highly strung person, does not get upset about the mess.

The vexed question of narrative closure versus experiential openness or newness had preoccupied Frisch from early on in his career. In his *Diaries* from the

1940s, he had put it in Biblical terms: he wrote, "Thou shalt not make graven images," and related it to love, using the immutable image as a metaphor of the determining, confining, and limiting force that image-making by one lover has over the freedom of the other, nailing him or her down once and for all, disallowing change.[18] A truly free, developing and changing, partner would now threaten to break the mold, the *idée fixe* the other has constructed. Making an image and holding the partner to it is therefore the end of love because it is the end of the possibility of change, the end of open time to develop. Love for Frisch always describes a state of being alive, movable, and accepting of change in others, a state of constant becoming.

Later in life Frisch returns to the issue of fixed, pre-existing story versus the mutable nature of biography, this time explicitly his own. In *Montauk* (1976), a brutally honest yet also highly literary and reflective autobiographical account of key moments, personal as well as professional and intellectual, in his life, he states (in English in an otherwise German text): "My greatest fear: repetition."[19] While we cannot enter the philosophical questions that the problem of repetition opens up, Frisch provides us with intriguing aspects unique to the authorial prerogative. Understanding that as a human being he should not expect his life to remain free from, or be lived outside of, repetition, he realizes that as an author he has to face recurring events differently each time: "That's the good thing about being a writer...for the writer as a person; he has to work through certain facts differently when they return in his life—in order to remain a writer..." (p. 122, my translation). This link between life and text, experience and story we already know, is stated from yet another, more provocative perspective in a statement Frisch reports having made at the end of a reading and book signing session: "I tell the American public: life is boring; I only go through experiences when I write" (p. 12, translation mine). This statement links experience and writing about it in a manner that is closer to the psychoanalytical concept of "working through" than to Morson's concepts of narrative openness. Frisch's openness lies in the ultimate revisability of the narratives generated by human experience up to their complete replaceability. Morson is *a priori* less concerned with (auto)biographical writing than with social narratives, and openness for him lies in the mere appearance of moments of alterity. His openings are *in* the narrative; Frisch's openness is a function *of*

---

18      Max Frisch, *Tagebuch 1946–1949* (Frankfurt am Main, 1950, rpt. 1977), p. 31.

19      Max Frisch, *Montauk. Eine Erzählung* (Frankfurt am Main, 1975), p. 18. The title of this slender book derives from a weekend trip the author undertakes with a woman on the occasion of an American reading tour in 1974. It is within the time frame of this weekend that the author takes stock of his life.

the narrative. In the long run, in (auto)biographical writing, the forces of life and story collude: both are closed. The final version of a life story must bring to convergence the narrated subject and the narrator at the present moment for the story to be coherent and recognizable as one's own. Only the last word, death, cannot be pronounced by the narrator. But death also means, narratively as well as in reality, "the end of story." In *Wilderness of Mirrors* Frisch's closing device is an unidentified corpse drifting down the Limmat river in Zurich. Gantenbein? Enderlin? Svoboda? Or yet another of the novel's characters?

• • •

Anna Seghers's "Der Ausflug der toten Mädchen" ("The Excursion of the Dead Girls")[20] presents an entirely different example of an open time structure, in this case pathologically open time.[21] "No, from much further, from Europe," are the opening words uttered by the first-person narrator, a foreign woman, in a dialog with a Mexican man. They are in a village, hot, dry, rather primitive, and they have nothing in common. The youngish woman had only just recovered from an illness and still feels weak and a bit dizzy. Friends, she says "had saved her from the most pressing misfortunes but conjured up more hidden ones in the process" (p. 8). She is clearly not at ease. Across the distance she had noticed a slightly dilapidated *rancho* and inquired about it, but nobody had given her much to go by. So she sets out on the hot, dusty path toward it in spite of the heat, the sun, her own weakness and exhaustion. As she peers into the yard through the broken door in the crumbly adobe wall, she feels the air to be much fresher and concludes that there must be a fountain or a pool at the back. And to her surprise she hears a regular creaking noise, although she had been convinced that the place was empty: "The creaking got louder and I noticed in the bushes that were getting denser and more lush, a regular up and down motion as if from a swing or a bar. My curiosity was now aroused and I ran through the gate toward the swing. At that moment someone called out

---

20     Anna Seghers, *Der Ausflug der toten Mädchen* (The Excusion of the Dead Girls) (Berlin, 1947, rpt. 2005).

21     The author's real name was Netty Reiling (1900–1983). She was born in Mainz, Germany, into a Jewish family. She studied in Heidelberg and joined the Communist party in 1928. Under the Nazis she fled first to Switzerland, then to France, and went into exile in Mexico. In 1947 she returned to Berlin and moved to East Berlin in 1951. She subsequently rose to literary prominence in the GDR and served for many years as the Chair of the East German Writer's Union. The story discussed here was written in 1943–34 in Mexican exile and published in German in New York in 1947. It is a story of ca. 30 pages. Translations are my own.

'Netty'!" (p. 9). This is the narrator's childhood name (and obviously also Anna Seghers real name), and she is strangely touched to be called by it.

As the slight dusty shimmer in the air clears and she sees through the bushes, she notices two of her classmates from her school in Germany sitting on the swing. In a strangely realistic, believable way, akin to a dissolve in a movie, the scene changes to Germany ca. 1910 and the gaggle of girls that made up the narrator's class during a school excursion on the Rhine. The day of the excursion now becomes the new narrative present, but other levels of time emerge as co-present. The narrator provides brief biographical sketches of all thirteen of her classmates, skipping back and forth between the afternoon of the excursion and later moments, still in the future on that day, but recognizable as later by their content, among them the death of a boyfriend of one of the girls in World War I, the rise as a Nazi functionary of another classmate, the marriage to a Nazi by a third, the resistance to Nazi cooptation put up by another of these young women later. But the shifts in time are also marked linguistically by the dominant tenses in this passage: the future, future past, and subjunctive past. Nora, for instance, one of the girls, was clearing and decorating with flowers their teacher's chair in the café on the Rhine. She "would certainly have regretted this later," the narrator muses, "as she became the leader of the national socialist women's union of our town" (p. 15). Their teacher, we learn, had been Jewish. The narrator also mentions the later suicide of yet another of her classmates because a Nazi lover of hers had threatened to denounce her for having had an affair with a Jew, an instance of *Rassenschande* at that time and thus forbidden.

All these young women are dead, yet the reader sees them alive on the occasion of that excursion, gets a glimpse of their future development and fate, sees them live through World War I, the Nazi period and into World War II and the bombings of their city—while also retaining in the back of his or her mind the narrative present and Mexico as the locale of storytelling. It is the double perspective of seeing these characters as both young, innocent, girls with an open future before them and as already dead after short lives distorted by the war which is the central attraction of this story, including the moral implications of assessing these lives. In death they are all equal; yet they did make choices during their lives, and some felt guilty. In the story's double-take, one would/did betray their Jewish teacher; another would/did deny help to a former friend but others, inversely, would be come and were charitable and generous, yet still died alone and senselessly in the war. The blending of temporal levels that renders the historical period from *ca.* 1910 to 1945 as both past and future, is artfully done, neither as backshadowing, the attribution of meaning to events *ex post facto*, nor quite as foreshadowing, the connection of present events to future, as yet unrealized, meanings. We might call it sideshadowing, but not quite in Morson's sense because the shadow of alternative possibilities falls not on individual characters, hinting at freedoms of decision-making and action they might have, but on the story as a whole through the "real" present moment and place, the dusty *rancho* in the Mexican countryside. The narrator is the only survivor, the only one with any freedom of choice at all. What we find is thus a use of time that resembles most closely one of Morson's "four diseases of presentness," namely "hypothetical time" or "multiple time" (p. 189), the strange and unstable co-presence of various temporalities. They are symptomatic of the narrated time being out of joint, as well as hinting at possible psychological problems in the narrator. In light of how the day of the school trip ends for the narrator, I suggest we call it "trauma time."

On her way home from the excursion, the narrator, on that day in 1910 a young girl like the others in the group from whom she just took leave, has "an anxiety attack upon turning into my street, as if I had a sense that it was destroyed" (p. 35). Previously, as their excursion boat landed at the pier where the trip ends, she experienced a temporal disjuncture and it appeared to her that "we were not returning from a trip but from a journey of years" (p. 31). Now walking through the city streets toward her house "something hard oppressed [her] heart as if something senseless, something evil were in store" for her (p. 32). Time literally splits for her and she sees the city simultaneously as it was in 1910 and as the bombed-out shell at the end of World War II. As she approaches her house, the presence of her mother on the veranda calms her down, yet "it seemed unbearably hard for me to climb up the stairs" (p. 36). She feels herself to be in the sort of dream in which one wants to run, at least to move, and finds oneself unable to or only very, very slowly. Also, she notices that her mother is much younger than herself: "She stood there happy and straight, destined to a busy family life full of the usual joys and burdens of everyday life not for a horrible, cruel end in a far-away village where she had been banned by Hitler. Now she seemed to recognize me and waved as if I had been gone. She always smiled and waved after excursions" (p. 36). As the narrator enters the staircase, she feels utterly exhausted again; the temperature feels too warm, too, and the light too bright. She hears the cry of turkeys, which surprises her greatly; then she transitions back into Mexico, and the multilayered and multi-temporal vision ends— having brushed quickly and almost casually over the death and destruction that World War II wreaked on her home town, including the death of her mother in that "far-off village where she had been banished by Hitler" (p. 36). In real life, Anna Segher's mother was sent to the camps and disappeared.

This is Morson's "multiple-time syndrome," to be sure, but not only as a narratological but also as a psychological concept, the more so, as it clearly reaches beyond the narrator to the author herself. Multiple time here is trauma time. Scholars have, for some time now, engaged trauma and writing about trauma as a distinct field in literary studies.[22] Anne Whitehead in her work *Trauma Fiction*, for instance, outlines "the (necessarily provisional) range of literary devices which characterize the emerging genre of trauma fiction" (p. 4) and names, among other aspects "trauma's disruption of time or history" (p. 5). She does so in reference to Cathy Caruth's work, who speaks, invoking Freud, of trauma as "the wound of the mind—the breach in the mind's experience of time, self, and the world" (p. 4; see n. 23). Robert Lemelson, Laurence J. Kirmayer, and Mark Barad, in a clinically oriented article, also emphasize that "the temporal structure of the response to trauma is complicated by the dynamics of memory and narrative."[23] Trauma, itself not an event, but rather the syndrome of often delayed and unpredictable reactions to an event, distorts the patient's sense of time and space. This is what happens to Segher's narrator. The story closes with her remembering, still hovering emotionally between the past and the present, but physically back in Mexico, her old Jewish teacher's request from the distant past to write an account of the school trip: "I was going to do the assignment right away tomorrow or even tonight, after my tiredness was gone" (p. 38). It turned out to be a truly long-term assignment, but with the end of the story, it is now completed.

---

22   See, for instance, Anne Whitehead, *Trauma Fiction* (Edinburgh, 2004); Cathy Caruth, *Unclaimed Experience: Trauma, Narrative, and History* (Baltimore / London, 1996); Marke S. Micale and Paul Lerner, *Traumatic Pasts: History, Psychiatry, and Trauma in the Modern Age, 1870–1930* (Cambridge, UK, 2001).

23   Robert Lemelson, Laurence J. Kirmayer, and Mark Barad, "Trauma in Context: Integrating Biological, Clinical, and Cultural Perspectives" in *Understanding Trauma: Integrating Biological, Clinical, and Cultural Perspectives*, eds. Robert Lemelson, Laurence J. Kirmayer, and Mark Barad (Cambridge, UK 2007), p. 464.

CHAPTER 15

# Time and the Self in Virginia Woolf and Richard Powers

*James F. Knapp and Peggy A. Knapp*

> But a soul can read in itself only what is distinctly represented there; it cannot unfold all its folds at once, because they go to infinity.
>
> G.W. LEIBNIZ, *The Monadology*, 1714

When Leibniz writes in the *Monadology* of a soul seeking to know itself, he uses the metaphor of a vast fabric, infinitely folded.[1] Later he describes a process of understanding in which tiny details may be gradually brought to clarity out of something present but previously seen only as a confused welter of perceptions, not yet available to conscious awareness. By looking into Virginia Woolf's *Between the Acts*, and Richard Powers' *The Echo Maker*, we will discuss how the history of individual characters can be seen as embedded in a representation of time reaching back to the prehistoric and forward to a troublingly uncertain future. Self-understanding can only be sought in the infinite sensations of the present. We make no pretense of solving philosophical problems still debated after three hundred years, but we do believe that, thinking with Leibniz, we can better grasp the deeply reflective artistry of these novels.

In a work written before the Monadology, Leibniz describes the endless stream of tiny, mostly unnoticed, perceptions that confront each individual:

> Ces petites perceptions sont donc de plus grande efficace qu'on ne pense. Ce sont elles, qui forment ce je ne say quoi, ces gouts, ces images des qualitiés des sens, claires dans l'assemblage, mais confuses dans les parties; ces impressions que les corps environnans font sur nous, et qui enveloppent l'infini; cette liaison que chaque estre a avec tout le reste d l'univers.[2]

---

1  *Pacidus Palalethi*, in Gilles Deleuze, *The Fold: Leibniz and the Baroque*, trans. Tom Conley (Minneapolis, 1993), p. 6: "The division of the continuous must not be taken as of sand dividing into grains, but as that of a sheet of paper or of a tunic in folds, in such a way that an infinite number of folds can be produced, some smaller than others, but without the body ever dissolving into points or minima."

2  "Nouveau Essais Sur L'Entendement," in *Gottfried Wilhelm Leibniz: Philosophische Schriften*, Band 6 (Berlin, 1962), pp. 54–55.

© KONINKLIJKE BRILL NV, LEIDEN, 2016 | DOI 10.1163/9789004312319_017

These tiny perceptions are therefore more effectual than one thinks. They make up this I-know-not-what, those flavors, those images of the sensory qualities, clear in the aggregate but confused in their parts; they make up those impressions the surrounding bodies make on us, which involve the infinite, and this connection that each being has with the rest of the universe.[3]

Here Leibniz links the past and the future of the "rest of the universe" with the fabric of individual perception as the soul discovers individual self-identity against a background of infinite confusion and infinite potential meaning. By an act of mind that Leibniz calls "apperception," which in modern parlance might be spoken of as "higher-order thought," insensible perceptions may be brought to consciousness and thus to understanding.[4] Although Virginia Woolf and Richard Powers represent consciousness artistically, as lived experience, they join Leibniz in reflecting deeply on how the world is present to each individual mind.

Even though the past and future of the entire universe is, for Leibniz, infinitely folded within each individual, the totality is present to the mind of God. When the individual soul seeks to represent the world, it can only do so by unfolding, within time, its own connection to past and future. Time is thus fundamental to our grasp of the outside world and of our specific identity within it. Time seemed an enigma to Saint Augustine, too, as he pondered the nature of his own selfhood in the *Confessions*—past and future have no being, and the present, which passes in a flash, cannot be extended and therefore has no space. Saint Augustine devoted the entire Book XI of the *Confessions* to the contrast between the human sense of sequence and the simultaneity of God's vantage point. We commonly think of three times: past, present, and future, but neither the past nor the future are actually located anywhere. The past exists only in memory of that which is no longer here, and the future only in anticipation of that which is not yet here; both must be regarded as features of

---

3   "Preface to the *New Essays*," in *Discourse on Metaphysics and Other Essays*, trans. Daniel Garber and Roger Ariew (Indianapolis / Cambridge, 1991), p. 55. In this essay, we do not intend to enter current philosophical debates about the complex evolution of Leibniz's thought over his long career, but we do want to use his provocative and often richly metaphoric notions to begin our own reflection on time and perception in two important novels.

4   Rocco J. Gennaro, "Leibniz on Consciousness and Self-Consciousness," in *New Essays on the Rationalists*, eds. Rocco J. Gennaro and Charles Huenemann (Oxford / New York, 1999), pp. 353–371. For Leibniz on apperception, see "The Principle of Nature and of Grace, Based on Reason," in Leibniz, G.W., *Philosophical Papers and Letters*, ed. and trans. Leroy Loemker (Dordrecht / Boston, 1969), p. 637.

a mental present. And the present can be divided into smaller and yet smaller increments that cannot be extended, so that it flies by even as it is being contemplated.[5]

Leibniz and Augustine pose the related problems of time and perception in philosophical terms, but literary narrative must also face such problems in presenting a world of sensations and events that has to be grasped and acted upon within the story. In recent years, the discussion of perception has shifted from "soul" or mind to brain, introducing a different vocabulary. The discourse of the brain locates specialized parts of the brain that connect with others to notice and interpret Leibniz's tiny perceptions and organize them into the apperceptions that allow us to understand, however well or poorly, and act on them. Two key components of our current understanding of the brain's structure and functioning have been electronic scanning technologies and clinical study of cases in which brains have been impaired by accident or illness. The scientific/ medical terms, of course, do not solve the paradoxes that philosophical discourse poses, but they do restate and, perhaps for our generation, clarify them. However, the problem for literary artistry remains that of presenting in narrative form the intricate fusion of past memory and anticipation of future events in the present consciousness of created characters. Their perceptions begin in a welter of tiny, unorganized details from the world outside as well as from the inner world of thought that includes memory and anticipation. To seem "real" to readers, characters must have enough consistency to suggest individual identity, but not so much as to preclude change and surprise.

Among recent novels that make use of the scientific lexicon is Richard Powers' *The Echo Maker* (National Book Award, 2006), a mystery story heavily invested in exploring the mental working of its central characters, one of whom has sustained serious brain injury in a car crash and several who try to help him. Gerald Weber, the expert from New York called to Nebraska to consult on the case of Mark Schluter, the injured man, has been likened to Oliver Sachs; through Weber the vocabulary of neuroscience comes to underlie the complex turns of the plot. All the major characters come to regard their seemingly self-chosen personas as the result of discontinuous "fugue-like" brain activity. Powers' *The Echo Maker* is about the "soul seeking to know itself" (Leibniz), one of the perennial subjects of the novelistic tradition. In one sense, the mystery, the case, is solved, but the philosophical problem that is its largest reference—how minds burdened with the past and pregnant with the future turn inner and outer worlds into apperceptions that lead to action—is not.

---

5   *The Confessions of St. Augustine*, trans. Henry Chadwick (Oxford, 1991).

Virginia Woolf, long regarded as a leading figure among twentieth-century Modernists most sensitive to the ephemerality of memory and sensation out of which the consciousness of self emerges, clearly speaks to the philosophical problem of apperceptions and action. Analogies with the artistic practices of impressionist and post-impressionist painting were brought early to criticism of her work, as was the notion of "stream of consciousness." While time and change were certainly part of impressionist art and its discussion (changing light on the face of Monet's Rouen Cathedral series, for example), the emphasis was overwhelmingly on the phenomenology of the present. To use the most traditional of metaphors, time was a river that flowed over the individual, presenting sensation and event (including personal memory) that was always felt, and sometimes understood. It is this metaphor of time as a stream always experienced in a brief and vanishing present that we want to complicate by counterposing Leibniz's metaphor of the crashing sea:

> Et pour juger encore mieux des petites perceptions, que nous ne saurions distinguer dans la foule, j'ay coutume de me servir de l'example du mugissement ou du bruit de la mer dont on est frappé quand on est au rivage. Pour entendre ce bruit, comme l'on fait, il faut bien qu'on entende les parties, qui composent ce tout, c'est à dire le bruit de chaque vague, quoyque chacun de ces petits bruits ne se fasse connoître que dans l'assemblage confus de tous les autres ensemble, et qu'il ne se remarqueroit pas si cette vague, qui le fait, estoit seule.[6]

> In order better to recognize these tiny perceptions that cannot be distinguished in a crowd, I usually make use of the example of the roar or noise of the sea that strikes us when we are at the shore. In order to hear this noise as we do, we must hear the parts that make up this whole, that is, we must hear the noise of each wave, even though each of these small noises is known only in the confused assemblage of all the others, and would not be noticed if the wave making it were the only one.[7]

In his metaphor, Leibniz emphasizes the plenitude of detail that the world presents to us, detail which is effectual even before it is consciously perceived. But unlike the river, which may be taken as figuring the ephemeral sensations of Augustine's almost non-existent moment, the metaphor of the sea asks us to think about listening, which is an act of the mind and part of Leibniz's larger

---

6  "Nouveau Essais Sur L'Entendement," p. 54.

7  "Preface to the *New Essays*," p. 54.

inquiry into the raising of perception from nearly insensible detail into conscious awareness. Moreover, Leibniz's tiny perceptions are always "plein de l'avenir, et chargé du passé" (filled with the future and laden with the past),[8] a notion of engagement with time and history far richer than sensory impressionism's narrow focus on the present and fully in keeping with the artistic projects of Woolf and Powers.

To represent the emergence of consciousness out of tiny perceptions weighted by the past and the future involves an enormously complex play of difference. For each individual, the particular details that become distinct enough to be perceived, to shape behavior, and occasionally to inform a higher state of self-reflection and self-understanding may reflect gender, age, social position, or historical situatedness. Illness or injury may disrupt this enormously delicate relation between self and world, and for the neuroscientists of Powers' novel *The Echo Maker*, like Freud a century earlier, mental illness or injury provided one avenue to understanding the normal process of identity formation. But experimental artists like Woolf and Joyce were no less concerned to make visible the differential emergence of many selves, and the difficulty of maintaining those selves over lifetimes. The village pageant of English history that forms the subject of Woolf's *Between the Acts* is only a literal token of the novel's intense awareness of time and perception for a gathering of men and women for whom self-understanding is as deeply threatened as it is for the central character of *The Echo Maker*. Like the brain-injured hero of Powers' novel, Woolf's characters all feel the ground of their selfhood to be destabilized, whether by war, aging, changing gender roles, or the disruptions of capitalist development. Perhaps the metaphoric image of water in Leibniz's writings that best captures the almost infinite multiplicity of this sense of self is neither the river nor the ocean, but simply the pond. In the *Monadology*, he describes the unimaginable plenitude that resides in even this most modest body of water:

> 67. Chaque portion de la matière peut être conçüe comme un jardin plein de plantes; et comme un Etang plein de poissons. Mais chaque Rameau de la plante, chaque membre de l'Animal, chaque goutte de ses humeurs est encore un tel jardin, ou un tel étang.
>
> 68. Et quoique la terre et l'air interceptés, entre les plantes du jardin ou l'eau interceptée entre les poissons de l'étang, ne soit point plante, ni poisson; ils en contiennent pourtant encore, mais le plus souvent d'une subtilité à nous imperceptible.[9]

---

8   "Nouveau Essais Sur L'Entendement," p. 55.
9   *Vernunftprinzipien der Natur und der Gnade; Monadologie* (Hamburg, 1956), p. 58.

67. Each portion of matter can be conceived as a garden full of plants, and as a pond full of fish. But each branch of a plant, each limb of an animal, each drop of its humors, is still another such garden or pond.

68. And although the earth and air lying between the garden plants, or the water lying between the fish of the pond, are neither plant nor fish, they contain yet more of them, though of a subtleness imperceptible to us, most often.[10]

However distant *The Echo Maker* may seem from the late modernism of *Between the Acts*, both novels regard the self as raised to distinctness out of a pool of infinite detail, where past and future do not simply flow past us like a stream, but always remain to be grasped, in the teeming depths of something more like a pond.

### *Between the Acts*

Virginia Woolf's final novel opens with a rather pointless and uncomfortable talk between Isa Oliver and Mrs. Haines, whose husband the unhappily married Isa fancies. When a bird outside "chuckled," Mrs. Haines asks if it is a nightingale and is told that it couldn't be a nightingale, that "It was a daylight bird, chuckling over the substance and succulence of the day, over worms, snails, grit, even in sleep."[11] The bird is imagined to have an intensely real inner life in dreams, one that exactly mirrors the reality of its outer, daytime world. A few pages later, Mrs. Swithin (Lucy), the elderly aunt of Isa's husband, is awakened very early by birds outside her window and picks up her favorite reading, an "Outline of History," to pass the hours before the rest of the household is up. Before long she has entered an inner world of her own and later stands at the window and sees "rhododendron forests in Piccadilly," populated by a variety of prehistoric creatures. This long past time is so vivid to her that when the maid enters, she has some difficulty returning to the present: "It took her five seconds in actual time, in mind time ever so much longer, to separate Grace herself, with blue china on a tray, from the leather-covered grunting monster who was about, as the door opened, to demolish a whole tree in the green steaming undergrowth of the primeval forest" (p. 9). As with most elderly women and men, the past is important to Lucy, who often speaks of her personal memories, but who also lives in some fuller sense in that past. The inward experience of it

---

10    *The Monadology* in Garber and Ariew, p. 78.
11    *Between the Acts* (New York, 1941), p. 3.

is very real to her and, mediated by a book in this case, it can even be a past reaching far back beyond her own lifetime.[12] Virginia Woolf often structures her narratives in such a way that moments of the past and the present intersect, as they do for Lucy. Her understanding of the problem of time was sophisticated, and contemporary interest in the topic was well known to her early in her career as a writer.[13] The novels of Virginia Woolf have long been seen as central to the Modernist exploration of inwardness, of the burden of the past in a time of uncertain futurity, and especially, in her writing, of the tiny perceptions (as Leibniz put it, the "indiscernibles") that link our inner lives to the reality of the world around us. Her representation of the structuring of perception within time remains vital to fiction's aesthetic nearly a century later.

*Between the Acts* is set at Pointz Hall, an historic house in the countryside, where a village pageant will take place in the afternoon. The narrator quotes an 1833 guide book that describes its fine setting, and before the guests begin to arrive, the family sits out in a sunny garden to admire the view, as they often do. At several key points in the novel, Woolf has a character reflect on the landscape, and always with a keen sense that what is perceived is not only seen from a particular point of view, but serves some vital self-interest of the perceiver. What Lucy sees is a view that reinforces her earlier reverie on the distant past. She sees a landscape that has not changed in centuries, which is nostalgic and comforting. She says its beauty is sad because it will be there "when we're not," even though other characters see a landscape threatened by cheap real estate development. But the notion she takes from the view is reassuring to her because she assumes the future will be much like the past, ignoring the reality of the present. But Giles Oliver, her nephew and Isa's husband, is far more intensely aware of the present, and for him it is not the past but the future that lies folded into the landscape he perceives. On the verge of war, he sees in the same view a "Europe, bristling with guns, poised with planes. At any moment guns would rake that land into furrows; planes splinter Bolney Minster into smithereens and blast the Folly" (p. 53). The very serenity of the countryside they both see is for Giles deeply disturbing because he perceives in it only a future destruction whose images are as real to him as are the dinosaurs of

---

12    Jane Duran, responding to Simone de Beauvoir's *Old Age*, discusses how age difference determines the reality of past, present or future for the characters in *To the Lighthouse* in "Virginia Woolf, Time and the Real," *Philosophy and Literature* 28 (October 2004), 300–309.

13    Ann Banfield, *The Phantom Table: Woolf, Fry, Russell and the Epistemology of Modernism* (Cambridge, UK, 2000) and "Time Passes: Virginia Woolf, Postimpressionism, and Cambridge Time," *Poetics Today* 24/3 (Fall 2003), 471–516.

Lucy's ancient forest. His greater awareness of the very real events in Europe's present means that in moments such as this, Giles lives emotionally in future time. Woolf is quite explicit in suggesting what it is these characters are doing when they gather the perceptual confusion of a distant landscape into distinct ideas and strong emotions. When Oliver, Lucy's brother, regards the landscape in an earlier scene in the novel, the narrator allows him a moment of self-consciousness about his own apperception, using a metaphor which emphasizes that every distinct perception is an act of mind: "he surveyed the landscape—flowing fields, heath and woods. Framed, they became a picture. Had he been a painter, he would have fixed his easel here, where the country, barred by trees looked like a picture" (p. 13).

Later, when Lucy and Mrs. Sands prepare sandwiches for a supper to be served during the pageant's interval, they both see bread and ham being cut, but their shared perception leads to two strikingly different moments of inner reverie:

> The cook's hands cut, cut, cut. Whereas Lucy, holding the loaf, held the knife up. Why's stale bread, she mused, easier to cut than fresh? And so skipped, sidelong, from yeast to alcohol; so to fermentation; so to inebriation; so to Bacchus; and lay under purple lamps in a vineyard in Italy, as she had done, often; while Sands heard the clock tick; saw the cat; noted a fly buzz; and registered, as her lips showed, a grudge she mustn't speak against people making work in the kitchen while they had a high old time hanging paper roses in the Barn. (p. 34)

Lucy looks at bread but sees a distant past of classical myth and Mediterranean ease, while cook's perception, structured by her own class position as a servant in a big house, looks at ham and remains resolutely in the present, her lips signaling anger and resentment at the conditions of her work. Thoughts that could barely be called conscious emerge in radically different forms within a single shared setting.

Isa's struggle to understand her present self is both deeper and more problematic than either Mrs. Sands' focus on immediate perception or Lucy's easy escape into a past of the mind. Like Lucy, Isa is drawn to books as a possible mediation, a way to frame her lived experience of time. Upon entering the library, she says "The library's always the nicest room in the house," but her remark (or thought) is punctuated as a quotation, a statement she has heard, perhaps wants to believe, but cannot truly feel. She sees the authors and titles of the literary canon (*The Faerie Queene*, Keats, Shelley, Yeats, Donne), then turns to biographies of famous men, to county histories, and finally to science.

But while she compares herself to a person with a raging toothache, looking for a cure, she finds no book in which to lose herself, as Lucy has done. "Book-shy she was, like the rest of her generation" (p. 19). But it is more complicated than that. Looking at the titles, she muses: "There they were, reflecting. What? What remedy was there for her at her age—the age of the century, thirty-nine—in books?" Seeking a personal "remedy," she identifies her own selfhood with the thirty-nine numbered years of her life, and those with the thirty-nine years of the century. The reality of "her age" is thus imagined in the most objective, timeline-of-history kind of terms, and it is this reality that she evidently expects books to reflect, though she knows that Keats and Shelley, and even *The Antiquities of Durham*, cannot do this.

That books fail her, however, does not mean that the inward consciousness she hopes to feed is not textually mediated: "For her generation, the newspaper was a book." When her father-in-law puts down his copy of the *Times*, she picks it up and reads: "'A horse with a green tail...' which was fantastic. Next, 'The guard at Whitehall...' which was romantic..." (p. 20). Her reading is at first enabled by literary expectations, but as she reads on she learns that the story is about a girl who is abducted and sexually assaulted by troopers, who drag her to their barracks. She concludes "that was real." But then Lucy comes in to put away the hammer and nails with which she had been mounting a placard for the pageant, and within a page, Isa begins to blend newspaper story and household tool into private fantasy: "The same chime followed the same chime, only this year beneath the chime she heard: 'The girl screamed and hit him about the face with a hammer.'" But this momentary fantasy is not allowed to become lengthy reverie, as in the next line, Mr. Oliver is turning the pages of his newspaper to find the weather forecast, which he then proceeds to read aloud. Woolf's juxtaposition here makes the obvious but very important point that the newspaper, by its very nature, is radically discontinuous. Every instance that Isa might term "real" is immediately set against celebrity, house fires, soap ads, and weather forecasts. In this sense the newspaper does reflect the incoherence of modernity felt by Isa, by Miss LaTrobe later in the novel as she attempts to stage her pageant, and by the audience as it tries to make sense of what it has just seen.

We do not argue that Isa's structure of feeling is caused by her reading a newspaper. We do propose that there is a congruence between the newspaper's textual discontinuity and the failure of so many of Woolf's characters to feel history as continuous. After the pageant is over, Lucy makes a simple remark that assumes a traditional sense of progressive time and causation: "We made more this year than last, he said, but then last year it rained" (p. 214). Isa's response suggests how little she can feel the truth of Lucy's kind of time:

"'This year, last year, next year, never...' Isa murmured." At the final scene of the pageant, the audience realizes that "the hands of the clock had stopped at the present moment. It was now. Ourselves" (p. 186). Here it would seem that Virginia Woolf nods to St. Augustine, when he writes that "nothing can be seen except what is present" (*Confessions,* XI.18.23).

Although Lucy and the others spend a great deal of time talking and thinking about past and future, they also feel that time is somehow nonexistent. Before the pageant begins, there is a kind of still point in the novel, when the empty dining room is described: "Empty, empty, empty; silent, silent, silent. The room was a shell, singing of what was before time was; a vase stood in the heart of the house, alabaster, smooth, cold, holding the still, distilled essence of emptiness" (pp. 36–37). After the pageant, the family returns to small tasks, and Woolf returns to the newspaper's discontinuity, but this time with an additional sense of the ephemerality of the Real it presumes to give us: "The paper crackled. The second hand jerked on. M. Daladier had pegged down the franc. The girl had gone skylarking with the troopers. She had screamed. She had hit him.... What then?" (p. 216). Ezra Pound called literature "news that stays new," and the harsh judgment of the press his quip implies is confirmed when the morning paper is brought in to the drawing room and described as "the paper that obliterated the day before." Its radical destruction of the past is made perfectly clear.

The daily paper is only one manifestation of the unsettling modernity which Isa and the other characters see all around them and which they struggle throughout the novel to accommodate to an inner perception of self shaped, to greater or lesser degrees, by a rapidly vanishing past. For Lucy, the past is large, and still very moving to her, while for the younger people, the past may be less affecting, since they perceive the present in much greater detail. The men and women on the street who look up to watch an airplane skywriting in *Mrs. Dalloway* might have seen an insubstantial and endlessly changing future of consumption in which even the sky may be overwritten by advertising. But more likely they only saw something wonderful, even beautiful. In the same novel, Peter has a similarly startling experience of change when he returns from many years in India to discover daylight savings time. The recent innovation of seasonal time change also figures in the trivial chatter of the audience in *Between the Acts* ("That's one good the war brought us—longer days"). That the clock time, which had regulated lives since early in the industrial era could be changed again, altering the felt experience of a summer evening, contributed to the feeling that modernity was changing every aspect of life, oftentimes speeding it up. The audience again free-associates as the play continues: "Each flat has its refrigerator...It took my mother half the morning to order dinner...

We were eleven. Counting servants, eighteen in family…Now they simply ring up the Stores" (p. 159). In this casual chat, Woolf shows how her characters familiarize social change by placing it in the context of the past, as memory allows the new to be felt as personal experience.

The historical pageant that serves to bring such diverse individuals together is an attempt by Miss LaTrobe, the director, to forge some kind of collective historical memory. Woolf's narrative devotes as much time to the crowd's running commentary on the play as it does to the actual content of the pageant. The comments, presented in a free indirect style in which individual speakers are seldom identified, reveal how often a word, a line of a song, or some detail of an actor's costume may call forth private memories and associations. Just as an "outline of history" invariably contains some overall structure or concept (the story of human progress, perhaps, or the rise and fall of great empires), an historical pageant may be assumed to have some underlying theme or idea, or at the very least, some principle of selection for what would otherwise be seen as a series of unconnected vignettes. This is very much the expectation of the audience, as one voice after another attempts to suggest what it all means. If there is any controlling idea, however, it is lost on the viewers: "I thought it brilliantly clever…O my dear, I thought it utter bosh. Did *you* understand the meaning? Well, he said she meant we all act all parts…. He said, too, if I caught his meaning, Nature takes part…. Then there was the idiot…. Also, why leave out the Army, as my husband was saying, if it's history? And if one spirit animates the whole, what about the aeroplanes?" (p. 197). The woman who repeats an earlier objection about Miss LaTrobe's failure to include the Army reminds us that History is conventionally thought to be about military victories and defeats, while another voice picks out a specific detail (the idiot) that seems to fit no scheme at all. And the final comment breaks the plane of the artwork entirely, expecting that a squadron of planes that happened to fly over the estate in close military formation must be seen as part of the play as well. Since Miss LaTrobe did make the outdoor setting central to her production, the comment is not entirely misplaced.

Vivid images are often said to be "etched in memory," and whatever the textual trace in book, newspaper, theatrical performance or personal remembrance, the characters of *Between the Acts* have only the scraps and orts of what they see and what they remember with which to construct the present. After the pageant is over Lucy spends some time gazing into a lily pond very much like Leibniz's fish pond of infinite detail. She notices that the lilies are folding up and then sees a jagged leaf that makes her think of Europe. This leads her to look at other leaves, naming them "India, Africa, America. Islands of security, glossy and thick" (pp. 204–205). By this time the outward reality of the pond has merged with her own inward need, and she continues her private herme-neutic of the fish pond, now finding the fish and noticing how varied they are ("silver; pink; gold; splashed; streaked; pied.") Her reverie finally leads her back to the climax of the play, when mirrors had been turned to the audience, which responded, "It was now. Ourselves." Lucy again murmurs "Ourselves," and con-tinues to gaze into the pool: "she followed the fish; the speckled, streaked and blotched; seeing in that vision beauty, power, and glory in ourselves" (p. 205).

Near the end of the pageant, Isa has a similar experience of noticing a tiny detail out of which she unfolds the future. Swallows (or possibly martins) sud-denly fly through the trees and land on a wall: "Yes, perched on the wall, they seemed to foretell what after all the *Times* was saying yesterday. Homes will be built. Each flat with its refrigerator, in the crannied wall. Each of us a free man; plates washed by machinery; not an aeroplane to vex us; all liberated; made whole..." (pp. 182–183). Her reverie is cut short by a "cacophony" of jazz which seems to belie her utopian vision, already impossibly compounded of genuine yearning, fear of invading German warplanes, day-old newspaper hype of development, and a half-remembered phrase from Tennyson. In an essay responding to the suggestion that he might be grouped with recent "systems novelists," Richard Powers describes this new genre in a way which makes it sound very far removed from the art of Virginia Woolf: "It is exhaustive and exhausting. It works by accumulation, changes of magnification and gauge, tangled networks of data, overloads of information-retrieval that attempt to map large-scale ecological, social, political, and scientific wholes." Woolf offers a few deft lines and details, not an encyclopedia, and yet Powers' conclusion could not be a more apt description of the temporal discontinuities that bur-den the present for Woolf's characters: "For these are the conditions under which the modern self emerges. We cannot know who we are without knowing where we are."[14]

### The Echo Maker

Woolf's weaving together of the force of past perceptions and anticipations of the future characterizes the fleeting and unstable present moment of her char-acters' inner lives. In like manner, Richard Powers explores the cacophony of inwardness from a twenty-first century perspective in *The Echo Maker*. While Woolf's fiction coheres around the historical pageant presented at Pointz Hall,

---

14     "Making the Rounds," in *Intersections: Essays on Richard Powers*, eds. Stephen J. Burn and
        Peter Dempsey (Champaign / London, 2008), p. 306.

Powers structures his narrative on the stages of Mark Schluter's response to a debilitating brain injury resulting from a truck crash he can no longer remember. Powers' chief narrative strategy is his nuanced handling of point of view in evoking inner lives, differing in style from one character to another, and including even the sand hill cranes who saw the accident. *The Echo Maker* is a many-faceted mystery story: What caused the improbable truck crash that injured Mark Schluter? Who left the note found on Mark's bed that seems the only clue: "I am No One / but Tonight on North Line Road / GOD led me to you / so You could Live / and bring back someone else"? What could trigger his reintegration with his past and reunion with the sister who has given up her own career to tend him? What explains the devoted persistence of Barbara Gillespie, Mark's nurse's aide? And implicated in all of these: how does the brain use the incidents and perceptions of a fleeting present to produce the narrative of identity that mediates selfhood?[15]

Karin Schluter, Mark's older sister and only living relative, leaves her consumer relations job in Sioux City in the middle of the night to watch over her brother's recovery in their home town of Kearney, Nebraska. Her selfhood seems rooted in their childhood bond, but when Mark recovers speech but no memory of the accident, he does not—or will not—recognize her. To her brother she is a cleverly designed robot or trained surrogate, part of an elaborate conspiracy against him. Danny Riegel, a committed (even saintly) environmentalist working for the Buffalo County Crane Refuge and school friend of both Schluters, offers Karin a lover's support and shelter. When Mark is diagnosed with Capgras Syndrome, Danny supplies Karin with Gerald Weber's well-known books on the brain and its disorders. Weber travels to Nebraska at Karin's urging to be a professional outside expert, but becomes increasingly involved in the case emotionally. Barbara Gillespie, Mark's obviously over-qualified nurse's aide, continues to devote herself to his recovery after his release from the rehabilitation facility that followed his hospitalization. She seems to have a strange rapport with Mark as well as an unexplained connection with Weber. Although Mark's traumatic loss of his factual memory of the

---

15    Bruno Latour finds Powers's open-ended approach to character a sign of its true realism: "What, then, is more realistic? To act as if continuity of existence was an unproblematic given? Or to show that it can be a highly variable gradient, which can be intensified or attenuated as the story unfolds? For Powers, *being* is never given; it always has a *temporal* character." See "Powers of the Fasimile: A Turing Test on Science and Literature," in *Intersections: Essays on Richard Powers*, in Burn and Dempsey, p. 267. To neglect its temporal dimension would be to reduce to a mere concept the rich tangle of tropes and metaphors that constitute the true realism of literary characters (p. 281).

crash and emotional connection with Karin structures the story, all five of these intertwined lives are laden with the past as memory and (mostly fearful) anticipations of the future, sometimes only slowly brought to consciousness as apperceptions. The echo makers of the title are the sand hill cranes that migrate to the Platte River near the North Line Road. The birds, given their name by the Anishinaabe clan because of their loud calls, are accounted for at the beginning of each of the five sections of the novel, sometimes from their own vantage point.

The mystery of time for the cranes (past present and future) is posed from an observer's angle on the first page of the book: the migrating flock converges "on the river at winter's end as they have for eons, carpeting the wetlands. In this light, something saurian still clings to them: the oldest flying things on earth, one stutter-step away from the pterodactyls" (p. 3). But the next page suggests an inner life: "This year's flight has always been. Something in the birds retraces a route laid down centuries before their parents showed it to them. And *each crane recalls the route to come.*"[16] The only witness to the nearly fatal crash of Mark Schluter's truck on a straight, little traveled, road may have been this immense gathering of birds. A year later, a crane whose mate has been shot (it is open season) seems to have retained the image of the crash: "She grows jittery again, remembering that ancient incident, the trauma of last spring. *Something bad once happened here, as loud and deadly as the new fatal wrong*" (p. 274, italics added). The birds can't testify ("What does a bird remember? Nothing that anything else might say" [p. 443]), but they haunt its prose as fleeting white images that also pose the problem of perception, apperception, and memory. The otherness of their long history—a stutter-step from the pterodactyls—their strange momentary kinship with the characters (mirror neurons? whose?), and their future along the Platte River in Nebraska is woven into the story of the fates linked by the accident.

Theories about the temporal relation between inward states and the world outside are of necessity under scrutiny in a novel about an injured brain. Just as Woolf uses books and newspapers to display her characters' thoughts—and her motif of discontinuity—the mental lives in *Echo Maker* are mediated by Weber's books. From the outside—the surface of the pond in Leibniz's figure—Mark's could be a case like many of Weber's, which have impelled him to write about the "endless surprise folded inside the most complex structure in the universe," with its billions of cells and inter-cellular connections. "Any given brain can put itself into more unique states than there are elementary particles in the universe" (pp. 93–94). Karin, reading Weber's book about extraordinary

---

16    *The Echo Maker* (New York, 2006), p. 4 (italics added).

mental conditions, hopes that the analysis of Mark's almost unique state, Capgras brought on by trauma, will lead to his cure. After meeting Mark, Weber appears on TV giving a general account of conscious activity that "seduces" Karin to hope all over again:

> Consciousness works by telling a story, one that is whole, continuous, and stable. When that story breaks, consciousness rewrites it. Each revised draft claims to be the original. And when disease or accident interrupts us, we're often the last to know. (p. 185)

Mark's inner life, as Karin intuits, is neatly described by this perspective, Weber's scientific position, a position he sees as part of the larger fabric of institutionalized work to which he contributes. His encounter with Mark, though, will force Weber to look beneath this surface (which, though not illusory is very general) in order to see the multitude of "petite perceptions" that swarm beneath it. This second part of his inquiry will cost Weber dearly (of which more later).

Karin's point of view is presented for the most part from a third-person account of her actions and thoughts, although those thoughts are often deep doubts minutely scrutinized. A notable exception is Karin's response to a public debate she and Danny attend on how to deal with a development plan that would impact the crane's migration habitat, which is Kearney's chief and only tourist attraction. She knows the seriousness of the stakes for Danny and the birds whose cause he represents when the development uses yet more of the water in the shallow river. She also knows that Robert Karsh, chief among the developers and another of her high school boyfriends, will present a slick, sophisticated defense of the project, likely to win the struggle with Danny's "guilt and facts." These are conscious judgments, but suddenly Karin feels the debate as a contest inside herself; in "a weird, fleeting *fugue* moment, she felt privately responsible for the whole contest" (p. 346, italics added). The single word "fugue" signals a discontinuity that goes beyond rational conflict, areas within the brain simultaneously singing different songs.[17]

The cacophony of the speakers and their motives striving to control the Development Council's deliberations matches Weber's account of the inner workings of all minds, and his influence has impelled Karin to imagine her own mind as like the public she sees before her:

---

17    Douglas R. Hofstadter, whose *Gödel, Escher, Bach: An Eternal Golden Braid* (New York, 1979, rpt. 1999) comes to mind often when reading novels by Powers, subtitles his book "A Metaphorical Fugue on Minds and Machines in the Spirit of Lewis Carroll."

How many brain parts had Weber's books described? A riot of free agents; five dozen specialties in the prefrontal bit itself. ...And they all had a mind of their own, each haggling to be heard above the others. ...A wave moved through her, a thought on a scale she's never felt. No one had a clue what our brains were after, or how they meant to get it. If we could detach for a moment, break free of all doubling, look upon water itself and not some brain-made mirror...For an instant, as the hearing turned into instinctive ritual, it hit her: the whole race suffered from Capgras. Those birds danced like our next of kin...[but we merely find them] a strange spectacle to gaze at from a blind. (pp. 347–348)

This internal monologue presents a moment in Karin's inner life compressing her neuroscience book learning, her direct experience with Mark's illness, and her long history with the cranes. As she sits beside Danny listening to the highly emotional hearing, a public event, she experiences its voices as a metaphor for the competing functions of the various parts of the brain. Then, in a sudden powerful mental "wave," the metaphor is suffused with her insight into Mark's Capgras as the general human condition in which the workings of intellect are disconnected, so that we no longer recognize emotionally what our reason tells us. Mark knows who his sister is, but fails to "see" her as kin; so it is with the developers and the birds.[18]

This passage works out three of the major strands of the novel's unfolding plot, positing them in a seeming instant of inner time. From Danny's point of view, this hearing is crucial to his attempt to save this segment of the Platte for the migrating cranes. In terms of historical time, his is the long view; he knows, and the novel (not to mention the real world) justifies his knowing, that the ecological effects of the proposed development will be disastrous for the migratory pattern of the birds. And Danny is no hypocrite; he never wavers from his self-effacing, eco-friendly style of life. Karsh's plan was obviously self-defeating in the long, and even the medium-long, run, since the birds could not come without the water, and the tourists would not come without the birds. Leibniz's pond is again relevant. Seen as a surface, the pond looks homogenous, but looking into its depths reveals a world of fishes and plants, each made up of still more complex entities. Seen as a public debate, the hearing is predictably contentious, but graspable, pitting the noble Danny against the opportunistic Karsh, settling into a near parable of a self-defeating capitalist

---

18    The intricate way this passage weaves together themes, events, and mind-sets justifies Ellen Siegelman's emphasis on the musicality of *The Echo Maker* in "Echoes of Memory, Echoes of Music," *Jung Journal* (2007), 46–54.

exploitation of nature. Seen as many minds, each teeming with a network of perceptions struggling to become apperceptions, it seems hopelessly complex. Seen against the eons-long history of life on earth, it looks like an evolutionary chapter that bodes no good for the survival of birds or humans.

Karin had built her conscious identity by attempting to save herself and her brother from their disastrous upbringing by unloving, nearly demented parents. Her long, apparently futile, struggle to reach Mark, though, gradually drains her of her sense of a continuous willed self, which reaches its nadir when Mark overdoses on the medication Weber has prescribed and she has agreed to: "She is nothing. No one. Worse than no one" (p. 407). Her musing here echoes the note on Mark's bed, which begins, "I am No One." She also seems to have internalized Weber's "Consciousness works by telling a story, one that is whole, continuous, and stable. When that story breaks, consciousness rewrites it. Each revised draft claims to be the original" (p. 185). At this low point, she doubts her usefulness to Mark, to Danny, and to the Refuge, although she throws herself into her work there with a furious energy. Her conscious grasp of the ecological facts she faces in this work surprises her by awakening her senses to the colors and scents of the natural world and connects her with what she had learned from Weber: "When one brain part is overwhelmed, another takes over" (p. 407).

While Karin's inner life is seen only fleetingly in this way, the interior monologues that reveal the inner lives of Mark and Weber are often and more elaborately presented. Mark's inner life is, of course, the very matter under scrutiny, for both the mystery story and the medical case history. His inner reasoning is often given as third-person narrative, which is meaningful in itself, since his problem involves a severing of one brain function from another. It sounds like this:

> They're after Mark Schluter's ass: this much is obvious. A man would have to be a vegetable to miss that much. Setting him up in some kind of experiment, some of it so hokey that even a child still stuck on Santa would snicker. But some of it so complex he can't even start figuring it. (p. 199)

It must have been planned before they removed his memory on the operating table; it must have caused the accident in the first place.

> The secret is out there in the empty fields. ...He needs a witness, but nobody saw what happened that night except the birds. Catch himself one of those cranes, one that was there, alongside the river. Find him a

sand hill, and swear it in. Scan its brain. ...The real deal is hidden behind
doubles. He has only one clue. One solid thing beyond doubt: the note.
The words from the person who found him, the one spectator to that
night's events, before the weirdness set in. (p. 199)

The inner life of none of the other characters is presented with this intensely
subjective flow masquerading as plain fact and logic.

When Mark tries to find the note writer he succumbs to another syndrome:
Fregoli's Delusion, that different people are the same person in disguise, in his
case the unshakable impression that most people he meets going door to door
asking for handwriting samples are really the same person. He decides it must
be "Danny Riegel, the birdman of Kearney" (p. 255). Not that Danny has been
continuously on Mark's mind. When he first found out that Karin was living
with Danny, it took him an effort to remember his close high school friend: "He
wrestled with the name, fitting the long past into the five seconds of fleeting
present where he now lived" (p. 216). (Saint Augustine's account of time
through the lens of neuroscience!) Karin begs Weber to come back to Nebraska,
hoping that he can see a way out of Mark's newly complicated condition. Mark
comes to hope so too; even though *she* is Kopy Karin, he trusts Weber.

Weber's own life becomes more and more deeply involved with the
Schluters. The mixed reviews of his latest book cause him a more devastating
disappointment than one would expect from so respected and confident a sci-
entist. Bothered by jet lag, his conference keynote address in Sydney goes
badly, and he suspects "the whole invitation might have been a setup," his audience filled with the "smiling faces of a species that hunted in packs" (p. 231).
His self-doubt fuels a growing need to cure Mark, to prove that his knowledge
of the brain, though presented to a "popular" readership, deserves medical
respect and does not merely exploit neurological oddities for his own fame.

Whenever he is in Nebraska, Weber calls his wife Sylvie every night. On his
second visit, after he has told Mark about his own brother and eaten the
reheated lasagna from his "surrogate sister," offered in a sort of bonding ritual,
Sylvie notices a change in her husband's tone. "Are you okay? You sound different, somehow. Your voice is very...I don't know. Unfolded. Like a philosopher or
something" (p. 305). Something has indeed unfolded within Weber's own complex neural map (or Leibniz's proposed folds of the soul). His books had
acknowledged that the self we acknowledge, "the whole, willful, embodied,
continuous, and aware" self that it is the job of consciousness to construct has
been deconstructed by science (p. 381). Yet he had not felt that discontinuity
emotionally until his bafflement with Mark's case and the dismissive review of
his book converge to undermine his sense of self: "Mark Schluter had gradually

dismantled his most basic sense of acquaintance, and nothing would ever seem familiar or linked again" (p. 355). Like Karen, Weber finds that Mark's crisis has triggered his own.

Registered in free indirect discourse, he faces his students in a university lecture hall:

> Their manner had changed this semester, turned sardonic. They had passed around public indictments of him by email and instant messenger. They still wrote down every word he said, but more now to catch him out, to root out charlatanism, their pens angled in challenge. (p. 357)

Here Weber's inner life is almost as out of sync with external realities as Mark's. His students seem mocking or hostile: "Bhloitov was furious....Miss Nutfraddle seemed ready to call the attorney general on her Blackberry" (p. 363). When another student, whom he calls "Young Silvie" leaves without asking an after-class question, Weber takes her leaving as desertion (perhaps even that of the real Sylvie): "*Come back. It's me. Still here*" (p. 366). The reader knows that Bhloitov is not furious because he does stay after class, humbly questioning Weber, worried that perhaps his father suffers from one of the syndromes introduced in the lecture. The young man's self-effacing question, however, does not bring Weber back to the "reality" we see in the scene. Instead he contemplates retirement, unable to face the "pens angled in challenge" to his authority.

Weber has always written and taught that the stories he tells about brains weirdly damaged or diseased are just extreme cases of our shared condition in which "the job of consciousness is to make sure that the distributed modules of the brain seem integrated" (p. 363), but his loss of confidence in his own continuous willed life is here badly shaken. He feels himself "eroded" by the mystery of the accident; he overreacts to bad reviews and begins to find his usual classroom case histories hollow. He seriously considers retirement. But like Karin, who found her physical senses oddly vivified by her loss of confidence in her purposeful self, Weber surprises himself by falling in love with Barbara, not because he knows her, but because she seems to him an urgent unsolved mystery. Always seen from the outside, Barbara acts with obvious resolve and intelligence far above her current job description, but reveals nothing of her own motives. No glimpse of her inner life appears behind her guarded dialogue; she is perhaps the most puzzling feature of this enigmatic narrative landscape. From their first meeting, Weber has experienced mere tinglings (tiny perceptions) of kinship with Barbara, and the erosion of his own confident self possession brings this sense of kinship to his conscious mind as he deduces that a similar disintegration must have undone *her*. His curiosity about her merges with his own feelings of self-estrangement. Something overwhelming seems to be at stake for her in Mark's recovery, and since that is now true for Weber as well, he goes back to Nebraska a third time in spite of the damage it inflicts on his marriage to the forbearing Sylvie, whom he knows so well. Weber has all along regarded the human mind as "the most complex structure in the universe," its workings always springing surprises, but his dispassionate professionalism deserts him as he contemplates and makes this third journey. He experiences his own mental equipment as a "riot of free agents" that his consciousness struggles to make into a narrative of self—he can't explain his state to himself any more than to Sylvie. And the reader must remain puzzled until almost the very end of the novel to see whether the ordinary world of circumstance and the inner worlds of the five main characters can be folded into a satisfying literary narrative.

• • •

Virginia Woolf and Richard Powers give us memorable insight into that crucial and fragile process through which consciousness of world and self are shaped out of a plenitude of perceptions that would otherwise overwhelm. For both authors, apperception is the consciousness of time, and it begins in confused perceptions like those of the crane whose mate was shot the year before and now senses something vaguely disturbing. If the crane were able to bring her uneasiness to more distinct consciousness, she might understand it as loss of the familiar past and anticipation of change in the future. At a still higher level, a character like Danny would know the cranes as a species to be caught between past and future, a past largely unchanged over countless generations and a future of likely oblivion. Gazing at the plenitude of an English landscape, Lucy fails to see it ravaged by future bomb craters, but she does see its distant past of tree ferns and dinosaurs. For aging English lady and vanishing North American bird alike, past and future may be experienced in the present, sometimes with startling clarity. It is as if, within consciousness, time were folded into the present, visible and inescapable.

The crane continues on its migratory flight, and Lucy slices ham for sandwiches. But for Mark Schluter the infinite detail of memory is terrifying because it cannot be reconciled with his perception of the present. The process through which the sense of self is continuously reassembled within the consciousness of past and present is broken. When the characters in these novels achieve a higher level of thought, there is often some kind of textual mediation, whether Lucy's Outline of History, or Isa's romantic poetry or newspaper "reality." For

Mark, it is video games, which offer a language of danger and uncertainty, of wariness, objects or opponents to be overcome, and a kind of time that is endlessly replayable. The past is not only there to be touched at certain moments: in the cyclical time of the game, it is always present.

When the perception of nature comes to clarity for a character like Danny, he understands what he sees in terms of environmental destruction and the narrative he writes is about preservation. For Weber, the mediating text is professional, the discourse of neuroscience. His perceptions must become data, and his data must issue in theory and the narrative of individual case study. But in this instance, the narrative can no longer contain the overwhelming detail of Mark's life. *Between the Acts* was written under the shadow of impending war and in a landscape threatened by modernity's ruthless development. *The Echo Maker* is also burdened by war, this time in Iraq, and development still threatens the landscape. As early as the twelfth century, in the *Lais* of Marie de France, narrative fiction had explored the inner lives of men and women coming to consciousness in a vanishing present laden with the past and pregnant with the future. Read together, Woolf and Powers show that art, like human consciousness, is always new, but always old as well.

CHAPTER 16

# Temporality and Control in Sondheim's Middle Period: From *Company* to *Sunday in the Park with George*[1]

*Raymond Knapp*

### Revisiting the Past in the Musical's Present

Stephen Sondheim's musicals from the early 1970s marked out new territory in many ways. In professional terms, they established him as an accomplished Broadway composer and lyricist, with *Company* (1970), *Follies* (1971), and *A Little Night Music* (1973) receiving mostly favorable critical notices and multiple Tony Awards, including Best Original Score for each. Even if these shows did not enjoy exceptionally long runs,[2] they launched his extraordinarily fruitful working relationships with producer-director Harold Prince and orchestrator Jonathan Tunick, and remained for many years the cornerstone of Sondheim's career and reputation. Thematically, too, these musicals set the terms for Sondheim's Broadway career. All three shows explore intricately interwoven sets of personal relationships while making anxious gestures toward an ever-receding past. Thus, *Company*'s not-so-young urban professionals cling to a

---

1   An earlier version of this essay, "Layered Temporalities in Sondheim's First Maturity: From *Company* to *Pacific Overtures*," was presented as part of "Time: Sense, Space, Structure: An Interdisciplinary Symposium" (The Claremont Colleges and Graduate University, February 22–24, 2007). A later version expanded the scope of the essay, "Bobby and Ben, Sweeney and George: On the Vagaries of Musical Control in Sondheim's 'Personal' Shows," presented as part of "Sondheim's Musicals: A Lecture and Panel Discussion" (co-hosted by the Department of Music and the Department of Drama, University of California, Irvine, March 9, 2007). To a much smaller extent, this essay also draws on earlier work of mine, *The American Musical and the Formation of National Identity* (Princeton, 2005), *The American Musical and the Performance of Personal Identity* (Princeton, 2006), and "Marking Time in Pacific Overtures: Reconciling East, West, and History within the Theatrical Now of a Broadway Musical," in *Musicological Identities: Essays in Honor of Susan McClary*, eds. Steven Baur, Raymond Knapp, and Jacqueline Warwick (Aldershot, UK, 2008), pp. 163–76.

2   Of Sondheim's shows, only *A Funny Thing Happened on the Way to Forum* (1962) had an initial run of over two years. Among his other shows, only *Into the Woods* and *Company* had initial runs over 700 performances (but less than two years), the same level of success enjoyed by *West Side Story* and *Gypsy*, for which Sondheim wrote only the lyrics.

© KONINKLIJKE BRILL NV, LEIDEN, 2016 | DOI 10.1163/9789004312319_018

1960s-styled sophistication; in *Follies*, aging members of an earlier generation of American urban sophisticates look to their pasts with rueful disappointment; and, in *A Little Night Music*, their Old-World counterparts from a still earlier generation try with similar pathos to reclaim the lost opportunities of their youth.

Sondheim's progression backwards and outwards continued through the 1970s with *Pacific Overtures* (1976) and *Sweeney Todd* (1979), each distinctively steeped in regret for a past that cannot be overturned. Although Sondheim's collaboration with Prince ended after the disastrously short run of *Merrily We Roll Along* (1981), regret over the past has since remained a consistent theme in his shows. Indeed, the fateful *Merrily* was itself a story of betrayed ideals, conveyed within a Pinter-like reverse chronology.[3] Sondheim's next show, *Sunday in the Park with George* (1984), written and directed by James Lapine, returned from the far-flung nineteenth-century settings of Sondheim's shows from the late 1970s to the American present in two big strides, starting in late-nineteenth-century Paris for the first act, but shifting to contemporary New York for most of the second, as part of a generational reconsideration of legacy. This restoration of a contemporary perspective on the past, expressed dramatically within a two-act structure, curiously foreshadows a similar shift in Sondheim and Lapine's next collaboration, *Into the Woods* (1987), where a more oblique shift in dramatic sensibilities between the two acts again takes us from a temporally remote European outlook—that of the traditional fairytale—to something like the American present, when the first act's linked fairytales go awry in the second.

These musicals provided Sondheim with repeated opportunities to explore, lament, and sometimes challenge the oppressive sway of historical time, indulging a tendency that has continued with extraordinary dramatic success in his two original shows of the 1990s. Thus, *Assassins* (1990) brings together more than a century's worth of American assassins and would-be assassins into the semblance of an alternative, disenfranchised community, whereas *Passion* (1994), through the shared singing of letters, overturns the tyrannies of time and space within a more personal sphere, thereby collapsing the writing and delivery of those letters into an experienced simultaneity.[4]

---

3  *Merrily* ran for only 16 performances, although the show—especially its score—has acquired fervent admirers. There have been many successful revivals, which often rearrange the plot in chronological order and/or include new songs. It has been especially successful in London, where the 2000 production won the Olivier Award for Best Musical, but it has never been revived on Broadway. Because of this complicated history, and because its success has come later than the period I am considering, I have not included *Merrily* in my discussion.

4  I discuss *Into the Woods* and *Passion* in *The American Musical and the Performance of Personal Identity*, pp. 150–163, 303–309. I discuss *Assassins* in *The American Musical and the Formation*

Musicals provide a particularly flexible and effective medium for the dramatic temporal confrontations that these manipulations all entail, stemming in part from the convention of the reprise, a traditional response to the felt need to replay something of the past, in affirmation, as part of an effective conclusion to a musical narrative. Indeed, going back at least as far as *The Merry Widow*, reprises in operettas and musicals have often ostensibly recalled music whose evocative reach into the past exceeds the show's actual narrative span, and whose suggestive power provokes those in the musical's present to recapture an otherwise forgotten past (as, for example, in *Kiss Me, Kate*'s "Wunderbar").[5] Because music lives both in memory and in the nowness of actual performance, it routinely collapses temporality in this manner, allowing and sometimes virtually demanding us to experience ("taste") the past in ways that other artifacts—photographs, narratives, poetry—do not, while also remaining elusively indeterminate, evoking yet refusing either to recover or to re-create the past in any more substantial way. Inherently, then, the temporal themes of Sondheim's musicals seem built-in not only to the musical as a genre, but also—and even more so—to the ways music works in the "real world": able to recall the past to us with aching vividness, yet always reminding us—in a way musicals have traditionally refused to acknowledge—that the recalled past cannot simply merge with present realities.[6]

### The Rhythms of Life

But there is another, more immediate way in which the music in Sondheim's shows manages temporality. Since music, among the arts, has the most intimate relationship to the flow of time itself, it often seems actually to regulate time, through its meter, rhythms, and forms. The reality, of course, is quite different: time is the master that music only *seems* to harness. Music must in every particular honor time as it inflects it and colors our experience of it; thus, even as music seems to set the terms for our experience of time, it must itself

*of National Identity*, pp. 162–176, and *"Assassins, Oklahoma!*, and the 'Shifting Fringe of Dark Around the Campfire,'" *Cambridge Opera Journal* 16 (2004), 77–101.

5   I discuss *The Merry Widow* in *The American Musical and the Performance of Personal Identity*, pp. 20–31; regarding shared or remembered songs in musicals, see also pp. 37–39 ("Ah! Sweet Mystery of Life" in *Naughty Marietta*), p. 108 ("Come What May" in *Moulin Rouge*), pp. 273 and 414, n. 11 ("My Ship" in *Lady in the Dark*), pp. 281–82 ("Wunderbar" in *Kiss Me, Kate*), and pp. 327–328 ("You Were Dead, You Know" from *Candide*).

6   Regarding the "real-life" dimension of this device, see Theodor Reik's *The Haunting Melody: Psychoanalytic Experiences in Life and Music* (New York, 1960).

conform to the terms it sets, with every move it makes having either to fit into established temporal patterns or risk collapse into chaos. And it is, once again, in the space between the appearance and reality of music's relationship to time that Sondheim finds purchase for his own rich engagement with this dimension, wherein music, as experienced in the "now," serves as a persuasive metaphor for the rhythms of life itself.

In Sondheim's shows of the 1970s and early 1980s, this more immediate dimension of music's engagement with temporality is variously articulated through rhythm and rhyme, layered musical interactions, and a quasi-operatic/ cinematic use of characteristic musical themes and textures. At issue for the characters involved in these temporal manipulations is *musical control,* through which each conveys his or her relationship to both the world and to the persona she or he has either opted or been chosen to perform. For characters in a Sondheim show, especially during moments of personal crisis, it is how they relate to their musical environments that counts, as conveyed not only through successful performance, but also, potentially, through estrangement or, in later shows, through a kind of psychic domination. It is these characters' expressed level of musical control that reveals most vividly their level of what we might term "real-world competence," by which I mean both their ability to cope with or manage the world, and their level of comfort with themselves and with their situations.

In the remainder of this essay, I will consider this dimension across the decade and a half beginning with *Company*—what may be well understood as Sondheim's "Middle Period"—partly to trace a particular shift in Sondheim's temporal manipulations, from a tendency to use musical control as an oppressive foil for the alienated or life-challenged individual (*Company, Follies, A Little Night Music*), to a tendency to establish, through operatic or cinematic deployment of music, a kind of psychic dominance (*Sweeney Todd* and *Sunday in the Park with George*). The shift is in some ways ironic, as my specific situational examples will demonstrate. For the early shows, key moments of crisis isolate the individual through temporal dislocation, as an emblem of profound alienation, whereas, for the later shows, the very opposite of alienation—that is, the complete *identification* between the central character's subjectivity and the world he inhabits—proves at least equally devastating. In *Pacific Overtures*, standing midway between these two poles as a kind of watershed, both approaches come into play, as individuals and cultures blend into each other within an ongoing musical play of oppression, alienation, dominance, and loss. In all of these shows, musical control is the key element in an ongoing quest to master temporality, a quest that most often fails, but which, through failure, dependably offers redemption.

## Alienation and Lost Chances: *Company, Follies,* and *A Little Night Music*

*Company,* 1970. A man stands apart from his circle of married friends, bemused and somewhat bewildered by the weird dance of married life. It doesn't help much to know that his friends envy him his single lifestyle, which enables him to partake more easily than they can of the new "life-style" freedoms made available in the 1960s. He knows that he's genera-tionally displaced from that scene, caught in between, all too aware that he should also be married, but unable to join in the dance, unable to commit to a life of absurd, incomprehensible pushing and circling. Finally, he wrenches himself away with a desperate cry, stopping the music to ask a starkly direct question, "What do you get?"

The music of *Company* is an eclectic mix drawn from the 1960s, including ele-ments of rock, minimalism, Broadway pastiche, the Swingle Singers, and, somewhat improbably, the Mahler revival—the latter having been accomp-lished, across the decade, largely through the efforts of Sondheim's first Broadway collaborator, Leonard Bernstein.[7] The climax of *Company*, described above, brings together and merges two of these elements, minimalism and Mahler. Minimalism, as Robert Fink has demonstrated, has served well as a representation of modernity; its motivic repetitions, for example, can evoke not only trancelike spiritual states, but also, on other occasions, the empty, bustling repetitions we encounter in our daily lives, or machinery, or saturation advertising.[8] In *Company*, door chimes, busy signals, car horns, and the main character's name—Bobby—combine to create tapestries of modern life in the manner of minimalism, creating a bustling musical activity that Sondheim and Tunick use to simulate a range of sonic backgrounds, such as phone chat-ter, the indistinct babble of a cocktail party, city streets, and the like. And it is, in the end, this repetitive and omnipresent soundscape that Bobby rebels against, in a scenario in which he, as the alienated, immobile subjective center, overwhelmed by the constant motion around him, calls it to a grinding stop.

---

7   Regarding the musical styles that influenced *Company*, see my extended discussion in *The American Musical and the Performance of Personal Identity*, pp. 293–303; see also Dan J. Cartmell's "Stephen Sondheim and the Concept Musical" (Ph.D. dissertation, University of California, Santa Barbara, 1983), pp. 159–170, and Stephen Banfield's *Sondheim's Broadway Musicals* (Ann Arbor, 1993), pp. 151–157, 161, and 170.

8   See Chapter 2 of Robert Fink's *Repeating Ourselves: American Minimal Music as Cultural Practice* (Berkeley / Los Angeles, 2005).

As it happens, this situation is nearly identical to that of the third movement of Mahler's "Resurrection" Symphony, Bernstein's particular favorite; in fact, this very movement, the scherzo, provided the basis for Luciano Berio's late-'60s tribute to Bernstein, with its repetitive, ever-circling motion providing a medium against which he could layer a variety of musical and sonic allusions, including, with the help of Samuel Beckett and the Swingle Singers, the indistinct babble of a cocktail party.[9] Mahler provided three slightly different programs for this movement, which I conflate here:

> A spirit of negation seizes the symphony's hero. The always-stirring, never-resting, never-comprehensible pushing that is life becomes horrible, like the motion of dancing figures in a brightly lit ballroom, into which he peers from outside in the dark night, without hearing the music, so that the circling motions of the couples seem absurd and pointless. He despairs. The world and life begin to seem unreal, as if deformed by a concave mirror. He wrenches himself away with a horrible cry of disgust.[10]

---

9     For more on the parallels between Mahler and *Company*, see my *The American Musical and the Performance of Personal Identity*, pp. 296–297.

10    Here are the three programmatic descriptions Mahler provided for the scherzo of his Second Symphony ("Resurrection"):

    1. "When you wake out of this sad dream, and must re-enter life, confused as it is, it happens easily that this always-stirring, never-resting, never-comprehensible pushing that is life becomes *horrible* to you, like the motion of dancing figures in a brightly-lit ballroom, into which you are peering from outside, in the dark night—from such a *distance* that you can *not* hear the *music* they dance to! Then life seems *meaningless* to you, like a horrible chimera, that you wrench yourself out of with a horrible cry of disgust" (quoted from Carolyn Abbate's *Unsung Voices: Opera and Musical Narrative in the Nineteenth Century* [Princeton, 1991], p. 124, emphasis removed and original emphasis restored). The German reads: "Wenn Sie dann aus diesem wehmütigen Traum aufwachen, und in das wirre Leben zurückmüssen, so kann es Ihnen leicht geschehen, daß Ihnen dieses unaufhörlich bewegte, nie ruhende, nie verständliche Getriebe des Lebens *grauenhaft* wird, wie das Gewoge tanzender Gestalten in einem helle erleuchteten Ballsaal, in den Sie aus dunkler Nacht hineinblicken—aus so weiter *Entfernung*, daß Sie die *Musik* hierzu *nicht* mehr hören! *Sinnlos* wird Ihnen da das Leben, und ein grauenhafter Spuk, aus dem Sie vielleicht mit einem Schrei des Ekels auffahren!" (as given in *Gustav Mahler Briefe, Neuausgabe*, ed. Herta Blaukopf [Vienna, 1982], p. 150).

    2. "The second and third movements are episodes from the life of the fallen hero. The Andante tells of his love. What I have expressed in the Scherzo can only be described visually. When one watches a dance from a distance, without hearing the music, the revolving motions of the partners seem absurd and pointless. Likewise, to someone

Mahler's music, like Sondheim's minimalist textures, creates a seemingly unstoppable, obliviously cycling and recycling welter of motion, against which the solitary subjective center finally rebels, in a despairing cry simulated by an extended dissonant chord.[11]

In *Company*, the repeated "Bobby" figures, deriving originally, along with other repeated motives, from door chimes and busy signals, and later used to simulate car horns in one of their many reappearances throughout the show, suggest the accumulated pressure Bobby feels from his married friends. This recurring music is Bobby's version of Mahler's "never-resting, never-comprehensible pushing," whose oppressive effect on Bobby was made even more vivid in the 2006 Broadway revival directed by John Doyle, since the ensemble cast also provided the instrumental ensemble support, on stage, in an imaginative adaptation of Tunick's original scoring by Mary Mitchell Campbell.

---

who has lost himself and his happiness, the world seems crazy and confused, as if deformed by a concave mirror. The Scherzo ends with the fearful scream of a soul that has experienced this torture" (as given in Henry-Louis de La Grange's *Mahler* 1 [Garden City, NY, 1973], pp. 784–785). The German reads: "Der zweite und dritte Satz, Andante und Scherzo, sind Episoden aus dem Leben des gefallenen Helden. Das Andante enthält die Liebe. Das im Scherzo Ausgedrückte kann ich nur so veranschaulichen: Wenn du aus der Ferne durch ein Fenster einem Tanze zusiehst, ohne daß du die Musik dazu hörst, so erscheint dir Drehung und Bewegung der Paare wirr und sinnlos, da dir der Rhythmus als Schlüssel fehlt. So mußt du dir denken, daß einem, der sich und sein Glück verloren hat, die Welt wie im Hohlspiegel, verkehrt und wahnsinnig erscheint. —Mit dem furchtbaren Aufschrei der so gemarterten Seele endet das Scherzo" (Natalie Bauer-Lechner, *Gustav Mahler in den Erinnerungen*, eds. Herbert Killian and Knud Martner [Hamburg, 1984], p. 40).

3. "A spirit of disbelief and negation has seized him. He is bewildered by the bustle of appearances and he loses his perception of childhood and the profound strength that love alone can give. He despairs both of himself and of God. The world and life begin to seem unreal. Utter disgust for every form of existence and evolution seizes him in an iron grasp, torments him until he utters a cry of despair" (La Grange, p. 785). The German reads: "Der Geist des Unglaubens, der Verneinung hat sich seiner bemächtigt, er blickt in das Gewühl der Erscheinungen und verliert mit dem reinen Kindersinn den festen Halt, den allein die Liebe gibt; er verzweifelt an sich und Gott. Die Welt und das Leben wird ihm zum wirren Spuk; der Ekel vor allem Sein und Werden packt ihn mit eiserner Faust und jagt ihn bis zum Aufschrei der Verzweiflung" (Alma Mahler, *Gustav Mahler: Erinnerungen und Briefe* [Amsterdam, 1940], p. 262).

11   I discuss the Mahler scherzo in Chapter 3 of my *Symphonic Metamorphoses: Subjectivity and Alienation in Mahler's Re-Cycled Songs* (Middletown, 2003), where I identify Mahler's technique as an example of absolute music deployed as a "topic."

The "Bobby" music and the elements it derives from are introduced within the first installment of the ongoing party of Bobby and his circle, with which the show opens.[12] After Bobby's outcry near the end of the show, which brings the final installment of this onslaught to a halt, he continues with "Being Alive," which has been described—by Sondheim himself—as a cop-out, since Bobby, as presented in the show, seems hardly capable of the redemptive release suggested over the course of the song.[13] But 1970 audiences much preferred this sugar-coating to the starker alternative Sondheim provided in his earlier attempt—the bitter "Happily Ever After"— which "Being Alive" replaced in out-of-town previews. Perhaps, though, the substitution was not just about Broadway's need for uplift. The dimly-felt associations of Mahler's "Resurrection" may also have played some part in forcing Sondheim's hand, since, in Mahler, the hero's "cry of disgust" yields fairly soon to a much more forgiving "ever after" than might have been expected. Or, perhaps, the sensibilities of the New York Philharmonic and Broadway audiences are closer than the former would like to suppose.

*Follies*, 1971. The famous Weismann Theater, the former venue for a series of revues between the wars, is to be torn down. Dmitri Weismann has invited former Follies performers and their associated others to one last party. The two central couples, former law students Ben and Buddy, and their former-chorus-girl wives Phyllis and Sally, have come to the reunion in order to revisit and perhaps reclaim their youth. But they are nearly destroyed in the process, for it turns out that the theater is haunted by their younger selves, waking memories that lead them to reconsider unhappy life choices, roads not taken, and ancient betrayals of friends, lovers, and their own desires. In the climactic sequence of numbers in *Follies*, the hell the four have put themselves and each other through becomes the basis for brilliantly funny vaudeville pastiche until

---

12    Ambiguously, the main action either keeps returning to the opening birthday party or takes place across and between a series of such parties. As Hal Prince notes, after describing the device of a recurring birthday celebration as "Pinteresque in feeling," "I am certain they were one [party]. I wouldn't be surprised if George Furth believes they were four. It doesn't matter" (Prince, *Contradictions: Notes on Twenty-Six Years in the Theatre* [New York, 1974], pp. 149 and 150).

13    Sondheim's comment is quoted in Craig Zadan, *Sondheim & Co.*, 2nd ed. (New York, 1986), p. 125. Sondheim's understandable regrets aside, it must be said that much depends on performance. Raúl Esparza's Bobby, in the 2006 Doyle revival, projected more depth earlier on in the show, and enough anguish in the early stages of the song to make this "copout" resolution absolutely compelling.

Ben—fallen somewhat from his days as a UN diplomat—breaks down during a suave "sophisticated" number, "Live, Laugh, Love."

As with Bobby in *Company*, Ben's breakdown is crucial to his ultimate release, and is conveyed in musical terms. But Ben's breakdown, unlike Bobby's, is presented directly as a failure of performance. As is typical for the vaudeville era of Tin Pan Alley, the "sophistication" of the song is conveyed rhythmically, requiring the singer to perform a casual mastery of sometimes complex cross rhythms; perhaps the quintessential example of this type is Irving Berlin's "Puttin' on the Ritz," and—as here—the sophistication is there meant to be seen as a façade, as was later exploited to hilarious effect when the song was used in Mel Brooks's *Young Frankenstein* as a vehicle for the monster's public debut. In "Puttin' on the Ritz," seven-beat patterns superimpose on a regular four-beat meter, contributing, in *Young Frankenstein*, to the ungainly grace of the number even before the monster's breakdown.[14] In Ben's "Live, Laugh, Love," the superimposed patterns are three beats rather than seven, repeated three times to carry the title triad of live, laugh, and love (stressed syllables in bold):

**Me**, I **like** to **live**,
**Me**, I **like** to **laugh**,
**Me**, I **like** to **love**.

But Ben's performance of happy-go-lucky sophistication, like the monster's in *Young Frankenstein*, is a sham, and he can't keep it up. As the number progresses, he misses a cue, then begins to trip over the rhythmic figure, which becomes a rut he can't seem to escape, sounding much like a broken record. Finally, he breaks away from the music, distracted by the words the music forces on him: "Me, I like to love me." But the music insists on continuing, and we start to hear it as though from his perspective, as an "always-stirring, never-resting, never-comprehensible pushing," an oppressive mode of living that leaves him in bewildered isolation, until he is rescued by the woman

---

14    In its original version (that is, before the better known version Fred Astaire performed in *Blue Skies*, dir. Stuart Heisler, 1946), "Puttin' on the Ritz" was performed with a black backup chorus, and made explicit racialized references (*Puttin' on the Ritz*, dir. Edward Sloman, 1930). Thus, the original chorus begins, "If you're blue and you / Don't know where to go to, / Why don't you go where / Harlem sits, / Puttin' on the Ritz. / Spangled gowns upon a bevy / Of high browns from down the levee, / All misfits, / Puttin' on the Ritz." For more on the song and its use in *Young Frankenstein*, see my "Music, Electricity, and the 'Sweet Mystery of Life' in *Young Frankenstein*," in *Changing Tunes: The Use of Pre-existing Music in Film*, eds. Phil Powrie and Robynn Stilwell (Aldershot, UK, 2006), pp. 105–118.

his lifelong performance of sham sophistication has hurt the most, his wife Phyllis.

Ben's breakdown vividly recalls a similar breakdown near the end of an earlier Sondheim show, *Gypsy* (1959). Although it was Jules Styne who composed the music for *Gypsy*, it was Sondheim, with librettist Arthur Laurents, who compiled the final number, "Rose's Turn," in which Madame Rose loses control while performing a parody of her daughter's act. Like Ben in *Follies*, she breaks down during a three-beat melodic phrase repeating against an established duple meter. And like Ben in *Follies*, Rose is distracted from maintaining her rhythmic groove by rhythmically reiterated words that carry unexpected meaning ("Mama..."). In both cases, the rhythmic problem serves as a figure for the character's increasing discomfort with performing an alien identity. Rhythmic crisis thus both represents and merges with identity crisis, with rhythm and identity each becoming not only a matter of performance, but also a test of basic human competence. And yet, in both cases, it is the breakdown, the failure of that test, that signals a deeper humanity, a valuable core that might redeem a seemingly wasted life.

> *A Little Night Music*, 1973. It is midsummer in turn-of-the-century Sweden. Fredrik, a lawyer, is nearly a year into his marriage to eighteen-year-old Anne, who remains a virgin, while also coming to dominate the erotic imagination of her nineteen-year-old stepson Henrik, a divinity student. Potential redemption from this impossible situation appears in the guise of actress Desirée Armfeldt, who was once Fredrik's lover and is now, some fifteen years later, on tour—and who happens to have a fourteen-year-old daughter named Fredrika. It is a classic operetta situation, set in an exotic European setting, the long twilight of the northern summer offering to rekindle a forgotten love, to rescue Fredrik and Desirée by allowing them to rediscover their one true romance. But, while it may be turn-of-the-century Sweden on stage, that stage is on Broadway, 1973, making the redemption that operetta offers seem remote and faintly ludicrous—this is, after all, only a year before *Young Frankenstein*'s send-up of Victor Herbert's "Ah, Sweet Mystery of Life," from *Naughty Marietta* (1910). Even in operetta terms, redemption seems more wistful than plausible, given Desirée's involvement with a married, jealous dragoon, and Fredrik's unshakable infatuation with Anne, who seems in complete control of the household.

We know Anne is in control of the household, including its rampant sexual tensions, through song—more specifically, through the combination song the three sing early on in the show, "Now, Later, Soon." Combination songs, a staple of the American musical stage going back at least to Gilbert and Sullivan's *H.M.S. Pinafore* (1878), are typically about compatibility, even when compatibility seems unlikely. Thus, in a combination song, separate groups or individuals first sing separate establishing songs, often quite different in character and even opposed in sentiment—but then the songs are sung together and, as if by magic, the combination works perfectly. The list of Broadway examples is long, including, for example, "An Old Fashioned Wedding" in *Annie Get Your Gun* (1966 revival), and a number of paired songs in *The Music Man* (1957). Closer at hand is the pastiche example Sondheim wrote for *Follies*, in which the young couples—the ghosts of Ben and Phyllis, and Buddy and Sally—parody the wise-cracking optimism of the Great Depression era in "You're Gonna Love Tomorrow" and "Love Will See Us Through."

But combination songs don't always demonstrate compatibility, or at least an untroubled compatibility. In "Tradition," the opening number from *Fiddler on the Roof* (1964), the only thing that the mamas, papas, sons, and daughters seem to agree on is the title word; when they combine, chaos ensues, each part oblivious to the others until they come together on the word "tradition." More subtle is the opening duet in a much older musical comedy, Mozart's *Marriage of Figaro* (1786). With great economy, Mozart establishes character difference in traditionally gendered ways. Figaro *measures*: as the scene opens, he is counting out the dimensions of the room he and Susanna will occupy, while she admires her new hat. Are these characters compatible? It would seem so; in fact, Figaro's measurements provide the perfect bass-line for Susanna's melody as she continues. But coexistence is not the only issue here; it's really about who's in charge—and that turns out to be Susanna, as we soon discover when *he* joins *her* tune, rather than the reverse.

Similarly, Sondheim sets up all three of his players—Fredrik, Henrik, Anne—with highly characteristic music. The lawyer Fredrik, in the patter-song "Now," considers a variety of alternative means for seducing his wife, enumerating a series of nested A's and B's while Anne babbles seeming nonsense, until he finally puts each alternative aside in favor of an afternoon nap. While he sleeps, Henrik, caught between his libido and Martin Luther, broods into his cello, taking his theme from Anne's admonition, "Later." And Anne, having managed each male in turn, is alluringly both maternal and girlish in "Soon," a quasi-lullaby in which she shifts keys and tempo capriciously, and gracefully negotiates whimsically intricate rhythms. And, when the three songs come together, they do mesh, for a time—allowing for Fredrik's drowsier pace, since he's singing in his sleep. But as the song continues, through force of will and gentle persuasion, it is Anne's song that the others bend to, unable to resist submerging their own identities into hers. Anne's seeming mastery is, however, itself merely a setup for her own imminent fall: in his sleep, Fredrik's final sung word is not "Anne," as before, but "Desirée."

Much later, in Act II, Fredrik's "life-incompetence," as revealed in this early combination song, has more serious consequences, when he cannot at first join Desirée in "Send in the Clowns." Indeed, his failure seems built into the song, which comes across as somehow broken despite its wistful lyricism, as it abjectly fails to "perform" one of the most traditional of operetta gestures: the sung waltz as a remembered, shared space for the reunion of old lovers (see discussion above). The song acknowledges its failure throughout, both through verbal expression, as Desirée seeks relief from an impossible situation, and through its halting, breathless melody, especially tailored for the vocal incapacities of Glynis Johns, the first Desirée. Later, at the end of the show, when Fredrik is virtually forced to accept his one chance at happiness, he finds himself at last able to join her in the song, able to accept brokenness as a necessary step toward rescue and redemption.[15]

### Laying Claim to the World: *Sweeney Todd* and *Sunday in the Park with George*

*Sweeney Todd*, 1979. It is early in the nineteenth century, London. Since we're also on a Broadway stage, we have Big Ben keeping time for us, even though the famous chimes wouldn't be built until 1858. But keeping time is not really the issue in *Sweeney Todd*, for we are in the timeless space of storytelling, concerned in particular with the tale of a brooding, embittered man who returns to London from a penal colony in Australia to avenge his wrongs and reconnect with the remnants of his family. And in this timeless realm, his mood seems to match perfectly the fog-ridden, almost subterranean world he reenters. Thus, in a kind of symbiosis, either he provides the musical soundtrack for this darkly foreboding world, or else the murky music of this murky place calls him out from the abyss, invoking his literal presence that we might once again attend his tale, and know that he and his world still survive.

The kind of musical fabric heard throughout *Sweeney Todd*, in which a continuous, often formless musical flow carries referential motives, such as the *Dies*

---

15      For more on "Send in the Clowns," and on *A Little Night Music* more generally, see my *The American Musical and the Performance of Personal Identity*, pp. 50–64.

*irae* and Big Ben, has been likened to not only Wagner, but also classic film scoring, and the music of horror films. Music serves in all three of these contexts to represent two things in a seemingly unmediated way: the sensibilities of a single mind or worldview, and the world itself. Within this kind of texture, leitmotivs seem to represent, from one perspective, the emergence of individual thoughts and feelings within a flow of consciousness, and from the other, the deep connection between objects we perceive directly and the world's essence, itself invisible except through such manifestations. This ability of music—and especially "background" music, or underscoring— to evoke simultaneously an individual consciousness and the world's essence resonates strongly with both nineteenth-century thought about the relationship between consciousness and the world (deriving from the German Idealism of Kant and Fichte) and the nearly simultaneously emergent view of music as a conduit between the two.[16]

In *Sweeney Todd*, the dominant consciousness is, of course, Sweeney's. Although it does not encompass the whole world, it sets its tone, or at least merges with it to such an extent that we cannot fully separate them. This is why, after many of the passages of dialogue in which the music stops, it resumes as underscoring for a sudden moment of awakening on Sweeney's part, with him thereby reasserting his dominion over the world's tone. Moreover, such moments represent significant turns in his—and the show's—psychological drama. In providing access to Sweeney's inner life, the music of *Sweeney Todd* fuses with and expresses *his* sense of the world above all else. In effect, when the opening "Prologue" conjures Sweeney from his grave, it does so by re-creating the world in his image, with that image being above all a musical one, sustained through an incessant flow of leithmotiv-saturated music. With such a fusion between Sweeney's sensibilities and the world he inhabits, it seems virtually inevitable that his death does not truly finish him. As in count- less horror films—which work essentially the same way, and often just as carefully coordinated with music—the dominant malevolent presence cannot die, so deeply intertwined is it within the world we have come to know. Even after his literal death, Sweeney's ongoing music asserts that the world has fully become *his* world, and once again conjures him from the grave with the force of inexorable fate.

---

16    For a discussion of the nineteenth-century basis for this cinematic trope, see my "'*Selbst dann bin ich die Welt*': On the Subjective-Musical Basis of Wagner's *Gesamtkunstwelt*" (*19th-Century Music* 29 [Fall 2005], 142–60). For more on this dimension of *Sweeney Todd*, see my *The American Musical and the Performance of Personal Identity*, pp. 331–342.

But Sweeney's control over the world's tone is anything but guaranteed. The first act sees him enter and fuse with London's subterranean world, establish himself, and launch his paired projects of revenge and family reunion, all with growing confidence. Near the end of the first act, however, when both projects seem to be nearing successful completion, with the evil Judge Turpin's throat beneath his razor, victory is snatched from him. His ensuing rage is staged as an operatic "mad scene," and, as he spins out of control in "Epiphany," it is specifically musical discourse that he fails to manage. Even as he settles from rage into a grim determination to slaughter whomever comes within reach of his blade ("They all deserve to die"), he continues to interrupt himself with wails of anguish over his lost wife and daughter, interspersed chaotically with bits of macabre humor derived from London's Music Halls. By the end of "Epiphany" he has brought his anguish under control, but only at the price of silencing his music, so that it is Mrs. Lovett, his lone confidante, who must start up the music again ("A Little Priest"). Although this situation is partly a matter of plotting—since baking Sweeney's victims into pies is her idea, not his—it is also logical: his world—his music, his life rhythms—are all spent, leaving a void for her to fill. But she finds herself unable to fill that void alone, and requires his musical participation before "A Little Priest" can truly get underway. Again there is a local logic: the song is a waltz song, straight out of the music hall, and—as with the Tango—"it takes two." But the logic is also of the specific world of this musical, whose music is fundamentally Sweeney's.

> ***Sunday in the Park with George*, 1984.** It is 1884, Paris, the city of light. A painter, Georges Seurat, attempts to reproduce a sense of unprocessed light on canvas, composing his images from thousands of dots of color, most famously in his monumental painting, then in progress, "Sunday Afternoon on the Island of La Grande Jatte." His obsessive manner of painting, dot by dot, requires tremendous concentration, and reflects his compelling desire to merge art and reality to an unprecedented degree. To this end, light, color, objects, trees, animals, people, all serve as raw materials for his art, to be reduced to dots of color and light, and rearranged like furniture to become, through design, a balanced composition. His art—his quest "to bring order to the whole"—becomes his reality; his art becomes, for him and those swept up in it, the *only* reality, a figurative *tableau vivant* that is enacted literally on stage.

In the first act of *Sunday in the Park with George*, as in *Sweeney Todd*, the musical fabric is dominated by the music of a single visionary character, with the dominant power of his vision—as betokened by a sonic environment, a musical ethos that emanates from his very life force—becoming one with the world's essence. True, George is an artist, not the mass murderer that Sweeney Todd becomes, and his "victims" live on as art rather than becoming the ingredients of meat pies. But the stakes, for the protagonist, are on the same order. We see George, like Sweeney, enact his vision only at the expense of what matters most to him, and, ultimately, he thereby becomes one of his own victims. Sweeney, after mistakenly murdering his beloved wife, is murdered in turn. And George, immersed in his painting in Act I while turning his models into the equivalent of cardboard cutouts, is in Act II reincarnated as an artist who must, as part of the "Art of Making Art," turn himself into a cardboard cutout, repeatedly. In this Sondheimian universe, ruled by obsessions that take the dramatic form of similarly obsessive musical textures, the reality created by an individual can be imposed on the world only by, in effect, consuming the world. And, inevitably, whoever exerts that forceful will is also part of the world that gets consumed.

*Sunday in the Park with George* problematizes the relationship of art and life, not only on these levels, but also across time, so that art's power to redeem is not short-changed, but presented alongside its capacity to absorb the living into the sometimes waking death of artistic obsession. It is this ambivalence that allows the show a redemptive theme denied *Sweeney Todd*, where the obsession is all-consuming and leaves behind no redemptive trace. But the crucial parallels remain: George, like Sweeney, seeks a vital human connection and is granted ample opportunity to secure it early on. Sweeney does not recognize his wife and places his reunion with his daughter in conflict with his thirst for revenge in which he drowns. The quest for revenge—in this context most palpably a refusal to let go of the past—satisfies for him a deeper need than mere human connection, providing a consuming connection to something that seems more deeply embedded in the world as he conceives it. George, too, finds himself having to choose between his connection with Dot—his lover and principal model—and connecting the dots of his painting, a choice that serves as a figure for the choice between life and art. Thus, the early song, "Color and Light," presents George and Dot as completely disconnected, yet obviously poised for connection at every turn, unconsciously aping each other's every thought and gesture.

Structurally, "Color and Light" assumes important features of the love duet: each of the lovers sings alone and at some length before they come together in overlapping phrases that echo each other verbally and musically, thence joining in a concluding phrase of shared devotion:

*Dot*:      So you want him even more.                                                      And you drown
*George*:                                     But the way she catches light...

| | | |
|---|---|---|
| *Dot*:    inside his eyes... | | I could look at him forever. |
| *George*: | And the color of her hair... | I could look at her forever. |

Moreover, this series of echoes brings into focus more elaborate echoes across their earlier, individuated contributions. Dot's movements in applying makeup, following the pointillistic music, directly imitate George's movements in applying paint to canvas. Their first sung contributions are identically framed, with his initial "More red" becoming, for Dot, "More rouge," and with her arrival echoing his, concluding in the title phrase, "color and light." Each ruminates, while the other sings, on the other's inexplicable concentration, his on his painting and hers on her reflection in the mirror.

But these tokens of potential connection seem designed above all to underscore the disconnect between George and Dot that we actually see and hear, climaxing poignantly in the third-person syntax of their mutual arrival ("I could look at him/her forever"). During the first two spans, George sings to the figures in his painting, then Dot sings of an imagined career as a Follies girl. During the third span, each talks in counterpoint to the other's singing, making it clear that their singing, even in parallel, does not point to their eventual union, but rather reinforces how different their respective worlds are. Thus, "More red" may parallel "More rouge," but we are meant to notice the difference between paint and makeup, the one applied to canvas and the other to a human face. Through most of the scene George and Dot are separated, literally and figuratively, by the painting, which is presented on a scrim that George works on from behind, on a ladder, while Dot sings from a contrived model's pose from which she breaks free only once (in this recalling the opening number, "Sunday in the Park with George").[17] And the structure of the song also serves to undermine rather than facilitate their union: in the first two spans, each retreats back to his or her respective world after arriving on the outward-gesturing "color and light." Most of the third span is spent with one speaking while the other sings, so that their coming together in the quoted phrases seems almost by happenstance; and the song continues, past this moment of confluence, with George alone, deliberating on what he should do, until the unfinished hat in the painting decides for him. The scene ends with his being (according to the stage direction) "consumed by light";[18] whereas before, painter and painting were held in balance, each plainly visible, at the

---

17    In the first song, "Sunday in the Park with George," she is posing for the painting in progress; her dressing-table pose anticipates Seurat's later painting, "Jeune femme se poudrant ou la Poudreuse" (1888–1890).

18    James Lapine and Stephen Sondheim, *Sunday in the Park with George* (New York, 1991), p. 40.

end of the song he disappears into the painting, an invisible presence behind the scrim.

Throughout the first act, George's relationship to those he is painting displays a similar ambivalence between art's need for life and life's need for art, with Dot's position being, naturally, the most fraught. While she is with George, Dot appears mainly in poses, dancing free only occasionally. That her name derives from his art seems like mere verbal playfulness, hardly capable of penetrating the surface of the drama. But the derivation does have a deeper point, since his obsession with his painting repeatedly brings him directly into confrontation with her, most overtly during the third span of "Color and Light":

> *George* [as he paints]: Blue blue blue blue / Blue still sitting / Red that
> perfume Blue all night / Blue-green the window shut / Dut dut dut
> Dot Dot sitting / Dot Dot waiting / Dot Dot getting fat fat fat
> More yellow / Dot Dot waiting to go / Out out out but
> No no no George / Finish the hat finish the hat /
> Have to finish the hat first / Hat hat hat hat
> Hot hot hot it's hot in here...

Thus, she is not only the artist's model, but also the (living) light with which he works.

When in the next scene George, now separated from Dot, forms his most vital connection of the show, it is to the boatman's dog Spot (whose name derives, like Dot's, from his manner of painting). Although George's "conversation" with Spot in "The Day Off" presents within the show as comic shtick, George's extended, empathetic channeling of Spot's vocality ("Ruff! Ruff! / Thanks, the week has been / Rough!") comes across as genuine connection, in part because it is based directly on a shared capacity for acute observation, even if it is based as well on the convenient fact that Spot cannot himself converse, and demands nothing of George. But, like the sadly vivid parallels between George and Dot in "Color and Light," George's capacity and need for connection here serve mainly to underscore the denial of both in relation to his human subjects, whose "higher calling" as components of his art precludes George's granting them an autonomy sufficient to support a genuine connection.

## Competing Temporalities, Competing Worlds: *Pacific Overtures*

In January, 1976, John Weidman's play about the opening up of Japan in the nineteenth century finally opened on Broadway, just in time to kick off the

American Bicentennial. As if the ironies of this timing weren't already deep enough to undermine the show's potential for success on Broadway, Hal Prince determined to tell the story from the Japanese point of view, to use an all-Asian, mostly male cast, to approximate as closely as possible Kabuki theatrical conventions, and to recruit Stephen Sondheim to turn the whole thing into a musical. None of this was easy, especially recruiting Asians on Broadway in the 1970s—no one even showed up for the first casting call—or convincing Sondheim to work on a project so close to the Brechtian theatrical mode he despised. Nevertheless:

> *Pacific Overtures*, 1976. In July, 1853, Commodore Matthew Perry and a fleet of four ships approach Japan to initiate trade agreements. Manjiro, a fisherman rescued from shipwreck several years earlier by Americans, has returned to warn his countrymen, and he joins forces with Kayama, the local governor, to outwit the Americans and protect Japan and its traditions. The Americans leave, but they soon return, followed by the English, Dutch, Russians, and French. Japan succumbs.

The arrival of the foreign admirals in Act 2 is staged as a single musical number, "Please Hello," a combination song that plays out, metaphorically, as a gang rape, but is presented as comedy according to the conventions of Broadway pastiche. Thus, the American admiral enters to a march reminiscent of Sousa, the English admiral to an intricately rhymed patter song à la Gilbert and Sullivan, the Dutch to a wooden-shoe number, and so on. As the stage accumulates admirals, overwhelming the Japanese Lord Abe, the combination song becomes a cacophony of competing would-be rapists, a nightmarish conflation of tunes like that near the end of *Follies*, and similar in effect to the incomprehensible pushing of the "Bobby" music in *Company*. But here there is no redemption for the victim of this sonic oppression. Lord Abe cannot call for the music to stop, nor is he rescued, and the music continues to its conclusion—a kickline of Admirals singing about "détente"—oblivious of the haplessly protesting victim. And the audience applauds, delighted at the spectacle of colorfully uniformed imperialists tearing themselves to shreds in a kind of feeding frenzy, and of Asian actors pre-empting the clichés of Broadway and turning them ferociously on their head.

Across the next two numbers in the show, "Bowler Hat" and "Pretty Lady," Japan's powerlessness to resist the Western onslaught is given poignant voice. "Pretty Lady," in particular, echoes the situation in "Please Hello," but leaves more room for us to identify with the silent victim, a samurai's daughter mistaken for a geisha. We are invited to hear Western tokens of complex beauty—imitative counterpoint over a ground bass—from her perspective, as we see the three, rather innocent British sailors circling her, confusing her, overwhelming her, and in the end, through their song, obliquely lamenting her—and Japan's—fate. And this catharsis, in turn, takes us to the harshly exhilarating finale, "Next," a machine-driven roller-coaster ride through the disasters and triumphs of twentieth-century Japan.

But there is, buried within *Pacific Overtures*, a happier alternative, a more hopeful blend of East and West. Much earlier, Kayama and Manjiro exchange "Poems" according to Japanese traditions, with Manjiro converting every poetic image that Kayama offers of his beloved wife, Tamate, into an image of America, most poignantly with the image of the moon:

> *Kayama*: Moon,
> I love her like the moon,
> Making jewels of the grass
> Where my lady walks,
> My lady wife.

> *Manjiro*: Moon,
> I love her like the moon,
> Washing yesterday away,
> As my lady does,
> America.

Within Manjiro's moment, the image expresses simple exuberance at the prospect of bringing America to Japan, but his words also resonate ominously with the larger context. Perhaps the moon, long a symbol of intoxicating madness, has blinded him to the fact that the most relevant "yesterday" America has recently undertaken to wash away is Japan itself. And Kayama, too, has been blinded by his improbable success, with Manjiro's help, in "outwitting" the Americans; he sings blithely of his wife, who we will soon learn has claimed herself, in suicide, as America's first victim. Yet, despite this overpoweringly sad undertow, the song's hopeful gesture toward cultural rapprochement transcends the show's tragic arch. What makes the song particularly hopeful is its easy blend of perspectives, with its singers learning from each other as they go, borrowing and re-elaborating each other's ideas until, finally, there is no alternative but to merge into one voice.

There is, importantly, a half-full / half-empty way to parse this situation. Here, as with all of the shows I have examined, the point of establishing characteristic temporal spheres for specific characters, or for the world as it appears to those characters, is to create the experience of alienation, of an unbridgeable gap between the subjective center and the promised happiness of connection to the world or significant others. In the case of *Pacific Overtures*, as I have argued elsewhere, musical numbers convey distinctively different and incompatible modes of being in time, the one circular, the other teleological.[19] In terms of my argument here, this incompatibility becomes absolute; one must prevail, and it is pre-ordained that the West's goal-directed sense of time must preempt Japan's circular sense of time—a mode of being that the West likes to call "timeless." And indeed, in the strange and oddly beautiful montage of "Bowler Hat," late in Act 2 (between "Please Hello," and "Pretty Lady"), we hear and see how Kayama, who figuratively represents Japan, has in accommodating the West become alienated from himself.

Such situations of alienation resonated powerfully in the early 1970s, especially on Broadway, whose golden age seemed to be slipping away as surely as youth, and as surely as the earlier world of operetta had. And such themes, of irreparable loss and desperate yearning, are central to Sondheim's artistic personality. Redemption, in this case, however, comes from the very soul of Broadway, and its reprise-driven prejudice to believe that what has been lost might be reborn in song. Thus, it matters that the first-act number, "Poems" is, among other things, a Broadway song, a genre with the capacity to outlive its dramatic moment, however fleeting, however overturned by later events, to remind us of other possibilities whenever it is recalled, whenever it is brought back into the present by the simple act of singing.

### Invocation, Failure, and the Potential for Redemption

The 1970s began and ended for Sondheim with shows in which the entire cast calls out the name of the protagonist in rhythmic repetition, within a melodic configuration laden with symbolism. The "Bobby" motive in *Company* derives from a conflation of the busy signal and the door chime, whereas "Sweeney" arises from the *Dies irae*, the Requiem Mass's invocation of Judgment Day. In both, the ritualized incantation, whether derived from the connective signposts of everyday life or of religion, is meant to conjure the presence of the one named.

---

19    See my "Marking Time in Pacific Overtures: Reconciling East, West, and History within the Theatrical Now of a Broadway Musical." For more on the show's perspective between "East" and "West," see also my *The American Musical and the Formation of National Identity*, pp. 268–281.

As I have argued above, in Sondheim's shows from the first half of the decade, music tends to create an alienating schism between the main character and both the world he has come to inhabit and the persona he has learned to perform; in *Sweeney Todd* and *Sunday in the Park with George*, music creates an equally disturbing *bond* between the title character and the world he has learned to dominate through sheer force of will. Yet, despite this crucial difference, not only are the *ends* nearly the same for the protagonists of these five shows, with the world each inhabits ultimately becoming his enemy, but also are the *means* similar for creating that situation, in musical terms. In *Company*, especially, it is the music that attaches to Bobby's name that comes to oppress him, and no more so than in the 2006 Broadway revival, in which the ensemble cast also provides the instrumental accompaniment, making the music more obviously the basis for the central drama. In *Follies*, all the principal characters call out to each other, too easily conflating their past selves with the rueful adults they have become; in the end, they call out, not for each other, but for an irretrievable past. And the plot of *A Little Night Music* truly gets underway only when Fredrik's sleeping subconscious calls out to Desirée at the end of "Now," displacing his callout to Anne in the waking version, which signals a shift in the object of his romantic obsession away from his destructively tantalizing, yet unattainable young wife, to his true beloved, if at that point only in memory. (To be sure, both invocations are emblems of his midlife crisis, and of his desperate need to reclaim his youth.)

All of these characters are in the end redeemed in one way or another, and what facilitates that redemption in each case is failure—a failure marked, specifically, by each character's disabled capacity to engage productively with his musical environment. Until that critical failure and the denouement of attendant personal crisis, they all have quite a lot in common with the protagonists of *Sweeney* and *Sunday*, especially with regard to their principal ongoing mode of failure: in their different ways, all five men fail, fundamentally, to connect. The potential for Bobby to connect romantically extends to nearly every female character in the show (and at least one male), yet his every encounter with them, including his sexual episode with April, ends in disconnect. Ben has predicated his career, and inadvertently his life, on enforcing a disconnection with his one true love, Sally. And Fredrik finds himself initially incapable of responding to Desirée's "Send in the Clowns," extending an entrenched romantic failure to act well on his own behalf. But Sweeney's failure to connect, with either his family or Judge Turpin's throat, pushes him over the edge. And, while the show's repeated invocations of its title character do not oppress him as they do Bobby in *Company*—quite the opposite, they give him life—those repeated invocations are even more obviously oppressive than *Company*'s "Bobby" music, and in the end Sweeney is absorbed fully into the bleakness of the world he has created in his image, which is in an important sense the only real connection he is able to forge. Similarly, George's connection is fundamentally and in the end *only* to the world he inhabits, in his case the world of art.

The terms of redemption, for those who achieve it, are critical. Bobby—who nearly wasn't allowed by his creators into the circle of the redeemed—finally realizes and fully accepts the fact that failure is the price of admission, although he seemingly has known all along. Thus, what makes "Being Alive" convincing—when it *is* convincing—is the ease with which a musical phrase's brutally honest rejection of commitment can become a plea for company, as the initial "Someone to hold me too close" transmutes into "Somebody hold me too close" (with the music's urgent repetitions at this point carrying an implied "please"). That the music does not need to change much for the meaning to change diametrically reassures us that both sentiments—both the realist rejection of company and the desperate need for it—are part of Bobby's life-rhythms.[20] Ben, who like Bobby nearly misses his chance for redemption, also earns it through failure, which allows him to accept Phyllis's generous, comforting acceptance of him, and thereby come to terms with his life choices. And, if Fredrik is generically pre-ordained to reconnect with Desirée—thanks largely to the underlying sweetness of operetta—he must nevertheless first fail utterly, and on all fronts; only then will he qualify for his share of the benevolence bestowed by the smiling Summer Night.[21] But Sweeney and George do not fail, and so are not and cannot be redeemed, but must instead leave redemption to a later generation, to Sweeney's daughter Joanna and George's great-grandson. To be sure, George Sr. is allowed at least a foretaste of redemption through an important shift from the musical to the visual, from the temporally measured rhythms driven by a living force to the timelessness of the frozen image, which might be reawakened, if only in the imagination.

---

20   This kind of musical shape-shifting returns with particular effectiveness in *Into the Woods*; see my discussion in *The American Musical and the Performance of Personal Identity*, pp. 150–163, esp. pp. 155–156.

21   Both *A Little Night Music* and the Ingmar Bergman film on which it is based (*Smiles of a Summer Night*, 1955), employ the conceit of the summer night's three smiles, which redeem, in the film, young love, fools, and the sad and lonely. In the musical, the beneficiaries shift, becoming the "young, who know nothing," the "fools who know too little," and the "old, who know too much." By leaving out the sad and lonely, the musical implicitly argues that the most important category is the central one, and Mme. Armfeldt confirms this at the end of the show when she observes, "The smile for the fools was particularly broad tonight."

The quest for musical control, as a figure for the desire to control one's life trajectory, including the past, is in Sondheim's musicals either doomed to fail or—should it succeed—to become a mortal trap. It is this trap that *Pacific Overtures* is the first of his shows to point to; *Pacific Overtures* argues that the impulse toward control is destructive at its root, and places that argument at the center of its narrative of cultural effacement. Yet, as argued above, *Pacific Overtures* also points to the redemptive capacities of Broadway, if more obliquely: destruction is never complete in a world in which the past may so routinely be recuperated in a reprise. On Broadway, at least, the past may be had for a song.

CHAPTER 17

# Remembering the Future

*Michael Cole*

Expectation refers to the future, and memory to the past. On the other hand the tension in an act belongs to the present: through it the future is transformed into the past. Hence an act may contain something that refers to what has not yet come to pass.

> SAINT AUGUSTINE, *Confessions*

Yet all experience is an arch wherthrough /
Gleams that untraveled world whose margin fades /
Forever and forever when I move.

> TENNYSON, *Ulysses*

We are all accustomed to the notion of remembering as the summoning up of past experiences in the process of dealing with the present. The study of memory, conventionally understood, has been one of psychology's most productive growth industries since the 1960s. However, as both Augustine and Tennyson suggest, the relations between past, present, and future in human experience are a good deal more complicated than common wisdom leads us to believe. Twenty years ago, Gerald Edelman[1] opened his book with a quotation from Edwin Boring, professor of Psychology at Harvard in the first half of the twentieth Century, who began his career when introspection was still accorded a place in an increasingly behavioristic academic psychology.[2] Boring pointed out that our common sense ideas about events occurring in the present are really based on memory of the past.

> To be aware of a conscious datum is to be sure that it has passed. The nearest actual approach to immediate introspection is early retrospection. The experience described, if there be any such, is always just past; the description is present. However, if I ask myself how I know the description is present, I find myself describing the processes that made up the description; the original describing is past: "Experience itself is at

---

1   G. Edelman, *The Remembered Present: A Biological Theory of Consciousness* (New York 1989).
2   See E.G. Boring, *The Physical Dimensions of Consciousness* (New York, 1933, rpt. 1963).

the end of the introspective rainbow. The rainbow may have an end and the end may be somewhere; yet I seem never to get to it."[3]

Edelman summarized a vast array of evidence from the neurosciences to substantiate his theory about what sort of organism human beings must be if the phenomenal present is "really" the past. I am less concerned with the technical adequacy of Edelman's neurological model than I am with the fact that remembering the present, if somewhat odd, is nonetheless a broadly recognized feature of human experience.

What, then, of memory for the future? Whether we look to the ideas of St. Augustine on the future as expectation, Miller, Galanter, and Pribram on plans,[4] or the Russian physiologist, Nicolas Bernstein[5] on the organization of living movement, one message repeats itself: The present is a dynamic evolving, trajectory that not only integrates current sensory input with prior experience, but also "calculates" an "imagined future" that then "feeds back" to complete a fundamental, transformational cognitive cycle characteristic of human thought. David Ingvar used the memorable phrase, "memory for the future" to highlight the complex ways in which what we normally think of time in relation to memory actually takes place in a distinctly non-linear fashion. He summarized a wide variety of evidence that plans, ambitions, and "sets" are normally remembered in great detail, just as memories of the past can be reconstructed. In addition, he summarized the neuropsychological evidence that memory for the future is selectively lost owing to lesions of the prefrontal and frontal cortices. Ingvar referred to these structures as the "neuronal substrate of the future."[6]

Of course, in one sense we all take for granted the existence of a "memory for the future." I can speak coherently, for example, of my memory of what I will be doing (plan to do) this weekend. Research on the selective disturbance of planning functions as a result of prefrontal and frontal lobe lesions has been well known for a long time.[7] Previously I did not think of such phenomena as memory for the future. It was only when I recently happened upon a reference to Ingvar's article, while ruminating about cultural mechanisms of cognitive development, that the idea of future memory began to seem like a necessary property of human thought. To understand why memory for the future is a

---

3   Ibid., p. 228.
4   G.A. Miller, E. Galanter, and K. Pribram, *Plans and the Structure of Behavior* (New York, 1960).
5   N.A. Bernstein, *The Coordination and Regulation of Movement* (Oxford, 1967).
6   D.H. Ingvar, "Memory for the Future," *Human Neurobiology* 4 (1985), 127–136 (130).
7   A.R. Luria, *Traumatic Aphasia: Its Syndromes, Psychology, and Treatment* (The Hague, 1970).

particularly interesting idea when thinking about human experience of time, I need to sketch a few of my ideas about culture and the role of culture in creating and recreating human beings.

### Culture as the Specific Medium of Human Life

My notions of culture have undergone a good many changes over the years as my personal experience and reading warred with each other in a search for coherence. As discussed in more detail elsewhere, I gradually came to develop a way of thinking about culture, and the mediation of human experience through culture, that derives its inspiration from the work of the Russian cultural psychologists Lev Vygotsky and Alexander Luria.[8] Central to their formulations is the notion that human beings live in an environment transformed by the artifacts of prior generations, extending back to the beginning of the species.[9] The basic function of these artifacts is to coordinate human beings with the physical world and each other. Cultural artifacts are simultaneously ideal (conceptual) and material. They are ideal in that they contain, in coded form, the interactions of which they were previously a part and which they mediate in the present.

They are material because they exist only insofar as they are embodied in material artifacts. This principle applies with equal force whether one is considering language/speech or the more usually noted forms of artifacts that constitute material culture. The American anthropologist White explained, "An axe has a subjective component; it would be meaningless without a concept and an attitude. On the other hand, a concept or attitude would be meaningless without overt expression in behavior or speech (which is a form of behavior). Every cultural element, every cultural trait therefore, has a subjective and an objective aspect."[10]

The special characteristics of human mental life are precisely those characteristics of an organism that can inhabit, transform, and recreate an artifact-mediated world. As the Russian philosopher, Ilyenkov puts it, "The world of things created by man for man and therefore things whose forms are reified

---

8    See M. Cole, *Cultural Psychology: A Once and Future Discipline* (Cambridge, MA, 1996).

9    C. Geertz, *The Interpretation of Cultures* (New York, 1973); E. Ilyenkov, "The Concept of the Ideal," *Problems of Dislectical Materialism* (Moscow, 1977); M. Sahlins, *Culture and Practical Reason* (Chicago, 1976); M. Wartofsky, *Models: Representation and Scientific Understanding* (Dordrecht, 1979).

10   L. White, "The Concept of Culture," *American Anthropologist* 61 (1959), 227–251 (236).

forms of human activity...is the condition for the existence of human consciousness."[11] The special nature of this consciousness follows from the dual material/ideal nature of the systems of artifacts that constitute the cultural environment. Human beings live in a double world, simultaneously "natural" and "artificial."

The special temporal characteristics of human psychological processes that accompany this view of human nature as created through processes of cultural mediation in an environment saturated with historically accumulated systems of artifacts were described in particularly powerful language by Leslie White:

> With words man creates a new world, a world of ideas and philosophies. In this world man lives just as truly as in the physical world of his senses. ...This world comes to have a continuity and a permanence that the external world of the senses can never have. It is not made up of the present only but of a past and a future as well. *Temporally, it is not a succession of disconnected episodes, but a continuum extending from infinity to infinity in both directions.*[12]

## The Future in the Present

In recent years we have seen some interesting suggestions for how people routinely import the future into their everyday interactions with others through the rhetorical strategies they use to create mutual understanding. For example, when people use rhetorical strategies that posit particular future states as the given content of an argument, the scholars of ancient Rome and early modern Italy referred to this process as "prolepsis," meaning the representation of a future act or development as being presently existing. Ragnar Rommetveit pointed out that ordinary human discourse is often proleptic "in the sense that the temporarily shared social world is in part based upon premises tacitly induced by the speaker."[13] Through prolepses, what is said serves..."to induce presuppositions and trigger anticipatory comprehension, and what is made known will hence necessarily transcend what is said."[14]

---

11    Ilyenkov, p. 94.
12    L. White, "On the Use of Tools by Primates," *Journal of Comparative Psychology* 34 (1942), 369–374 (372) (italics mine).
13    R. Rommetveit, *On Message Structure: A Framework for the Study of Language and Communication* (New York, 1974), p. 87.
14    Ibid., p. 88.

Stone and Wertsch[15] used prolepsis in this manner to characterize the way in which teachers seek to induce children's understanding of how to complete cognitive tasks with which they are having difficulty; in effect, the teachers presuppose (a least hypothetically) that the children understand what it is they are trying to teach as a precondition for creating that understanding.

### Social Organization of the Future in the Present

A particularly convincing illustration of how prolepsis operates in everyday interaction to create the non-linearity of cultural time with respect to human experience is especially visible in the interactions that take place at the birth of a child. In this first encounter between generations, parents make visible how the cultural past greets the newborn as its cultural future; the palpable constraints in place in adulthood are transformed into palpable constraints at birth so that "future structure from the past" is transformed into material constraints on the process of organism-environment interaction at birth. This illustration comes from transcripts collected by the British pediatrician, Aiden McFarlane, who tape recorded the conversations that took place between parents at their children's birth.[16] He found that the parents almost immediately start to talk about and to the child. Their comments arise in part from phylogenetically determined features (the anatomical differences between males and females) and in part from cultural features they have encountered in their own lives, including what they know to be typical of boys and girls. Typical comments include "I shall be worried to death when she's eighteen" or "It can't play rugby," said of girls.

Putting aside our negative response to the sexism in these remarks, we see that the adults interpret the phylogenetic-biological characteristics of the child in terms of their own past cultural experience. In the experience of English men and women living in the mid-twentieth century, it was "common knowledge" that girls do not play rugby and that when they enter adolescence, they will be the object of boys' sexual attention, putting them at various kinds of risk. Using this information derived from their cultural past and assuming cultural continuity (for example, that the world will be very much for their daughter as it has been for them) parents project a probable future for the child. In this way, parents represent and enact the future in the present, an

---

15    C.A. Stone and J.V. Wertsch, "A Social Interactional Analysis of Learning Disabilities," *Journal of Learning Disabilities* 17 (1984), 194–199.

16    A. McFarlane, *The Psychology of Childbirth* (Cambridge, MA, 1978).

example of prolepsis. Moreover, if less perhaps obviously, the parents' purely *ideal* recall of their past and their imagining their child's future become a fundamental *materialized* constraint on the child's life experiences in the present. This rather abstract, non-linear process of transformation is what gives rise to the well-known phenomenon that even adults totally ignorant of the real gender of a newborn treat the baby quite differently depending upon its symbolic/cultural gender. Adults literally create different material forms of interaction based on conceptions of the world provided by their cultural experience. They bounce boy infants—those wearing blue diapers— and attribute "manly" virtues to them while they treat girl infants—those wearing pink diapers—in a gentle manner and attribute beauty and sweet temperaments to them.[17]

I believe the process illustrated by MacFarlane to be universal, but I know of no recordings equivalent to Macfarlane's from other cultures. However, an interesting account of birthing among the Zinacanteco of South-central Mexico appears to show the same processes at work. In their summary of developmental research among the Zinacanteco, Greenfield, Brazelton, and Childs (1989) report a man's account of his son's birth at which the son

> was given three chilies to hold so that it would...know to buy chili when it grew up. It was given a billhood, a digging stick, an axe, and a [strip of] palm so that it would learn to weave palm.[18]

Baby girls are given an equivalent set of objects associated with adult female status. The future orientation of differential treatment of the babies is not only present in ritual, it is coded in the Zinacantecan saying, "For in the newborn baby is the future of our world."

### The Future in the Present in Early Childhood:
### An International Case

To give some flavor of the ways in which the proleptic cultural organization of experience present at birth continues to provide specific patterns of

---

17  J.Z. Rubin, F.J. Provezano, and Z. Luria, "The Eye of the Beholder: Parents' View on the Sex of a Newborn," *American Journal of Orthopsychiatry* 44 (1974), 512–519.

18  P.M. Greenfield, T.B. Brazelton, and C.P. Childs, "From Birth to Maturity in Zinacantan: Ontogenesis in Cultural Context," in *Ethnographic Encounters in Southern Mesoamerica: Celebratory Essays in Honor of Evan Z. Vogt*, eds. V. Bricker and G. Gossen (Albany, 1989), p. 177.

culturally mediated experience that shape environmental influences on children's experience, I provide an example from a later period of development that involve comparisons of Japanese and American preschool educational practices. Joseph Tobin and his colleagues conducted a comparative study of preschool socialization in Hawaii, Japan, and China.[19] They recorded classroom interactions that they then showed to teachers and other audiences in all three countries to evoke their interpretations and basic cultural schemata relevant to the preschool child. Only the Japanese and American data is discussed here.

When Tobin and his colleagues videotaped a day in the life of a Japanese preschool, young Hiroki was acting up. He greeted the visitors by exposing his penis and waving it at them. He initiated fights, disrupted other children's games, and made obscene comments. When American preschool teachers observed the videotape, they disapproved of Hiroki's behavior, his teacher's handling of it, and many aspects of life in the Japanese classroom in general. Starting first with the overall ambience of the classroom, Americans were scandalized by the fact that there were 30 preschoolers and only one teacher in the classroom. How could this be in an affluent country like Japan? They could not understand why Hiroki was not isolated as punishment.

The Japanese had a very different interpretation. First, while teachers acknowledged that it would be very pleasant for them to have a smaller classroom, they believed it would be bad for the children, who "need to have the experience of being in a large group in order to learn to relate to lots of children in lots of kinds of situations."[20] When asked about their ideal notion of class size, the Japanese teachers generally named 15 or more students per teacher in contrast with the 4 to 8 preferred by American preschool teachers. When Japanese preschool teachers observed a tape of an American preschool, they worried for the children. "A class that size seems kind of sad and under populated," one remarked. Another added, "I wonder how you teach a child to become a member of a group in a class that small."[21]

Members of the two cultures also had very different interpretations of the probable reasons for Hiroki's behavior. One American speculated that Hiroki misbehaved because he was intellectually gifted and easily became bored. Not only did the Japanese reject this notion (on the grounds that speed is not the same as intelligence), they offered a different interpretation. Hiroki, they believed, had a dependency disorder. Owing to the absence of a mother in the

---

19      J.J. Tobin, D.Y.H. Wu, and D.H. Davidson, *Preschool in Three Cultures* (New Haven, 1989).
20      Ibid., p. 37.
21      Ibid., p. 38.

home, he did not know how to be properly dependent and consequently how to be sensitive to others and obedient. Isolating Hiroki, they reasoned, would not help. Rather, he needed to learn to get along in his group and develop the proper understanding in that context.

Tobin and his colleagues comment on the Japanese view of their preschool system and Hiroki's behavior as follows:

> ...Japanese teachers and Japanese society place [great value] on equality and the notion that children's success and failure and their potential to become successful versus failed adults has more to do with effort and character and thus with what can be learned and taught in school than with raw inborn ability.[22]

The Japanese who watched the tape disapproved of the promotion of individualism that they observed in tapes of an American classroom, believing that "A child's humanity is realized most fully not so much in his ability to be independent from the group as his ability to cooperate and feel part of the group."[23] One Japanese school administrator added:

> For my tastes there is something about the American approach [where children are asked to explain their feelings when they misbehave] that is a bit too heavy, too adult like, too severe and controlled for young children.[24]

There are many interesting implications to be drawn from these brief observations. In the present context my purpose is to relate them to the situation such children encounter as adults, in particular, the situation that Japanese boys face should they pursue a career in the American pastime of baseball. This point is made in a fascinating account of the fate of American baseball players who play in the Japanese major leagues (R. Whiting, 1989).[25] Despite their great skill, experience, and physical size, American ballplayers generally have a very difficult time in Japan. There are many reasons for their difficulties, but crucial to their difficulty is a completely different understanding of keys to success in this team sport, a difference that mirrors differences in preschool

---

22    Ibid., p. 24.
23    Ibid., p. 39.
24    Ibid., p. 53.
25    R. White, *You've Gotta Have Wa* (New York, 1989).

education in the two cultures to an amazing degree. The title of the book, *You Gotta Have Wa*, pinpoints one key difference. *Wa* is the Japanese word for group harmony and, according to Whiting, it is what most clearly differentiates Japanese baseball from the American game. Although American ballplayers maintain that individual initiative and innate ability are the key ingredients to success, the Japanese emphasize that "the individual was nothing without others and that even the most talented people need constant direction."[26] Linked to the emphasis on group harmony is an equivalent emphasis on *doryoku*, the ability to persevere in the face of adversity as the key to success; Americans emphasize individual talent. Whiting pointed out that the ideals of *wa* and *doryoku* are cornerstones not only of Japanese baseball, but of Japanese business as well.

> *Wa* is the motto of large multinational corporations, like Hitachi, while Sumimoto, Toshiba, and other leading Japanese firms send junior executives on outdoor retreats, where they meditate and perform spirit-strengthening exercises, wearing only loin-cloths and headband with doryoku emblazoned on them.[27]

Despite their acknowledged talent, American players, whose understanding of the sources of success, the cultivation of which can clearly be seen in their preschool education, are generally unable to submit to the Japanese way of doing things. In a remark that echoes poignantly on the Japanese disapproval of the American emphasis on verbalizing and valuing personal feelings over group harmony, one American ballplayer who had a long and acrimonious public dispute with his manager who was led to ask in desperation, "Don't you think that's going too far? What about my feelings? I have my pride you know." To which the manager replied, "I understand your feelings; however, there are more important things."

Here again we see how culture operating on young children creates an effect conditioned not by present necessity, but by deep beliefs about "how things will work" in the child's future, that serves as a conceptual schema, a cultural model, for how they treat children in the present. Cultural differences in behavioral organization in the present may have relatively minor consequences in the present life of children, but major effects in terms of the long-term organization of their behavior.

---

26    Ibid., p. 70.
27    Ibid., p. 74.

### The Temporal Nature of Images: Picturing the Future in the Narrative Present

My examples so far have been taken from neuroscience and psychology. As a final example, I turn to literature of the complex temporal organization of culturally mediated human experience. I draw upon a passage in which Fyodor Dostoevsky illustrates the important principle that human experience of an image, such as a painting or photograph, requires the person viewing the image to place it within a temporal sequence in the act of making it interpretable. The following passage is taken from Dostoevsky's *The Idiot* in a scene where Prince Myshkin is asked to describe a picture of an execution he has witnessed. To follow the ensuing exposition, readers should first imagine for themselves a picture of a man about to be guillotined.

At the start of the story we the readers, and the women who have provoked the Prince into telling the story, have only a vague image of the picture that Prince has in mind. This image of the end product, the picture, is the starting point for experiencing the temporal expansion of (the imagined) prior context.

> "It's literally a minute before death," the prince began with perfect readiness, carried away by the recollection and to all appearance immediately forgetting everything else, "that very moment when he [the prisoner] has just mounted the ladder and has just stepped onto the scaffold. Here, he glanced in my direction; I looked at his face and understood everything... But, then, how can you recount it! I would like terribly, so terribly, for you, or someone else, to draw it. It would be best if it were you! I thought at the time that a picture of it would do good. You know, here it is necessary to imagine everything that has gone before, everything, everything. He had been living in prison and expecting his execution not for another week at the least; he had been counting on the usual formalities, that an order would have to be forwarded somewhere and would only come back in a week. But here, suddenly, by some chance, the business was shortened."

Then the Prince describes the interchange between the executioner and the prisoner.

> At five o'clock in the morning he was sleeping. It was in late October; at five o'clock it was still cold and dark. The prison superintendent came in quietly, with a guard, and carefully touched him on the shoulder; the latter sat up, leaning on his elbow— he sees the light. "What is it?"—"The execution is at ten o'clock." Just roused from sleep, he did not believe it,

began to argue that the order wouldn't be ready for a week, but when he came fully to his senses, he ceased arguing and fell silent—that's how they told it—then he said: "But it's hard it should be so sudden..."—and fell silent again, and didn't want to say anything more.

When Prince Myshkin imagines for us what the prisoner must have been thinking, we move with Myshkin into the prisoner's conscious experience, still remote from the event, but powerful because we also have, as it were, the prisoner's projected encounter with the inevitability of our own non-existence, our absence in time. The closer that moment comes, the more rapid and sensory-soaked the description becomes.

> I think that here, too, it must seem that there is still an eternity left to live, while they're taking him along. I fancy he must have thought along the way: "There's a long time left, three streets more; I shall pass by this one, then there'll still be that one left, and then that one, with the baker on the right... It'll be a long time before we get to the baker's!" There are crowds of people all around, shouting, clamor, ten thousand faces, twenty thousand eyes—you have to bear it all, but the main thing is the thought: "There's ten thousand of *them*, and they're not executing any of them, but they're executing me."

Then the ascent up the ladder as a scene of procedural "preliminaries"...

> There is a little ladder leading to the scaffold; here, before the ladder, he suddenly began to cry, and this was a strong and manly person, a great villain, they say, he was. The priest never left him for a moment, and rode with him in the cart and kept talking all the while—I doubt that he heard: he'd start to listen, but after two words he'd no longer comprehend. So it must have been. At last he began going up the ladder; his legs were tied together so that he could only move with tiny steps. The priest, who must have been an intelligent man, left off speaking and only kept giving him the cross to kiss. At the foot of the ladder he was very pale, and when he got to the top and stood on the scaffold, he suddenly became white as paper, as white as writing paper.

Once on the scaffold, the Prince/Prisoner—the two are one now—begins to externalize the prisoner's internal experience of a future with no future. It is through these proleptic imaginings that we, with Prince Myshkin, imaginatively experience the future execution.

His legs must have growing weak and wooden and he must felt nauseous—
as though something were choking his throat and it almost tickled—have
you ever felt that when you were frightened, or in very terrifying moments
when all your reason remains but has no more control? It seems to me
that if one is faced with inevitable destruction, for instance, a house is
falling on you, then you would suddenly feel a terribly longing to sit down
and close your eyes and wait—come what may! Well, in that very
moment, whenever this weakness began, the priest would hastily, with a
rapid gesture and in silence, put a cross all of a sudden to his very lips, a
little silver cross, four-pointed—he kept putting it up frequently, every
minute. And the moment the cross touched his lips, he would open his
eyes and seem to come alive again for a few seconds, and his feet would
move. He kissed the cross greedily, made haste to kiss it, as if he was in
haste not to forget to grab some kind of provision, just in case, but I doubt
that he was cognizant of anything religious at the time. And so it went
until he was laid on the plank...

From this moment the "narrative future," the imagined event, becomes a felt
future experience *now* for the Prince and for the reader.

It's strange that people rarely faint at these last moments! On the contrary,
the head is terribly alive and must be working with such mighty, mighty,
mighty force, like an engine at full speed; I imagine there's a continual
throbbing of various thoughts, all unfinished, and perhaps even absurd,
such irrelevant thoughts. "That one there keeps looking—he has a wart
on his forehead, see, one of the executioner's bottom buttons is rusty"...
and in the meantime, you know everything and remember everything;
there is this one point that you simply can't forget, and you can't lose consciousness, and everything moves and turns about it, this one point. And
to think that it is like that up to the very last quarter of a second, when
your head is already lying on the plank, and waits, and...*knows*, and suddenly hears above how the steel starts to glide! You'd hear that for certain!
If I were lying there, I should listen on purpose and hear! It may only be
one tenth of a moment, but you'd hear it for certain!

And it's just because Prince Myshkin has located his own thoughts second by
second with the prisoner's that we can experience a future event, for us as well
as the prisoner,—a terrifying unforgettable but unavoidable future. As if to
prolong our discomfort, Doestoevsky carries his thought experiment a step
future, placing death, for a moment, in the past.

> And just imagine, it's still argued that, maybe, when the head flies off, then for one more second, it may be that it knows that it's flown off— what a concept! And what if it's five seconds!...

Then, at the end of his story, the Prince returns to summarize his image of the painting. We return to compare our image of the execution, now that the Prince has brought us into the very heart of the events that gave rise to the image of an execution. This time, when the Prince provides a distilled image of the paining, the beginning point of our experience of the execution becomes a new end-point. Through Doestoevsky's act of literary imagination, we are at one with the prisoner. Hence, the Prince's summary of the picture, terse as it may be, now immediately expands into the harrowing event that it synoposizes.

> Paint the scaffold so that only the last step can be seen distinctly and up close; the criminal has stepped upon it: his head, his face white as paper, the priest holds up the cross, the former greedily holds out his blue lips and looks, and—*knows everything*. The cross and the head—that's the picture, the face of the priest, the executioner, his two attendants, and a few heads and eyes below—all that might be painted in the background, in a haze, for embellishment... That's the picture.[28]

### Final Remarks

My reading of contemporary neuroscientific research on the relation of conscious experience to the workings of the brain, no less than research on the nature of conversational understanding or the way in which adults organize the experience of their children urge upon me the idea that cultural mediation of human experience introduces a fundamentally non-linear element into the temporal relations organizing human consciousness. A great deal remains uncertain about the complex systems properties that arise as a result of the cultural mediation of consciousness, and I pretend to no overall understanding of all the factors at work. But I believe that a deeper understanding of the role of time in human experience is not achievable unless the kinds of phenomena I have presented here are taken systematically into account.

---

28    Fyodor Dostoevsky, *The Idiot*, new rendition by Anna Brailovsky, based on the Constance Garnett translation (New York, 2003), pp. 67–70 (italics hers).

# CHAPTER 18

# "Real Time". On the Whereabouts of Time in New Media: The Case of Webcams

*Ike Kamphof*

## Here and Now—New Media and the Challenge of Presence

We live our lives as embodied beings situated in space and time. We move and act in the presence of other people, and of the objects, animals and plants that constitute our environment. Our being-in-the-world always has a particular shape and mood, intimately tied up with the relationship that we, as sensual and intentionally acting beings, have with our environment. A sense of presence in space and time, and of our co-presence with other living beings and with things is sometimes explicit, but more often only peripheral.

While the quintessential experience of presence may be a feeling of an—always specific—"here and now," and the main parameter for co-presence being "physically" and "actually" proximate to others, experiences of presence and co-presence are not limited to the here and now. Deeply immersed in an artwork, I am "in the moment" but also elsewhere. Feelings of co-presence certainly do not require that we share both space and time with the person, object or event that we become aware of.

Upon entering my room, I may sense that somebody has been there, and experience a somewhat uncanny co-presence in location, though right now there is no trace of the intruder. A more imaginative elaboration of being-with somebody in place, but not in time, is part of many tourist experiences, as when I delight in the idea that right now I am standing in exactly the same place at Waterloo where Napoleon beheld his last, ill-fated battle.

More in line with the focus of this essay, our feeling of co-presence with others can also be based on sharing time, though spatially we may be on opposite sides of the globe. Far away from home, I can be deeply comforted by the thought that at this precise moment my family is having dinner together. I can even feel related to events that I have no direct acquaintance with, but that must be happening now. In this vein I muse that at this minute, when I feel elated about life, somebody else on this earth is suffering, or indeed is dying.

These examples suggest that the criteria for who is, and for what is, experienced as present or co-present are flexible, determined by actual and physical proximity, but also by imaginative intensities that bridge distances in space and

© KONINKLIJKE BRILL NV, LEIDEN, 2016 | DOI 10.1163/9789004312319_020

time. Nevertheless, the "here" and the "now" remain the reference points for the range of experiences that we normally accept as feelings of presence and co-presence. People whose experiences stretch this normality too far are felt to be oversensitive, while those whose reactions show little awareness of the things and people in their immediate environment are seen as dreamers or as cold.

At present this normality is being strongly challenged by the development of electronic media. Photography and television, each in its own way, make the remote present, often with great emotional force, in the intimacy of our living rooms. The telephone and, more recently, the internet bring people together in so-called "real time," but in a space that is not easily described. Its specific qualities are merely hinted at by terminology like "virtuality" or "cyberspace."

The pervasiveness of electronic media in our culture calls up vitally important questions about the kind of presence these media provide and what this may entail for our fundamental orientation in, and sensitivity to, the world. Is the presence of the world that is offered by TV, captivating as it may be, at the same time a diminished and superficial presence? Can one really speak of co-presence when physical bodies stay in separate places and the "meeting" takes place in the fleetingness of what is presented on a computer screen? And what about people who consider those at the other end of their mobile phone more present than the people sitting around them?

The question of what may be happening to the "here" and the "now" in relation to each other gives rise to special concerns. While the overcoming of distance by the media reduces the importance of local space, the "now" gains prominence where information is transmitted with increasing velocity. For French philosopher and critic of technology Paul Virilio, it is indeed speed and the—real or professed—immediacy of media technologies that characterizes our present culture. "Real time," Virilio claims, is replacing "real space." He exhorts us to come to terms with this development immediately.[1]

The sense of urgency in Virilio's work stems from the supposition that a reshaping of our sense of being-in-the-world by the new technologies is not merely an aesthetical issue, but has grave implications for our ethical engagement with the world.[2] His fears are shared by another critic of new media. In "The Scene of the Screen," Vivian Sobchack analyses the ways in which our

---

1   See esp. Paul Virilio, *The Vision Machine* (Bloomington, 1994), *Polar Inertia* (London, 1999), *The Information Bomb* (London, 2000). A useful collection of essays is *The Virilio Reader*, ed. James Der Derian (Oxford, 1998). The essay that forms the starting point of investigations here is Paul Virilio, "The Visual Crash," in *CTRL[SPACE]: Rhetorics of Surveillance from Bentham to Big Brother*, eds. Thomas Y. Levin, Ursula Frohne. and Peter Weibel, (Karlsruhe, 2002), pp. 108–113.

2   I use the term "aesthetic" here in its technical sense as referring to the process of perceptual reception.

existential, embodied presence to the world, to ourselves and to others has been reconfigured in conjunction with the subsequent development of three visual media: photography around the 1840s, followed in the 1890s by cinema and by electronic media in the 1940s.[3] With the latter, an "absolute present" takes hold on our culture. Sobchack refers here not only to real time media, but to the possibility of direct access in general, through interactive engagement with all kinds of data, live or recorded, that are, as it were, always presently available through electronic media. Real time media form a poignant case of this present of instant access. Like Virilio, Sobchack is anything but positive about the recent developments and points to the imminent danger of a loss of physical and moral gravity that could well cost us our future.[4]

Although the supposedly intimate relation between our perceptual orientation in, and responsiveness to, the world and the involvement and responsibility that we may feel is a complex issue, many will recognize these premonitions as their own. These concerns are as serious as they are difficult to investigate. Ultimately they require an interdisciplinary approach that would bring together the extensive research being carried out in engineering departments and laboratories, studies in neuroscience and analyses of aesthetic response in the field of the humanities. The questions of what presence and co-presence are, how they can be evoked in artificial and telepresent environments, and what that might mean for our culture and way of being together, could then be placed in a historical and ethical setting with the help of philosophical aesthetics and studies in visual culture.[5]

The project I undertake here has a more limited scope. It is situated in a subsection of phenomenological aesthetics that provides a fruitful starting point to approach the issues faced. I follow Virilio's call to investigate the basic shape of our experiences of presence and co-presence in the light of real time technologies and focus on one example of these that is targeted by

---

3  Vivian Sobchack, "The Scene of the Screen," in *Carnal Thoughts: Embodiment and Moving Image Culture* (Berkeley, 2004), pp. 135–162.

4  Ibid., pp. 158, 159.

5  For an overview of technical research on presence, see http://www.ispr.info/. Much of this research seems to ignore the historical aspect and cultural embedness of response. For a history of the dynamic between the creation of illusions and response, see Oliver Grau, *Virtual Art: From Illusion to Immersion* (Cambridge, MA, 2003). A further complication with experiences of presence is that the awareness of presence is itself a reflective experience, which makes explicit what is normally only peripheral. In the experience of new media, presence can become more explicit, precisely because it is both different and similar than physical and actual proximity.

Virilio in his essay "The Visual Crash": the fast proliferation of live cameras on the internet.[6] This choice is motivated by methodological concerns—as will become clear later—as well as by the striking popularity of this relatively simple technology.[7]

Through webcam sites users have direct access to live views of places as different as a research station in Antarctica, a bus stop in the town of Mantova, Italy, the pyramids of Giza, an osprey nest or a waiting room of the division of motor vehicles of the department of administration in Fairbanks, Alaska.[8] All of these, and many other places are just a click away. Together they form, as Campanella has aptly remarked, "something of a grassroots global telepresence project."[9] As such they seem a fitting illustration of Virilio's claims on the new pre-eminence of real time over real space. I will focus on how time is displayed in the presentation of these camsites, and how it shapes actual practices of viewing that develop around this new medium.

I will not attempt to answer the question to what extent our aesthetic and ethical sensibilities are indeed being reshaped at present.[10] Instead I will point out some methodological problems with analysing new media in the light of this question. These problems specifically pertain to time. As much as I sympathise with the forebodings brought forward by Virilio and Sobchack, pointing out these problems does entail a critique of their approach. My subsequent analysis of time and webcam viewing aims to complement their work, by sketching out a few lines that warrant consideration in, among others, further ethnographic research of emerging viewing practices.

---

6    Virilio's focus in "The Visual Crash" is less on our sense of presence, and more on global surveillance and journalism. However, concerns with presence are central to Virilio's work on real time and telepresence technologies in general.

7    It is impossible to say how many webcam sites are available for public viewing on the net. The main webcam portals, such as earthcam.com or worldlive.cz list over 2000 sites. The actual number may run into hundreds of thousands. EarthCam claims to receive, on average, over 200.000 unique visitors daily (earthcam.com, Press Release [New York, July 17, 2007]). Popular sites may receive as many as 3000 hits a day.

8    http://www.martingrund.de/pinguine/; http://digilander.libero.it/strafat/; http://www.pyramidcam.com/; http://www.friendsofblackwater.org/camhtm.html; http://www.state.ak.us/dmv/FBKwebCam.htm.

9    Thomas J. Campanella, "Eden by Wire. Webcameras and the Telepresent Landscape," in *The Robot in the Garden: Telerobotics in the Age of the Internet*, ed. Ken Goldberg (Cambridge, MA, 2000), pp. 22–46, 23.

10   I address this question in Ike Kamphof, "Webcams to Save Nature. Online Space as Affective and Ethical Space," in *Foundations of Science* 16/2–3 (2011), 259–274.

Before elaborating on these points, I will present a short sketch of Virilio's and Sobchack's main concerns.[11] The gist of their concerns resounds in much that is brought forward on the subject, whether in academia or the popular media.

## Disembodiment and the Fickleness of Electronic Presence

Though they come to their conclusions by different paths, Sobchack's and Virilio's criticisms of new media are similar in their main lines. Finding their basic inspiration in Maurice Merleau-Ponty's phenomenology of the body, both point to the marginalization of the lived body and its situated engagement in a concrete lifeworld shared with others, which accompanies new media.[12] As Virilio states, we now have the possibility, through real time connections, of being present at locations other than where we are physically located. We are leaving our full-bodied self behind for an indeterminate presence, the presence of an anyone, anywhere.[13] Sobchack speaks of the experience in which electronic media engage their spectators "as so diffused as to belong to *no-body.*"[14]

Where Virilio compares telepresence directly with real-life experience, Sobchack comes to her point by a detour through cinema. In cinema what is seen on the screen appears as the result of an animated, intentional seeing. As the camera moves through the visual field, spectators are confronted with a seeing that seems in the process of coming into being, signifying and developing towards a future. Film reflects and makes visible at the same time, the way a human subject sees, while being engaged with the world. Cinema thus affirms the human, embodied mode of seeing and presence to the world. In contrast, electronic presence is dispersed in a centerless network of information, randomly accessed by quasi-disembodied beings. Thus, both authors link

---

11    In the case of Sobchack's essay, "The Scene of the Screen," it must be mentioned that this stands out in her work—which mostly focuses on film—in being much more generalizing. This other work provides extremely valuable suggestions for investigations into our response to electronic imagery. See esp. Vivian Sobchack *The Address of the Eye. A Phenomenology of Film Experience* (Princeton, 1992), and the essays collected in *Carnal Thoughts* (see n. 3).

12    Sobchack, "The Scene of the Screen," pp. 161, 162.

13    Virilio, "Polar Inertia," pp. 130–131. Cf. "Conversation between Paul Virilio and Hans-Ulrich Obrist," June 8, 1991, *Urbanaria*, the website of the first annual exhibition of the Soros Center for Contemporary Arts – Ljubljana, online: http://www.ljudmila.org/scca/urbanaria/txt/e/virilio.htm, p. 4.

14    Sobchack, "The Scene of the Screen," p. 152.

electronic media to the disembodiment of presence and engagement, which is at the same time seen as a departicularization and diluting of the presence of the human user and of his or her engagement with the content provided.

Disembodiment on the side of the subject is mirrored on the side of the object by dematerialization and virtualization of what is presented in new media. Sobchack speaks of electronic imagery as constituting an alternative world, consisting of copies that appear to lack an original ground.[15] As a result the imagery looks thin and cannot hold our attention, a point also suggested by Virilio.[16] The networked present of electronic media goes together with "instant stimulation" and volatile "impatient desire."[17] Virilio also emphasizes how the acceleration and fragmentation of our interaction with the environment in and through electronic media takes away the meaning of local space and time. Moreover, the instant dissemination of imagery across the world erases the possibility of interpreting and reflecting on what is perceived.[18]

In short, both authors expect disembodied and dematerialized presence to result in a scattered engagement with a discontinuous network. They suggest that this shallow and erratic engagement is a reflection, even a consequence, of the discontinuities and thinness of the network of contents. In that network, unevaluated units of information come to viewers as if by coincidence. Electronic media and real time technology thus encourage viewers to be incessantly on the lookout for the unexpected. Narrative, suspense and history as forms and expressions of sustained experience are supplanted by a mode characterized by reiterated short-lived surprise. Although in line with popular suspicions that new media are turning viewers into sensation seekers characterized by an ever shorter attention span and hovering in a constant "now" of superficial curiosity, there are, nevertheless, problems with this position. These problems pertain to popular opinion as much as to scholarly conjectures in this vein.

### Where is Time—and When are We?

My methodological concerns here can be summarized in two questions. First, where do we locate time when we speak about the way media affects our perception and sense of presence? Is it a matter of the hardware and software developed—that is, of technology—or also of the use individuals or groups of

---

15    Ibid., p. 154.
16    Paul Virilio, *The Vision Machine*, pp. 64–66, 72.
17    Sobchack, "The Scene of the Screen," p. 153.
18    "Conversation between Paul Virilio and Hans-Ulrich Obrist," p. 8.

people make of it? How important are different applications of what could technically be considered the same medium, and what about the subject matter presented? To what extent does time, as located in the living bodies of viewers, still cut through our experience? Second, where does the analysis of the temporal structures of a given medium situate itself with regard to time? Let me try to elucidate these two points by commenting on some of the choices made by Sobchack and Virilio.

To start with the first question, Sobchack's phenomenological approach is sophisticated enough to avoid the trap of a simple technological determinism. She emphasizes the interplay between technology and culture as technologies develop, historically informed by their own materiality, but also by their political, economic and social context. Technology co-constitutes culture, but is itself also an expression of cultural values. Moreover, technologies are not merely used as instruments, but also lived by the human beings who create them. She rightly concludes that the relationships between technology and those making up a specific culture at a given historical moment are dynamic and reversible.[19]

However, her subsequent discussion divides the various media along lines that seem to be precisely technological—*in casu*, photography, cinema and electronic media. This leads to a view of media as much more homogeneous than they actually are. Speaking of electronic culture in this manner, she refers to media as diverse as television, VCR and DVD players and recorders, the computer and mobile phones. It is true that all these media provide instant access to content and invite interaction that is anchored in the now of that access—that is, if one counts zapping channels as the main use of television. Yet focussing on a general characterization of this sort hides the variety that exists in modes of interaction or even within one mode that affect the experience of viewers in different ways. These can hardly be said to be all random and scattered. A serial on television is not only seen in the present, but invites its fans to a repeated investment of time that connects the present of access to the past and the future. Playing online games or writing a weblog demand from users a commitment in time and social engagement.

The criticism of an overly general and too technologically inspired notion of media holds even more for Virilio's quite abstract treatment of real time technology, which shows little concern for differences in display, subject matter, or use. Virilio tends to focus on military technologies and practices as seminal to our culture. He ignores the possibility that, even when certain structures have come into being for military purposes, actual uses may be

---

19    Sobchack, "The Scene of the Screen," p. 137.

different from, or even subvert, the original purposes. The internet, born from the military Arpanet, is a case in point. Though Virilio aims to address our critical sense, and in that pre-supposes that we have a choice in the matter, his work often exhibits a technological—or possibly a cultural—determinism.

I propose a different approach and will start here at the opposite end of the spectrum: a single application of electronic media, the webcam sites mentioned before. For these sites, real time and a reiterated temporal present are certainly defining aspects. Still, one wants to know how real time shapes this particular medium and its use—or, I would say, its possibilities of use. An overly general approach to media tends to obscure the various options that exist side by side and spring forth from the dynamic interplay between technologies, cultural forms and different users. Close examination of a single medium, its various levels, and the practices that form around these, can yield results and questions that are useful for the study of other media.

A final remark on this first methodological point concerns phenomenology as a methodology for analysing presence. While offering vital insights into the ways we live with technologies, phenomenology cannot help but presuppose a somewhat abstracted "we," situated in an archetypical situation of viewing.[20] It is indispensable in mapping the outlines of the kind of presence propounded by a specific medium, and that is an important part of what my own analysis aims to do. However, our experience, mediated and unmediated, has multiple layers and we partake of different temporalities, often at the same time. It is coloured by culture, class, and gender, and by personal habits and sensibilities. As a result there is always a certain leeway. Phenomenology cannot have the final word about what users actually do. This requires empirical investigation into existing viewing practices, such as that provided by ethnographic research.[21] Here, I will complement, and confront, my phenomenological sketches with observations on emerging viewing practices.[22]

• • •

---

20 For insightful contributions of phenomenology to the study of technology and media, see also Don Ihde, *Technology and the Lifeworld: From Garden to Earth* (Bloomington, 1990) and *Bodies in Technology* (Minneapolis, 2002).

21 A general introduction into the application of ethnography to new media environments is Christine M. Hine, *Virtual Ethnography* (London, 2000).

22 These observations stem from the study of three forums/guestbooks, conducted in 2006: www.tnacso.net/cont/jailcam.php (discontinued November 27, 2006); http://www .infotecbusinesssystems.com/wildlife/?ch=Wildlife&sh=eaglecam (continued in other format after April 2006); http://www.Africam.com; and from my own experiences as a webcam enthusiast.

As pertains to the second methodological question, our present situation, traced through by accelerating cycles of the development and decay of new technologies and the cultural possibilities these entail, seems aptly character-ized by Peter Lunenfeld as a state of future/present, exemplified in judgments like: by the time you learn this program, it will have become obsolete.[23]

To come to terms with this continually evolving situation, several options for positioning our thinking about new media present themselves. A theory may find its roots and yardstick predominantly in the past, rejecting the "advances" taking place around us by pointing to the loss of established modes of experience and treasured values that they entail. This kind of theorizing risks blinding itself not only to fruitful aspects of current developments, but also to the persistence of past experience and ideals in the present. Sobchack's account of electronic culture, being strongly focussed on what has been lost of photographic and especially cinematic culture, often suffers from this nostal-gia. So do frequent admonitions that point to the loss of community, sustained by face-to-face contact, that supposedly accompanies new media.[24]

A second possibility, exemplified by much of Virilio's work, boldly orients the analysis of the present situation to the future. Flecked with verbal forms like "is becoming," "will become," and "will have become," Virilio extrapolates current technical possibilities, drawing them to a logical conclusion in a future state where they reach their ultimate realisation. To others this future may appear as a longed-for technological utopia; to Virilio it appears as nothing less than apocalyptic.

Speculation about what may occur and looking back to the past are useful tools for an evaluation of new media, if only for the critical distance they provide from the present. However, both must be based on careful examination of actual practices. These are more differentiated than analyses founded on dominant log-ics and tendencies allow, a layered and murky mix of past, present and future practices, experiences and values, of technical possibilities and just as real prac-tical and material constraints.[25] There is little reason to expect the future to be

---

23    Peter Lunenfeld, *Snap to Grid: A User's Guide to Digital Arts, Media and Cultures* (Cambridge, MA, 2000), pp. 27–28. My account follows two of the options presented by Lunenberg.

24    For an overview of the debate on traditional and virtual communities, see David Bell, "Community and Cyberculture," *An Introduction to Cybercultures* (London, 2001), pp. 93–111.

25    I do not intend to reject the value of work like that of Virilio. An examination of general logics is necessary for outlining the interpretative framework for the investigation of new media. In analyses that ignore this, the framework remains implicit and thus hidden. Critical scrutiny of presuppositions then becomes difficult. The larger picture painted here with the help of Virilio and Sobchack forms the common background to my analysis, both where I follow their suggestions and where I criticize them.

different in this respect. A focus on actual practices and on the unavoidable materiality of technology is central to the analysis that I will present here.

### Sharing the World in Real Time: The Different Shapes of Webcam Viewing

In less than 15 years, webcams have developed from a technology for specialists to an extremely popular medium with a wide range of applications. They are used in relatively restricted settings for chatting, videoconferencing, online performance, prostitution, and surveillance purposes; but they are also used in highly public venues for traffic or weather information, and for tele-visiting distant places as different as landscapes, city squares and open offices. As was mentioned, it is this last application that will concern me here.[26] Put generally, webcam sites of this kind display imagery, and occasionally sound, that is uploaded in real time from a camera located in some real place to a server, through which this imagery becomes accessible to users by way of the internet. Usually the camera is static, although some sites offer minimal possibilities for viewer control, such as turning the cam or zooming in and out. Recorded imagery is uploaded continuously (videostream) or at regular intervals with update times varying between a graceful few seconds to many hours. Needless to say, the claim to provide a real time connection, vital to these sites, requires update times to be as short as possible.

Firm belief in the unstoppable progress of technology may suggest that videostream cams hold the future. However, in the light of constraints in space for data storage and in bandwidth, this is anything but certain. All the more so, because webcam sites are frequently set up by private individuals with limited resources. The most satisfying sites hover between videostream and update times no longer than a few minutes. Although, on the face of it, webcam sites

---

26   The first webcam was the Trojan Room Coffee cam, established by scientists from the computer laboratory of the University of Cambridge, Massachusetts in 1991, meant solely for internal use. It displayed to colleagues all over the building, the amount of coffee available in the kitchen next to the laboratory. In 1993 the site was moved to the internet and became an instant hit with viewers all over the world. It is no longer online. The most famous cam was the one student Jennifer Ringley installed in her college dorm from 1996–2003. Many followed her example of a life online. The focus on cams for televisiting places allows the analysis here to remain close to the basic characteristics of the medium and the kinds of presence it propounds. In the use of webcams for self-exposure or for prostitution—the most extensive kind of use—the time of the performance superimposes itself on the fundamental structures of the real time connection to other places.

appear to be another medium of visual representation, in line with painting, photography, cinema and television, it would be a mistake to consider them that alone. They are also media of connection, and this precisely in their real time operation.[27] Though the view may not be spectacular, there is something oddly touching about seeing the sun shine on a patch of dirty snow next to an ugly research station in Antarctica, right now. The feeling of co-presence effected between me as a viewer, in the privacy of my office, and this small event in a remote corner of the world, is as much a function of vision as it is of a virtual "touch." Webcams do not just show users objects and events in another physical location: they link users to these objects or events, uniting them, as it were, in the same temporal space.[28]

This connectivity is felt even when the visible information is minimal, when, for instance, the camera lens is damaged or temporarily covered with raindrops or ice. One can even say that the impossibility of establishing a connection, due to technical difficulties, is still experienced as a connection, though negatively—as a missing one. As such webcams realise, materially and on a global scale, a presence and co-presence that was only latent before and dependent on imaginative extrapolations of our being here and now. Moreover, they offer contiguity in time as something to be explored and enjoyed as such.[29]

In what follows, I will survey the contours of the co-presence made possible by this new real time medium and how these contours are sketched out in various ways by the interplay of the technology, the levels of presentation of camsites, their subject matter, and actual viewing practices.

---

27   Arguably, webcams are not representational media at all. The imagery is taken as a means to perceive a remote environment, much like a telescope.

28   Where Virilio repeatedly speaks of the Renaissance perspective of real space as being replaced by the perspective of real time that unites the globe, this is an important aspect of what he means.

29   This tele-connectivity is emphasized by sites that offer the possibility of some minor type of interaction. A website, broadcasting a scene from a living room in Mesa, Arizona, gave users the option to turn on a variety of lamps in the remote place. Simple as it may be, at its inception in 2005, the cam site received several hundreds of visitors a day, leaving the owners baffled by the incessant flickering of the lamps (www.clickiton.com). A similar site allowed users to blow bubbles on a patio in south Florida (www.andieandmike.org/bubblecam-push.stm). A grimmer version of interactivity was conceived by John Underwood, who intended to offer viewers the possibility to shoot live animals on a Texas ranch (http://news.bbc.co.uk/2/hi/technology/4022147.stm). The plan was not carried out.

## Surfing the World by Will Power: The Cam Portal

Webcams are most easily found and accessed through "directories" or "portals," such as earthcam.com or worldlive.cz, that collect links to cams and order them in categories according to subject matter or geographical location. By clicking on a hyperlink users access a specific website that, in turn, contains a link to a live cam. With their lists of links—supplemented, on occasion, by maps of the world that users can zoom in on—portals portray the places that they connect to as all equally available. Access here comes closest to Sobchack's idea of content being spread out in a simultaneous network. Closing the website of a single cam brings one back to the portal.[30]

If the metaphor of travel, frequently invoked by cam portals, makes any sense, the suggested routine for a journey is that of an effortless quickstep-like clicking back and forth through online content. Though taking place within the framework provided by the portal, the itinerary of this trip, as it unfolds in time, depends almost completely on the choices individual users make, motivated by their curiosity, specific interests and sensibilities, or their boredom. There is no pre-set path or pace to be followed, nor is the end of the personal itinerary given beforehand. The journey is brought to an end either from the outside, by the call of practical duties, or by an upsurge of the recurrent demands of the body, or from the inside, because time and travel halt in something that strikes the viewer: a sudden modest "epiphany."[31]

One can see in this logic of choice and epiphany, indications for the prevalence of restless desire, aimlessly wandering in the present, and the imperative of superficial surprise. Virtual travel by means of webcam portals shares many features with web browsing in general. Tara McPherson describes how the "now" of web browsing is attached to the smooth motions of the cursor over

---

30  The idea suggested by names like EarthCam and WorldLIVE and by earthcam's promotional tag as the place "where the world watches the world" is exceedingly overstated. A quick look at the worldmap of available cams shows how little of the world and of specific areas is viewable online.

31  I borrow the term "epiphany" from Espen Aarseth's analysis of artworks or discourses "whose signs emerge as a path produced by a non-trivial element of work" (Espen J. Aarseth, "Aporia and Epiphany in 'Doom' and 'The Speaking Clock.' The Temporality of Ergodic Art," in *Cyberspace Textuality: Computer Technology and Literary Theory*, ed. Marie-Laure Ryan [Bloomington, 1999], pp. 31–41 [32]). Aarseth calls these works "ergodic" as opposed to "narrative" works. The trajectory users devise while browsing the net can also be called "ergodic."

the screen. She characterizes the activity, directed by our predilections, as "volitional mobility."[32]

While the prolonged presentness of web travel, which is more about keeping up the "momentum than about the moment," can be linked to the gliding cursor, I would add that the will resides in the hand holding the mouse and in the click of the right finger that transfers this "hold" to the screen.[33]

With the webcam portal, the analogy between physical mobility and virtual travel by swerve of cursor and click of mouse gains a special flavour. The suggestion that we visit "real" places in real time transforms surfing the web into a means of conquering the material resistances that come with physical travel. In our Western culture, which conceives time in terms of a scarce good, to be managed and used efficiently, these resistances tend to appear as a "loss of time." Virtual travel offers the opportunity to master time and relish the full time of visiting, now and without waste.

The portal thus encourages viewing as browsing through current content, interspersed, perhaps, with screen grabs in which more remarkable views are swiftly appropriated and stored. The sites speak of "peaking in" and "checking out," terms that hint at both speed and power. The fun is stumbling on small coincidences, where the time of the trip and that of real events meet for a triumphant moment.

• • •

Looking at this picture—part real, partly belonging to the ideology of the net— from the perspective of actual viewing practices, two observations modify the unitary image suggested by the phenomenology of the portal.

1

Anyone with some experience in browsing will be familiar with the fact that presence of content is never guaranteed on the internet; rather, it is intermittent. Webcam sites are no exception here. One clicks an icon, and a website may appear, or it may not. One more click, and after some waiting—the uploading of cam views is often rather slow—live images may appear. A fair number of times they do not, due to technical difficulties with the camera, the

32    Tara McPherson, "Reload. Liveness, Mobility and the Web," in *The Visual Culture Reader*, ed. Nicholas Mirzoeff (London, 2002), pp. 458–470, 460.

33    Ibid., p. 462.

hardware or software, or limitations in bandwidth and internet congestion. Sometimes the cam has simply been discontinued. Though technology is often pictured as immaterial and transparent, it is anything but that, remaining firmly placed in a physical world. When evaluating technology, I suggest we take these malfunctionings seriously. Instead of ignoring them as soon-to-be-remedied side effects, we should recognize them as inherent in the current state of technology. Though this state changes over time, technology always has such a current state, where its ideal potential shows up as "blemished" by material and practical constraints.

In the black holes that punctuate the virtual presence, the real, physical world reasserts itself. This is true for all frustrated browsing, but particularly pertinent in the case of webcam viewing, where it refers directly to the reality of the missing or disturbed content. Waiting for updates, partly discoloured and choppy views, sleeping or dead links—all of these remind viewers of the physical world and of the very real distances separating a remote camera and my desktop.

Though the black holes do not always appear visually in such a way that they can become a meaningful part of experience, more elaborate camsites often have forums or guestbooks attached.[34] These constitute an intrinsic and vital part of the medium. Interestingly, almost as much of the communication on these forums is about the hows and whys of absences of connection and visibility, about the hazards of camera maintenance, as it is about what there is to see. One even finds discussions of absences of view whose causes lie at the users' end of the connection, that result, for instance, from the recurrent needs of the body, necessitating one to shop for food, cook, sleep, and make one "miss out" on views. In these discussions, that come across as being about trifles to a hasty glance, much of what is lacking in material and physical presence on the net returns or is compensated for.

Waiting times are of particular interest when evaluating presence. As a viewer, situated behind my computer at home or in an office, my presence never completely coalesces with the electronic presence the screen holds out to me. My eyes do not fall together with the mechanical eyes of the cam. At the margins of my view, I am aware of my body, situated in a familiar room. I may register the weather outside, the time of day, the outlines of neighbouring

---

34    Often sites do also "present" absence, as when a black screen hosts an explanatory line like "webcam not available after dark" (http://www.walt-n-ingrid.com/webcam/front-yard; accessed 2007-02-09) or "network cable (between webcam + station) broken after big storm. But camera is OK!" (http://www.martingrund.de/pinguine/pinguincam1.htm; accessed: 2007-01-30).

houses and sounds of what is going on in the street. I see what is on the screen, but also the screen itself and my own physical desk. At times a colleague or family member may call for my attention. Waiting times make explicit the many small gaps that separate me from a smooth electronic existence and forcefully bring my experience back to where it is physically located, in front of a black screen.

Potentially, these interruptions foreground the body as the central seat of experience, the node to which I return after my adventures in the electronic world, and to which these adventures have to be tied in order to become part of one lived history. They offer the space for realigning my presence as an embodied being, here and now, to my mobile presence on the net. Or better, they *can* do this, because the extent to which this actually happens depends on what viewers do in their waiting time. Do they quickly change to other electronic tasks—in order not to lose time perhaps? Or do they take that time to reflect on real or electronic experiences? In what ways does increased distance to the screen, which may be waiting for a connection while the user roams around the room doing other chores, merge electronic spaces and times with real space and time?

2

Portals invite browsing, but viewers are not obliged to stay within that mode. Epiphanies in their very structure come to us as given from outside, and can summon forth more prolonged attention. It is not easy to differentiate the superficial shock of the new from a more profound involvement or to decide what stimulates the one rather than the other. The structure of content is one factor, another the sensibilities of different viewers, and still another the viewing situation—think here of the way the world can suddenly appear in a fresh light when we are sick or otherwise stopped in our ordinary activities. The relationship between superficiality of content and attention span does not only work in one direction, as might be suggested by forebodings about new media. Faster interaction can make content appear more shallow, where sustained attention tends to thicken it. This is a lesson we learn from art and aesthetic appreciation. In this respect we are only just beginning to familiarize ourselves with the net.

One observation on webcam viewing is of specific importance here. Regular visitors to webcam sites tend not to stay within the overwhelming totality of present content offered by the portal. They make the net smaller and more habitable by creating lists of favourite cams that are the result of personal choice, but also of being struck or "touched" by something.[35] Supported by the personal campage option on portal sites and by special viewing software, favourites provide the possibility of repeated visits to the same site. In this way, specific sites and the places they connect to can become part of little daily rituals—like always visiting a favorite cam with the morning coffee. "Volitional mobility" gives way to other temporalities; a quick pick turns into attachment or captivation.

### The Webcam Site: Sharing Time

Looking at the screen on the level of single websites, I will concentrate on the actual cam view and its movement in accordance with the dictates of real time, for which the surrounding content of banners, ads and textual information makes up the stable backdrop. The core of the cam view is usually the time stamp, placed right within the image at the upper or lower rim, giving the date and time—the hour, minutes and often seconds—at the place depicted. Here the medium explicitly points to itself as providing a real time connection. In the case of imagery that is updated, the website will mention the refresh rate.

For the user, the time stamp is the point of reference for establishing the connection between two places, not by simply erasing local time, but by offering an opportunity to correlate the local time of the place viewed and that of the place of viewing, while uniting them in the real time of the medium. The time stamp also facilitates a quick check of the veracity of the connection by allowing the local time to be situated within the (often vaguely comprehended) framework of international time zones. Given the ample opportunity for hoaxes on the net, this check is vital. Another check is provided by clues to natural time within the image: the changes in the light between night and day, and with the passing of the seasons, which also help users to orient themselves and the places visited in global space.

The time stamp, which indicates the moment the picture was taken, high-lights the indexical status of the imagery. As with documentary photography and film, the affective force of the index is played out between the presence,

---

35    The term "touch" is used by J.-F. Lyotard for the experience of the "now" of events as something miraculous that cannot be articulated without loosing this initial presence. Art recalls this "now," implicated in any perception as fundamentally receptive. It is also present in more colloquial experiences of wonder that draw us out of ourselves and commit us to something that came from elsewhere. See esp. the essays collected in J.-F. Lyotard, *The Inhumann Reflections on Time* (Cambridge, 1991).

now, of the trace of some real object or event and the pastness of what was in front of the camera, even if seconds ago. Yet webcams tend to overrule the inevitable pastness marking a single image by pointing forward to continually refreshed imagery, on the verge of arriving just now. The broken nature of updates and the choppy movement of most videostreaming reminds viewers that the real is never truly in the image. At the same time, the persistence of updating and streaming, marked by the running time stamp, compels one to stay with the imagery, including its black holes, and through these with the remote reality in an ongoing presentness. The latter is never fully achieved but perpetually promised. This seems to be one of the sources of viewers' addiction to cams.[36]

As a rigidly real time medium, the webcam relentlessly follows the linear movement from one moment to the next of objective time: around the clock, day after day, month after month. Staying with a place through the webcam does not necessarily mean being attentive to what there is to see all the time—which is, in any event, humanly impossible. Supported by special software, avid viewers keep connections to favorite cams open in a corner of their screen while doing other things. In this way connections to specific places can become part of the daily lives and experience of viewers. The frequent and quite emotionally tinted messages that appear on forums upon the impending closure or even temporary malfunctioning of a cam suggest that this is indeed the case for certain users.[37]

Another feature that ties viewers to single cams and the places they connect to is the already-mentioned forum, where registered users actively communicate with others about what there is to see. Particularly around cams that offer views of wild or domestic animals, small virtual gatherings occur where people share a hybrid mix of real and virtual environments.

The continuous hovering in a present that moves forward linearly without selection and change of pace has distinctive consequences for the way subject

---

36    Addiction is frequently mentioned by viewers on forums. As McPherson also mentions, the "now" of the net tends to stimulate ongoing immersion: one doesn't want to miss anything ("Reload. Liveness," p. 464). In the case of webcams, the idea that there is always something else "to be with," does not just tie viewers to online content, but to real events that are by nature always evolving.

37    When I first started viewing cams and one of my personal favourites, a cam in the Djuma game reserve, threatened to be closed in June 2005, I was surprised by my own emotional response. I was even more surprised to find the accompanying messageboard flooded with passionate pleas, goodbyes and expressions of gratitude (http://www.djuma.co.za/vuyacam.php). Something similar happened when the "jailcam" closed in December 2006 (http://www.tnacso.net/cont/jailcam.php).

matter appears on cams. Webcams not only tend to show the ordinary course of events—traffic flows, animals foraging or drinking, people waiting in line or doing deskwork, the change of light during the day—they also suck any prospect for drama out of the subject by their unremitting, mechanical recording and transmitting. The "immediate" or "authentic" quality of this presentation of imagery should be seen in juxtaposition with the edited material that we are used to in film and especially on television. We have come to know live reporting, with its quick selection and instant construction of small stories, only too well. The "real" in webcam sites is the unedited, undramatic, and untold flow of things in time. As a result, what there is to see appears as contingent and often very ambiguous as to its meaning.

This way of seeing is indeed not the seeing of human beings, intentionally engaged with their environment, but a machine vision. While certain users find this simply boring, others are fascinated by precisely this and challenged to other modes of attentiveness than those addressed by narrative and the immediate creation of meaning. The indeterminate shape and significance of events portrayed in this way becomes the object of intense discussion on the forums, leading to a multifaceted seeing-together.[38]

## Two Modes of Sharing Time: Spotting and Dwelling With

Not all cam sites stay with the mentioned indeterminate vision. On the basis of the specific subject matter of sites, two main ways of viewing can be distinguished that structure the time of viewing and engagement with other places in different ways. One I will call "spotting," and the second, "dwelling" with a remote place. Spotting adds a certain excitement to cam viewing by superimposing a more human time over the mechanistic time of the cam. While in principle any cam site allows for spotting, with viewers devising arbitrary goals and waiting for chosen events to materialize, certain sites actively stimulate spotting in the presentation of their subject matter. Thus viewers are invited to

---

38　What I am suggesting, in response to Sobchack's evaluation of electronic presence as dehumanized in comparison to that of cinematic presence, is that, for human beings, who never fall together with this presence but see their own seeing at the same time, the alienating character of machine vision can challenge users to experiment with other ways of seeing. On the forums many viewers express relief over the liberation from the authoritative position of the editor in other visual material. The ambiguity of scenes viewed can lead to a "seeing together." Discussing various possibilities, viewers do not naïvely take the cam for neutral. Instead they debate the limits of view and exchange perspectives, interpretations and additional knowledge on the places and events viewed.

catch wild animals that often come to drink at a waterhole, to watch for the milkman or the bus in some suburban street, to catch a suspect being brought into jail or to discover pretty girls on a tropical beach.

While the goal is pre-determined, and the technological set-up that allows for the sightings suggests a certain mastery over events to be observed, the actual events, and their being seen, remain dependent on circumstances that are not under the control of technology or of viewers. Again the emphasis is on coincidence, but this time without any direct power over content. All the set-up offers is a chance to witness the event. When this happens, it comes as a reward for time and patience spent, but remains a stroke of luck, arriving, as it were, from elsewhere.

The most elaborate site of this kind is Africam.com, which has a forum where users list and discuss animal sightings and communicate observations and information on animal behavior. At its inception, the Africam team experimented with means to guarantee "spectacular" content by showing animals in an enclosure, or by leaving bait out for predators. Viewers' responses quickly taught them that the camsite could only work when animals were spotted "on the animals' terms."[39]

Somewhat different are the many sites where a camera is pointed at a bird nest. While viewers are encouraged to watch for signs of hatching, this does not preclude their becoming attentive—in the long waiting process leading up to that moment—to less significant and defined occurrences, as the birds guard their eggs, day by day, week after week, with only minor movement. Many of these trifles are debated on accompanying forums. This stimulates a more intense involvement of viewers with the place, the animals, and with each other, which ends with the young leaving the nest and the gathering around them saying their slightly sad goodbyes.[40]

This brings me to the second way of viewing, following or dwelling with what emerges, no matter how undefined and uneventful. Generally not equipped with sound, what the webcam shows of real objects and events is movement, or behavior in the case of people or animals. But here movement and behavior are nascent, barely surfacing as such. A person scratches in a vague response to the cleaning person passing by; an animal shudders, not even centre-screen; cars cross an intersection marking out diverse trajectories; somebody waiting shifts a leg; the water in a pond slowly turns from a transparent blue to murky

---

39    http://www.africam.com/. Peter Armitage, *The Show Must Go On* (Pinetown, 2003), pp. 25–26, 58–59.

40    The most elaborate example was the eaglecam (www.infotechbusinesssystems.com/wildlife/eaglecam; as online in the spring of 2006).

brown as clouds pass by. As a viewer, one recognizes and registers these events, too ordinary and too non-descript to be named, in a sphere that belongs to us as feeling and embodied beings, a realm where they hover without taking on explicit meaning.

In this kind of viewing, time itself becomes the subject matter of the cam. Time here is the embodied being-in-time of people, animals—and also, I would say, of traffic, buildings, earth and water, and most exquisitely, of light. This is highlighted by cams that focus on typically contemporary ways of passing time: sitting in a waiting room, standing at a bus stop, doing deskwork, hanging around a shopping mall. Some sites keep records of views from the past 24 hours, which compete with Monet's paintings of the Rouen cathedral in their subtle variations of colour and light.[41]

Here the connection in real time opens up an interesting possibility. By uniting us in real time with a place viewed, what happens in the remote place and what happens here, where my body is situated and is living its time, permeate each other. The co-presence effected in this embodied following what happens in another place could be called a kind of empathy. Empathy is not the same as sympathy, nor does it lead straightforwardly to ethical engagement with places and the objects and people in them. It is a fundamental way of orienting ourselves in the world and feeling our way towards people and objects co-present with us.

The connection between our sensual being-in-the-world and ethical feelings of care and responsibility is a pertinent problem in the light of new media. Common sense tacitly recognizes a close relationship between the two in the use of words like "consideration" and "sensitivity" for both a more neutral and an ethical alertness to issues and to other people. A basic openness to the world as undetermined, and possibly indeterminable, seems closely related to an ethical attitude that, if not actively caring, at least refrains from prejudice.

In her evocative essay, "The Passion of the Material. Toward a Phenomenology of Interobjectivity,"[42] Sobchack posits aesthetic appreciation and ethical sensibility as materially grounded in our embodied being-in-the-world. She especially emphasizes our shared material being with the objects and people around us, and the way this shared being colors two different experiences for which we use the term "passion": suffering and devotion.

---

41   See http://www.state.ak.us/dmv/AFOwebCam.htm; http://digilander.libero.it/strafat/; http://www.pyramidcam.com/gallery2006/gallery.php?page=&sort=&perPage=&album=365.

42   *Carnal Thoughts*, pp. 286–317.

In suffering, we are confronted with the vulnerability we share with all material objects. We feel what it is like to be treated as only an object, a feeling that forms the basis for an ethical attitude towards our environment. Devotion to the things and people around us expands our being to other things and people. We reach out to experience their being as our own. This devotion reverberates in aesthetic care and appreciation.

If we accept this account, a crucial question emerges: what motivates us to live in tune with this assumed immanent aesthetic and ethical potential of our being? Sobchack does not address this question, but strongly suggests that a reflective awareness of our shared material being is vital. She also discusses how this awareness can be evoked by the way the eye of the camera treats material being in certain films. What I have described here as visually dwelling with others—with people, animals, objects, air and light—invited by the meticulous and relentless progression of the webcam, is very close to this.

The webcam addresses our perception on the level of an empathic, physical recognition of other beings, but it can also make this empathy explicit and turn it into a ground for sympathy and care. This results from the choppiness and the graininess of the views that intensifies our seeing and gives the hovering image a visual quality that seems as vulnerable and material as we are. More importantly, it is an effect of time—of the slow, mechanical pace of the medium and its indifference to meaning—that turns the slightest gestures and changes of scene into astonishing events, emerging just now.

Experiences like this may be too rare and too contemplative to play a major role in the development of ethical feelings that could span the globe. But dwelling with other places and people may be ethically relevant in a more mundane way. As I have suggested, sharing time with places, and the animals, objects and people in them, weaves these places into our daily experience, visually and haptically. Sustained relationships with a place through a webcam can generate affective bonds that are akin to the ones that arise between us and the various beings in our neighborhood. These bonds go deeper than conscious decision. The house we live in, the roads we ride, the people we greet on the way to work are deeply engrained in our daily existence. They shape the rhythm of our own embodied being and in that way sustain us. We often only feel this when they fall away, and we are lost or homesick.

Material commonalities and harmonies are realized here practically, within the fabric of our experience.[43] The extent to which this shared being,

---

43    Important questions here belong to the field of neuroscience. For instance, to what extent are phenomena like limbic resonance (that give a neurological basis to the phenomeno- logical observation that our being is not enclosed within the confines of individual bodies

as maintained through media, will pan out in actual caring towards the new neighbors in our globalizing environment is still in its experimental phase. The involvement testified to on forums and the passionate overtones in many messages are significant in this respect. It is also not difficult to come up with observations where recognition of the vulnerability of others through media results in indifference or invites aggression.[44] In this all too real turmoil, I would like to close with a few practical suggestions.

### Looking Back and Looking Forward

In my admittedly sketchy investigation of webcam sites, I hope to have made clear that these sites allow for and actively encourage at least three main kinds of viewing: browsing and quickly grabbing; spotting, while being actively oriented to the chance of witnessing some desired occurrence; and dwelling with what happens, intensely attentive or partly engaged in other things.

These three ways of viewing imply three different ways of sharing time and being co-present with events, people and objects in other places. They take shape in the interplay of the development of hardware and software and the specific sensibilities of users. But they also refer back to established viewing practices: those of travel and tourism, of viewing art and watching movies or television—but also, it seems, to an age-old living together with objects and people around us.

Closer study is needed of the effects of telepresence on our existential experience of presence, the care we feel for our environment and the responsibility that we show towards it. This requires that the problem be placed in a wider context. Electronic media are as much the cause of change as they are expressions of it. Arguably, it is precisely because our relationships with our immediate environment have become more "volitional" and possibly more superficial that they can be replaced by electronic communication. Yet many more times we are not satisfied with this and find ways to merge electronic and real life presences—as when people feel the urge to visit the "real" place of the webcam or to meet the members of an electronic forum "in the flesh."

---

and minds) adversely affected by electronic communication? At present, I do not know of any studies in this area.

44    A telling internet phenomenon, in this respect, are flame wars, conflicts fought by text-messaging in which people show very little constraint as to what is said and how.

   Remaining closer to the investigation undertaken here, it has become clear
that viewing experiences of one and the same media application differ greatly
in the kind of involvement they allow. These differences have been established
here as differences in temporality. From the phenomenological and ethnographic perspective sketched, three groups of questions suggest themselves for
further enquiry"

1.  *Waiting Time.* What do people do in the time lapses that break seamless elec-
    tronic presence? When and under what conditions does multitasking—the
    involvement in different activities at the same time—offer possibilities
    of cross-overs and connections between various real and electronic pres-
    ences of one physical body? When does it disperse our being into so many
    hopelessly different atomic existences?
2.  *Chatting Time Away.* In what ways is the physical, historical existence of
    viewers part of what may look like an endless conversation over trifles?
    How does this make people present to each other and to the place about
    which they share their "views"?
3.  *The Time of Touch.* What factors determine when viewers surf, spot, lurk
    or are drawn in and touched? How does visiting camplaces and fellow
    viewers "for real" affect viewers' experiences in this respect? When does
    emotional engagement with a place move beyond spectatorship to turn
    into actual sensual and ethical involvement, so that "there and now"
    becomes at the same time, "here and now"?

I would hope that these further investigations not only, as Sobchack and Virilio
suggest, save us a future, but also provide us with a deeper understanding of
our present involvement in the world, with its current and its timeless traits.

# CHAPTER 19

# No-Time in Non-Places

*Leonard Michael Koff*

We begin with Aristotle, a good place to begin. His understanding of time grounds all subsequent discussions of the nature of time, including discussions among those who argue for, or theorize, or imagine, or experience no-time in a non-place.

For Aristotle, time does not exist without change: "So just as there would be no time if there were no distinction between this 'now' and that 'now,' but it were always the same 'now': in the same way there appears to be no time between two 'nows' when we fail to distinguish between them."[1] Indeed, for Aristotle, time depends on change occurring or on change that has occurred, both of which mean that an interval of time has elapsed, whether or not we realize this.[2]

---

1   Aristotle, *Physics*, trans. Philip H. Wicksteed and Francis M. Cornford (Cambridge, 1929, rev. and rpt. 1967) iv, ch. 11, 218$^b$ (p. 383). Aristotle's definition of time describes what is now called the A-series theory of time, where time is *tensed*—experienced as past, present, or future— and *moves*—said to be always passing in a "forward" direction. The B-series theory of time holds that events stand in "the unchanging relations of temporal precedence or simultaneity to each other" (see Robin Le Poidevin, *The Images of Time: An Essay on Temporal Representation* [New York, 2007], p. 63: [The B-series theory says] "that time does not pass: there is only temporal order"). As Le Poidevin explains, "if there really is an A-series, if time really is passing, then B-series relations are simply a product of A-series positions... To hold that there is an A-series is not to assert the unreality of the B-series." There is, however, "no plausible way to invert this and represent the A-series as supervening on the B-series in...a simple way. Passage [of time] does not simply emerge from order... It would, perhaps, be possible to hold that A-series passage and B-series order were completely independent of each other, but this would be a deeply unattractive position" (p. 63). We are left, therefore, with often irreconcilable semantics of time that entail confusion, for example, about what we mean when we say that we perceive events *as present*: we might mean that we are present *as events pass*—that we perceive events passing—or that we remember events so that we perceive a memory image *as present* in which events pass. See also M. Joshua Mozersky, "The B-Theory in the Twentieth Century," in *A Companion to the Philosophy of Time*, eds. Heather Dyke and Adrian Bardon (Chichester, 2013), pp. 167–182.

2   Not realizing this seems to create a "now of indeterminate duration," not a present and, therefore, "no time" in Aristotle's sense, because there seems to be no change in such a "now": time is the measure of change. See *Physics*, iv, chap. 11, 218$^b$ (p. 383): "For when we experience no changes of consciousness, or, if we do, are not aware of them, no time seems to have

© KONINKLIJKE BRILL NV, LEIDEN, 2016 | DOI 10.1163/9789004312319_021

But Aristotle's discussion of time is not a view that makes no-time—time without change—impossible to postulate or experience. And indeed Sidney Shoemaker suggests how the idea of time without change might be argued for.[3]

> When we have been asleep, we are prepared to allow that a good deal of time has elapsed since a given event occurred even though we were not ourselves aware of any change during the interval, for in such cases it is plausible to hold that our belief that an interval of a certain duration has elapsed is founded on the inductively grounded belief that changes did occur that we could have been aware of had we been awake and suitably situated.[4]

There are, as Shoemaker implies here, logically conceivable circumstances "in which the existence of changeless intervals *could* be detected."[5]

---

passed, any more than it did to the men in the fable who 'slept with the heroes' in Sardinia, when they awoke; for under such circumstances we fit the former 'now' on to the later, making them one and the same and eliminating the interval between them, because we did not perceive it." On Aristotle's understanding of change, see *Physics*, v.

3   Aristotle's understanding of a "now of indeterminate duration" provides the context for Shoemaker's argument that there can be no-time, which is not the same as "no time" in Aristotle's sense. See Aristotle's rhetorical question in *Physics* iv, chap. 10, 217$^b$–218$^a$ (p. 373): "Some of it is past and no longer exists, and the rest is future and does not yet exist; and time, whether limitless or any given length of time we take, is entirely made up of the no-longer and not-yet; and how can we conceive of that which is composed of non-existents sharing in existence in any way?" Aristotle's question implies that a "now" without a discernable past or future means that time, as a measure of change, does not exist: without boundaries, there is, in Aristotle's sense, no change. But Aristotle's argument here, implied by his rhetorical question, is an argument that for Aristotle isn't true. The argument is meant to catch those who, for Aristotle, reason incorrectly. A "now of indeterminate duration" has a past that, yes, "no longer exists," but that had to exist, and a future that does not "yet exist," but will exist. There always is, for Aristotle, a past and a future, hence a present, hence change, hence time, the measure of change. See also iv, chap. 10, 218$^b$ (pp. 375, 377): "Again, if simultaneity in time, and not being before or after, means coinciding and being in the very 'now' wherein they coincide, then, if the before and the after were both in the persistently identical 'now' we are discussing, what happened ten thousand years ago would be simultaneous with what is happening to-day, and nothing would be before or after anything else." For Shoemaker's argument that there are ways to perceive what occurs or has occurred in this one continuous "now," see "Time without Change," in *The Philosophy of Time*, Oxford Readings in Philosophy, eds. Robin Le Poidevin and Murray MacBeath, (Oxford, 1993, rpt. 2009), pp. 63–79.

4   Shoemaker, p. 65.

5   Shoemaker, p. 79 (italics his).

If it is possible for there to be changeless intervals, then it may seem compatible with my total experience that any number of such intervals, each of them lasting billions of years, should have elapsed since I had my last meal, despite the fact that the hour hand of my watch has made one revolution and the fact that my lunch is still being digested.[6]

As Shoemaker says, "we are all possessed of a 'sense of time,' an ability to judge fairly accurately the length of intervals of time, at least of short intervals, without using observed change as a standard," and thus

> one can tell whether the second hand of a clock is slowing down without comparing its movements with those of another clock, and if one hears three sounds in succession one can often tell without the aid of a clock or metronome how the length of the interval between the first and the second compares with that of the interval between the second and the third.[7]

These considerations suggest for Shoemaker that "it is logically impossible for someone to know that nothing, including the state of his own mind, is changing," and although it is logically impossible "for someone to be aware of the existence of a changeless interval during the interval itself...it does not...follow from this that it is impossible for someone to be aware of the existence of such an interval before or after its occurrence."[8]

---

6 Shoemaker, p. 66.

7 Shoemaker, p. 66. Shoemaker suspects this—that "there will be a constant change in my cognitive state as the interval progresses"—that "led Aristotle to stress change 'in the state of our minds' in his discussion of the relationship between time and change." Shoemaker adds that "it does not seem to be true to say, as he [Aristotle] does, that one must *notice* a change in the state of one's own mind in order to be aware of the passage of time" (pp. 66–67, italics his).

8 Shoemaker, p. 67. See pp. 67–68: "...it is logically impossible that anyone should know, at any given time, that the then current state of the universe is such as to make impossible the existence in it of life and consciousness, yet most of us believe that we have very good reasons for thinking that the universe has been in the very remote past, and will again be in the very remote future, in just such a state." Indeed, "people should have well-grounded beliefs about when in the past such intervals have occurred and when in the future they will occur again, and that they should be able to say how long such intervals have lasted or will last... I think that the sorts of grounds there could conceivably be for believing in the existence of changeless intervals are such that no sound argument against the possibility of such intervals can be built on a consideration of how time is measured and of how we are aware of the passage of time."

I won't rehearse here in any detail Shoemaker's theoretical example of conceivably detectable no-time.[9] Rather, I want to extend Shoemaker's argument from the conceivably detectable to the actually detectable and place us in the same position of observation with respect to no-time as Shoemaker does, that is, at no-time's beginning and ending boundaries. Moreover, I want to argue that no-time, which can occur within the flow of time itself, is more than either an empty or "frozen" interval: no-time can be a projection that brackets time itself as "time out of time,"[10] set apart by our making it into an object of awareness, a present memory, a memory image; no-time can also be, in theological contexts, a metaphysical discovery. Thus the content of all the examples of no-time, imagined and metaphysical, that I want to look at can be said to be experienced within no-time itself, as well as observed, as Shoemaker argues, at no-time's beginning and ending boundaries.[11]

---

9    Shoemaker's argument for detecting changeless intervals through whatever we use as our clock, intervals he assumes are part of the flow of time—as opposed to detecting unchanging things or substances (Shoemaker is interested in ordinary becoming [changes of things or substances that "can only occur while the subject of change exists" (p. 64)])—entails the following: postulating circumstances of detection analogous to the ones used to argue for change during an interval in which someone is asleep. Shoemaker's circumstance is this: that there could occur a total freeze of existence (pure changelessness) in one region of a three-region world that someone in one or the other or both unfrozen regions detected because freezes were occurring at different, but overlapping intervals in each of the three regions, but that in a certain year there was a total freeze in all three regions at the same time.

10   See Le Poidevin, *Images of Time*, esp. pp. 104–122: "Psychologists have long recognized that time perception is not directly analogous to perception of objects and their properties, for the temporal features of events do not present themselves for inspection in the way in which spatial properties do" (p. 120). On the puzzle of time perception, see esp. pp. 97–122. For "a number of considerations—the problematic nature of time perception, the difficulty of locating temporal properties in space and time, the fact that models of time perception do not include them as causal factors—point to the conclusion that time order and duration are *acausal* properties of events" (p. 120, italics his). On Le Poidevin's description of a chronometric explanation of time order and duration, where these are also mind-independent, see pp. 116–120. In the context of time perception, "the fact that psychological models of order and duration perception make [time] perception indirect does not itself show that these properties are mind-dependent, any more than the indirect nature of depth perception (involving as it does comparison by the brain of the input from the two eyes) shows depth to be mind-dependent" (p. 121).

11   I am not attempting to reconcile the A-series theory of time (as flow) with the B-series theory (as indivisible intervals that can be present as if out of the flow of time itself), though I want to see no-time as dependent on perceptions of memory images, constructed or discovered. Shoemaker makes clear that his argument for no-time does not take on the

No-time, as I mean it, is an indivisible static instance, situated in a flow of time, that is in some relationship to the flow of time itself—interrupting it or tangential to it or approaching simultaneity with it. Its content has duration, of a certain duration, which can point to the past or the future, or the past and the future as a composite instance of time, an instance perceived as present now.[12]

---

problem of relating the A-series theory of time and the B-series theory. Rather he wants to see no-time as possible in an A-series theory of time. That theory of time entails our being present as events occur, although what we are receiving are not strictly speaking events; rather in the A-series theory of time we are perceiving what Carl L. Becker, for example, calls—the term is not original with him—the "specious present" (see *The Heavenly City of the Eighteenth-Century Philosophers* [New Haven / London, 1932], pp. 119–122). See William James, *The Principles of Psychology* (New York, 1983), p. 609: "Time...considered relatively [sic] to human apprehension, consists of four parts, viz., the obvious past, the specious present, the real present, and the future. Omitting the specious present, it consists of three...nonentities—the past, which does not exist, the future, which does not exist, and their conterminous, the present; the faculty from which it proceeds lies to us in the fiction of the specious present." This is the Augustinian position. See Augustine, *Confessions*, trans. Henry Chadwick [Oxford, 1991] bk. xi. 27 (p. 236): "When time is measured, where does it come from, by what route does it pass, and where does it go? It must come from out of the future, pass by the present, and go into the past; so it comes from what as yet does not exist, passes through that which lacks extension, and goes into that which is now non-existent." On the specious present, see Barry Dainton, "The Perception of Time," in *A Companion to the Philosophy of Time*, pp. 389–409. The B-series theory of time entails being present at a memory image—perceiving an event—all or part of which we "make present." See Le Poidevin, pp. 62–68, on the A-series theory and the B-series theory of time memory: "For the A-theorist, our changing beliefs about a given event [occurrence] track the changing facts of reality... For the B-theorist, in contrast, our changing beliefs about an event do not track *changing* facts, but rather the same fact from different perspectives... " (pp. 64–65, italics his).

12  On the relationship with respect to music between present theories and memory theories in our effort to "make sense of our experience of succession, persistence and change," see Ian Phillips, "Perceiving Temporal Properties," *European Journal of Philosophy* 18/2 (2008), 176–202. On music without the impositions of time, see *Give My Regards to Eighth Street: Collected Writing of Morton Feldman*, ed. B.H. Friedman (Cambridge, MA, 2000), esp. p. 12: "When sound is conceived as a horizontal series of events all its properties must be extracted in order to make it pliable to horizontal thinking... In order to articulate a complexity of such close temporal ordering one might say differentiation has become here the prime emphasis of the composition. In a way, the work resulting from this approach can be said not to have a 'sound.' What we hear is rather a replica of sound... " Feldman's compositions are thus without forward motion, without temporal/rhythmic imposition. They are "notices" one tunes into as they occur. Indeed, Feldman's *Second String Quartet*, for example, is 5 hours long. For Feldman, imposing time on music falsifies it, working against the nature of sound itself, sound suddenly heard and then remembered.

No-time is maker-dependent that may be, or that in fact is in certain contexts—theological, for example, or utopian or dystopian—maker-independent. The content of no-time can be changeless or seen as changing (within a static indivisible instance) or changed; its content can take us out of time's flow, as Aristotle understands the "flow of time," and arrest that flow, conceiving it or reconceiving it in some valuable or, in some cases, necessary way. No-time can seem like—can be experienced as—an interval of time, but not in Aristotle's sense of interval as defined in the *Physics*.[13] We can, while we are awake, create no-time intentionally, that is, create circumstances in which no-time exists, or discover no-time, both of which entangle us in no-time's presence. In short, we can experience no-time without having to regard it as a kind of time having change, if only we could be aware of it as having change, or realize this later: we can experience no-time without having to argue for it as changeless with a

---

13    See n. 1 and 2 above. On two conceptions of a static instance, see also Le Poidevin, esp. pp. 129–132. Both are extrapolations: the first derives "from a process of dividing an interval into smaller and smaller parts." Thus, "if time is (as we assume) continuous, this process has no end"; the second derives not from dividing an "interval into smaller and smaller parts"—an instance is not "part of an interval at all, but as an indivisible boundary between two parts of an interval" (p. 129) with a certain duration. See Le Poidevin, pp. 130–132, on Zeno's two paradoxes in light of each notion of an instance: the Achilles and the Tortoise paradox, where there is no static instance and Achilles can never overtake the Tortoise who has a head start in their race; and the Arrow paradox where the arrow is "at rest at each moment of its flight, and so is at rest throughout the period of its flight" (p. 130). With respect to Achilles and the Tortoise, as Le Poidevin says, "instants do not need to enter the discussion at all. The central idea that generates the problem is that of intervals that can be divided indefinitely" (p. 131). With respect to the Arrow, Le Poidevin says, we can "distinguish between moving *in* an instance, which we concede is impossible, and moving *at* an instant. An object is in motion *at* a given instance if the object is in a different position at any instant before or after that instant. Properly understood, the static account of change, far from leading us to Zeno's Arrow, offers a way out of it" (p. 131, italics his). The Arrow paradox, Le Poidevin says, "depends on treating the moment in question as genuinely indivisible, not composed of earlier and later parts. It might be durationless, or, if time is only finitely divisible, then we could conceive of each instant as having a very short (but not infinitesimal) duration. In this second case, there would be no need to assert, with Aristotle, that 'Time is not composed of indivisible nows.' The essential point about each instance, whether or not it has a duration, is that it be indivisible, for then it cannot contain motion" (p. 130) and can be *changeless* in the sense I mean it; that is, no-time can be changeless or changing or changed; it can contain arrested change, as it were, reflecting an idea of changelessness that points backwards or forward in time, or backwards and forward, as conceived or reconceived time. For an analysis of the connection Aristotle makes between motion and the "now," see Michael F. Wagner, *The Enigmatic Reality of Time: Aristotle, Plotinus, and Today* (Leiden/Boston, 2008), esp. pp. 149–271.

certain "frozen" content that necessarily excludes us from knowing no-time's content from within no-time itself.

With respect to visualizing no-time, there is a close-up of a handless clock in one of the most famous scenes from Ingmar Bergman's *Wild Strawberries* (1957) meant to mark Professor Isak Borg's personal no-time, the nightmare of impending death: a "weird" and "unpleasant" morning walk through empty streets and ruined houses. We are in this scene inside what Aristotle might call a "now of indeterminate duration," though Bergman has permitted us to observe its changing psychologically content. In the film, we hear the handless clock tick, which are Professor Borg's heartbeats, and when he touches the back of an enigmatic figure in an overcoat and hat, who, when touched, falls to the ground bleeding, church bells toll sounding the last changed moment of the professor's life. Moreover, there are large glossy eyes below Borg's handless clock that tell us that the clock is looking down at the professor as he looks up at the clock. Furthermore, when Borg takes out his own pocket watch, it too is handless. Here is no-time as the living Freudian unconscious, as no-time both created and discovered.

Fig. 19.1 is not a dream; it is a contemporary public instance of no-time, finite as well as infinite. The figure shows a cross section of a modern airport during the day where there are no people in view on the ground floor, or no people yet, but where there is an escalator to several floors where we can see the possibility of people. Fig. 19.2, also during the day, is the photo of empty seats in front of an airport snack bar where whoever might have been there is not there now, or who-ever is going to be there is not there yet: people buying food, eating, waiting to board planes. Fig. 19.3 shows an empty passageway in a modern airport con-structed to evoke a cathedral aisle; indeed the architect may have intended the "cathedral passageway" to comfort any experience of passing through a place emptied of existents by recalling, for those passing through the "aisle," histori-cally full Christian time before one reaches a directional sign near the end of the passageway. Fig. 19.4 shows the tracks of a city tram: the haunting photo antici-pates the tram's appearance—it has not yet arrived—or its appearance again—it has already passed; the image is a moment out of time for an unspecified static length. Fig. 19.5 shows a clock below a cross that's on, from the angle of the photo, a monumental public building—the angle makes the building almost tyrannical. The juxtaposition of cross and clock suggests that these two instruments for tell-ing time are in some ironic relationship with each other: here indeed is no-time waiting to begin as real time, standing against the cross of eternal time that starts "of its own accord" the very moment we bring the cross to consciousness. The clock is a stopped object and, as such, a non-existent we make exist by imagining it starting. All these photos are images of no-time, shown in black-and-white, that reflect, the more we stare at them—black-and-white enhances the sensa-tion of no-time in a non-place—the existence of changeless, or changing, or changed existence, in places where time is about to be born, or has already been born and is just out of sight. What these photos show, beyond what believe we actually see or anticipate seeing in them, is the experience of existence emptied for a time, but not permanently emptied, of existents.[14]

As to why no-time is a duration one wants to experience, there are various answers, and indeed when one enters no-time, it is not necessarily an interval from which one may want to exit.[15] Because no-time is a condition in the world of change, that is, a duration of arrested time and, as a result, a non-place, no-time can seem to return us to longed-for states of being. Is no-time an entrance into the transcendent or the possibly transcendent? Or does the idea or the experience, the willed experience, of no-time awaken our sense of transcen-dence lost? Indeed, the timeless and the placeless have a pull that relieves us of engagement with the real world because we have achieved freedom from it, a state of being in which we are free to imagine otherwise, an otherwise some-times desired as a foretaste of perfection, sometimes felt as an experience of power over time itself. No-time can seem salvific in the middle of existence and sometimes dreaded in the course of life as a state of alienation, a state that is solitary and confining, or indeed a state we fear because it may come upon us.[16]

---

14      Let me note that the "melting clock" in Salvador Dali's *The Persistence of Memory* (1931) is
        a visual representation of "no time" rather than no-time as understood here. According to
        Dali, the several soft watches were inspired by the surrealist perception of a Camembert
        melting in the sun.

15      For an important reformulation of the simple question, what is time?, see Adrian Bardon,
        *A Brief History of the Philosophy of Time* (Oxford, 2013), p. 175: "*Time is not so much a 'what'
        as a 'how,' and not so much a question as an answer*. Time as we know it in experience is a
        matter of how we adaptively organize our own experiences; in a physical and cosmologi-
        cal context, it is a matter of how we can most successfully model the universe of occur-
        rences. As such, time is an answer: a solution to the problem of organizing experience and
        modeling events" (italics his). For an amusing take on this issue, see Roger Zelazny, *The
        Guns of Avalon* (New York, 1972), p. 17: "The most difficult thing about Time, I have learned,
        is doing it."

16      On no-time as oblivion, see W.G. Sebald, *The Rings of Saturn*, trans. Michael Hulse
        (Frankfurt-am-Main, 1995, rpt. London, 1998), esp. p. 24: "For the history of every individ-
        ual, of every social order, indeed of the whole world, does not describe an ever-widening,
        more and more wonderful arc, but rather follows a course which, once the meridian in
        reached, leads without fail down into the dark... There is no antidote...against the opium
        of time. The winter sun shows how soon the light fades from the ash, how soon night
        engulfs us. Hour upon hour is added to the sum, how Time itself grows old. Pyramids,
        arches and obelisks are melting pillars of snow. Not even those who have found a place

Our wanting to be in no-time requires, of course, that we puzzle out real-life connections to ourselves as well as to social and psychological configurations, and this seems of value. In this way, no-time not only awakens our memories of the before, but forces us to consider our social and ethical situatedness, our now, and our relationship with the other we would leave behind in the timeless. Entering no-time avoids what Emmanuel Levinas calls the "face of the other" in the non-place that no-time creates and reflects: entering no-time is socially and ethically, indeed metaphysically complex.[17]

• • •

Integral to Aristotle's argument about the nature of time is the assumption that there is an all-time, an idea of eternal time that philosophers in the high and late medieval ages, and medieval authors, call *absolute time*, engaged as they are by Aristotle in what Augustine calls the practice of finding Egyptian Gold.[18] As Pierre Duhem puts it, Aristotle "attempted to affirm

---

amidst the heavenly constellations have perpetuated their names: Nimrod is lost in Orion, and Osiris in the Dog Star. Indeed, old families last not three oaks. To set one's name to a work gives no one a title to be remembered, for who knows how many of the best of men have gone without a trace? The iniquity of oblivion blindly scatters her poppyseed and when wretchedness falls upon us one summer's day like snow, all we wish for is to be forgotten." Cf. the effects of the accidents of weather on our sense of freedom from time that Patrick Modiano describes in *Dora Bruder*, trans. Joanna Kilmartin (Berkeley / Los Angeles / London, 2015), p. 48: "Perhaps it was one of those mild, sunny winter days when you have a feeling of holiday and eternity—the illusory feeling that the course of time is suspended, and that you need only slip through this breach to escape the trap that is closing in on you."

17 For Levinas, no-time is finally not a meaningful, that is, ethical human concept since it is an idea of time without our presence in it; for the face of the other, for Levinas, changes our experience of time itself. See Rudolf Bernet, "Levinas's Critique of Husserl," in *The Cambridge Companion to Levinas*, eds. Simon Critchley and Robert Bernasconi, (Cambridge, UK 2002), p. 94: "Only the 'instant' linked to the unexpected upsurge of the other can change my life to the point of forcing or allowing me to re-commence [past, present, and future] from the beginning. Such a fragmentation of my existence into a multiplicity of discontinuous instants is testimony to my essential dependence on the other" where forgiveness and hope enable my time to re-commence. See Levinas, "Diachrony and Representation," in *Time and the Other*, trans. Richard A. Cohen (Pittsburgh, 1987), esp. pp. 119–120. Cf. Gaston Bachelard, *Intuition of the Instant*, trans. Eileen Rizo-Patron (Evanston, 2013), who, in his critique of Bergson, anticipates Levinas's idea of an ethical instant.

18 Saint Augustine, *On Christian Doctrine*, trans. D.W. Robertson, Jr (New York, 1958), 2, xl, 60 (p. 75), on assimilating aspects of the pre-Christian world that anticipate Christianity.

the existence of a unique time, one that is the same in all places, on land and on sea, that would be the same in other worlds if, *per impossibile*, there existed other worlds. He seems not to rest until he had rejoined the Pythagorean doctrines by means of the proposition that the clock that defines the unique and true time is the sphere of the fixed stars."[19]

Dante's description of heavenly time and place, which rests on Christian philosophy in the context of Aristotle, is an accommodation to our earthly coordinates; our ascent from the peak of a medieval theological fiction (from atop Mount Purgatory) enables us to understand—and experience—the absolute time and the spaceless space of a heaven created, as Dante explains, through Neoplatonic emanation.[20] For Dante, *Paradiso* is an eternal place beyond Eden; its physical geography is familiar, constructed by the biblical text, reflecting our fully human desire for transcendence from physical and bodily coordinates of ordinary space and time. Eden in Dante valorizes the natural elements of our world—bird song, flowers, colors, and so on—and sees them as figuring a perfected state of being. But the structure of Paradise can only be explained as a gesture to a human, that is, a Ptolemaic understanding of time and space and the universe-creating descents and assents of the Many from and into the Neoplatonic One:

> Dentro dal cielo de la divina pace
> si gira un corpo ne la cui virtute
> l'esser di tutto suo contento giace.
>     Lo ciel seguente, c'ha tante vedute,
> quell'esser parte per liverse essenze,
> da lui distratte e da lui contenute.
>     Li altri giron per varie differenze
> le distinzion che dentro da sé hanno
> dispongono a lor fini e lor semenze.
>     Questi organi del mondo così vanno,

---

19    Pierre Duhem, *Medieval Cosmology: Theories of Infinity, Place, Time, Void, and the Plurality of Worlds*, ed. and trans. Rober Ariew (Chicago / London, 1985), p. 310. The Pythagoreans argue for "the existence of a time transcending the world of bodies, by which all movements and all changes of visible nature could be measured" (p. 310).

20    See Robert M. Durling in *The Divine Comedy of Dante Alighieri: Paradiso*, trans. Durling (Oxford, 2011), note to 2.106–148 (p. 62, italics his): "God's creative power is *transmitted downward*, increasingly diversified, through the heavenly spheres governed by the angelic intelligences (the 'movers')."

come tu vedi omai, di grado in grado,
che di sù prendono e di sotto fanno.

Within the Heaven of God's peace there
turns a body under whose power lies the being
of all that it contains.
The next heaven, which has so many sights,
divides that being into different essences, separated
by it, yet contained by it.
The other spheres through various
differences dispose the distinctions held within
them to their ends and to their sowings.
These organs of the world thus descend, as
you can see, by degrees, for they take from
above and fashion below.[21]

The 'Heaven of God's peace,' as Robert Durling explains, "is the Empyrean...
beyond space and time, identified by Beatrice as located in 'the mind of God.'"
Within the Empyrean lies the *primum mobile*, the first moveable sphere, and
within in lies the eight concentric spheres—the sphere of fixed stars and those
of seven planets; all these together "revolve around the stationary earth." It is
from the sphere of the fixed stars, to which the *primum mobile* transmits "God's
undifferentiated creative power,"[22] and which transmits that power to the
lower planetary spheres, that soul or mind knows what is called "absolute
time." It is the time the soul uses to recognize time on the "the stationary earth."
Creation in Dante is creation conceived a descending power, which is descending love that has intellectual content, embodying itself (em-mattering itself
through Neoplatonic emanation [the matter of the universe includes us]), and
we, as we desire to ascend to God, answer His descending grace. This is a perfected physical, intellectual, and metaphysical system, experienced emotionally; it is all there is, literally and virtually.[23]

---

21    *The Divine Comedy of Dante Alighieri: Paradiso*, 2.112–123 (pp. 50, 51).
22    Durling, notes to 2.112–114, 115–117, (p. 62).
23    See Durling, p. 745: "The sphere of the fixed stars ['tante verdute' (so many sights)] is...the
      first step in the derivation of multiplicity from the 'simple light' of God's power. In contain-
      ing all the multiplicity of the universe virtually, the sphere of the fixed stars follows a pat-
      tern similar to that attributed to the Logos, but here there is a material embodiment of the
      power, which descends through the cosmos from high to low in the mediation character-
      istic of medieval Neoplatonism." On Plato's idea of time "as the moving image of eternity"
      and Plotinus' Neoplatonic time, where "eternity and time...correspond to two sorts of

For Aristotle, as well as for Dante, as we've seen, knowing time by means of eternal time—absolute time—requires our presence: "If...we are to render *time* intelligible, we must come to understand the constitutive role we play in the objective order of time. We cannot have any understanding of what time is like 'in itself,' totally independent of our apprehension or experience of it. For the reality of time is partially constituted by the soul's measuring activities."[24] This "does not mean that the measurement is subjective or that any measurement a soul makes is correct." What is meant is "that we cannot give an adequate account of time without including in the account a soul which is measuring the changes. It is from the perspective of a soul which endures through time, which lives in the present, remembers the past, and anticipates the future, that change is measured."[25]

But in responding to Aristotle, Christian philosophers define something called *potential time* that does not need our apprehension to exist and, in this way, separates the existence of time as change from our apprehension of change. According to Pierre Duhem, "Even if heaven stopped, time would continue to be and to measure the movements of others bodies. Moreover, even if all movement were to stop, time would still exist and would measure the universal rest... If heaven moves actually, potential time coincides with actual and positive time which measures the movements of heaven; if heaven is immobile, potential time continues to exist—it is then the time that would measure the movement of heaven if heaven moved. We know this time independently from the movement of heaven; therefore, if heaven were immobile, we could, using this potential time, measure the duration of heaven's rest."[26]

---

reality: one eternal and intelligible; the other temporal and perceptible, and equated with the natural universe" (p. 276), see Wagner, *Enigmatic Reality of Time*, esp. pp. 275–363.

24    Jonathan Lear, *Aristotle and the Desire to Understand* (Cambridge, 1988), p. 79 (italics his). See also pp. 76–77: "It is change...as well as our recognition of it that grounds our recognition of a before and after and the interval which the distinct nows mark. This recognition— the marking of distinct nows—itself recognizes the reality of time and is also a *realization* of time itself. For time is nothing other than a number or measure of change" (italics his).

25    Lear, pp. 78–79. See also p. 75: "Aristotle does not believe that one cannot give account of the objective reality of time without including in the account an enduring soul which *could* experience the change" (italics his).

26    Duhem, p. 295. See Johannes Duns Scotus, *Opera Omnia* I, lib. II, dist. II, quaest. XI (Hildesheim, 1968), p. 324: "Et ita dico quod, quando iste motus caeli non erit, poterit tamen alius motus mensurari per tempus hujus motus primi caeli, une quantum scilicet motus ille posset fieri cum tanta parte illius motus, si esset; et nunc et cum tanta parte quietis cum quanta pars motus posset esse (Therefore I assert that if the movement of the first heaven is not, one can still measure all other movements by means of the time

The complement of potential time is the *absolute clock*: As Duhem puts it, "Man is free to choose as he wishes the continuous movement that will serve him to define time, the movement that will become his clock. This movement will therefore be uniform by convention, and it will be the same for all movements regulated by the observer, by means of the first movement...nature does not impose upon man the clock he must use; the choice of a clock is the result of an arbitrary convention in which one considers only reasons of fitness and convenience. It seems that all the ancient philosophers embraced the first view of time. The Pythagoreans and later the Neoplatonists all admitted, under various forms, the existence of a time transcending the world of bodies, by which all movements and all changes of visible nature could be measured. According to them, a uniform movement was any movement that accomplished equal distances in durations marked as equal by the divine clock. Since they were divine beings, the celestial bodies can only accomplish uniform rotations. Each one of them, and in particular the orb of the fixed stars, is therefore a visible clock regulated exactly by the clock that marks the perfect time in the world of ideas."[27] Thus Ockham, according to Duhem, "relates all local movements to an absolutely fixed reference conceived by our reason," so that he "measures time with a purely ideal clock that our mind constructs as soon as it perceives any movements or change whatever. It is by relating time (the time Duns Scotus called *potential time*) to this ideal clock, that man recognizes the uniformity of the diurnal movement and...observes or constructs visible clocks."[28]

---

marked off by the movement of the first heaven—meaning that one can know in which part of the celestial movement the considered movement can be accomplished—if celestial movement existed; presently it is accomplished during a part of the celestial rest equal to what can coexist with such a part of celestial movement)." On the idea of potential time, see Neil Lewis "Space and Time," in *The Cambridge Companion to Duns Scotus*, ed. Thomas Williams (Cambridge, 2003), esp. pp. 89–93. See also Richard Cross, *The Physics of Duns Scotus: The Scientific Context of a Theological Vision* (Oxford, 1998). On angelic time as eviternity, that is, time that entails "some kind of extended duration [without before and after] in the absence of any physical process," an analogue of no-time as the "possibility of absolute time" (for Bonaventure and Olivi) and as the "counterfactual accounts of relativistic time" (for Scotus), see Richard Cross, "Angelic Time and Motion: Bonaventure to Duns Scotus," in *A Companion to Angels in Medieval Philosophy*, ed. Tobias Hoffmann (Leiden / Boston, 2012), esp. pp. 126–147. Eviternity is an "all-at-once," an idea of time in the absence of change that logically lies between eternity (God's timelessness, an unextended point), on the one hand, and, on the other, the time of mutable material beings (a line), which has duration, of a certain duration, and is a process.

27    Duhem, pp. 309–310.
28    Duhem, p. 320 (italics his).

Unlike Christian cosmologists, modern ones have the concept of *time zero*, an idea of no-time rather than the idea of potential time or an absolute clock: "In the finite case...as we consider the cosmos at ever-earlier times, it's accurate to picture the entirety of space continually shrinking. Although the mathematics breaks down at time zero itself, it's correct to envision that at moments ever closer to time zero, the universe is an ever-smaller speck... [But] if space is truly infinite in size, then it always has been and always will be. When it shrinks, its contents are squeezed ever closer together, making the density of matter ever larger. But its overall extent remains *infinite*. After all, shrink an infinite tabletop by a factor of 2 and what do you get? Half of infinity, which is still infinite. Shrink by a factor of 1 million and what do you get? Infinity still. The closer to time zero you consider an infinite universe, the denser it becomes at every location, but its spatial extend remains unending."[29]

Modern cosmologists do understand time as change, where time in our world, with our gravitational field, depends on the absence of a particular symmetry: "Things in the universe must *change* from moment to moment for us even to define the notion of *moment to moment* that bears any resemblance to our intuitive conception. If there were perfect symmetry between how things are now and how they were then...time as we normally conceive it wouldn't exist."[30] Moreover, the end of time—the "big crunch"—is "not just the end of

---

29    Brian Greene, *The Hidden Reality: Parallel Universes and the Deep Laws of the Cosmos* (New York, 2011), p. 26 (italics his). See also p. 66, on the fluidity of time: "Whereas everyday experience convinces us that there is an objective concept of time's passage, relativity shows this to be an artifact of life at slow speeds and weak gravity. Move near light speed, or immerse yourself in a powerful gravitational field, and the familiar, universal conception of time will evaporate."

30    Brian Greene, *The Fabric of the Cosmos: Space, Time, and the Texture of Reality* (New York, 2004), p. 226. See pp. 128–129: "Our experiences...teach us" that time "seems to have an intrinsic direction, pointing from what we call the past toward what we call the future, and things appear to change—milk spills, eggs break, candles burn, people age—in universal alignment with this direction" (italics his). And based on classical physics that his essay assumes, Greene comes to argue that time has a "direction to the way things unfold *in* time" (p. 143, italics his). See p. 175 on entropy and gravity: "*The future is indeed the direction of increasing entropy. The arrow of time—the fact that things start like this and end like that but never start like that and end like this—began its flight in the highly ordered, low-entropy state of the universe at its inception*" (italics his). On the three arrows of time, thermodynamic ("the direction of time in which disorder or entropy increases") psychological ("the direction [of time] in which we remember the past but not the future"), and cosmological ("the direction of time in which the universe is expanding rather than contracting"), see Stephen W. Hawking, *A Brief History of Time: From the Big Bang to Black Holes* (New York, 1988), pp. 143–153 (145). See note 15 for Adrian Bardon on the nature of time.

matter. It is the end of *everything*. Because time itself ceases at the big crunch, it is meaningless to ask what happens next [or what happened before the big bang]... There is no 'next' for anything at all to happen—no time even for inactivity nor space for emptiness... [The universe] will disappear into nothing at the big crunch, its glorious few zillion years of existence not even a memory."[31]

### On Platonic Space

In the *Timaeus*, Plato writes about the beginning of the universe in ways analogous to modern cosmologists, where time is a conceptual marker for an expanding material universe. For Plato, "Platonic space," more accurately called a matrix, "is what *must* be there in the beginning, even before the act of creation occurs... For whatever comes to be must 'come to be in a certain place.' Compared with such spatial necessity, time is secondary in status—merely a 'moving *image* of eternity,' that is devised by the Demiurge to keep track of the circular motions of the heavens...time is a distinctly *late* addition to the scene of creation."[32] That matrix "is not...strictly material in character.

---

31   Paul Davies, *The Last Three Minutes: Conjectures about the Ultimate Fate of the Universe* (New York, 1994), p. 123 (italics his). See p. 64 on black holes and gravity: "Theory suggests that at the center of the black hole, gravity rises without limit. Because the gravitational field manifests itself as a curvature, or warping, of spacetime, the escalating gravity is accompanied by a spacetime warp that also rises without known limit. Mathematicians refer to this feature as a spacetime singularity. It represents a boundary, or edge, of space and time through which the normal concept of spacetime cannot be continued. Many physicists believe that the singularity inside a black hole genuinely represents the end of space and time, and that any matter that encounters it will be completely obliterated. See Hawking, *A Brief History of Time*, p. 713; Hawking is rethinking the notion of the obliteration of matter in black holes.

32   Edward S. Casey, *On the Fate of Place: A Philosophical History* (Berkeley/Los Angeles/London, 1997), p. 32 (italics his). See *Timeaus*, trans. Donald J. Zeyl (Indianapolis/Cambridge, 2000), 52a (p. 41): " ...we must agree that that which keeps its own form unchangingly, which has not been brought into being and is not destroyed, which neither receives into itself anything else from anywhere else, nor itself enters into anything else anywhere, is one thing. It is invisible—it cannot be perceived by the senses at all—and it is the role of understanding to study it. The second thing is that which shares the other's name and resembles it. This thing can be perceived by the senses, and it has been begotten. It is constantly borne along, now coming to be in a certain place and then perishing out of it. It is apprehended by opinion, which involves sense perception," and 37d (pp. 23–24): "So, as the model was itself an everlasting Living Thing, he [god] set himself to bringing this universe to completion in such a way that it, too, would have that character

Although it takes on material qualities, it is not itself composed of matter. As exhibiting or reflecting these qualities, it is more like a mirror of the physical than a physical thing itself. It has no qualities of its own, for, if it did, it could not be altogether receptive of the qualities of the things that occupy it, nor would it reflect them faithfully... Thus we cannot even characterize the receptive matrix as aqueous—as we are certainly encouraged to do [for example] at the beginning of the [Acadian] *Emuma Elish* and in Genesis... As preelemental, Space or the Receptacle is 'a nature invisible and characterless.' Yet the Receptacle is neither a void nor placeless."[33] In its "literal sense of 'uterus' or 'womb,' the matrix is the generatrix of created things: their *mater* or material precondition. As such, it is the formative phase of things—things that will become more fully determinate in the course of creation... Vis-à-vis the generative matrix, the task of creation becomes that of crafting and shaping, ultimately of controlling, what is unformed or preformed in the matrix itself. Creation becomes a matter of mastering matter."[34] To master "is not to bring into being in the first place but to control that which has already been brought into existence. It is still a matter of creation, at least in that sense of creation inherent in the Hebrew word *bará* used in 1 Genesis: a word whose cognate meanings include 'to carve' (for example, the tip of an arrow) or 'to cut up' (for example, a carcass). What is now at stake is not creation *ex nihilo*—an action

---

to the extent that it was possible. Now it was the Living Thing's nature to be eternal, but it isn't possible to bestow eternity fully upon anything that is begotten. And so he began to think of making a moving image of eternity: at the same time as he brought order to the heavens, he would make an eternal image, moving according to number, of eternity remaining in unity. This image, of course, is what we now call 'time.'" See also 39d-e (p. 26): "And so people are all but ignorant of the fact that time really is the wanderings of these [heavenly] bodies, bewilderingly numerous as they are and astonishingly variegated. It is nonetheless possible to discern that the perfect number of time brings to completion the Perfect Year at that moment when the relative speeds of all eight periods have been completed together and, measured by the circle of the Same that moves uniformly, have achieved their consummation. This, then, is how as well as why those stars were begotten which, on their way through the heavens, would have turnings. The purpose was to make this living thing as like as possible to that perfect and intelligible Living Thing, by way of imitating its sempiternity." For a close textual analysis of the Platonic concept of the space, see Kenneth Sayre, "The Multilayered Incoherence of Timaeus' Receptacle," in *Plato's Timaeus as a Cultural Icon*, ed. Gretchen J. Reydams-Schils (Notre Dame, 2003), pp. 60–79.

33    Casey, pp. 32–33.

34    Casey, p. 24 (italics his).

we have discovered to be as rare as it is problematical—but creation *ex datis*, 'out of the given.'"[35]

What is fascinating here is that Platonic space, which is *present* before matter or time, evokes, as a description of the universe, powerful emotions, so that imagining it creates a particular kind of longing to be in it—to be *there* just before the beginning of creation or just after its end, or at the beginning and the end of creation—because both states of being would return us to a "first state." Indeed there is a modern a-religious discourse of return to the "before creation" or the "after creation" as salvific reach so as to create a supermodern space void of existents. Here is the idea of space and time we can inhabit at will in the middle of life.

### Experiencing No-Time in Supermodern Space

For Marc Augé, we inhabit supermodern space when, for example, "we are driving down the motorway, wandering through the supermarket or sitting in an airport lounge waiting for the next flight to London or Marseille." Such places, he writes, "have the peculiarity that they are defined partly by the words and texts they offer us: the 'instructions for use,' which may be prescriptive ('Take right-hand lane'), prohibitive ('No smoking') or informative ('You are now entering the Beaujolais region')."[36]

---

35   Casey, pp. 23–24. Cf. Yeshayahu Leibowitz, *Accepting the Yoke of Heaven: Commentary on the Weekly Torah Portion* (New York, 2006) on a Jewish view of creation that places it outside time in every sense in which human and divine time are conceived: "The first verse in the Torah presents the great principle of faith: the world is not God—as opposed to any other outlook which exists in human culture: as opposed paganism, whose gods belong to the world and are to be found in the world; as opposed to Christianity, which at first blush received from Judaism the concept of the transcendent God, but immediately attired it in a human form which existed in the world; and as opposed to atheism, for which the world is the totality of being, or, in other words, that it is God. As opposed to all of these, the Torah says, 'In the beginning, God created heaven and earth'; the world is not God, and God is not the world—God is beyond the world, beyond any reality to which man's concepts are bound, and beyond the needs and the interests which stem from man's existence in the world" (p. 14).

36   Marc Augé, *Non-Place: Introduction to an Anthropology of Supermodernity*, trans. John Howe (London/New York, 1995), p. 96. See also p. 94: "Clearly the word 'non-place' designates two complementary but distinct realities: spaces formed in relation to certain ends (transport, transit, commerce, leisure), and the relations that individuals have with these spaces. Although the two sets of relations overlap to a large extent, and in any case officially (individuals travel, make purchases, relax), they are still not confused with one

Sometimes these [instructions] are couched in more or less explicit and codified ideograms (on road signs, maps and tourist guides), sometimes in ordinary language. This establishes the traffic conditions of space in which individuals are supposed to interact only with texts, whose proponents are not individuals but "moral entities" or institutions (airports, airlines, [the] Ministry of Transport, commercial companies, traffic police, municipal councils); sometimes their presence is explicitly stated ("this road section financed by the General Council," "the state is working to improve your living conditions"), sometimes it is only vaguely discernible behind the injunctions, advice, commentaries and "messages" transmitted by the innumerable "supports" (signboards, screens, posters) that form an integral part of the contemporary landscape.[37]

Listen to Augé describing the experience that texts point out in the supermarket and at the automated teller machine:

The customer wanders round [sic] in silence, reads labels, weighs fruit and vegetables on a machine that gives the price along with the weight; then hands his credit card to a young woman as silent as himself—anyway, not very chatty—who runs each article past the sensor of a decoding machine before checking the validity of the customer's credit card.

There is a more direct but even more silent dialogue between the cardholder and the cash dispenser: he inserts the card, then reads the instructions on its screen, generally encouraging in tone but sometimes including phrases ("Card faulty," "Please withdraw your card," "Read instructions carefully") that call him rather sternly to order.[38]

Supermodern space, abstract, or rather abstracted, is, in a transient way, eternal. Texts and tags define it, then organize it as well as the movements within it. Supermodern space contains the idea of objects, the idea of matter, by literally containing pointers to objects and matter. Such space may be alienating, but only for a short duration; it then becomes comforting just because it can

---

another; for non-places mediate a whole mass of relations, with the self and with others, which are only indirectly connected with their purposes. As anthropological places create the organically social, so non-places create solitary contractuality."

37  Ibid., p. 96.
38  Ibid., pp. 99–100.

"fabricate the 'average man,' defined as the user of the road, retail or banking system."[39]

> No doubt the relative anonymity that goes with...temporary identity can even be felt as a liberation, by people who, for a time, have only to keep in line, go where they are told, check their appearance. As soon as his passport or identity card has been checked, the passenger for the next flight, freed from the weight of his luggage and everyday responsibilities, rushes into the "duty-free" space; not so much, perhaps, in order to pay at the best prices as to experience the reality of his momentary availability, his unchallengeable position as a passenger in the process of departing.[40]

The user of a non-place, "in contractual relations with it (or with the powers that govern it)"[41]—and "always required to prove his innocence"[42]—is free from connections that anchor, that impinge.

> A person entering the space of non-place is relieved of his usual determinants. He becomes no more than what he does or experiences in the role of passenger, customer or driver. Perhaps he is still weighed down by the previous day's worries, the next day's concerns; but he is distanced from them temporarily by the environment of the moment. Subjected to a gentle form of possession, to which he surrenders himself with more or less talent or conviction, he tastes for a while—like anyone who is possessed—the passive joys of identity-loss, and the more active pleasure of role playing.[43]

Unlike supermodern space, continuous space "bristles with monuments— imposing stone buildings, discreet mud shrines—which may not be directly functional but give every individual the justified feeling that, for the most part, they pre-existed him and will survive him. Strangely, it is a set of breaks and discontinuities in space that expresses continuity in time."[44] The non-place is a parody of the return to Eden, for there is no history there, no indelible

---

39   Ibid., p. 100.
40   Ibid., p. 101.
41   Ibid., p. 101.
42   Ibid., p. 102.
43   Ibid., p. 103.
44   Ibid., p. 60.

connection between past and present, no indelible projection between present and future. History is "transformed into an element of [textual] spectacle."[45]


## Personal Space and History

For Georges Perec, personal space, each room in the Parisian apartment house, 11 Rue Simon-Crubellier, that his novel describes, is timeless and placeless, but it also contains history as the material and immaterial record of real time and place.[46] Perec writes about the layers of history in personal space, both deliberately there and accidentally there, as well as layers of apprehension and memory that are there, too. Perec's space is museum-like: an unintentional museum.[47] In describing who and what reside, or resided, in each room of Perec's Paris apartment house at his moment of apprehension, Perec gives us people and things literally and figuratively there, alive there *now* so that the full presence of what *was* there once is there now, too. The result is a timeless, but historically thick interior and internal landscape.

> Not much is left of these three rooms in which Gaspard Winckler lived and worked for nearly forty years. His few pieces of furniture, his small workbench, his jigsaw, his minute files have gone. On the bedroom wall, opposite his bed, beside the window, that square picture he loved so much is no longer: it showed an anti-chamber with three men in it. Two were standing, pale and fat, dressed in frock-coats and wearing top hats which seemed screwed to their heads. The third, similarly dressed in black, was sitting by the door in the attitude of a man expecting visitors, slowly putting a pair of tight-fitting new gloves on over his fingers.[48]

Winckler's rooms contain objects that have remained there since his death, including pictures of people from an earlier day who were never there except as Winckler's wish for visitors; the idea of anticipating visitors reveals Winkler's consciousness still present for Perec. Moreover, as Perec explains, Winkler

---

45    Ibid., p. 103–104.

46    Georges Perec, *Life: A User's Manual*, trans. David Bellos (Paris, 1978; Boston, 1987).

47    On timelessness and historical memory as experienced in a museum, where "classification [collection] manages time," see Susan A. Crane, "The Conundrum of Ephemerality: Time, Memory, and Museums," in *A Companion to Museum Studies*, ed. Sharon Macdonald (Malden, MA / Oxford, 2008), pp. 98–109 (101).

48    Perec, pp. 5–6.

spent his days creating jigsaw puzzles using paintings another resident of 11 Rue Simon-Crubellier, a man named Bartlebooth, who occupies himself now—he outlived Winkler—re-creating his own life in art. If Bartlebooth's paintings are one remove from his experience, Winker's puzzles are two: they are "puzzled" presences of Barlebooth's painted experience able to be played as a game again and again.

In a postmodern sense, no one has left his or her room in Perec's apartment house. Indeed, no one's consciousness has been obliterated either: historical memories remain, as well as wishes for memory—the desire to assemble (to paint), to disassemble (to "cut up"), to reassemble (to complete the puzzle of experience again and again as if the arc of experience were both a continuous mystery and the affirming and clarifying goal). Valène, for example, another resident of Perec's apartment house,

> had the feeling that time had been stopped, suspended, frozen around he didn't know what expectation. The very idea of the picture he planned to do and whose laid-out, broken-up images had begun to haunt every second of his life, furnishing his dreams, squeezing his memories, the very idea of this shattered building laying bare the racks of its past, the crumbling of its present, this unordered amassing of stories grandiose and trivial, frivolous and pathetic, gave him the impression of a grotesque mausoleum raised in the memory of companions petrified in terminal postures as insignificant in their solemnity as they were in their ordinariness, as if he had wanted both to warn of and to delay these slow or quick deaths which seemed to be engulfing the entire building story by story: Monsieur Marcia, Madame Moreau, Madame de Beaumont, Bartlebooth, Rorschach, Mademoiselle Crespi, Madame Albin, Smautf. And himself, of course, Valène himself, the longest inhabitant of the house.[49]

Valène's sense of time stopped and of place disembodied, his idea of a "shattered building laying bare the racks of its past," the idea of "the crumbling of" a "present," the sense of the "unordered amassing of stories grandiose and trivial" is not, of course, in Augé's sense, no-time in a non-place, but rather a species of it just because rooms in Perec's apartment house speak historical consciousness that continues to exist. Distinctions of time and place do not obtain there, but history does—random, flattened, and cut up into all-time and all-place.

---

49   Ibid., p. 127.

## Imagining No-Time in Fabulist Space

In this context, where places have thick history imagined as a present and random conjunction, which is history without a thesis, Italo Calvino describes entire cities, peopled and busy, constructed out of pure metaphysical play.[50] Calvino describes at least eight civic entities that exhibit species of timelessness, each an example of a city imagined as, for example, *infinite*, or *expanding*, or *celestial*, or *instantiated*, or a city as a *twin of itself*, or *tentative*, or *confounding*, or *anticipated*. Fabulated cites, for example:

· Like Esmeralda, an *infinite* city, where "a network of canals and a network of streets span and intersect each other." The inhabitants "are spared the boredom of following the same streets every day... Combining segments of the various routes, elevated or on ground level, each inhabitant can enjoy every day the pleasure of a new itinerary to reach the same places. The most fixed and calm lives in Esmeralda are spent without any repetition... A map of Esmeralda should include, marked in different colored inks, all these routes, solid and liquid, evident and hidden."[51]
· Like Olinda, an unchanged *expanding* city "that grows in concentric circles, like tree trunks which each year add one more ring. But in other cities there remains, in the center, the old narrow girdle of the walls from which the withered spires rise, the towers, the tiled roofs, the domes, while the new quarters sprawl around them like a loosened belt. Not Olinda: the old walls expand bearing the old quarters with them, enlarged, but maintaining their proportions on a broader horizon at the edges of the city..."[52]
· Like Andria, a *celestial* city, "built so artfully that its every street follows a planet's orbit, and the buildings and the places of community life repeat the order of the constellations and the position of the most luminous stars: Antares, Alpheratz, Capricorn, the Cepheids... Though it is painstakingly regimented, the city's life flows calmly like the motion of the celestial body and it acquires the inevitability of phenomena not subject to human caprice." The inhabitants who feel themselves "part of an unchanging heaven, cogs in a meticulous clockwork take care not to make the slightest change in" their city or their habits.[53]

---

50    Italo Calvino, *Invisible Cities*, trans. William Weaver (New York, 1972; rept. 1974).
51    Ibid., pp. 88–89.
52    Ibid., p. 129.
53    Ibid., p. 150.

- Like Eudoxia, an *instantiated* city, knowable in its already completely woven image, "laid out in symmetrical motives whose patterns are repeated along straight and circular lines, interwoven with brilliantly colored spires, in a repetition that can be followed throughout the whole woof... In Eudoxia, which spreads both upward and down, with winding alleys, steps, dead ends, hovels, a carpet is preserved in which you can observe the city's true form... All of Eudoxia's confusion, the mules' braying, the lampblack stains, the fish smell is what is evident in the incomplete perspective you grasp; but the carpet proves that there is a point from which the city shows its true proportions, the geometrical scheme implicit in its every, tiniest detail."[54]
- Like Beersheba, a city as its own *twin*, a double projection of itself, "one celestial and one infernal," but "the citizens are mistaken about their consistency. The inferno that broods in the deepest subsoil of Beersheba is a city designed by the most authoritative architects, built with the most expensive materials on the market, with every device and mechanism and gear system functioning, decked with tassels and fringes and frills hanging from all the pipes and levers. Intent on piling up its carats of perfection, Beersheba takes for virtue what is now a grim mania to fill the empty vessel of itself... Still, at the zenith of Beersheba there gravitates a celestial body that shines with all the city's riches, enclosed in the treasury of cast-off things: a planet a-flutter with potato peels, broken umbrellas, old socks, candy wrapping, paved with tram tickets, fingernail-cutting and pared calluses, eggshells. This is the celestial city, and in its heavens [sic] long-tailed comets fly past, released to rotate in space from the only free and happy action of the citizens of Beersheba, a city which, only when it shits, is not miserly, calculating, greedy."[55]
- Like Zoe, a *tentative* city whose existence is always in doubt but that exists because it is doubted: "In every point of this city you can, in turn, sleep, make tools, cook, accumulate gold, disrobe, reign, sell, question oracles... The traveler roams all around and has nothing but doubts: he is unable to distinguish the features of the city, the features he keeps distinct in his mind also mingle. He infers this: if existence in all its moments is all of itself, Zoe is the place of indivisible existence. But why, then, does the city exist? What line separates the inside from the outside, the rumble of wheels from the howl of wolves?"[56]
- Like Zaira, a *confounding* city, mysterious and playful because it speaks to our allegorical turn that would reveal meaning, a city of indelible historical

---

54    Ibid., p. 96.
55    Ibid., pp. 112–113.
56    Ibid., p. 34.

relationships that "does not tell its past, but contains it like the lines of a hand, written in the corners of the streets, the gratings of the windows, the banisters of the steps, the antennae of the lightning rods, the roles of the flags, every segment marked in turn with scratches, indentations, scrolls."[57]

· Like Chloe, a city that hasn't begun, a city *anticipated*: "In Chloe, a great city, the people who move through the streets are all strangers. At each encounter, they imagine a thousand things about one another; meetings which could take place between them, conversations, surprises, caresses, bites. But no one greets anyone; eyes lock for a second, then dart away, seeking other eyes, never stopping... A voluptuous vibration constantly stirs Chloe, the most chaste of cities. If men and women began to live their ephemeral dreams, every phantom would become a person with whom to begin a story of pursuit, pretenses, misunderstandings, clashes, oppressions, and the carousel of fantasies would stop."[58]

Calvino's eight cities mirror the mode of postmodern fabulation where possible places in possible coordinates are brought into being. Fabulation has no bounds. Moreover, as play places, fabulated cities are conceptually and indeed emotionally irresistible because they speak to aspects of the freedom and the mystery of places conceived as embodied varieties of timelessness: cities freed of time so that they address issues of existence itself, cities in continuously interactive nexus with their inhabitants and their inhabitants' ideas of existence. For Calvino, fabulated cities are in re-imaginable time and geography.[59]

### And Then There's Las Vegas

For Hunter Thompson, Las Vegas, too, is a city of timeless experience. The real Las Vegas is in fact a fabulation we can actually visit: this makes it irreal. At base,

---

57    Ibid., p. 11.

58    Ibid., pp. 51–52.

59    For real cities that history and myth make timeless, see Cynthia Ozick, "The Butterfly and the Traffic Light," in *The Pagan Rabbi and Other Stories* (Syracuse, NY, 1995), pp. 209–210: "Jerusalem, that phoenix city, is not known by its street names. Neither is Baghdad, Copenhagen, Rio de Janeiro, Camelot, or Athens; nor Peking, Florence, Babylon, St. Petersburg. These fabled capitals rise up ready-spired, story-domed and filigreed; they come to us at the end of a plain, behind hill or cloud, walled and moated by myths and antique rumors. They are built of copper, silver, and gold; they are founded on milkwhite stone; the bright thrones of ideal kings jewel them... It is different with places of small repute or where time has not yet deigned to be an inhabitant. It is different especially in America." For a magisterial examination of the theories of time's literal embodiment, see Marvin Trachtenberg, *Building-in-Time: From Giotto to Alberti and Modern Oblivion* (New Haven / London, 2010).

for Thompson, Las Vegas speaks to disruptions in the human body's sense of identity. Thompson's hilarious and frightening vision of America's mid-twentieth-century drug culture is situated in the disparate landscape of a city that is the projection of interior disintegration.

> This is the main advantage of ether: it makes you behave like the village drunkard in some early Irish novel…total loss of all basic motor skills: blurred vision, no balance, numb tongue—of all connection between the body and the brain. Which is interesting, because the brain continues to function more or less normally…you can actually *watch* yourself behaving in this terrible way, but you can't control it.[60]

The experience of no-time inside the casinos makes of that space a kind of psychological free-for-all that includes a technology that amazed Thompson back then: the casino space can be televised with a spatial grandeur that rivals divine presence illuminated in manuscripts in a much earlier day.

> This madness goes on and on, but nobody seems to notice. The gambling action runs twenty-four hours a day on the main floor, and the circus never ends. Meanwhile, on all the upstairs balconies, the customers are being hustled by every conceivable kind of bizarre shuck. All kinds of funhouse-type booths… Stand in front of this fantastic machine…and for just 99¢ your likeness will appear, two hundred feet tall, on a screen above downtown Las Vegas…Jesus Christ. I could see myself lying in bed in the Mint Hotel, half-asleep and staring idly out the window, when suddenly a vicious nazi [sic] drunkard appears two hundred feet tall in the midnight sky, screaming gibberish at the world: "*Woodstock Uber Alles!*"[61]

For Thompson, Vegas is a place of extraordinary and accessible, warped, and parodic transcendence.

### No-Time as Willful Escape

There is, of course, a darker side to no-time and non-place because their call can be so powerful that some of us want to enter no-time at any cost. For Thomas de Quincey,

---

60    Hunter S. Thompson, *Fear and Loathing in Las Vegas: A Savage Journey to the Heart of the American Dream* (New York, 1971), p. 45 (italics his).
61    Thompson, pp. 46–47 (italics his).

A man who is inebriated, or tending to inebriation, is, and feels that he is, in a condition which calls up into supremacy the merely human, too often the brutal, part of his nature: but the opium-eater (I speak of him who is not suffering from any disease, or other remote effects of opium) feels that the diviner part of his nature is paramount; that is, the moral affections are in a state of cloudless serenity; and overall is the great light of the majestic intellect.[62]

Opium, de Quincey explains, "communicates serenity and equipoise to all the faculties, active or passive: and with respect to the temper and moral feelings in general, it gives simply that sort of vital warmth which is approved by the judgment, and which would probably always accompany a bodily constitution of primeval or antediluvian health."[63] When he was under the influence of opium, "the sense of space, and in the end, the sense of time, were both powerfully affected. Buildings, landscapes, &c. were exhibited in proportions so vast as the bodily eye is not fitted to receive. Space swelled, and was amplified to an extent of unutterable infinity... I sometimes seemed to have lived for 70 or 100 years in one night; nay, sometimes had feelings representative of a millennium passed in that time, or, however, of a duration far beyond the limits of any human experience."[64]

---

62    Thomas de Quincey, *Confessions of an English Opium-Eater and Other Writings*, ed. Barry Milligan (London, 2003), p. 47.

63    De Quincey, p. 46.

64    De Quincey, p. 76. See also de Quincey's short story, "Savannah-la-mar," in *Great Short Stories of the Masters*, ed. Charles Neider (New York, 1989), p. 49, which imagines the reverse, not the experiences of the infinitely large, but the infinitely small: "The time which *is* contracts into a mathematical point; and even that point perishes a thousand times before we can utter its birth. All is finite in the present; and even the finite is infinite in its velocity of flight towards death" (italics his). Moreover, the instrument that measures the passage of time is not the hourglass of sand, but an hourglass of drops, as if the passage of time were being mourned. Cf. Henry James, "The Great Good Place," in Neider, p. 191, who imagines George Dane entering no-time as an experience of consciousness, but not as a dark experience, not as drug-induced, nor in a dark place, but as the place created as respite for the consequences of fame, the weight of the experience of being everything in the awareness of others. James' no-time is imagined as historical travel that the mind would mercifully live in: "One by one he touched, as it were, all the things it was such rapture to be without. It was the paradise of his own room that was most indebted to them—a great square fair chamber, all beautified with omissions, from which, high up, he looked over a long valley to a far horizon, and in which he was vaguely and pleasantly reminded of some old Italian picture, some Carpaccio or some early Tuscan, the representation of a world without newspapers and letters, without telegrams and photographs, without the dreadful fatal too much." See also the description in Julian Barnes, *The Sense*

### Leaps of Time: Time Travel and Theological Parody

De Quincey's experiences create for him the nightmare of time travel.

> A thousand accidents may, and will interpose a veil between our present consciousness and the secret inscriptions on the mind; accidents of the same sort will also rend away this veil; but alike, whether veiled or unveiled, the inscription remains for ever; just as the stars seem to withdraw before the common light of the day...which is drawn over them as a veil—and that they are waiting to be revealed when the obscuring daylight shall have withdrawn.[65]

But if time travel with the aid of hallucinogens renders the place from which one comes and the place to which one travels one space, time travel can also be imagined theologically. In the twelfth-century, Walter Map in his *De Nugis Curialium* (*Of Courtiers' Trifles*) narrates a tale that is, from our perspective, perhaps the radiating episode of time travel.[66]

The tale of Herla, king of "the most ancient Britons" describes Herla's encounter with a "pygmy" king ("in respect of his low stature, not above that of a monkey" [*pigmeus uidebatur modicitate stature, que non excedebat simiam*]) with a "huge head" (*capite maximo*). His visage fiery red, (*ardenti facie*), he had a "long red beard reaching to his chest" (*barba rubente prolixa pectus contigente*),

---

*of an Ending* (New York, 2011) of a gesture that refuses the idea of time as an external dimension: wearing "watches with the face on the inside of the wrist. ... It made time feel like a personal, even secret, thing" (p. 6). For perhaps the most evocative description of a memory *in the present*, acutely sensitive to the bodily conjunction of past and present, see Marcel Proust, "Filial Sentiments of a Parricide," in Neider, pp. 348–349: "If you look at someone while his mind is intent upon bringing back something from the past, restoring it to life for an instant, you will see that his eyes go suddenly blind to the surrounding objects which they reflected an instant before. 'Your eyes are blank, you are somewhere else,' we say; however, we see only the external signs of the phenomenon that takes place in the mind. At such a moment the most beautiful eyes in the world no longer touch us with their beauty; they are, to change the meaning of a phrase of Well's, no more than 'machines to explore time,' the telescopes of the invisible, which become at best measures to gauge one's advancing age... At the moments when my eyes met hers, I had a sense of the supernatural; her gaze, by some feat of resurrection, firmly and mysteriously joined the present to the past."

65    De Quincey, p. 77.
66    Walter Map, De *Nugis Curialium*, ed. and trans. M.R. James, rev. C.N.L. Brooke, and R.A.B. Mynors (Oxford, 1983), 1, 12 (pp. 26–31).

a hairy belly, and "legs declined into goats' hoofs" (*crura pedes in caprinos degenerabant*). The pygmy king himself mounted on a goat says that he and Herla both are "closely connected...in place and descent" (*loco michi proximus et sanguine*) and that Herla, a royal, deserves that his wedding "should be brilliantly adorned by my presence as a guest"—Herla is about to marry the daughter of the King of the Franks. In this courtly context, the pygmy king makes a pact with Herla that brings him into the pygmy's realm. When he attends Herla's wedding, bringing lavish gifts that celebrate, in the context of Map's work written during the reign of Henry II and Queen Eleanor, Herla will reciprocate precisely one year later.[67]

But when Herla journeys to the pygmy's kingdom, he enters "a cave in a high cliff" with "a light which seemed to proceed not from the sun or moon, but from a multitude of lamps, to the mansion of the pygmy [*quod non uidebatur solis aut lune sed lampadarum multarum ad domos pigmei transeunt*]." The lamps are the lights of a nether world. And when Herla returns home, "laden with gifts and presents of horses, dogs, hawks, and every appliance of the best for hunting or fowling," arriving "once more at the light of the sun," he finds that his queen has died and that he was seen, as a shepherd relates, to have "disappeared in company with a pygmy at this very cliff, and was never seen on earth again, and it is now two hundred years since the Saxons took possession of this kingdom, and drove out the old inhabitants." Herla thought that "he had made a stay of but three days [*qui per solum triduum moram fecisse putabat*]," the elapsed time of Christ's descent into hell.

Map's courtier's tale has its origins in stories describing the trickery of an elder race, such as dwarves, which is the stuff of folklore. As such, Map's tale is a parody, in the guise of folklore, of courtesy politics at a twelfth-century court. It is also a parody of a descent into hell in which Map's view of court politics is set. The pygmy in this parody of royal politics and theologically transformative rites of passage is a king who is also a satyr-like devil—Pan himself, as Map says—whose appearance suggests something finally too sexual for a wedding, though Herla does not respond to what the pygmy king's appearance points to as his dangerous nature. Herla's return to this world is a species of resurrection, where he brings gifts to an upper world in which he, when he arrives, is only a memory: a story. In addition to the appropriate gifts that Herla brings back from the pygmy's realm, he is given a small "bloodhound" (*sanguinarium*), being enjoined that "on no account must any of his train dismount until that dog leapt from the arms of his bearer [Herla himself]." But some in Herla's

---

67    The full nature of Map's connection with Henry II is still not quite clear ("He was a clerk the court of Henry II and Queen Eleanor" [p. xxiii]). See Map, Introduction, p. xiii–xxii.

train, "forgetting the pygmy's orders, dismounted before the dog had alighted, and in a moment fell into dust." Death is one Herla's gifts from the underworld. Map's tale of Herla is a dark, witty, cautionary political parable about generosity and trust.

It is, however, the idea of time lapse that marks Map's tale of time travel for us. Herla travels "forward in time" in the pygmy's realm and returns to the world above two hundred years latter. The tale's theological allusions, which connect the tale to Christ's own time travel, thus give to the events of Map's story a resonance that makes time travel fraught with personal, indeed anti-Christian danger: the treacherous pygmy king must have held Herla captive and figuratively had him executed. Moreover, the tale rests in politically significant time travel, which the parody of theological descent and resurrection makes a universal political experience. Map's tale can be viewed as an etiology of realpolitik. Christ's return after three days creates a new world in resurrected time, the beginning of a new time; Herla's return means his kingdom has been destroyed.[68] Herla is a parodic Christ.

Actual time travel is, of course, not possible, though we may desire it for a variety of reasons. The primary [human] experience

> is motion, and we move neither in space nor in time, but in space-time. It makes no sense to speak of motion in space independently of time, or of motion in time independently of space. What constitutes motion is only a change of position in space during a given interval of time. Even before we grapple with relativity, we can say that space and time are unbreakably linked with each other. Motion itself would be impossible if we had to decide the direction in time as well as in space. Elementary time abstracted from the fact of motion—independently of whether life

---

68  On language that points to the experience of no-time, see John Perry, "Temporal Indexicals," in *A Companion to the Philosophy of Time*, pp. 486–506. The dates used to denote divine time are not, of course, like dates on a human calendar to which Perry contrasts temporal indexicals such as now, today, tomorrow, yesterday, last month, a year ago, past, future, and present, all of which refer to periods of human time. We should add that the discussion of time in this essay, including duration in the real world, assumes classical physics: "We do not ascribe various durations to the different parts of space, but say that all endure simultaneously. The moment of duration is the same in Rome and in London, on earth and the stars, and throughout the heavens..." (see *Newton: Philosophical Writings*, ed. Andrew Janiak [Cambridge, 2005], p. 26). The implication here is that no-time, as I am describing it, assumes God-given classical physics, as does the theological time of Herla's "sleep." Relativistic physics and the modern science fiction informed by it are beyond the scope of this essay.

invents such clever things as memory—is not a rigid parameter but merely a way of specifying change and motion.[69]


## The Saint's Body

But just as time travel catches our escapist imagination, medieval saints' lives do, too, even today if perhaps not for reasons of devotion. Both time travel and medieval saints' lives speak to our yearning for, and our desire for proof of, transcendence as a species of miraculous no-time, where matter becomes *matter eternalized* in our very presence. My example of a saint's life comes from Chaucer in a tale set in late antiquity when Christianity was under siege. The *Second Nun's Tale* speaks to the power of the Christian body, first dressed for a chaste marriage and then as a continuously impervious being that does not burn in a "bath of flambes rede" (8.515)[70]—in water and fire at the same time—to convince Romans that Christianity is true: it can translate a body into the bodily matter of eternity, a resurrection visible here on earth.

The *Second Nun's Tale* first imagines the conversion of the pagan Valerian, describing the effect of Cecilia's assuming, in a worldly way, a chaste body, having put "an haire" (a hair shirt [133]) under her "a robe of gold" (132), her wedding dress. This is not, of course, a miraculous transformation of desire, but a willful one. She would be married in a real-world Christian body so that

---

69    Martin Bojowald, *Once Before Time: A Whole Story of the Universe* (New York, 2010), p. 226.
      See pp. 225–226, on the arrow of time: "In particular, the question of why one cannot sin-
      gle-handedly reverse the arrow time is independent of the enticing fantasy of time travel.
      During time travel in the usual sense…time does not run backward; it merely proceeds
      differently for some people. A person or a group of people is enclosed in a bounded region
      that, after a process rarely specified further, arrives in the past of its neighborhood.
      Neither for external bystanders nor for the time travelers does a step back in time occur;
      rather one region of space is merely separated from the outer world and subject to a modi-
      fied process of time. While time travelers do venture into their neighborhood's past, their
      own time keeps running forward. After all, they retain their memories of all their past, a
      part of which is now in the future of their neighborhood, and they do not grow younger,
      or even die when a time before their birth day is reached." On the distinction between
      mathematical and physical time, where time independent of matter is idealized and real
      physical time is relational, see pp. 218–219. On ideas of time travel, see Douglas Kutach,
      "Time Travel and Time Machines," in *A Companion to the Philosophy of Time*, pp. 301–314.
70    Citations of Chaucer are from *The Riverside Chaucer*, 3rd ed., gen. ed. Larry D. Benson
      (Boston, 1987).

she would be, for her husband, the visible sign of Christian sexual values. She thus presents herself at her wedding as a body sealed against male desire—and her own. Cecilia does not refuse marriage; rather she intends to live as a married virgin. She and Valerian live as a victorious couple, he a tamed man, a figure of Cecilia's Christian marital imagination. All this makes Cecilia ready as a practicing Christian for her "bath of flambes."

When Almachius, the Roman prefect, commands that everyone sacrifice to Jupiter on pain of punishment ("Whoso wol nat sacrifise, / Swape of his heed" [365–366]), Cecilia, brought to the place of sacrifice, refuses, encouraging the others brought with her to accept martyrdom. They do: "With humble herte and sad devocioun, / And losten bothe hir hevedes in the place, / Hir soules wenten to the Kyng of grace" (397–399). Maximus, who is Almachius subordinate officer, sees the souls of the others, and perhaps Cecilia (the text is unclear), ascend to heaven, and in recounting what he sees, his witnessing converts others. Maximus himself is beaten to death and Cecilia takes his body to be buried.

Then, for a second time, Almachius demands Cecilia sacrifice to Jupiter. Again, it is not altogether clear from the text whether or not she, too, was martyred in the presence of Maximus when the others were and then returned to earth. In any case, this time she refuses Almachius directly, explaining the power of her Christian faith to withstand apostasy—the model for their confrontation is Christ's confrontation with Pilate. And this time her refusal to sacrifice results in her being placed in a bath of flames.

> The longe nyght, and eek a day also,
> For al the fyr, and eek, the bathes heete
> She sat al coold, and feelede no wo.
> It made hire nat a drope for to sweete.
> But in that bath hir lyf she moste lete [give up],
> For he Almachius, with ful wikke entente,
> To sleen hire in the bath his sonde [messenger] sent. (519–525)

But Cecilia does not perspire amid boiling water. Her Christian body remains cold and true in water that cannot change her physical nature. She is in theological no-time, a place whose boundaries cannot be crossed, only witnessed; the assaults of earthly physical conditions have no affect on her. Cecilia only dies after her neck is "smoot" (526) three times, though her head is not severed; it bleeds for three days, the blood caught fountain-like by those surrounding her to whom she never ceases to "teche" (538). In this state—bleeding, impervious—she lives for three days, a sacred duration, and her being wounded imitates the wounding to which the sacred flesh of the savior was subjected.[71]

We should add that Chaucer also knows about another mystery of death where the blood of Arcite in the *Knight's Tale* is "clothered" (1. 2745). His death removes him from nature despite our best medicine: "Hym gayneth neither, for to gete his lif, / Vomyt upward, ne dounward laxatif," so that "Nature hath now no dominacioun" (2755–2756, 2758). Neither purgatives (to induce vomiting) nor laxatives can help him, and he exists where "Nature wol nat wirche" (2759). All that's left to say is this: "Fare wel phisik! Go ber the man to chirche!" (2760). Arcite's existence out of time doesn't, of course, make him a saint, though his condition is a dark example of no-time. The jocular tone in the *Knight's Tale* here is the tone of anxiety. Arcite is the parody of a saint, whereas Cecilia's imperviousness to the annihilation of matter is our hope, our theology.

## The Timeless as the World Itself

Paganism, too, has something of Christianity's hoped-for transcendence: the connection between body and spirit that the theology of our religious traditions writes out of our descriptions of experience. For Carlyle, however, the indivisible exists in the ongoing here and now.

> That great mystery TIME, were there no other; the illimitable, silent, never-resting thing called Time, rolling, rushing on, swift, silent, like an all-embracing ocean-tide, on which we and all the Universe swim like exhalations, like apparitions which are, and then are *not*: this is forever very literally a miracle; a thing to strike us dumb,—for we have no word to speak about it. This Universe, ah me—what could the wild man know

---

71    Cf. the martyrdom of the second-century Polycarp. He is set in the center of a circle of flames, like Cecilia, but without water, and his body doesn't burn. Polycarp is "not like a human being in flames but like a loaf baking in the oven"—he is imaged as a host—"or like a gold or silver ingot being refined in the furnace"—he is imagined as a crucible, fragrant with an odor like incense itself. From Polycarp's impervious being blood flows just as blood flows from Cecilia's neck (*Early Christian Writings: The Apostolic Fathers*, trans. Maxwell Staniforth [Harmondsworth, UK/New York, 1968], p. 161. See Caroline Walker Bynum, *The Resurrection of the Body in Western Christianity, 200–1336* (New York, 1995), pp. 45–46: "Martyred flesh had to be capable of impassibility and transfiguration, suffering and rot could not be the final answer. If flesh could put on, even in this life, a foretaste of incorruption, martyrdom might be bearable...[and] a sort of anesthesia of glory might spill over from the promised resurrection...."

of it; what can we yet know? That it is a Force, and thousand-fold Complexity of Forces; a force which is *not* we. That is all; it is not we, it is altogether different from us. Force, Force, everywhere Force; we ourselves a mysterious Force in the centre of that. "There is not a leaf rotting on the highway but has Force in it; how else could it rot?"[72]

For Carlyle we have, unfortunately, lost our visceral connection to what he calls the ocean of time that is, in Carlyle's grand defense of the spirituality of paganism, the eternal place of which we are part, as if we are creation's epiphenomenon, a mystery within mystery. "Breath" is Carlyle's metaphor, an allusion to the "breath of life" that God Himself breathes into His creation, though the idea of God as creator is not, in Carlyle, the God who is outside the human world. We are, though Carlyle veers from this possible analogy, white caps on the ocean itself.

> The world, which is now divine only to the gifted, was then divine to whosoever would turn his eye upon it. He stood bare before it face to face. "All was Godlike or God."[73]

The phrase "face to face" is Carlyle's pun on the phrase "face to face" from Exodus (33, 11) that describes how Moses was privileged to see God. For the pagan, "face to face" means being in a world whose objects are one with a god who is in them and everywhere. For Carlyle, the pagan represents the closest connection to the spiritual present in the material. This defines, in Carlyle, a notion of soul and body that is a continuous first state: an available no-time.

### Eternal Place in the Present: Sabbath Time

There is a Jewish version of the idea of the transcendence of matter, though it is one that occurs in the world of time and matter once a week. The Sabbath is set apart "for freedom, a day on which we would not use the instruments which have been so easily turned into weapons of [spiritual] destruction, a day for being with ourselves, a day of detachment from the vulgar, of independence of external obligations, a day on which we stop worshipping the idols of technical civilization, a day on which we use no money, a day of armistice in the economic

---

72    Thomas Carlyle, *On Heroes, Hero-Worship, and the Heroic in History*, Lecture 1: The Hero as
      Divinity. Odin. Paganism: Scandinavian Mythology (1840), p. 5 (italics his).
73    Carlyle, pp. 5–6.

struggle with our fellow men and the forces of nature..."[74] Moreover, as Abraham Heschel explains, "one must abstain from toil and strain on the seventh day, even from strain in the service of God."[75] Sanctifying time by dividing it between *this* day and the *other* days gives to Judaism its "architecture of time"—the phrase is Heschel's: "There is no quality that space has in common with the essence of God. There is not enough freedom on the top of the mountain; there is not enough glory in the silence of the sea. Yet the likeness of God can be found in time, which is eternity in disguise."[76]

For Heschel, "Time is the process of creation, and things of space are results of creation. When looking at space we see the products of creation; when intuiting time we hear the process of creation. Things in space exhibit a deceptive independence. They show off a veneer of limited permanence. Things created conceal the Creator. It is the dimension of time wherein man meets God, wherein man becomes aware that every instant is an act of creation, a Beginning, opening up new roads for ultimate realizations. Time is the presence of God in the world of space, and it is within time that we are able to sense the unity of all beings."[77]

To experience Sabbath time, then, we need to mark it off physically. Doing this is not creating the Sabbath—it was, according to Judaism, already created. Rather marking off Sabbath time makes it possible for eternity to be here for us. By dividing up the world into *this* kind of activity and *that* kind of activity, we make holiness present, a condition of being that the Sabbath, created before creation, calls us to join in the language of Jewish prayer as its groom; for the Sabbath is lonely: "Things are our tools; eternity, the Sabbath, is our mate. Israel is engaged to eternity."[78] And eternity needs to be wed in a place where it can be among us.

---

74   Abraham Joshua Heschel, *The Sabbath: Its Meaning for Modern Man* (New York, 1951), p. 28.

75   Heschel, p. 30. See also Heschel, p. 106, n. 8: "Rabbi Zera used to seek out pairs of scholars (engaged in learned discussion) and say to them, 'I beg of you do not profane it' (the Sabbath, by neglecting its delights and good cheer)."

76   Heschel, p. 16.

77   Heschel, p. 100. See also p. 99: "Space is exposed to our will; we may shape and change the things in space as we please. Time, however, is beyond our reach, beyond our power. It is both near and far, intrinsic to all experience and transcending all experience." See p. 75: "What is the Sabbath? *Spirit in the form of time.* With our bodies we belong to space; our spirit, our souls, soar to eternity, aspire to the holy. The Sabbath is an ascent to the summit. It gives us the opportunity to sanctify time, to raise the good to the level of the holy, to behold the holy by abstaining from profanity" (italics his).

78   Heschel, p. 48. See also p. 52 on the lonely Sabbath: "With all its grandeur, the Sabbath is not sufficient unto itself. Its spiritual reality calls for companionship of man. There is a great longing in the world. The six days stand in need of space; the seventh day stands in

As Judith Shulevitz explains, Nehemiah shut the gates that "made the Sabbath a geographical construct as well as a temporal one. The Sabbath, a day set apart, became a city enclosed, and a nation withdrawn into itself. Nearly a millennium later, the rabbis of the Talmud turned the Sabbath into a more modern space, a place both enclosed and open. They developed a body of laws whereby, by following certain prescribed procedures, a community could construct a boundary marking off a Jewish neighborhood for the duration of the Sabbath. In lieu of a gate, the laws called for a wall, but not the impermeable kind of wall that surrounded Jerusalem. This was not a wall of brick or clay or stone. This was to be a notional wall—a wall concept... a boundary marked by thin wires strung from pole to pole high above the head. This wall-like entity they called an *eruv*, which in Hebrew means 'mixing' or 'commingling,' since an *eruv* brings together entities otherwise kept apart on the Sabbath."[79] This created the rabbinic cave in the world, if you will, once a week. The rabbinic *eruv*, as Shulevitz puts it, sweeps away biblical restrictions that forbid Jews

> to walk very far or carry anything on the Sabbath from one domain to another...to carry [anything] from the private domain, or home, to the public domain, or street, or the other way around, or between two private domains, and so on... [The rabbinic *eruv* bundles] together assorted city spaces—apartment buildings and alleyways, courtyards and front yards—and [re-categorizes] them, within limits, as one large private domain... The *eruv* advanced the legal fiction that all the Jews in a neighborhood live in one...Jewish household... With an *eruv* in place, traditional Jews stroll though the streets on the Sabbath with a commanding ease, as if moving from room to room. Women carry their babies and push strollers; men carry their books and prayer shawls; guests carry wine to their hosts... An *eruv* [thus] delineates the contours of a Jewish

---

need of man. It is not good that the spirit should be alone, so Israel was destined to be a helpmeet for the Sabbath." See p. 59: "To name it [the Sabbath] queen, to call it bride is merely to allude to the fact that its spirit is reality we meet rather than an empty span of time which we choose to set aside for comfort or recuperation." See p. 60: "The idea of the Sabbath as a queen or a bride is not a personification of the Sabbath but an exemplification of a divine attribute, an illustration of God's need for human love; it does not represent a substance but the presence of God, His relationship to man." On the idea of marriage to the Sabbath, see pp. 51–52: "The Hebrew word *le-kadesh*, to sanctify, means, in the language of the Talmud, to consecrate a woman, to betroth."

79  Judith Shulevitz, *The Sabbath World: Glimpses of a Different Order of Time* (New York, 2010), p. 50.

space... The *eruv*, in other words, is a segregator and identity-enhancer and nation-builder. Its quasi-fictional walls were the state upon which the Jews imagined their way into the idea of community.[80]

Inside the Platonic cave, we only see illusions, reflections on the wall in front of us of ideas behind us, outside the cave. In the Platonic cave, we are chained to illusions until we turn to see where reality is, and walk outside. Inside the rabbinic cave, where there is the absence of instrumental activity, but not the body, we study Torah, the teachings of the God who created the world and its organizing, sanctifying laws, one of which is to observe the Sabbath and, in this way, create a foretaste of eternity. The Sabbath is created through reciprocal desire. The rabbinic cave does not reveal the illusions of the world nor give entrance to disembodied spirit if only we would leave it. The rabbinic cave is figuratively a marked off space where transcendence is present in matter that manifests it.

> A story is told about a rabbi who once entered heaven in his dream. He was permitted to approach the temple in Paradise where the great sages of the Talmud, the Tannaim, were spending their eternal lives. He saw that they were just sitting around tables studying the Talmud. The disappointed rabbi wondered, "Is this all there is to Paradise?" But suddenly he heard a voice: "You are mistaken. The Tannaim are not in Paradise. Paradise is in the Tannaim."[81]

Inside the rabbinic cave we remember and observe—memory creates the present and the future—God's redemption of the Israelites from Egypt, their redemption from slavery, from work—our redemption, a condition we have on the Sabbath when we read about creation and the creation of the Israelites in related cosmological and historical texts. The Jewish God is the god of history, of events in time that are the figurative interior of the rabbinic cave.

> To Rabbi Shimeon eternity was not attained by those who bartered for space but by those who knew how to fill their time with spirit. To him the great problem was *time* rather than *space*; the task was how to convert time into eternity rather than how to fill space with buildings, bridges and roads; and the solution of the problem lay in study and prayer rather than in geometry and engineering.[82]

---

80    Shulevitz, pp. 51–52.

81    Heschel, p. 75.

82    Heschel, p. 41 (italics his).

As Heschel puts it, "Eternal life does not grow away from us; it is 'planted within us,' growing beyond us. The world to come is therefore not only a posthumous condition, dawning up the soul on the morrow after its departure from the body. The essence of the world to come is Sabbath eternal, and the seventh day in time is an example of eternity."[83] We exist in duration, as Henri Bergson would put it,[84] recognizing the eternal as the duration in which time exists, as Bergson has put it.[85] We would marry it in the language of Jewish mysticism.

Without naming him, Heschel describes the distinction Bergson draws between duration and time as we experience it. Bergson makes this distinction using the clock, the instrument that marks for him our apprehension of time divided into past, present, and future.

> When I follow with my eyes on the dial of a clock the movement of the hand which corresponds to the oscillation of the pendulum, I do not measure duration, as seems to be thought; I merely count simultaneities, which is very different. Outside of me, in space, there is never more than a single

---

83    Heschel, p. 74.

84    Henri Bergson, *Time and Free Will: The Idea of Duration* in *Henri Bergson: Key Writings*, eds. Keith Ansell Pearson and John Mullarkey, Athlone Contemporary European Thinkers (New York/London, 2002), p. 60: "Pure duration is the form which the succession of our conscious states assumes when our ego lets itself *live*, when it refrains from separating its present state from its former states" (italics his). See, also Bergson, *The Creative Mind* in *Key Writings*, p. 266: "...the more we accustom ourselves to think and to perceive all things *sub specie durationis*, the more we plunge into real duration. And the more we immerse ourselves in it, the more we set ourselves back in the direction of the principle, though it be transcendent, in which we participate and whose eternity is not to be an eternity of immutability, but an eternity of life: how, otherwise, could we live and move in it? *In ea vivimus et movemur et sumus*" (italics his).

85    Bergson, *Creative Evolution*, trans. Arthur Mitchell (1911; rpt. New York, 1944) p. 5: "The apparent discontinuity of the psychical life is...due to our attention being fixed on it by a series of separate acts: actually there is only a gentle slope; but in following the broken line of our acts of attention, we think we perceive separate steps... A thousand incidents arise, which seem to be cut off from those which precede them, and to be disconnected from those which follow. Discontinuous though they appear, however, in point of fact they stand out against the continuity of a background on which they are designed, and to which indeed they owe the intervals that separate them... Our attention fixes on them because they interest it more, but each of them is borne by the fluid mass of our whole psychical existence. Each is only the best illuminated point of a moving zone which comprises all that we feel or think or will—all, in short, that we are at any given moment. It is this entire zone which in reality makes up our state." See also p. 7: "Duration is the continuous progress of the past which gnaws into the future and which swells as it advances."

position of the hand and the pendulum, for nothing is left of the past positions. Within myself, a process of organization or interpenetration of conscious states is going on, which constitutes true duration. It is because I
*endure* in this way that I picture to myself what I call the past oscillations
of the pendulum at the same time as I perceive the present oscillation.[86]

And these "so-called successive oscillations," even though "there will never be
more than a single oscillation, and indeed only a single position, of the pendulum, and hence no duration,"[87] become an entrance into—our method into, as
it were, for Bergson—the apprehension of the duration that is for him existence
itself: "Withdraw...the pendulum and its oscillation and there will no longer be
anything but the heterogeneous duration of the ego, without moments external
to one another, without relation to number," which is what Bergson calls "succession without mutual externality."[88] Succession itself exists "solely for a conscious spectator who keeps the past in mind and sets the two oscillations...side
by side in an auxiliary space."[89] Now "between this succession without externality and...externality without succession, a kind of exchange takes place, very
similar to what physicists call the phenomenon of endosmosis."[90] In that
exchange, an exchange of apprehension that we remember, pure duration presents itself.[91] Time as duration and time as succession perceived through instru-

---

86  Bergson, *Time and Free Will*, p. 63 (italics his).

87  Ibid., p. 63.

88  Ibid.

89  Ibid., See also Bergson, *Duration and Simultaneity* in *Key Writings*, p. 208: "Impersonal and
    universal time, if it exists, is in vain endlessly prolonged from past to future; it is all of a
    piece; the parts we single out in it are merely those of a space that delineates its track and
    becomes its equivalent in our eyes; we are dividing the unfolded, not the unfolding." See
    also Bergson, *Time and Free Will*, p. 60: "We set our states of consciousness side by side in
    such a way as to perceive them simultaneously, no longer in one another, but alongside
    one another; in a word, we project time into space, we express duration in terms of exten-
    sity, and succession thus takes the form of a continuous line or a chain, the parts of which
    touch without penetrating one another."

90  Bergson, *Time and Free Will*, p. 63. Endosmosis is the inward flow of a fluid through a
    permeable membrane toward a fluid of greater concentration. Bergson's description of
    the relationship between time as duration and time as succession is, in this instance,
    aqueous. His metaphor here is helpful because it places the mechanical movements of a
    pendulum inside consciousness, which is, for Bergson, fluid.

91  See Bergson, *Time and Free Will*, p. 64: "There is real space, without duration, in which
    phenomena appear and disappear simultaneous with our states of consciousness.
    There is a real duration, the heterogeneous moments of which permeate one another;
    each moment...can be brought into relation with a state of the external world which is

ments of measurement come into existence reciprocally: they depend on one another; they interpenetrate one another in consciousness.[92]

Bergson's distinctions thus speak to Heschel's own mysticism in which physical objects we observe in tensed time as having a past, as existing in the present, as continuing to exist in the future, obscure our wished-for and life-affirming presence in time as pure duration.[93]

---

contemporaneous with it, and can be separated from the other moments in consequence of this very process. The comparison of these two realities gives rise to a symbolical representation of duration, derived from space. Duration thus assumes the illusory form of a homogeneous medium, and the connecting link between these two terms, space and duration, is simultaneity, which might be defined as the intersection of time and space." See also Augustine, *Confessions*, xi, xx, 26 (p. 235) on the relationship between time and memory: "Perhaps it would be exact to say: there are three times, a present of things past, a present of things present, a present of things to come." See, too, xi, xxv, 36 (p. 242): "So it is in you, my mind, that I measure periods of time."

92   For a skeptical description of what we can really know when we observe time as a clock shows it, see G.E. Moore, *Some Main Problems of Philosophy* (New York 1953), who argues against the position that from an instance of elapsed time the soul or mind, as Aristotle would put it, can apprehend time *as an idea* and that, moreover, from this idea one can understand time in the real world as uniform and universal, either experienced or known from observing a watch. See pp. 206–207: "But in the case of Time it seems to me doubtful whether we ever directly apprehend any time at all, *either* the time of ordinary life *or* any other. Consider the cases in which, if at all, we do directly apprehend a lapse of time. We undoubted do directly apprehend changes of various sorts, both movements—that is to say changes of relative position in directly apprehended space—and other changes, which do not consist in movement: e.g., we may directly apprehend a light growing brighter, a colour changing, or a sound growing louder... I can observe now the second-hand of my watch actually moving; and you all know what it looks like to *see* the second-hand moving...whereas I *do not* see the hour-hand moving; and the difference which I express in this way is certainly a difference in what I directly apprehend. The case of the second-hand is...a case in which I do directly apprehend a change—in this case a movement—and it would seem that this, if any, must be a case in which I directly apprehend a lapse of time. No doubt it is very difficult to decide exactly *how much* of the movement I do directly apprehend, and where direct apprehension passes into memory: but there seems no doubt that I do directly apprehend *some* movement. But do I directly apprehend any length of time which the movement occupies? I am bound to say I cannot be certain that I do" (italics his).

93   Bergson, *Time and Free Will*, p. 64: "There is a real space, without duration, in which phenomena appear and disappear simultaneously with our states of consciousness. There is a real duration, the heterogeneous moments of which permeate one another; each moment, however, can be brought into relation with a state of the external world which is contemporaneous with it, and can be separated from the other moments in consequence of this

When gazing at reality while our souls are carried away by spatial things, time appears to be in constant motion. However, when we learn to understand that it is the spatial things that are constantly running out, we realize that time is that which never expires, that it is the world of space which is rolling through the infinite expanse of time.[94]

Because the language of Heschel's mysticism is expressed through—using— biblical events, themselves regarded as mystical, Heschel can put his understanding of pure duration and tensed time in, for example, the following way: according to the Book of Exodus, Moses beheld his first vision in the flame of fire: "He gazed, and there was a bush all aflame, yet the bush was not consumed" (3, 2). This tells us, for Heschel, where narrative is the mode of philosophical discourse in Judaism,[95] that time "is like an eternal burning bush [a mystical object rather than a material one]. Though each instant must vanish to open the way to the next one, time itself is not consumed," that is, "time is the process of creation, and things of space are results of creation."[96] Experiencing duration for Heschel is recognizing the source of what is continuously being created. Here is a mysticism that depends on the existence of the material, a mysticism that does not understand the idea of duration without existents.

### The Emotions of No-Time

Let me conclude with two passages, the first from Jhumpa Lahiri's short story that describes a radiant understanding of eternal continuity; the second from Paul Harding's *The Tinkers* that imagines no-time as the end of time, which is, for Harding, full of everything there is. What is fascinating is that both Lahiri and Harding give us all time—and no-time—to grandfathers, as eternal a perception of time as exists in living human beings themselves. In the Lahiri, there is a moment of lived—and living—no-time for an actual grandfather who watches his grandson sleep, catching up the boy's future time in the memory of his own lifetime, his relived past.

He turned to face his sleeping grandson, the long lashes and rounded cheeks remind him of his own children when they were young. He was

---

very process. The comparison of these two realities gives rise to a symbolical representation of duration, derived from space."

94    Heschel, p. 97.
95    See Yoran Hazony, *The Philosophy of Hebrew Scripture* (Cambridge, MA, 2012).
96    Heschel, p. 100.

suddenly conscious that he would probably not live to see Akash into adulthood, that he would never see his grandson's middle age, his old age, this simple fact of life saddening him. He imagined the boy years from now, occupying this very room, shutting the door as Ruma and Romi had. It was inevitable. And yet he knew that he, too, had turned his back on his parents, by settling in America. In the name of ambition and accomplishment, none of which mattered anymore, he had forsaken them. He kissed Akash lightly on the side of his head, smoothing the curling golden hair with his hand, then switched off the lamp, filling the room with darkness.[97]

Lahiri sees the self, embodied in the grandfather the grandfather as well as the grandson, as a clock, the recorder of time and time's embodiments, the place where personal and personal historical awareness resides, which is history's residence in the particular.[98]

Harding's perspective is decidedly darker than Lahiri's, though virtually identical. When Harding's protagonist, a clockmaker, dies, his personal house dies, too, including the grandfather clock, his correlative, which he was repairing.

There he lay among the graduation photos and old wool jackets and rusted tools and newspaper clippings about his promotion to head of the mechanical-drawing department at the local high school, and then about his retirement and subsequent life as a trader and repairer of antique clocks. The mangled brass works of the clocks he had been repairing were strewn among the mess. He looked up three stories to the exposed support beams of the roof and the plump silver-backed batts of insulation that ran between them. One grandson, or another (*which?*) had stapled the insulation into place years ago and now two or three length of it had come loose and lolled down like pink wooly tongues.[99]

---

97    Jhumpa Lahiri, "Unaccustomed Earth," in *Unaccustomed Earth* (New York, 2008), p. 51.

98    On the self as clock, see also Paul Harding, *The Tinkers* (New York, 2009), pp. 33–34: "That was it, he realized; the clock had run down. All the clocks in the room had wound down— the tambours and carriage clocks on the mantel, the banjo and mirror and Viennese regulator on the walls, the Chelsea ship's bells on the rolltop desk, the ogee on the end of the table, and the seven-foot walnut-cased Stevenson grandfather's clock, made in Nottingham in 1801, with its moon-phase window on the dial and pair of robins threading flowery buntings around the Roman numerals. When he imagined inside the case of that clock, dark and dry and hollow, and the still pendulum hanging down its length, he felt the inside of his own chest and had a sudden panic that it, too, had wound down."

99    Harding, p. 10 (italics his).

The self as clock here suggests that at death the living self becomes the conscious place of all time's disintegrating constituents, both bodies and objects. Death for Harding is a species of no-time, of duration that holds time's "earthly things" as they are disappearing.

> There was the sky, filled with flat-topped clouds, cruising like a fleet of anvils across the blue. George had the watery, raw feeling of being outdoors when you are sick. The clouds halted, paused for an instant, and plummeted onto his head. The very blue of the sky followed, draining from the heights into that cluttered concrete socket. Next fell the stars, tinkling about him like the ornament of heaven shaken loose. Finally, the black vastation itself came untacked and draped over the entire head, covering George's confused obliteration.[100]

Both Harding and Lahihi have made no-time the time and place of reflection, as well as prophecy. For them, no-time *is* our existence: it includes our memory and our vision of the lives of those who follow us and those who won't.

---

100     Harding, pp. 10–11.

FIGURE 19.1    *Escalator at Tbilisi International Airport, Georgia*
PHOTO: GVILAVA

FIGURE 19.2    *Cafeteria in the Regional Airport, Santa Genoveva/Goiânia, of the Zona da Mata,*
*Goiânia, Brazil*
PHOTO:  REINALDO CÉSAR DE O. GUEDES

FIGURE 19.3
*The corridor leading from terminal 2 to*
*terminal 1, Copenhagen Airport*
PHOTO:  THIERRY CARO

FIGURE 19.4     *Tram tracks in Lisbon*
PHOTO: MATHIEU FORMISYN

FIGURE 19.5    *Akureyrarkirkja (Lutheran church of Akureyri), designed by Guðjón Samúelsson, completed in 1940, Akureyri, Iceland*

# Index

Printed in the United States
By Bookmasters